Space Exploration and Humanity

Space exploration and humanity

Space Exploration and Humanity

A Historical Encyclopedia

VOLUME 2

History Committee of the American Astronautical Society

Stephen B. Johnson, General Editor

Timothy M. Chamberlin, Michael L. Ciancone, Katherine Scott Sturdevant, and Rick W. Sturdevant, Section Editors

David Leverington, Technical Consultant

ABC-CLIO

Santa Barbara, California • Denver, Colorado • Oxford, England

Library of Congress Cataloging-in-Publication Data

Space exploration and humanity : a historical encyclopedia / History Committee of the American Astronautical Society ; Stephen B. Johnson, general editor ; Timothy M. Chamberlin . . . [et al.], section editors.

    p. cm.

  Includes bibliographical references and index.

  ISBN 978–1–85109–514–8 (hard copy : alk. paper) — ISBN 978–1–85109–519–3 (ebook)
1. Astronautics—History. 2. Astrophysics—History. 3. Astronautics—United States—History. 4. Outer space—Exploration—History. 5. Astronautics and civilization.
6. Astronautics—Historiography. 7. Astrophysics—Historiography. I. Johnson, Stephen B., 1959– II. American Astronautical Society. History Committee.
TL788.5.S636   2010
629.403—dc22   2010009664

ISBN: 978–1–85109–514–8
EISBN: 978–1–85109–519–3
14 13 12 11 10   1 2 3 4 5

This book is also available on the World Wide Web as an eBook.
Visit www.abc-clio.com for details.

ABC-CLIO, LLC
130 Cremona Drive, P.O. Box 1911
Santa Barbara, California 93116-1911

This book is printed on acid-free paper ∞

Manufactured in the United States of America

Portions of the glossary are reprinted from New Cosmic Horizons: Space Astronomy from the V2 to the Hubble Space Telescope (copyright (c) 2000 David Leverington) and Babylon to Vogager and Beyond (copyright (c) 2003 David Leverington). Both reprinted with the permission of Cambridge University Press.

Contents

VOLUME 1

About the Editors

Timothy M. Chamberlin is design editor for the *Tulsa World* newspaper in Tulsa, Oklahoma. His work includes coverage of the space shuttle *Columbia* accident and the future of spaceflight for *The State* newspaper in Columbia, South Carolina. He has developed a website for finding relevant information about current U.S. space policy and law, and has written about space advisory committees appointed by the president for *Space Times* magazine. He is an honorary member of the American Astronautical Society (AAS) and serves as editor of the AAS History Committee's newsletter *Explorer*. Tim holds an MS in space studies from the University of North Dakota, North Dakota. He serves as the section editor for the Human Flight and Microgravity Science section of this encyclopedia.

Michael L. Ciancone is an engineer at the NASA Johnson Space Center in Houston, Texas, where he provides technical support on human spaceflight safety to the Constellation Program. He is also a bibliophile who has maintained an active interest in pre-Sputnik rocket societies and spaceflight visionaries, as well as the cultural and social impacts of spaceflight. In connection with these interests, he chairs the American Astronautical Society (AAS) History Committee and coordinates the review panel for the annual Eugene M. Emme Award for astronautical literature. He is a member of the History Committee of the International Academy of Astronautics. Michael has written papers on spaceflight visionaries such as David Lasser and Luigi Gussalli, and has served as curator for an exhibit at the Western Reserve Historical Society on "Cleveland and Outer Space, The Cleveland Rocket Society (1933–37)." He is the author of *The Literary Legacy of the Space Age—An Annotated Bibliography of Pre-1958 Books on Rocketry & Space Travel* (1998) and the volume editor for the papers presented during the International Academy of Astronautics (IAA) History Symposium of the 2002 World Space Congress in Houston, Texas. He serves as the section editor for the Space and Society section of this encyclopedia.

Dr. Stephen B. Johnson is associate research professor at the National Institute for Science, Space, and Security Centers at the University of Colorado at Colorado Springs, Colorado, and is currently assigned to the NASA Marshall Space Flight

Center for the Ares I and Constellation programs. He acquired his doctorate at the University of Minnesota in the history of science and technology, and has written *The Secret of Apollo*, which won the 2002 Eugene M. Emme Award for astronautical literature, *The United States Air Force and the Culture of Innovation 1945–1965*, numerous journal articles in publications such as *Technology and Culture*, *History and Technology*, *Air Power History*, and *Journal of Industrial History*, and essays in books including *Critical Issues in the History of Spaceflight*, *The Societal Impact of Spaceflight*, and *The Business of Systems Integration*, and was the editor of *Quest: The History of Spaceflight Quarterly* from 1998 to 2005. Stephen has contributed to encyclopedias such as *The Readers Guide to the History of Science* and *American National Biography*. He is the general editor for this encyclopedia and the section editor for the Astrophysics and Planetary Science, Civilian and Commercial Space Applications, and Technology and Engineering sections.

Dr. David Leverington, now retired, is a writer on the history of astronomy and space research. Educated at the University of Oxford, UK, with a degree in physics, he was, in the 1970s, design manager of GEOS, Europe's first geosynchronous scientific satellite, and program manager, at the European Space Agency, of *Meteosat*, Europe's first meteorological satellite. Later he was engineering director in BAE Systems' Space Division, responsible, among other things, for *Giotto*, the spacecraft that intercepted Halley's Comet in 1986, the Photon Detector Assembly for the *Hubble Space Telescope*, the Envisat/Polar Platform, the Medium Energy Detector Assembly for *Exosat*, subsystems on the *ISEE-B*, *Ulysses*, and *Hipparcos* satellites, and remote sensing instruments that flew on the American *UARS* and *TIROS N* satellites. He was deputy CEO of British Aerospace Communications and on the management board of the UK's Earth Observation Data Centre. He has written three books: *A History of Astronomy from 1890 to the Present*, *New Cosmic Horizons*; *Space Astronomy from the V2 to the Hubble Space Telescope*; and *Babylon to Voyager and Beyond: A History of Planetary Astronomy*. David has been published in many journals including *Nature* and the *Quarterly Journal of the Royal Astronomical Society*, and is a contributor to Elsevier's *Encyclopedia of the Solar System, 2nd Edition*. He is included in the marquis' *Who's Who in the World 2006* and is a fellow of the UK's Royal Astronomical Society. He serves as technical consultant for this encyclopedia's Astrophysics and Planetary Science section and its meteorological and civilian Earth observation articles.

Katherine Scott Sturdevant, MA and PhD candidate, is professor of history at Pikes Peak Community College in Colorado Springs, Colorado. She has taught a wide range of undergraduate American History courses for 25 years and has been a scholarly historical editor for nearly 30 years. She authored *Organizing and Preserving Your Heirloom Documents* and *Bringing Your Family History to Life through Social History*, received state and national teaching excellence awards, and is active in many forms of curriculum innovation, public history publishing, interpretative training, source collection and preservation, historic preservation, and speaking for general audiences. She acted as the education editor for the early stages of this encyclopedia, ensuring at the outset that the level of writing and editing is appropriate for the target audience of high school seniors, college freshmen, and general readers.

Dr. Rick W. Sturdevant is deputy director of history at Headquarters Air Force Space Command, Peterson Air Force Base in Colorado Springs, Colorado. He acquired his PhD in 1982 from the University of California, Santa Barbara, California, and joined the U.S. Air Force history program in April 1984 as chief historian for Air Force Communication Command's Airlift Communications Division at Scott Air Force Base, Illinois. In 1985, he moved to Peterson Air Force Base to become chief historian for the Space Communications Division, and in 1991 moved to the Air Force Space Command (AFSPC) history office. In addition to producing classified periodic histories and special studies for AFSPC, Rick has published extensively on the subject of military aerospace history in such periodicals as *Space Times*, *High Frontier: The Journal for Space & Missile Professionals*, and *Journal of the British Interplanetary Society*, and essays in a number of books including *Beyond the Ionosphere: Fifty Years of Satellite Communication*, *To Reach the High Frontier: A History of U.S. Launch Vehicles*, and *Harnessing the Heavens: National Defense through Space*. He has also contributed to encyclopedias such as *Air Warfare: An International Encyclopedia* and *Encyclopedia of 20th-Century Technology*. Rick is editor of the International Academy of Astronautics and American Astronautical Society (AAS) history series. A recipient of the Air Force Exemplary Civilian Service Award and the AAS President's Recognition Award, he was elected an AAS Fellow in 2007. He serves as the section editor for the Military Applications section of this encyclopedia.

Military Applications

Military applications for artificial Earth satellites originated with conceptual studies in the years immediately after World War II. During the latter half of 1945, U.S. Navy Lieutenant Robert Haviland and Commander Harvey Hall from the Bureau of Aeronautics, inspired by a speculative report on spaceflight written in May 1945 by German rocket scientist Wernher von Braun, campaigned for creation of a high-priority satellite development program. Learning of the Navy initiative but refusing Commander Hall's offer of a joint program, Army Air Forces (AAF) Major General Curtis LeMay directed Douglas Aircraft Company's AAF-funded Project RAND to produce a feasibility study titled "Preliminary Design of an Experimental World-Circling Spaceship," which RAND finished at the beginning of May 1946. RAND released a dozen more detailed studies in February 1947 on the feasibility of a space program.

Meanwhile, Soviet engineers pondered similar questions. In 1945 Sergei Korolev discussed with captured German scientists the possibility of designing a spaceship. His compatriot Mikhail Tikhanravov initiated studies in 1947 on possible launch systems for a satellite. Their efforts culminated in March 1950 with what was likely the first detailed Soviet report on the requirements for launching a satellite, given the constraints of existing technology.

For the Soviet Union, the United States, and a few other nations interested in having their own spaceflight capabilities, sufficiently powerful launch vehicles posed the greatest immediate problem. As Theodore von Kármán, director of the AAF Scientific Advisory Group, asserted in a December 1945 report to AAF Commanding General Henry "Hap" Arnold, the German V-2 ballistic missile showed that a satellite was a definite possibility. The aspiring space powers confiscated V-2 technology and gathered up German rocket scientists at the end of World War II, using those resources to augment their fledgling rocket development programs. Driven primarily by Cold War requirements for delivering thermonuclear warheads over great distances, the superpowers developed intermediate-range and intercontinental ballistic missiles (ICBMs) during the 1950s. The Soviet Union produced the R-7 Semyorka; the United States developed the Jupiter, Thor, Atlas, and Titan. Those missiles doubled as the first space-launch vehicles and became the technological foundation for space boosters into the twenty-first century.

Military space history can be divided into several periods based on the maturation of satellite systems and their applications. From the late 1940s through the 1950s, conceptual and design studies predominated. The U.S. military focused primarily on intelligence collection and reconnaissance from space, along with closely related applications like meteorology and early warning. Not far behind were applications for early warning, communications, and navigation. Intense rivalry among the U.S. military services, each seeking to gain a dominant role in managing U.S. military space activities, led initially to costly duplication of research and development efforts. To achieve better control of the entire U.S. space program, President Dwight Eisenhower's administration initially created the Advanced Research Projects Agency (ARPA) under the Department of Defense (DoD), then supported the establishment of NASA to handle civil space programs. By comparison, the Soviet military, finding itself overwhelmed in the early 1960s with as many as 30 diverse proposals for space systems from different design bureaus, struggled to assert stronger control over the quest for space-based applications.

From 1960 into the mid-1970s, the two superpowers began using first-generation military satellites operationally for reconnaissance, intelligence collection, early warning, meteorology, navigation, and communications. While used operationally, some of those satellites were primarily test platforms or prototypes. During this period, the U.S. Air Force (USAF) launched GRAB electronic intelligence, Corona reconnaissance, Missile Defense Alarm System (MIDAS) early warning, Transit navigation, Defense Meteorological Satellite Program (DMSP), Vela nuclear detection, and Initial Defense Communications Satellite Program (IDCSP) and other types of military communications satellites. The Soviet Union countered with similar systems, but Soviet military capabilities in space generally lagged behind similar U.S. applications by several years to a decade. China and Europe also developed some military space capabilities, another decade or more after the United States and Soviet Union.

The technological evolution of military space capabilities continued during the 1970s and early 1980s, when many applications first achieved full operational

Satellite image of Beijing, China, taken on 27 May 1967. (Courtesy U.S. Geological Survey)

status. Defense Support Program (DSP) satellites provided early warning, and Defense Satellite Communications System (DSCS) II satellites aided command and control of U.S. strategic forces. For U.S. reconnaissance and surveillance, satellites such as KH-9 Hexagon, KH-11 Kennan, and Lacrosse entered the scene. In the Soviet Union, more advanced versions of Korund and Strela communications satellites appeared, and Yantar imaging satellites began augmenting older Zenit capabilities.

Technological advances had implications for military organization and force structure. Soviet planners hastened an effort that the Ministry of Defense had directed in November 1968 to integrate space forces into overall military planning. Following the USAF lead in 1982, the U.S. Army and Navy established operational space commands. In September 1985 the DoD created U.S. Space Command, which gained overall responsibility for operational employment of all three services' military space systems. Organizational arrangements also changed to accommodate the institutionalization of managerial and engineering processes, most notably systems engineering and integration, to effectively and efficiently coordinate production of supremely complex space systems. Compared to the Soviet Union, however, U.S. military planners during this period made little headway on integrating space with overall planning for contingency or wartime operations.

From the mid-1980s into the 1990s, both the U.S. and the Soviet defense establishments replaced their first operational military satellite systems with improved, next-generation capabilities. For space-based communications, particularly on U.S. satellites, that meant jam-resistant DSCS 3 platforms and Milstar satellites survivable through all levels of conflict, including nuclear war. Crosslinks between orbiting military satellites reduced reliance on ground stations for telemetry, tracking, and control. In the realm of navigation, it meant U.S. Global Positioning System (GPS) and Soviet Global Navigation Satellite System (GLONASS) satellite constellations capable of providing more precise location in three dimensions, instead of two, and timing to within billionths of second became the norm.

A major shift in military space applications occurred during the First Gulf War in 1991 and, shortly thereafter, the collapse of the Soviet Union. Satellites built and launched for strategic purposes associated with the Cold War suddenly became essential for theater or tactical purposes in the so-called "War on Terrorism." Early-warning satellites designed to detect ICBM launches from the Soviet Union spotted Scud missile launches from Iraq, and communications satellites intended originally for command and control of nuclear forces linked coalition land, sea, and air forces in the Persian Gulf theater with command centers in the United States. For the first time, nearly the whole range of space-based capabilities—reconnaissance, intelligence, surveillance, weather, navigation, and communications—originally identified in the May 1946 RAND report were used in warfare. Military leaders around the globe, including many in the United States and Russia, suddenly acknowledged space systems as what insiders termed a "force multiplier" that could significantly increase the combat strength and capabilities of army, navy, marine, and air units.

Consequently, during the 1990s, both Russia and the United States sought ways to provide their warriors with more space-based support. In 1992 Russia began a decadelong process to reorganize its military space forces as a separate service—the Russian Space Forces (VKS). Meanwhile, the USAF established a Space Warfare Center to inform leaders from all the military services about how space systems could support their combat operations and to hear from those same leaders what sorts of additional space support they would like to have. This resulted in the formulation of different, sometimes ingenious, ways to employ existing systems and in the identification of requirements for entirely new satellite systems, one example being space radar. If GPS enabled combat operations centers to track friendly or so-called "blue" ground, sea, and air forces, space radar held the promise of enabling those same centers to track enemy or "red" forces. Although U.S. military planners did not envision having an operational space radar constellation before 2015, officers and enlisted troops realized that such a system could provide almost total awareness of the evolving situation in a given battle space.

Other nations focused on independent procurement of military space systems, either to reduce their reliance on U.S. systems or to have parallel capabilities in the event of conflict with U.S. forces. Although proliferation of military satellites internationally had begun before the 1990s, widespread awareness of their importance to the success of U.S. military operations in Bosnia, Kosovo, Afghanistan, Iraq, and elsewhere between 1991 and 2003 prompted numerous countries or organizations to accelerate their quest for such systems. Technological advances associated with microsatellites and small, low-cost launch vehicles offered more countries affordable opportunities to achieve a modicum of independence with regard to space.

The proliferation of military satellite systems internationally focused on communications, reconnaissance, and surveillance. The United Kingdom continued improving its Skynet military communications satellites and began manufacturing North Atlantic Treaty Organization (NATO) communications satellites formerly produced in the United States. Italy's first military satellite, *Sicral 1*, went into orbit in February 2001. Israel launched its *Ofek 5* reconnaissance satellite in May 2002 and *Amos 2*, a satellite dedicated to military telecommunications, in December 2003. France orbited its *Helios 2A* imaging-reconnaissance satellite along with four Essaim electronic-intelligence microsatellites in December 2004, and it began launching Syracuse 3 satellites in October 2005 to satisfy its military communications requirements. China launched its first military communications satellite, *Feng Huo 1*, in January 2000 and an improved type of Fanhui Shi Weixing (FSW) film-recovery photoreconnaissance satellite in November 2003. Concerned about the threat from secretive North Korean nuclear and missile programs, Japan launched its first two military spacecraft in March 2003 to begin an Information Gathering Satellite (IGS) constellation. South Korea launched its first military communications satellite, *Mugunghwa 5*, in August 2006. India, which launched a *Technology Experiment Satellite* in 2001 that subsequently relayed high-quality images of the war in Afghanistan and of Pakistani troop movements along the border, planned to have a space-based Military Surveillance and Reconnaissance System operational by 2007. Meanwhile, Russia, which faced severe economic constraints after the collapse of the Soviet Union, maintained its military space operations at a much reduced level.

By 2006 it was universally recognized that civil or commercial satellites could have military applications and vice versa. In terms both of design and capabilities, military, civil, and commercial satellites were becoming more alike. Furthermore, in a growing number of instances, such dual-use capabilities extended beyond national boundaries to users in specific regions or around the globe. This became most apparent with space-based navigation, where GPS, operated by the USAF and originally intended to serve U.S. military forces, could be used by anyone in the world with a commercially available GPS receiver. Similarly, Russia had intended GLONASS for military use but, in the 1990s, acknowledged its potential civilian uses, and the European Union's planned Galileo commercial satellite navigation system offered the prospect of supporting military operations.

In the realm of meteorology, the technology associated with separate U.S. military and civil polar-orbiting satellites converged. Seeking higher performance and cost savings through a single, integrated system, the United States assigned primary responsibility for telemetry, tracking, and control of both DoD and Department of Commerce polar-orbiting meteorological satellites to the National Oceanic and Atmospheric Administration's Office of Satellite Operations in June 1998. For the first time in U.S. history, on-orbit control of military satellites resided with a civil organization.

With respect to reconnaissance and surveillance, where data from space once was collected exclusively by defense and intelligence organizations for defense and intelligence organizations, declassification of Corona satellite imagery proved useful for civil entities monitoring environmental change. U.S. forces purchased commercial imagery from French SPOT (Satellite Pour l'Observation de la Terre) satellites and from U.S. remote-sensing satellite companies like DigitalGlobe (named Earth Watch before September 2001) and GeoEye (formed by the acquisition of Space Imaging by ORBIMAGE in January 2006). Israel and India had dual-use observation satellites. South Korea's multipurpose *Arirang 1* satellite had been producing poor-resolution imagery since 1999; its *Arirang 2*, launched in July 2006, could capture extremely high-resolution, multispectral images, both for making geographic surveys and for tracking North Korean military activity. France and Italy planned to begin construction of a shared, dual-use observation system near the end of 2006.

Communications represented a fourth dual-use area. Rapidly escalating demands for satellite channels and greater bandwidth to support systems like unpiloted aerial vehicles, precision-guided munitions, and tracking of friendly forces compelled the military to rely increasingly on commercial satellite communications. Military leaders preferred to use dedicated military satellites like those in the Ultrahigh Frequency Follow-On System, DSCS, or Milstar for command and control purposes, but limited capacity and transmission speed on those platforms sometimes compelled commercial augmentation. Growth in requirements for transmitting so-called "general purpose" information only increased military reliance on commercial satellite systems. By one estimate, the total demand for U.S. military communications in 2007 would be 9.5 Gbps, with commercial vendors meeting between 1.8 and 5.5 Gbps.

Countries and Organizations with Dedicated Military Satellites Indicating Year of First Successful Launch

Country or Organization	Communications	ELINT	Navigation	Reconnaissance	Meteorology	Early Warning	Nuclear Detection	Ocean Surveillance
Chile	1995			1995				
China	1984	1975	2000	1975				
France		1995		1995*				
Germany				2006				
Israel				1988				
Italy	2001							
Japan				2003				
NATO	1970							
South Korea	2006							
Soviet Union/ Russia	1964	1967	1967	1962		1972		1965
Spain	2006							
United Kingdom	1969			2005				
United States	1958	1960	1960	1960	1962	1960	1963	1971

*with Germany, Italy, and Spain

Human spaceflight marked one area into which the most powerful military establishments put considerable conceptual, design, research, and development effort for more than a half century but failed to produce a military spaceplane. During the 1930s–40s, Austrian engineer Eugen Sänger and his German wife, Irene Bredt, conceived and designed a piloted, hypersonic spaceplane—the "Silverbird" antipodal bomber. Walter Dornberger, Wernher von Braun's military superior during World War II, immigrated to the United States in the late 1940s and began working for Bell Aircraft Company, where he propounded the Sänger-Bredt concept. In April 1952 Bell proposed to the USAF a piloted bomber missile (BOMI) and, two years later, the USAF contracted with that company to study the feasibility of developing such a system for bombardment and reconnaissance. By early 1956 BOMI had evolved into three separate studies—a rocket bomber (Robo), a reconnaissance spaceplane (Brass Bell), and an experimental boost-glide vehicle (Hywards)—only to be consolidated under the Dyna-Soar (X-20) program six days after the launch of *Sputnik* in October 1957. That program ended in 1963 without producing more than a factory mock-up, but wind-tunnel testing of a modified X-15 aerospaceplane design with thin delta wings for sustained hypersonic cruise continued into the late 1960s.

Further U.S. technological research and development for winged spaceflight, sponsored at least in part by the DoD, continued for decades. The X-24A and X-24B piloted, rocket-powered lifting-body research conducted by the USAF and NASA from 1966 to 1975 contributed significantly to the design of NASA's Space Shuttle. Despite a directive to use the Shuttle for all national-security space launches, USAF plans in the early 1980s for a dedicated military Shuttle fleet never came to fruition. The USAF briefly contemplated a Trans-Atmospheric Vehicle (TAV) in the early 1980s, but yielded by late 1985 to political pressure for a jointly sponsored and funded USAF, Navy, NASA, ARPA, and Strategic Defense Initiative Organization program to develop a National Aero-Space Plane (NASP), for which the USAF had overall management responsibility. Viewed by people in President Ronald Reagan's administration as the potential progenitor of all piloted space transportation systems in the post-Shuttle era, NASP—otherwise referred to as the X-30—yielded significant progress in several critical technologies but failed to fly before budget woes led to the program's cancellation in 1994.

Meanwhile, the Soviet Union had begun studying the Sänger-Bredt idea after engineer Aleksey Isayev discovered documentation on the piloted, antipodal-bomber project at Peenemünde in May 1945. By spring 1947, design work on such a bomber to deliver nuclear weapons had begun under Mstislav Keldysh's leadership at Design Bureau NII-1 in Moscow. Although that project fizzled in 1950, designers Pavel Tsybin and Vladimir Myasishchev perpetuated the quest for a reusable spaceplane in the late 1950s with their short-lived, respective Gliding Cosmic Apparatus (PKA) and M-48/VKA-23 projects. In 1959 designer Vladimir Chelomey initiated studies for Raketoplan. Insufficient technical progress and Soviet politics, however, resulted in termination of Raketoplan and transfer of that project's resources in the mid-1960s to Chief Designer Gleb Lozino-Lozinskiy's Spiral program. The latter finally succumbed in 1978 to technical problems and insufficient funding.

Like the United States, the Soviet Union eventually opted for a lifting-body design and began development of the Buran shuttle in 1976. The only orbital launch of Buran, without humans on board, occurred in 1988. Russian President Boris Yeltsin officially terminated the costly program in June 1993 before a piloted flight could occur. For several years thereafter, Russian engineers pursued a Multi-purpose Aerospace System (MAKS) based on a small Molniya spaceplane launched from atop an AN-225 carrier aircraft but, like previous endeavors, it never became operational.

Capsules and space stations proved to be better means of achieving human space-flight for military purposes. In 1958 the U.S. Army proposed Project Adam to launch a human in a capsule atop a Redstone booster, and the USAF competed with a similar Man-In-Space-Soonest (MISS) program using a capsule atop a Thor or, later, Atlas booster. The following year, Wernher von Braun's team at the Army's Redstone Arsenal, presented Project Horizon, which would have used spent boosters for Earth-orbiting space stations and lunar habitats. For strictly military purposes, those ideas coalesced in the mid-1960s under the USAF Manned Orbiting Laboratory (MOL) program, which included a "Blue Gemini" capsule and a pressurized Titan III upper-stage orbital station. When President Richard Nixon's administration can-celed MOL in 1969, some U.S. "military man in space" advocates hoped to conduct defense-related reconnaissance and surveillance from NASA's proposed space station, but policy making at the highest levels of government eventually led to a decision ban-ning such activities onboard the *International Space Station*.

Soviet activity with respect to military capsules and space stations closely paral-leled U.S. activity. As early as September 1957, Soviet scientists and engineers had begun assessing the reentry temperatures and thermal protection required for return vehicles of various shapes and, by January 1958, had come to favor a stubby, conical capsule over a winged-glider design. In January 1965 Vladimir Chelomey's OKB-52 (Experimental Design Bureau) undertook development of an Almaz Autonomous Piloted Orbital Station (APOS) similar to the proposed USAF MOL. The following year, Almaz APOS and a competing Soyuz R program were canceled in favor of Almaz OPS. Unlike the APOS design, which included an attached, three-seat "Vozvrashemui Apparat" (Return Apparatus or VA) reentry capsule and no docking port, the Almaz OPS dispensed with the VA capsule and added a docking port. The first successful Almaz OPS military space station flight, *Salyut 3*, occurred in June 1974 and tested numerous kinds of reconnaissance sensors. A second Almaz military space station, *Salyut 5*, went into orbit in June 1975 and, during 409 days of operation, was visited by the crews of *Soyuz 22* and *Soyuz 24*. With reentry and burnup of *Salyut 5* in August 1977, the story of the world's only operational military space station ended. Like its U.S. counterparts, the Soviet Union concluded that reconnaissance by space crews was too expensive compared to using advanced robotic systems.

As the number of countries employing satellites for military purposes expanded dramatically during the 1990s–2000s, U.S. strategists became increasing alarmed.

The collapse of the Soviet Union had left the United States standing unequivocally alone as the world's premier space power. But, ironically, U.S. space supremacy made attacks on U.S. space systems by lesser powers more probable during periods of tension or open warfare. Therefore, U.S. military space leaders began emphasizing principles of "space control," which included situational awareness, defensive counterspace operations, and offensive counterspace operations.

The USAF renewed its attention to improving space surveillance through acquisition of better terrestrially based radar, electro-optical, and passive electronic sensors, along with space-based sensors and improvements in space-catalog software, to enhance detection of on-orbit threats. New X-band radar systems augmented older Ballistic Missile Early Warning System (BMEWS) and PAVE PAWS radars, which had space surveillance as a secondary mission. The USAF Space Command deactivated Baker-Nunn optical cameras, which had been in use since the late 1950s, and sought to upgrade Ground-based Electro-Optical Deep Space Surveillance (GEODSS) camera systems that used 1970s technology. Responsible for keeping track of all artificial objects in Earth orbit, the USAF, during the late 1990s, implemented the most significant improvement in space cataloging since the late 1950s—the High Accuracy Catalog. In March 2004 the USAF contracted with a Boeing–Ball Aerospace team for development and initial operation of a Space-Based Space Surveillance (SBSS) system.

Discussion intensified regarding placement on U.S. military satellites of threat-detection and defensive systems. Military scientists and engineers pursued ways both to locate and counter an enemy's attempt to jam signals from U.S. navigation and communications satellites. To jam enemy satellite signals without permanently damaging the satellite itself, the USAF contracted with Northrop Grumman Corporation to develop a ground-based, transportable Counter Communications System that became operational in 2002. A vigorous debate arose about whether space-based weapon systems, defensive and offensive, were appropriate or necessary. In 2006 practically everyone agreed that space already had been "militarized" for decades, but large numbers questioned whether it should be "weaponized."

Arguments about weapons in space had historical roots buried deep in science fiction and surrounding the earliest years of actual spaceflight following the launch of *Sputnik*. Successful test flights of Soviet and U.S. ICBMs in the late 1950s made it clear to the world that all-out war between the superpowers would involve thermonuclear warheads passing through space to targets halfway around the globe. Some doomsday theorists even proposed launching nuclear warheads from Earth-orbiting platforms or lunar emplacements. Other proposals in that early period aimed to develop direct-ascent or orbital interceptors to inspect and, if necessary, destroy enemy satellites. During the 1960s the USAF actually deployed an operational anti-satellite (ASAT) system using nuclear-tipped Thor missiles, and the Soviet military developed a fractional-orbit nuclear delivery system. With mounting apprehension about the possibility of a Multiple Orbital Bombardment System (MOBS) or a Long-duration Orbital Bombardment System (LOBS), both superpowers began building antiballistic missile systems. The 1972 ABM Treaty, which recognized the limitations of missile defense systems at that time, possibly forestalled a new, extremely expensive arms race.

Nonetheless, driven partially by technological advances and by strategists who saw weaponization of space as inevitable, the issue resurfaced in the 1980s. President Ronald Reagan's Strategic Defense Initiative included investigation of technologies for space-based lasers, kinetic-kill vehicles, and other types of antimissile or antisatellite systems. The USAF actually tested an experimental air-launched ASAT weapon that destroyed its target. Although none of those weapon systems were deployed operationally, U.S. plans for a national missile defense system continued into the twenty-first century and resulted, eventually, in the United States officially withdrawing from the ABM Treaty in June 2002, thereby clearing the way for deployment of ground- and sea-based interceptor missiles. Both NASA and the USAF experimented successfully with spacecraft for close-up, on-orbit inspection of other spacecraft, a function not far removed from an on-orbit ASAT capability.

According to the Outer Space Treaty of 1967, nations should treat space as a uniquely fragile, global commons open to everyone and belonging to no one. Consequently, all who seek access to outer space ought to feel secure in their rights to use it on a sustainable basis, free from space-based threats by others. In the early years of the twenty-first century, however, a majority of experts perceived significant erosion in this concept of space security. As societies and economies around the world became increasingly dependent on harmonious support from civil and commercial space systems, trends in military space activity created ominous overtones. Competing civil, commercial, and military interests rendered the dynamics of space security unclear.

Rick W. Sturdevant

MILESTONES IN THE DEVELOPMENT OF MILITARY SPACE APPLICATIONS

1932 1 December: German Army Ordnance recruits Wernher von Braun.

1934 19 December: First flight of A-2 rocket.

1936 6 January: Planning of Peenemünde test site begins.

1937 4 December: First A-3 test flight.

1942 10 March: Johns Hopkins University established Applied Physics Laboratory.

3 October: First successful A-4 test.

1944 1 July: Army Ordnance founds Jet Propulsion Laboratory.

7 September: First V-2 launched at London.

1945 2 May: von Braun surrenders to U.S. troops.

9 July: White Sands Proving Ground created.

18 July: Soviet engineers set up Institute Rabe in Bleicherode, Germany, to reassemble V-2s.

3 October: Committee for Evaluating the Feasibility of Space Rocketry created by U.S. Navy.

1946 16 April: First V-2 flight from White Sands.

2 May: Project RAND report on satellite feasibility.

November: Soviet Union establishes Council of Chief Designers for ballistic missiles.

1947 27 July: Kapustin Yar selected as Soviet missile test site.

18 October: First Soviet V-2 test flight.

1948 14 May: RAND becomes independent, nonprofit corporation.

1950 24 July: First rocket launched from Cape Canaveral.

1952 4 November: National Security Agency established.

1954 January: Soviet R7 intercontinental ballistic missile (ICBM) program approved.

10 February: Teapot Committee recommends acceleration of U.S. ICBM program.

27 November: U.S. Air Force issues requirement for reconnaissance satellite.

1955 4 January: Titan ICBM program recommended.

20 May: U.S. National Security Council (NSC) issues NSC 5520, calling for scientific satellite program to support intelligence programs.

1956 7 May: Thor intermediate-range ballistic missile program approved.

1957 11 June: First Atlas ICBM test flight (failed).

12 September: North American Air Defense Command established.

4 October: *Sputnik* launched by Soviet military.

17 October: Baker-Nunn camera filmed *Sputnik* rocket body in orbit.

1958 7 February: Advanced Research Projects Agency established.

21 July: MITRE Corporation chartered.

16 December: First rocket launched from Vandenberg Air Force Base (AFB).

18 December: Launch of *SCORE*, world's first communications satellite.

19 December: USAF Interim National Space Surveillance and Control Center established.

1959 28 February: *Discoverer 1*, first polar-orbiting satellite, launched.

1960 26 February: *MIDAS 1* launched.

13 April: First Transit satellite launched.

3 June: The Aerospace Corporation formed.

22 June: Launch of *GRAB 1*, world's first electronic intelligence (ELINT) satellite.

18 August: Launch of *Discoverer 14*, first photoreconnaissance satellite.

4 October: Launch of U.S. Army's *Courier 1B*.

1961 1 January: Ballistic Missile Early Warning System at Thule AB, Greenland, declared operational.

31 January: First successful Samos launch.

1 July: Space Detection and Tracking System becomes operational.

6 September: National Reconnaissance Office established .

1962 26 April: First Zenit 2 photoreconnaissance satellite launched.

23 May: Department of Defense (DoD) cancels Advent program.

23 August: First successful launch of Defense Meteorological Satellite Program (DMSP) satellite.

1963 21 March: First launch of lifting reentry vehicle, M-12 Raketoplan model.

9 May: Successful West Ford launch and needle dispersion.

1 August: Program 505 Mudflap antisatellite (ASAT) declared operational.

17 October: First pair of Vela satellites launched.

10 December: Dyna-Soar program canceled; Manned Orbiting Laboratory (MOL) authorized.

1964 1 June: Program 437 ASAT becomes fully operational.

21 December: First Quill radar reconnaissance satellite launched.

1965 23 April: *Molniya 1* communications satellite launched.

6 May: First completely successful Lincoln Experimental Satellite launch.

1966 16 June: Launch of first seven Initial Defense Communications Satellite Program satellites.

1967 1 July: DODGE satellite launched.

27 October: First test of complete Soviet Istrebitel Sputnikov coorbital ASAT.

30 October: First launch of Soviet Tselina ELINT satellite.

23 November: First successful launch of Tsiklon navigation satellite.

1969 9 February: *TACSAT 1* launched.

10 June: MOL program canceled.

25 August: Soviet Fractional Orbital Bombardment System becomes operational.

21 November: *Skynet 1A* satellite launched.

1970 20 March: *NATO 2A* satellite launched.

6 November: First Defense Support Program satellite launch.

1971 15 June: First launch of KH-9 Hexagon.

2 November: Launch of first two Defense Satellite Communications System (DSCS) 2 satellites.

14 December: First White Cloud launch.

1972 26 May: Antiballistic Missile Treaty and Strategic Arms Limitation Talks (SALT) 1 agreement signed.

19 September: Launch of Soviet *Oko 1*.

1974 13 December: First successful launch of Yantar 2K satellite.

26 December: First Parus satellite launched.

1975 22 December: Soviet Raduga first launched.

1976 2 June: First launch of Satellite Data System satellite.

15 December: Tsikada satellite first launched.

19 December: KH-11 Kennan first launched.

1978 24 January: Radioactive debris from Radar Ocean Reconnaissance Satellite *Kosmos 954* spreads over northern Canada.

9 February: First Fleet Satellite Communications launch.

22 February: First Global Positioning System (GPS) Block 1 satellite launch.

1979 18 June: SALT 2 treaty signed.

1982 17 May: First launch of Potok satellite.

1 September: USAF Space Command established.

12 October: First Soviet Global Navigation Satellite System launch.

30 October: Launch of first DSCS 3.

1983 23 March: President Ronald Reagan announced Strategic Defense Initiative.

1 October: Naval Space Command established.

1985 13 September: *Solwind P78-1* destroyed by air-launched ASAT.

23 September: U.S. Space Command established.

25 October: First launch of Soviet Luch satellite.

1988 7 April: Army Space Command activated.

2 December: Lacrosse satellite first launched.

1989 14 February: First GPS Block 2 satellite launch.

1991 16 January: First Gulf War begins.

31 July: Strategic Arms Reduction Treaty (START) 1 signed.

1993 3 January: START 2 signed.

3 September: First successful launch of Ultrahigh Frequency Follow-On satellite.

1994 7 February: First launch of Milstar satellite.

1995 27 April: GPS becomes fully operational.

7 July: *Helios 1A* reconnaissance satellite launched.

1997 6 June: First launch of Araks satellite.

17 October: Mid-Infrared Advanced Chemical Laser successfully targets *MSTI 3.*

1998 1 October: USAF transfers operational control of DMSP to National Oceanic and Atmospheric Administration.

2000 1 May: GPS selective availability discontinued.

2001 7 February: Italy launches *Sicral 1.*

24 March: Russian Space Forces established as separate service.

2002 24 May: Strategic Offensive Reductions Treaty or Moscow Treaty signed.

9 August: First U.S. joint *Space Operations Doctrine.*

2003 18 March: Second Gulf War begins.

3 June: USAF becomes DoD executive agent for space.

24 November: National Geospatial-Intelligence Agency created.

2004 2 August: USAF publishes *Counterspace Operations Doctrine.*

2006 16 December: *TACSAT 2*, first launch in U.S. joint-services project to develop satellites specifically for tactical applications.

19 December: *SAR-Lupe 1*, first German military radar imaging satellite launched.

24 December: *Meridian 1*, first launch in new class of Russian government and military communications satellite.

2007 11 January: First successful Chinese ASAT test.

8 March: Orbital Express mission launched.

10 October: First launch of Wideband Global System satellite.

Rick W. Sturdevant

Bibliography
Center for Defense Information. http://www.cdi.org/.
Center for Nonproliferation Studies. http://cns.miis.edu/.
Teresa Hitchens and Tomas Valasek, *European Military Space Capabilities: A Primer* (2006).
Stephen B. Johnson, "The History and Historiography of National Security Space," in *Critical Issues in the History of Spaceflight*, ed. Steven J. Dick and Roger D. Launius (2006), 481–548.
RussianSpaceWeb. http://www.russianspaceweb.com/.
Asif A. Siddiqi, *Challenge to Apollo: The Soviet Union and the Space Race, 1945–1974* (2000).
David N. Spires, *Beyond Horizons: A Half Century of Air Force Space Leadership* (1997).

ANTISATELLITE SYSTEMS

Antisatellite systems, developed by the United States, the Soviet Union, and China to defend against potential space threats, such as an orbital nuclear bombardment system. Antisatellite (ASAT) systems included direct-ascent and orbital devices.

Early satellite development created concerns about the need for ASAT capabilities. The increasing nuclear rivalry between the United States and the Soviet Union in bomber and ballistic missile delivery systems threatened to spill into space. Critical space-based reconnaissance and early-warning systems were also potential links that, if severed, could give a decisive strategic advantage to the opponent. An effective ASAT system might successfully defeat or defend those vital satellites.

Nations have several options for conducting ASAT operations. First, one might destroy the satellite. Second, a country might attack electronically the communications link between the satellite and its ground control station. Third, a nation could strike the satellite's ground control facilities through a conventional military attack or sabotage. Fourth, an opponent could try to use denial and deception techniques, such as camouflage, to counter a satellite's ability to conduct its mission.

The United States and the Soviet Union concentrated on developing a capability to destroy each other's satellites. These efforts focused on either a direct-ascent weapon or an orbital system. Direct-ascent systems rely on a booster, such as a rocket or aircraft, to fire a device into a trajectory for interception of the satellite. These weapons have the potential, depending on the booster's capability and location, to intercept targets in low Earth orbit or higher. Orbital systems, also called co-orbital ASATs, allow the ASAT apparatus to maneuver into position to destroy its target.

The United States developed four notable direct-ascent ASAT systems. First, the Army maintained a single, nuclear-armed Nike-Zeus on Kwajalein Island in the Pacific from 1963 to 1966. Originally designed as an antiballistic missile (ABM) system to shoot incoming warheads, the Program 505 Mudflap Nike-Zeus system possessed the same capabilities required for an ASAT weapon: detection, tracking, and interception of an object in space. The second operational ASAT capability evolved directly from a series of high-altitude nuclear blasts conducted over Johnston Island, south of Hawaii. Those tests confirmed that nuclear explosions could damage sensitive electronic components on satellites. As a result, the U.S. Air Force (USAF) initiated Program 437, which maintained two nuclear-armed Thor intermediate-range ballistic missiles on alert constantly during 1964–70 to conduct ASAT operations. A third direct-ascent system, the Air-Launched Miniature Vehicle (ALMV), involved a USAF F-15 interceptor aircraft launching a small, two-stage rocket that carried a heat-seeking Miniature Homing Vehicle to destroy a satellite through direct, high-speed impact. Although developed and successfully tested in 1985, the ALMV was never deployed operationally, and the program was canceled in 1988. Fourth, during the 1990s, the Army contracted with The Boeing Company to develop a Kinetic Energy (KE) ASAT ground-launched missile system.

American military leaders investigated other ASAT options. The USAF's first attempt to develop an ASAT capability was its satellite inspector (SAINT) program in 1956. Based on a co-orbital concept that required the inspector vehicle to maneuver itself within 50 ft of its target to conduct an analysis of the satellite, SAINT held the possibility of evolving into a satellite interceptor with a kill capability. Objections to weapons in space, cost, and technical concerns forced cancellation of the SAINT program in 1962. Exploring another option, in 1997 the Army and USAF tested ground-based MIRACL (Mid-Infrared Advanced Chemical Laser) lasers to see if they could temporarily blind a satellite. The lasers successfully affected a satellite in a 300-mile orbit.

The Soviet Union also developed several ASAT alternatives to counter U.S. satellites. Its co-orbital ASAT, Polet or Istrebitel Sputnikov (IS), could position itself in lethal range and explode a device that expelled metal shards into its target. This conventional weapon, which required about two orbits to intercept a satellite, allegedly became operational in 1971. Additionally two lasers based at Sary Shagan had sufficient power possibly to blind the optical systems of reconnaissance satellites. The Soviet Union had a direct-ascent ASAT potential with its nuclear-tipped, exo-atmospheric Galosh ABM missiles deployed around Moscow. Under the modified 1972 ABM Treaty, it could have maintained one site with 100 interceptors. During the late 1980s, the Soviet Union undertook development of an air-launched ASAT similar to the ALMV of the United States.

On 11 January 2007, China successfully tested a direct-ascent, kinetic-kill ASAT weapon that destroyed that nation's own *Feng Yun 1C* polar-orbiting meteorological satellite. The test created many thousands of new fragments, making it the worst-ever single debris event in space. It sparked global protests and prompted many experts to express concern that a space-arms race might result.

Some proposed multi-use space systems had perceived ASAT capabilities. For example, the U.S. Dyna-Soar and Space Shuttle were thought to have the ability to carry weapons or devices that could damage or destroy a satellite in orbit. The Soviet Union also believed the Strategic Defense Initiative, a ballistic missile defense system, could be used as an ASAT system. Similarly, the United States suspected during the 1960s that the Soviet Union's Raketoplan/Spiral space plane and Almaz space station had ASAT potential.

Clayton K. S. Chun

See also: Antiballistic Missile Systems, Russian Space Forces, United States Air Force

Bibliography
Clayton K. S. Chun, "A Falling Star: SAINT, America's First Antisatellite System," *Quest* 6, no. 2 (Summer 1998): 44–49.
Paul Stares, *The Militarization of Space* (1985).
Mark Wade's "Encyclopedia Astronautica". http://www.astronautix.com/.

Air-Launched ASAT (Antisatellite). *See* Air-Launched Miniature Vehicle.

Air-Launched Miniature Vehicle

Air-Launched Miniature Vehicle (ALMV) was the primary U.S. antisatellite (ASAT) program during the 1980s. Although the U.S. Air Force (USAF) tested components several times, Congress cut funding and imposed significant testing restrictions, lest the United States spark a space-related arms race. The system was never deployed.

National Security Decision Memorandum 345, signed in January 1977 by Brent Scowcroft, President Gerald Ford's national security advisor, called for development of a non-nuclear ASAT capability. To meet that requirement, the USAF awarded Vought Corporation a contract in 1979 for development of the ALMV, otherwise called the Miniature Homing Vehicle or Prototype Miniature Air-Launched Segment. A direct-ascent ASAT designed for high-altitude launch from an F-15 fighter, the ALMV system measured 18 ft long, weighed 2,700 lb, and included three main components: a Short-Range Attack Missile first stage, an Altair III second stage, and a miniature homing vehicle that used eight cryogenically cooled infrared telescopes to acquire the target satellite with which it would collide at high velocity.

Even before the ALMV was ready for testing, powerful political forces threatened its existence. In unsuccessful negotiations during 1978–79, the United States held it as a potential bargaining chip to offset the Soviet co-orbital ASAT. Congress restricted

Launch of a U.S. Vought ASM-135 ASAT missile, 13 September 1985. (Courtesy U.S. Department of Defense)

ALMV testing in the early 1980s. In September 1985, the USAF conducted its only full-system test, which resulted in successful, but controversial, interception and destruction of the still-functioning *Solwind P78-1* satellite. Faced with additional funding cuts and a congressional ban on further testing unless the president certified that Soviet leaders had broken their self-imposed moratorium on testing their co-orbital system, the USAF recommended terminating the program, which occurred in March 1988.

Peter L. Hays

See also: United States Air Force

Bibliography

Dwayne A. Day, "Arming the High Frontier: A Brief History of the F-15 Anti-Satellite Weapon," *Spaceflight* 46, no. 12 (2004): 467–71.
Global Security. http://www.globalsecurity.org/.
Paul B. Stares, *The Militarization of Space: U.S. Policy, 1945–1985* (1985).

Dyna-Soar

Dyna-Soar, a United States Air Force (USAF) program to develop an orbital space-plane for military operations. Since the 1930s, two fundamental reasons for replacing aircraft had been to increase range and to decrease vulnerability. Yet even with jet engines, planes in 1952 could not fly nonstop from the United States to the Soviet heartland; even with aerial refueling, a capability still in development at that point, the most advanced conventional bomber would require more than 10 hours to reach Moscow, Russia, from Omaha, Nebraska. Consequently, in April 1952 Bell Aircraft

Company pressed the USAF to develop a hypersonic boost-glider, which would fly much faster, higher, and farther than any existing intercontinental jet bomber or guided missile—a true aerospace bomber, one that could operate in both air and space. The person primarily responsible for Bell's boost-glide program was Walter Dornberger, former commander of Nazi Germany's rocket installation at Peenemünde, who envisioned a reusable, hypersonic, glide bomber boosted into an orbital trajectory by a missile. Because Dornberger's bomber-missile concept, known as "BOMI," offered the prospect of extending airpower doctrine into space, the USAF financed a five-year program to study the feasibility of hypersonic boost-glide technology.

The Dyna-Soar program evolved from consolidation of several feasibility studies— Hywards, Brass Bell, and Robo—on 10 October 1957. Before year's end, the USAF had approved a development plan for System 464L or Dyna-Soar, an acronym for Dynamic Soaring, a piloted, reusable spacecraft for reconnaissance, nuclear bombardment, or antisatellite missions. The Soviet Union countered in 1960 with its own Raketoplan development program that evolved after 1964 into Spiral.

Because the Soviet Union might have perceived an operational Dyna-Soar as a direct military threat, both the Eisenhower and Kennedy presidential administrations preferred to rely on robotic, unarmed reconnaissance satellites such as Corona. On 21 February 1962, the USAF deleted military subsystems from the Dyna-Soar program and, two days later, Defense Secretary Robert S. McNamara restricted program objectives to development of an orbital research system. The Department of Defense officially designated the Dyna-Soar glider as the X-20 on 26 June 1962.

Technical advances in NASA's human spaceflight program, lack of a clear operational requirement, pressure for a military space station instead of an orbital glider, and national policy opposing weapons in space converged, culminating in the cancellation of Dyna-Soar on 10 December 1963. At the same time that McNamara announced Dyna-Soar's demise, he authorized the USAF to develop a space station—the Manned Orbiting Laboratory. Although Dyna-Soar never flew, its legacy appeared in the Space Shuttle's design features and operational capabilities. Furthermore, wind-tunnel testing of Dyna-Soar models improved research methodologies for later programs, and the need for a bigger, better booster to launch Dyna-Soar resulted in the Titan III.

Roy F. Houchin

See also: Hypersonic and Reusable Vehicles, Manned Orbiting Laboratory, Space Shuttle

Bibliography

Robert Godwin, ed., *Dyna-Soar: Hypersonic Strategic Weapons System* (2003).

Roy F. Houchin, "Hypersonic Technology and Aerospace Doctrine," *Air Power History* 46, no. 3 (Fall 1999): 4–17. *Quest* 3, no. 4 (Winter 1994).

High-Altitude Nuclear Tests

High-altitude nuclear tests, accomplished by launching a rocket armed with a nuclear weapon to an altitude greater than 40 kilometers, were conducted by the United States and the Soviet Union during 1958–62.

The United States performed the world's first such test. Called Teak, it employed a rocket to send a 3.8-megaton device to an altitude of 76.8 km over Johnston Island in the Pacific on 1 August 1958. A series of three tests, code-named Argus, began on 27 August 1958 in the South Atlantic when the USS *Norton Sound* launched a rocket carrying a one- to two-ton nuclear package. The tests, which evaluated the effects of blast, heat, X rays, nuclear radiation, and electromagnetic pulse on satellites and ballistic missile systems, occurred in space at an altitude of 500 km. To map radiation from the Project Argus tests, the Defense Department's Advanced Research Projects Agency launched its *Explorer 4* satellite. Scientists reported that X rays and other phenomena related to high-altitude nuclear explosions could seriously damage the electronic systems in satellites and ballistic missiles.

On 9 July 1962, the United States conducted an experiment with a Thor missile carrying a 1.4-megaton payload. This Starfish Prime test, part of the Fishbowl series, occurred at an altitude of 400 km over Johnston Island and inadvertently crippled at least three Applied Physics Laboratory satellites and the first Telstar satellite, which was launched the day after the test. Furthermore, electromagnetic pulse from the explosion disrupted electrical components and communications 800 km away in the Hawaiian Islands. Four more Fishbowl tests followed at various yields and altitudes. Data collected throughout the 1960s by several spacecraft—*OGO 1*, *OGO 3*, *OGO 5*, *OV3-3*, and *1963-38C* (also called *Transit SE 1*)—substantiated that the Starfish Prime radiation flux lasted years. Although the vulnerability of all satellites, friendly and unfriendly, to high-altitude nuclear explosions became apparent, both the United States and the Soviet Union developed antisatellite capabilities (Program 437 and Galosh ABM, respectively) using nuclear-tipped missiles.

The Soviet Union, like the United States, conducted several high-altitude nuclear tests. Its first, on 27 October 1961, included a 1.2-kiloton device that exploded at an altitude of 150 km. Four more tests followed before the program ended in November 1962. All high-altitude nuclear testing ended with the Limited Test Ban Treaty of 1963. To monitor compliance with that treaty, the U.S. Air Force, beginning in October 1963, launched several pairs of Vela nuclear-test detection satellites, which subsequently proved invaluable for the scientific study of naturally occurring gamma radiation from deep space.

Clayton K. S. Chun

See also: Nuclear Detection, Nuclear Test Ban Treaty

Bibliography
V. N. Mikhailov, *Catalog of Worldwide Nuclear Testing* (1999).

Istrebitel Sputnikov. *See* Soviet Antisatellite System.

Kinetic Energy Antisatellite

Kinetic Energy Antisatellite (KE-ASAT), a ground-launched, direct-ascent antisatellite (ASAT) system consisting of missile and weapon-control subsystems. The former included a booster, shroud, Kinetic Kill Vehicle (KKV), and launch-support capability. A battery control center and mission control element constituted the weapon-control subsystem. By 2006 the United States had neither tested KE-ASAT in space nor declared it operational.

In 1989, the year after cancellation of the U.S. Air Force (USAF) Air-Launched Miniature Vehicle (ALMV) program, the U.S. Army initiated the KE-ASAT program. Like the ALMV, it would use kinetic energy (physically hit the target) to disable satellites in low Earth orbit, but KE-ASAT would minimize creation of space debris by employing a large Kevlar "fly swatter" mechanism to contain pieces of the target satellite. In 1990 Boeing won the KKV demonstration/validation contract and, in 1996, a follow-on development contract. Hover testing of a prototype in August 1997 validated the KKV concept.

KE-ASAT was criticized across the political spectrum. The Department of Defense did not specifically request KE-ASAT funding after President William Clinton's administration terminated it in 1993. Nonetheless, Congress revived the program in 1996 and funded it as a relatively small-scale, research-and-technology-demonstration effort, only to become disillusioned with its management and results. A General Accounting Office report in December 2000 described the program as in disarray. Additional criticism occurred when opponents discovered approximately $15 million annually continued to flow toward KE-ASAT development under the USAF account for space-control technology. In 2002 the Army Space and Missile Defense Command (SMDC) became responsible for the KE-ASAT program and contracted with Davidson Technologies Inc. of Huntsville, Alabama, for direct support. Work on the KE-ASAT concept continued in 2005 under an SMDC contract with Huntsville-based Miltec Corporation to build an Applied Counterspace Technology test bed at Redstone Arsenal.

Peter L. Hays

Bibliography
Center for Defense Information. http://www.cdi.org/.
Global Security. http://www.globalsecurity.org/.
Steven J. Lambakis, *On the Edge of Earth: The Future of American Space Power* (2001).

MIRACL, Mid-Infrared Advanced Chemical Laser

MIRACL, Mid-Infrared Advanced Chemical Laser, a megawatt-class, continuous-wave, chemical laser managed by the U.S. Army Space and Missile Defense Command (SMDC) at its High Energy Laser Systems Test Facility on White Sands Missile Range, New Mexico. Burning ethylene as fuel with nitrogen triflouride as oxidizer in a rocket-like engine to produce excited fluorine atoms as an exhaust product, MIRACL then injected deuterium that combined with the fluorine to produce excited deuterium fluoride molecules from which the laser's resonator mirrors extracted optical energy. In the two decades after 1980, when it first lased, MIRACL underwent nearly 160 tests and accumulated more than 3,600 seconds of lasing time.

The U.S. Navy procured MIRACL, developed by TRW Corporation, and its associated SeaLite Beam Director (SLBD), built by Hughes Aircraft, to demonstrate the potential of lasers for ship defense against aircraft and tactical missiles. Although Congress canceled the SeaLite program in 1983, the U.S. Air Force (USAF) began using the MIRACL and SLBD equipment in 1985 for antiballistic missile tests under President Ronald Reagan's Strategic Defense Initiative. Too large for operational deployment by the U.S. military, MIRACL nonetheless paved the way for development of smaller, high-energy lasers mounted on ground-mobile, airborne, or space-based platforms.

The potential use of high-energy lasers as antisatellite (ASAT) weapons became especially controversial. Critics feared a new arms race and further militarization of outer space. A congressionally imposed ban prevented testing MIRACL on an orbiting satellite during 1990–95, but its expiration cleared the way for experimentation. In October 1997 MIRACL successfully targeted the USAF *Miniature Sensor Technology Integration 3* (*MSTI 3*) satellite for 10 seconds at various power levels, never above 50 percent of the laser's maximum. Afterward a USAF spokesman said the test had allowed collection of information about the effects of laser beams on satellite components without permanently damaging *MSTI 3*.

Peter L. Hays

See also: Strategic Defense Initiative, TRW Corporation

Bibliography

Federation of American Scientists. http://www.fas.org/.
Steven J. Lambakis, *On the Edge of Earth: The Future of American Space Power* (2001).

Nike Zeus. *See* Program 505 Mudflap.

Polet. *See* Antisatellite Systems.

Program 437

Program 437, an antisatellite (ASAT) system the United States Air Force (USAF) operated during 1964–70. Crews maintained two nuclear-armed Thor intermediate range ballistic missiles on Johnston Island in the Pacific.

Program 437 originated from a series of high-altitude nuclear tests conducted by USAF and Atomic Energy Commission scientists in 1962 that explored the effect of electrically charged particles on satellites. Code-named Fishbowl, those tests used Thor missiles to demonstrate the deadly result of electromagnetic pulse from a nuclear explosion. Consequently when USAF leaders advocated adoption of nuclear-armed Thor missiles to defend against a multiple orbiting bombardment system and other enemy satellites, Secretary of Defense Robert S. McNamara ordered immediate development of the system, which became fully operational in May 1964.

The Thor ASAT system was under the control of the Aerospace Defense Command's 10th Aerospace Defense Squadron, which operated and maintained the system. Armed with a Mark 49 1.4-megaton nuclear warhead, the system had an effective kill range of about 5 miles to compensate for interception and timing errors, but crews intercepted test targets as close as 131 ft. Analysts needed several hours to track a satellite and conduct missile launch preparations. Crews would prepare both missiles for launch, one as the primary and the other for backup. Personnel from Vandenberg Air Force Base, California, provided logistical support. Department of Defense and USAF officers experimented with Thor as a satellite inspector. Limited capability, expensive operations, launch site vulnerability, and other concerns forced the USAF to terminate Program 437 on 2 October 1970. A typhoon on 19 August 1972 destroyed the Johnston Island launch facilities, which ended the program.

Clayton K. S. Chun

See also: Delta, United States Air Force

Bibliography

Clayton K. S. Chun, *Shooting Down a "Star": Program 437, the U.S. Nuclear ASAT System and Present-Day Copycat Killers* (2000).

Program 505 Mudflap

Program 505 Mudflap, the first U.S. operational antisatellite system, operated from 1 August 1963 to 23 May 1966. The program used a modified Nike Zeus antiballistic missile, the DM-15S, a direct descendant of the Nike Ajax and Nike Hercules family of surface-to-air missiles that defended the nation against the Soviet Union nuclear armed bomber threat during the 1950s. As the threat evolved from bombers to long-range missiles, the U.S. Army initiated Nike Zeus development in 1957 to counter a possible Soviet intercontinental ballistic missile (ICBM) attack by intercepting incoming warheads while still outside Earth's atmosphere. Although concerns about the system's survivability in a nuclear conflict and the reliability of its radar guidance subsystem led to abandonment of Nike Zeus as an antimissile weapon, the Army reoriented the program in the early 1960s to counter the emerging threat of a Soviet orbital bombardment system.

The Army's advocacy of Nike Zeus as a nuclear-armed antisatellite system proved fortuitous. When political sensitivities and technical concerns forced the U.S. Air Force (USAF) to scrap its SAINT (Satellite Interceptor) on-orbit interception and inspection program, national leaders turned to the Army. In April 1962 Secretary of Defense Robert S. McNamara asked the Army to expand its Nike Zeus antisatellite experiments. Encouraged by the results, McNamara approved Program 505 Mudflap on 1 May 1962.

Testing of a modified Nike Zeus, designated DM-15S, began in December 1962 at White Sands Missile Range, New Mexico. It had an extended range of 150 miles and could deliver a one-megaton warhead to an altitude of 151 miles. A DM-15S test missile fired from Kwajalein Atoll on 24 May 1963 demonstrated it could strike into space when it came within a predicted kill radius against a target Agena D upper stage. That led McNamara to direct on 27 June 1963 that a Nike Zeus antisatellite weapon be placed on alert at Kwajalein, and a single DM-15S became operational there on 1 August. The USAF rival Program 437, which had greater range but slower reaction time, combined with pressure to fund the Vietnam War and a controversy about service roles and missions, forced McNamara to cancel Program 505 in May 1966. That left the USAF in control of U.S. antisatellite operations.

Clayton K. S. Chun

See also: United States Army, White Sands Missile Range

Bibliography
Clayton K. S. Chun, " 'Nike-Zeus' Thunder and Lightning," *Quest* 10, no. 4 (2003): 40–47.

Raketoplan and Spiral

Raketoplan and Spiral, successive Soviet Union programs during the 1960s–70s to develop a military spaceplane capable of orbital maneuvering to intercept, inspect, and destroy enemy satellites and of landing at conventional airfields. Semyon Lavochkin's OKB-301 (Experimental Design Bureau) had worked during the early 1950s on the La-350 Burya intercontinental cruise missile for delivery of nuclear weapons, and plans called for its evolution into a piloted shuttle-type, recoverable vehicle. Before the Soviet Union canceled it in 1960, the Burya project supplied experienced cruise-missile designers like Chelomey with useful results. In July 1958 government officials had permitted Chelomey's OKB-52 to begin development of long-range, winged ballistic missiles and, within a year, Chelomey had begun exploring the idea of winged, piloted spaceships along two lines—a Kosmoplan (Space Glider) for interplanetary flights and a Raketoplan (Rocket Glider) for Earth-orbital flights.

Initially OKB-52 worked on several Raketoplan variants for piloted, suborbital missions, such as reconnaissance, bombing, and foreign-satellite identification. An official decree in May 1961, however, redirected the bureau's efforts toward a new, piloted variant of the Raketoplan for military missions in Earth orbit and deep space. By 1963 OKB-52 engineers had drafted a plan containing four variants: a single-seat, orbital antisatellite spaceplane; a single-seat orbital bomber for terrestrial targets; a seven-passenger, intercontinental ballistic spacecraft; and a twin-seat scientific model for circumlunar flight. Presumably, regardless of the variant, these vehicles would deorbit in a thermally protected container that would be discarded after atmospheric reentry, deploy wings, and use a turbojet engine to land on any airstrip.

The launch of M-12, a Raketoplan scale model, on 21 March 1963 marked the world's first test flight of a lifting reentry vehicle. Although OKB-52 engineers had begun constructing full-scale R-1 and R-2 test models before Premier Nikita Khrushchev's fall from power in October 1964, lingering political difficulties combined with technical challenges to cause suspension of the Raketoplan development program by January 1965. Nonetheless Chelomey's interest in winged, reusable spacecraft continued, leading him in 1975 to employ Raketoplan subscale test data in designing

764 | Military Applications • Antisatellite Systems

the LKS (light space aircraft) interceptor, a new spaceplane similar to the European Hermes. Not until September 1983, 14 months before Chelomey died, did Soviet officials successfully terminate his spaceplane proposal.

Meanwhile, in 1965 the Soviet Air Force had ordered OKB-52's research database transferred to Artem Mikoyan's OKB-155. The simultaneous movement of several of Chelomey's best Raketoplan designers from OKB-52 to Mikoyan's OKB-155 signaled takeoff of the Spiral spaceplane program under Chief Designer Gleb Lozino-Lozinskiy. Three major components comprised the Spiral system: a reusable, hypersonic, air-breathing launch aircraft (GSR); an expendable two-stage rocket (RB); and an orbital spaceplane (OS). An atmospheric flight-test version of the Spiral OS, the MiG 105-11 or Spiral EPOS (Experimental Passenger Orbital Aircraft), incorporated the airframe and some systems of the projected orbital version. Although technical problems and insufficient funding delayed the first EPOS flight until October 1976, construction and testing of several, subscale BOR (unpiloted orbital rocket) lifting bodies during the late 1960s provided valuable design data for the Spiral program. After the eighth EPOS test in September 1978, the Spiral program ended. In December 1981 Soviet officials ordered 48 senior members of the Spiral design team to join their former leader, Lozino-Lozinskiy, at NPO (Scientific-Production Association) Molniya to assist with development of the Buran space shuttle.

Rick W. Sturdevant

See also: Hypersonic and Reusable Vehicles

Bibliography
Peter Pesavento, "Russian Space Shuttle Projects, 1957–1994," *Spaceflight* 37 (May, June, July, August 1995).
Russian Space Web. http://www.russianspaceweb.com/.
Asif A. Siddiqi, *Challenge to Apollo: The Soviet Union and the Space Race, 1945–1974* (2000).

SAINT, Satellite Interceptor. *See* Antisatellite Systems.

Soviet Antisatellite System

Soviet antisatellite system, co-orbital IS—Istrebitel Sputnikov (satellite destroyer)—conceived by designer Vladimir Chelomey in 1959. The IS satellite would enter an orbit similar to that of its target, maneuver closer to the target, then detonate conventional explosives to destroy the target with shrapnel. During an April 1960 meeting, Soviet leaders discussed the possibility of developing a semiautomated, maneuverable spacecraft for antisatellite (ASAT) purposes, and an official resolution to proceed followed in July 1960. Undoubtedly it came in response to a combination of factors: their shooting down of Francis Gary Powers's U-2 reconnaissance aircraft in May; their concern about the impending capability of U.S. reconnaissance satellites to orbit over Soviet territory; and their perception of U.S. offensive intent behind

Soviet military space capabilities posed an ever-increasing threat to U.S. land, sea, air, and space forces in the 1980s. The Soviet Union operated and tested an orbital antisatellite weapon that was designed to destroy space targets with a multi-pellet blast. (Courtesy Ronald C. Wittmann/Defense Intelligence Agency)

the SAINT (satellite interceptor) program. Responsibility for overall design of the IS system went to Anatoli Savin's KB-1 bureau, which also was involved with designing systems for ballistic missile defense, with Chelomey's OKB-52 (Experimental Design Bureau) developing the spacecraft itself.

On-orbit testing began with Polet (flight) prototype launches in November 1963 and April 1964 of the control and propulsion subsystems required for maneuvering the IS spacecraft to its target. When development of the UR-200 launch vehicle originally intended for the IS lagged, further on-orbit testing was delayed pending manufacture of a substitute rocket, designated 11K67 (Tsiklon 2A). Consequently the first test of a complete IS satellite did not occur until 27 October 1967. One year later, for the first time, an IS satellite used onboard radar guidance to intercept a target satellite repeatedly before the target's destruction by a second IS, *Kosmos 252*, which generated 139 orbital fragments. In 1971 a new type of armored target satellite, with the designation DS-P1-M, allowed Soviet engineers to collect data on the number of shrapnel hits during as many as three attacks by IS interceptors. Before compliance with the Antiballistic Missile Treaty in 1972 caused a temporary cessation in IS testing, the Soviet Union had successfully performed seven interceptions and five detonations.

ANATOLI IVANOVICH SAVIN
(1920–)

Anatoli Savin, a participant in the Soviet Union nuclear project during World War II, later worked in the radio-electronics industry and participated during the 1960s in development of the Soviet strategic antimissile defense system. From 1973 to 1999, he headed TsNII Kometa, a leading manufacturer of Soviet "smart weapons." He directed research and development of such military space systems as the Istrebitel Sputnikov (IS) antisatellite program, Oko for strategic early warning, and MKRTs Legenda, which combined space-based electronic and radar reconnaissance capabilities for naval targeting. In the 1990s Savin became an ardent supporter of international cooperation for peaceful uses of outer space.

Peter A. Gorin

A resumption of IS testing in 1976 gave Soviet engineers an opportunity to demonstrate that during its first orbit after launch, the ASAT could intercept and destroy its target. In addition they doubled the altitude at which an IS could intercept and destroy its target. Testing of optical guidance, however, proved less satisfactory than radar for purposes of target interception. By July 1979 the Soviet Union considered its IS system operationally ready, but on-orbit testing continued into June 1982 at the rate of approximately one intercept per year. Much to the Soviet military establishment's consternation, Soviet Communist Party General Secretary Yuri Andropov announced a unilateral moratorium on ASAT testing in March 1983. Despite a failed attempt in May 1987 to launch a 90-ton Polyus spacecraft capable of carrying ASAT weapons and unconfirmed rumors of a MiG-launched satellite interceptor project, Andropov's edict effectively halted the Soviet ASAT program.

Rick W. Sturdevant

See also: Reconnaissance, Russia (Formerly the Soviet Union)

Bibliography

Asif A. Siddiqi, "The Soviet Co-Orbital Anti-Satellite System: A Synopsis," *Journal of the British Interplanetary Society* 50, no. 6 (June 1997): 225–40.

Paul B. Stares, *The Militarization of Space: U.S. Policy, 1945–84* (1985).

Thor ASAT (Antisatellite). *See* Program 437.

X-20. *See* Dyna-Soar.

BALLISTIC MISSILES AND DEFENSES

Ballistic missiles and defenses emerged during the last years of World War II as major military concerns. Even as Nazi Germany launched V-2 missiles strategically against

Great Britain and tactically against the Remagen Bridge, the Allies studied how to defend against the V-2, while at the same time beginning to develop their own ballistic missiles in response. Concluding they could not intercept and destroy V-2s in flight, they attacked the launchers. The limited success of that effort, however, left psychologically shaken civilians clamoring for more extreme measures. Consequently, the Allies considered aiming massive, timed artillery barrages toward anticipated V-2 trajectories to defend London, but the potential for unexploded ordnance falling back onto the city and the conclusion of the war in Europe rendered that discussion moot.

The seeming impossibility of defending against ballistic missiles made a huge impression on military planners in the United States, Soviet Union, United Kingdom, and France, who immediately after the war exploited the V-2 "rocket scientists" to rapidly develop their own ballistic missiles. Even as defense budgets dwindled in the late 1940s, the U.S. Army Air Forces (precursor to the U.S. Air Force) began projects to study the feasibility of long-range ballistic missiles capable of traveling 5,000 miles—MX-774—and antimissiles capable of intercepting and destroying missiles traveling 4,000 mph at an altitude of 500,000 ft—MX-794 Wizard and MX-795 Thumper. In May 1946 the War Department Equipment Board chaired by General Joseph Stilwell recommended that the United States use radar, computer, and guided-missile technologies to design defensive antiballistic missile (ABM) systems. Nine years later, Bell Telephone Laboratories used an analog computer to simulate 50,000 intercepts of ballistic missiles, thereby strengthening the position of U.S. ABM advocates. Secretary of Defense Neil McElroy assigned primary responsibility for ballistic missile defense to the U.S. Army in January 1958. By that time, the United States and the Soviet Union had begun flight testing their respective versions of an intercontinental ballistic missile (ICBM).

The Soviet Union followed an almost parallel course with respect to development of ballistic missiles and defenses. Under its Special Committee for Reactive Technology, a large organization emerged in the late 1940s to develop long-range ballistic missiles. By September 1953 progress toward development of an ICBM prompted senior military leaders to ask for, and receive, permission from the Central Committee of the Communist Party to study the feasibility of a Soviet ABM system. According to an official Soviet report, a computer-controlled V-1000 Griffon antimissile with a high-explosive warhead first intercepted and destroyed a ballistic-missile warhead in March 1961. A similarly successful test by the United States in 1962 placed both super-powers, which already had deployed operational ICBMs, on course to build operational ABM systems.

From the early 1960s onward, technological improvements rendered ballistic missiles more reliable, accurate, and lethal. Their proliferation throughout the world, placement of multiple independently targeted reentry vehicles (MIRVs) on ICBMs, and deployment of the Soviet Fractional Orbital Bombardment System prompted Soviet and U.S. leaders to seek commensurate advances in missile defense. When the latter proved too technically challenging to be operationally satisfactory, leaving ballistic missiles practically invulnerable in flight, both superpowers formulated offensive, rather than defensive, nuclear doctrines. This led to a costly offensive arms race and to the concept of Mutual Assured Destruction (MAD), where the United States and the Soviet Union both possessed sufficient nuclear weaponry to completely annihilate the other, regardless of which side launched a first strike. Meanwhile, despite the questionable effectiveness of ABM systems, both countries began deploying them at great expense. That generated the potential for dangerous defensive competition, plus escalation of the already intense offensive missile race and a heightened likelihood of strategic instability.

Consequently, in November 1969 U.S. and Soviet negotiators began talks in Helsinki, Finland, to limit offensive and defensive strategic, nuclear forces. The first series of Strategic Arms Limitation Treaty (SALT) talks, which ended in January 1972, resulted in agreement on two documents: an ABM Treaty severely limiting the deployment of missile defense systems, and an Interim Agreement on the Limitation of Strategic Offensive Arms. The latter provided the basis for a subsequent SALT II treaty that, although never ratified by the United States, met with Soviet and U.S. compliance into the mid-1980s.

By the 1980s U.S. experts became increasingly enamored with the possibility of achieving deterrence through the development of both ground- and space-based, primarily nonnuclear missile defenses instead of continuing to threaten the Soviet Union with assured destruction by ballistic missiles carrying thermonuclear warheads. That resulted in President Ronald Reagan's Strategic Defense Initiative and in renewed negotiations to limit strategic nuclear weapons. The Intermediate-range Nuclear Forces (INF) Treaty, signed at Washington, DC, in December 1987, called for unprecedented onsite inspections to verify compliance with its terms and resulted, by May 1991, in the elimination of all Soviet and U.S. nuclear, ground-launched ballistic and cruise missiles with ranges between 500 and 5,500 km, along with associated infrastructure. Signed in July 1991, the Strategic Arms Reduction Treaty (START) I limited the number of nuclear delivery vehicles—missiles and bombers—to 1,600 with no more than 6,000 warheads. START II, signed in January 1993 but never enforced, eventually was surpassed in May 2002 by the Strategic Offensive Reductions Treaty (SORT), which called for the United States and Russia to reduce their operationally deployed nuclear warheads to 2,200 and 1,700, respectively, by 2012.

The demise of the Soviet Union reduced the strategic threat, but enhanced dangers from rogue nations, such as Iraq and North Korea, and terrorist groups, such as Al Qaeda and Hezbollah. The First Gulf War in 1991 highlighted the increasingly urgent need for developing better capabilities for theater or tactical missile defense to counter threats from rogue nations or terrorist organizations. In the Second Gulf War in 2003, U.S. theater missile defenses successfully intercepted a number of Iraqi missile attacks. The terrorist threat was graphically demonstrated in the summer of 2006 by the launch of hundreds of ballistic missiles by Hezbollah against Israel. By that time, the United States had withdrawn from the ABM Treaty and had deployed an initial strategic defense system in Alaska to defend against a North Korean missile strike. The quest for reliable, effective strategic and theater missile defense systems continued, along with planning for new missile systems to replace aging models.

Rick W. Sturdevant

See also: Early Warning; Korolev, Sergei; Von Braun, Wernher

Bibliography

Clayton K. S. Chun, *Thunder over the Horizon: From V-2 Rockets to Ballistic Missiles* (2006).

Missile Defense Agency. http://www.mda.mil/.

Richard L. Russell, "Swords and Shields: Ballistic Missiles and Defenses in the Middle East and South Asia," *Orbis* 46, no. 3 (Summer 2002): 483–98.

System Planning Corporation, *Ballistic Missile Proliferation—An Emerging Threat* (1992).

Antiballistic Missile Systems

Antiballistic missile systems are generally composed of three basic components that together defend against a ballistic missile attack. Sensors detect enemy missile launches and begin tracking them, which includes discriminating between warheads and decoys. Command-and-control systems use the sensor data to direct the course of the antiballistic missile (ABM) battle, which involves weapons that destroy the attacking missiles or warheads. The technical difficulties and large expense associated with implementing these extremely complex systems sparked a decades long, continuing debate.

Two types of sensors, active and passive, detect and track incoming targets. Radars, the principal active form, transmit their own radio signal and detect objects that reflect it. Modern ABM radars use a phased-array concept where the transmitted beam is steered electronically to handle numerous targets and interceptors simultaneously. To ensure the required accuracy, ABM radars operate in higher frequency bands where short wavelengths allow detection and tracking of small objects over great distances. For example, X-band radars can "see" a golf ball 2,400 miles distant. Infrared sensors, the primary passive form, detect infrared radiation from a missile's exhaust. Beginning in the early 1970s, U.S. Air Force Defense Support Program (DSP) and Soviet Oko satellites employed infrared (IR)-sensing technology to detect missiles during powered flight. In the 1990s Russia improved its detection capabilities with the launch of Prognoz satellites, and the United States planned for a Space-based Infrared System to replace DSP. With the proliferation of ballistic missile technology and advances in IR detectors, nations such as France and Japan, and the multinational European Union, began studying the feasibility of obtaining their own launch-detection spacecraft.

The first ABM weapons, developed in the early 1960s by both the United States and the Soviet Union, were rocket-powered interceptors. Their inaccuracy, due to their being guided by radio signals from the ground, necessitated nuclear warheads to ensure a reasonable probability of destroying their targets. The U.S. Army's Nike-Zeus, designed to protect large urban areas, successfully intercepted an Atlas intercontinental ballistic missile (ICBM) reentry vehicle (RV) in July 1962. Nonetheless, reliance on mechanically driven radars made tracking a large number of warheads impractical and discrimination between decoys and actual warheads difficult. In 1968, the U.S. Army received its first Spartan, which added a third stage and updated subsystems to the Nike-Zeus. The exo-atmospheric Spartan successfully intercepted its first RV in August 1970.

Shortly thereafter, in December 1970, the short-range, endo-atmospheric Sprint destroyed an RV launched by a Minuteman ICBM. The Soviet Union mirrored the Spartan-Sprint combination during the 1960s by developing medium-range Galosh and, during the 1970s, short-range Gazelle ABM missiles.

During 1962–63, the Soviet Union began constructing the first ABM system. To protect Moscow, the original components of the A-35 system included a network of Dnestr-M radars on the Soviet Union's periphery for early warning, Dunai-3M and Dunai-3U radars for tracking and battle management, and nuclear-armed Galosh interceptors. When the Soviet Union refused to discuss ABM limitations in 1967, President Lyndon Johnson directed deployment of "Sentinel," a Spartan-Sprint system. Faced with political opposition against the deployment of Sentinel sites near most major U.S. cities, two years later, President Richard Nixon refocused Sentinel to protect ICBM bases instead of cities and renamed it "Safeguard." Despite technical progress, the enormous expense and questionable effectiveness encouraged leaders of the two superpowers to sign the ABM Treaty, part of the first Strategic Arms Limitation Treaty, in 1972. A 1974 revision limited each superpower to one fixed, land-based ABM site. The United States completed its single Safeguard site in northeastern North Dakota in October 1975, only to cease operations two months later. Although the Soviet Union began system improvements in 1978 to develop the upgraded A-135 system around Moscow, the upgrade remained incomplete and the system's operational readiness doubtful in 2005.

By the early 1980s, computer technologies, missile flight-control components, and infrared sensors had advanced such that onboard guidance systems could direct an interceptor to collide physically with its target and destroy it through the kinetic energy produced by the collision. The U.S. Army successfully demonstrated this in June 1984 and, henceforth, all U.S. ABM missiles used the kinetic-kill principle. That demonstration was part of the Strategic Defense Initiative (SDI), popularly termed "Star Wars," which President Ronald Reagan had announced in March 1983. The SDI Organization, renamed the Ballistic Missile Defense Organization in 1993 and the Missile Defense Agency in 2002, sought to develop new ABM system components to defend the United States against missile attacks. These included space-based, kinetic-kill interceptors such as Brilliant Pebbles in addition to land-based capabilities like the Terminal (originally Theater) High-Altitude Area Defense missile system. Research, begun around 1960, continued on directed-energy weapons such as lasers or particle beams that could destroy attacking missiles from platforms in space or the air, at sea, or on the ground. Considerable effort also went into designing computer software for the Battle Management/Command, Control, and Communications system that would direct the ABM response to incoming warheads. The Russians, meanwhile, pursued similar advanced technologies with especially heavy emphasis on lasers, but most experts assessed their progress as approximately a decade behind the United States.

Despite the questionable performance of the U.S. Army's Patriot missile against Iraqi Scuds during the First Gulf War in 1991, and less than satisfactory intercept tests over the Pacific Ocean throughout the 1990s, President George W. Bush announced his determination in May 2001 to deploy an operational ABM "test bed" within a decade. When first deployed, it would have limited operational capability but, as technical problems were overcome, system capability would expand and eventually reach fully operational status. Those who favored the president's decision reasoned that, in an era of complex systems, the line between operational testing and actual operations had diminished virtually to the vanishing point. Opponents argued that near-term prospects for building an ABM system with a high probability of destroying hostile missiles appeared too dismal to warrant the expense.

Undeterred by critics, President Bush notified Russia in December 2001 that the United States intended to withdraw from the 1972 ABM Treaty, because it hindered the ability of the United States to protect its people from future terrorist or rogue-state missile attacks. By early 2005, the U.S. Army had placed six interceptor missiles at Fort Greely, Alaska, and two at Vandenberg Air Force Base, California. Russia continued its ABM research and operations.

Rick W. Sturdevant

See also: Ballistic Missile Early Warning System, Defense Support Program, Oko

Bibliography
Donald R. Baucom, *The Origins of SDI: 1944–1983* (1992).
Missile Defense Agency, *Harnessing the Power of Technology: The Road to Ballistic Missile Defense from 1983 to 2007* (2000).
Paul Whitmore, "Red Bear on the Prowl: Space-related Strategic Defense in the Soviet Union," pts. 1 and 2, *Quest* 9, no. 4 (2002): 22–30; and 10, no. 1 (2003): 54–62.
Steven J. Zaloga, *The Kremlin's Nuclear Sword: The Rise and Fall of Russia's Strategic Nuclear Forces, 1945–2000* (2002).

Ballistic Missiles

Ballistic missiles, rocket-propelled devices designed to fly without reliance on aerodynamic surfaces for maneuvering and to deliver an explosive warhead to a target. While the military rocket made its operational debut in thirteenth-century China, it remained largely unchanged and ineffective as a military weapon until early in the twentieth century. Then, beginning in 1926 when American rocket pioneer Robert H. Goddard flight-tested the world's first liquid-propellant rocket, this device evolved rapidly into an intercontinental ballistic missile (ICBM) capable of delivering a multi-megaton thermonuclear warhead within 500 ft of an intended target. These powerful weapons came to threaten the existence of the human race, creating a "balance of terror" as the United States and the Soviet Union confronted each other across the decades of the Cold War.

About seven years after Goddard's flight test, the German Army initiated a major rocket development program. Headed by a brilliant young engineer and rocket enthusiast, Wernher von Braun, the German program sought to produce a weapon that would extend the range of traditional artillery and possibly serve as a terror weapon to undermine enemy morale. The German Army originally asked von Braun to

Test flight of an LGM-118A Peacekeeper Intercontinental Ballistic Missile at Vandenberg Air Force Base, California, 13 November 1985. (Courtesy U.S. Department of Defense)

develop a rocket that could deliver a one-ton warhead over a range of about 150 miles. In 1942, his team successfully tested a rocket meeting those specifications. Developed under the designation A-4 and popularly known as the V-2, this weapon made its operational debut in September 1944 when the first missile struck London. Even as Germany was developing the V-2 and using it against the Allies, it was devising plans for the world's first ICBM, which would have been used against New York City had the war continued long enough.

Following World War II both the United States and the Soviet Union used captured German rockets and German rocket experts to accelerate their own long-range missile programs. Both the United States and the Soviet Union also made critical progress in developing thermonuclear warheads small enough to be mounted on early ICBMs and powerful enough to compensate for guidance inaccuracies. Moreover, because the material for these warheads was relatively inexpensive, their advent ushered in an era of "nuclear plenty" that reinforced the Eisenhower presidential administration's "New Look" defense policy of relying on strategic nuclear rather than conventional forces. The Soviet Union, under Nikita Khrushchev's leadership, adopted a similar policy in the belief it strengthened both national security and the economy. At the same time, significant advances in inertial guidance produced substantial improvements in missile accuracies over intercontinental ranges. Consequently, the Soviet Union successfully tested the world's first ICBM in August 1957, and the United States duplicated that feat four months later. Fear of a "missile gap" favoring the Soviet Union spurred U.S. reconnaissance efforts that led to launch of the world's first photo-imaging satellite in 1960.

Within three months of their successful ICBM test, the Soviet Union used its R-7 rocket to launch *Sputnik*, the world's first artificial satellite. Within four months, the

United States used an Army Jupiter intermediate range missile to launch *Explorer 1*, its first satellite. Virtually all space launch vehicles from 1957 into the mid-1960s originated as, or were derived directly from, ballistic missiles.

Early long-range rockets like the R-7 and the Jupiter used liquid fuels such as alcohol and kerosene, which were mixed with liquid oxygen and ignited in a combustion chamber to produce thrust. Unfortunately, liquid oxygen posed a problem. Because its storage required an extremely low temperature, those early missiles could not be kept on alert for immediate launch. Some combinations of fuel and oxidizer, such as aniline and red fuming nitric acid, could be stored at room temperature, but their chemical characteristics made handling them difficult. Their hypergolic nature when combined posed a threat to missile technicians, as was evidenced by the accidental explosion of a Titan II ICBM near Damascus, Arkansas, during routine maintenance in September 1980 and the tremendous explosion of an R-16 ICBM at Baikonur in the Soviet Union in 1960 that killed dozens. Another disadvantage of the liquid-propellant ballistic missile was the complex system of lines and pumps required to combine propellant and oxidizer in its combustion chamber.

These difficulties prompted both the United States and the Soviet Union to develop highly stable solid propellants that allowed underground storage of missiles on operational alert for extended periods with minimal maintenance. Use of such propellants in smaller, shorter-range rockets and those used for jet-assisted takeoff of aircraft had provided the basis for development of long-range, solid-propellant rockets. A key breakthrough came with the discovery that asphalt could serve as the propellant in a solid rocket motor that used potassium perchlorate as its oxidizer. Secondarily, asphalt became the binder that held the propellant and oxidizer together. Solid propellants, which erased the need for the complex, fragile plumbing and pumping systems associated with liquid-propellant missiles, enabled the development of long-range missiles that could be carried aboard submarines and launched from beneath the water's surface.

By the mid-1960s both the United States and the Soviet Union were deploying substantial numbers of long-range, nuclear-tipped ballistic missiles in hardened, underground silos and on highly mobile, stealthy submarines. As the decade ended, however, the growing number of missiles deployed by the two superpowers caused increasing concern. Moreover, the lethality of these forces increased greatly when both countries acquired the capability to use a single missile to launch several warheads, each of which could be aimed at a different target. Along with those multiple independently targeted reentry vehicles (MIRVs), technicians developed decoys that simulated attacking warheads, thereby countering the ballistic missile defenses being built by the two countries.

The rapid expansion of ballistic missile forces on hair-trigger alert increased the possibility of a miscalculation that could result in a devastating nuclear war. To stabilize the situation, the United States and the Soviet Union began a series of negotiations in 1968 that extended into the twenty-first century. These included the Strategic Arms Limitation Talks (SALT) during 1968–79 and the Strategic Arms Reduction Talks (START) that began in 1982.

The SALT talks produced three major treaties: two in 1972 and one in 1979. The Anti-Ballistic Missile Treaty of 1972, in conjunction with a 1974 protocol, severely restricted the missile defenses each side could deploy. The signatories hoped to dilute incentives for further expansion of nuclear forces and, especially, to eliminate the possibility that one side might launch a surprise attack, then use its defenses to thwart whatever response its weakened opponent could mount. The second 1972 treaty, an interim accord on offensive systems, limited the number of strategic launchers on each side: 1,618 ICBMs and 950 submarine-based missiles for the Soviet Union compared to 1,054 ICBMs and 710 submarine-based missiles for the United States. In 1979, the SALT II agreement, though never officially implemented, restricted each side to 2,250 launchers of which 1,320 could be MIRVed.

By the early 1980s, MIRVing had dramatically escalated the number of warheads each side's ballistic missiles could deliver. Consequently, the United States pushed for a new series of talks not merely to "control" strategic missile forces but to reduce them. In 1985, those START talks merged with the three-part Nuclear and Space Talks, all of which included discussions on missile-defense issues and intermediate-range nuclear forces (INF).

Although the missile-defense negotiations failed to produce an agreement, the United States and the Soviet Union signed the INF Treaty in December 1987. It eliminated all missiles with ranges between 500 and 5,500 km, including the Soviet Pioner (North Atlantic Treaty Organization designation SS-20) and American Pershing II missiles. Moreover, in July 1991, leaders of the two countries signed the START I treaty, which reduced to 1,600 the number of delivery systems each side could have. It also cut to 6,000 the number of warheads each nation could have, a 41 percent reduction for the Soviet Union and 43 percent for the United States. A few months after that treaty was signed, the Soviet Union disintegrated, with the Russian Federation inheriting the bulk of Soviet nuclear-tipped missiles.

In January 1993 U.S. leaders and Russian Federation leaders signed the START II treaty, under which the two countries agreed to reduce their nuclear warheads to between 3,000 and 3,500 by 2003. The START III treaty of 2002 further lowered those limits to between 1,700 and 2,200 warheads by the end of 2012.

While the world's two leading nuclear powers worked to constrain their strategic force structures, missile technology proliferated around the globe. A few nations, most notably France, Brazil, Argentina, and India, undertook major indigenous ballistic missile programs. They, along with the Soviet Union, helped other nations develop their own missile programs. The most notable stream of proliferation issued from the Soviet Union and flowed into China and North Korea. By the early 1990s, those three nations had sold ballistic missiles to Pakistan, Syria, Iraq, Iran, and six other countries and had helped several of them develop their own ballistic missile technology bases. Consequently, at the beginning of the twenty-first century, almost 30 nations possessed ballistic missiles with ranges exceeding 30 km.

Because the Soviet Union was a major source of ballistic missile proliferation, it came as no surprise that the Scud-B, a variant of the Soviet R-11 rocket from the mid-1950s, was the most common long-range tactical missile in the world by the close of the twentieth century. Nor was it surprising that the Scud and its variants became the long-range missiles most commonly used in Middle Eastern conflicts. During the so-called "War of the Cities" during the last year of the 1980s Iran-Iraq War, for example, Iraq fired almost 200 Scud-variant missiles at Iranian cities, including Teheran. Later, the same type of Iraqi missile was fired at U.S. and coalition forces and at Saudi and Israeli cities during the First Gulf War.

As the twenty-first century dawned and the Cold War faded into memory, people breathed easier, knowing the world had avoided all-out nuclear war between the huge missile forces of the United States and the former Soviet Union. Nevertheless, a new shadow darkened the path of human progress as nuclear, chemical, and biological weapons, along with the missiles to deliver them, became widespread among so-called rogue nations.

Rick W. Sturdevant

See also: Antiballistic Missile (ABM) Treaty, Reconnaissance, Solid Propulsion, V-2

Bibliography
Clayton K. S. Chun, *Thunder over the Horizon: From V-2 Rockets to Ballistic Missiles* (2006).
Jacob Neufeld, *The Development of Ballistic Missiles in the United States Air Force, 1945–1960* (1990).
Michael J. Neufeld, *The Rocket and the Reich: Peenemunde and the Coming of the Ballistic Missile Era* (1995).
Steven Zaloga, *The Kremlin's Nuclear Sword: The Rise and Fall of Russia's Strategic Nuclear Forces, 1945–2000* (2002).

Fractional Orbital Bombardment System

Fractional Orbital Bombardment System (FOBS), a weapon-delivery system developed by the Soviet Union to insert payloads into orbit, then de-orbit nuclear warheads in reentry vehicles. The Soviet Union attempted 24 FOBS test launches during 1965–71 and deployed the system operationally during 1969–83. Both the second Strategic Arms Limitation Treaty in 1979 and the first Strategic Arms Reduction Treaty in 1991 prohibited FOBS.

The earliest, concrete proposal for this type system originated from Soviet Chief Designer Sergei P. Korolev, who began preliminary work in 1960 on Global Missile Number 1 (GR-1), part of a plan to develop the large N1 booster for the Soviet crewed lunar effort. By 1962–63 the Soviet Union had at least three major FOBS-type projects: the GR-1, another headed by General Designer Vladimir N. Chelomey, and one by Mikhail K. Yangel's design bureau. In early 1965, before full testing of any system, the Strategic Rocket Forces conducted a comparative analysis and selected Yangel's version. After 20 test launches, the first battalion of FOBS (R-36-O missiles) went on combat duty at Tyuratam in August 1969. The Soviet Union deployed only 18 of the missiles and began dismantling them in 1982, removing the last one from duty in February 1983. Estimates of the FOBS warhead explosive yield varied from 2 to 20 megatons, and assessment of its accuracy predicted it could hit within 3 to 5 km of its intended target.

Apparently, Soviet leaders thought FOBS would provide more strategic flexibility, as it could strike the United States from the south, the direction with the fewest ground-based early warning sensors. However, the development of U.S. space-based early warning with the Missile Defense Alarm System and the Defense Support Program negated any advantage, as they detected missile launches from anywhere on Earth. Secretary of Defense Robert S. McNamara publicly announced U.S. knowledge about the existence of the system in November 1967, but he downplayed its significance, arguing that because FOBS payloads were not in a sustained orbit, they did not violate the recently signed Outer Space Treaty forbidding nuclear weapons in space.

Peter L. Hays

See also: Defense Support Program, Military Space Policy

Bibliography

Asif A. Siddiqi, "The Soviet Fractional Orbiting Bombardment System (FOBS): A Short Technical History," *Quest* 7, no. 4 (Spring 2000): 22–32.

Paul B. Stares, *The Militarization of Space: U.S. Policy, 1945–1984* (1985).

Strategic Defense Initiative

Strategic Defense Initiative (SDI), often referred to as "Star Wars," a research-and-development program announced by U.S. President Ronald Reagan in March 1983 to explore largely space-based means of detecting, intercepting, and destroying strategic nuclear missiles launched by the Soviet Union. Pentagon reports in the early 1980s claimed the Soviet Union had improved the accuracy for its missiles and, if true, that threatened to undermine the existing U.S. strategy of Mutual Assured Destruction (MAD) by rendering U.S. land-based strategic missiles more vulnerable. In response to this perceived vulnerability and inspired by calls from retired U.S. Army Lieutenant General Daniel O. Graham and his High Frontier organization for a national missile defense system, President Reagan chose to initiate a long-term, technically daunting, extremely expensive program to create a complex, space-based missile defense system. In National Security Decision Directive 85, he affirmed the U.S. commitment to eliminate the threat from Soviet missiles. The president envisioned a system neither susceptible to offensive countermeasures nor easily overwhelmed by a mass raid of Soviet missiles.

In January 1984 National Security Decision Directive 119 formally established SDI and three months later the Strategic Defense Initiative Organization (SDIO) was created within the U.S. Department of Defense (DoD) with U.S. Air Force Lieutenant General James A. Abrahamson Jr. as its director. For its agenda and schedule, the SDIO initially turned to an October 1983 report from an Office of Technology Assessment panel headed by James Fletcher, former administrator of NASA. It began exploring technologies in five general areas: directed-energy weapons, such as lasers and particle beams; kinetic-energy weapons that physically hit a target; systems concepts and battle management; survivability and lethality; and systems for surveillance,

Artist's rendering of one of the Strategic Defense Initiative designs from Los Alamos National Laboratory. It shows a space-based particle beam weapon attacking enemy intercontinental ballistic missiles. (Courtesy U.S. Department of Energy)

target acquisition, tracking, and kill assessment. Researchers soon acknowledged that significant technological advances would be necessary before engineers could design small, lightweight, relatively inexpensive, reliable space-based missile "killers."

During an October 1986 conference between Soviet President Mikhail Gorbachev and President Reagan in Reykjavik, Iceland, Reagan rejected the idea of limiting SDI strictly to research in exchange for reductions in the Soviet nuclear arsenal. By summer 1987 the SDIO recommended to the DoD Acquisition Board a plan to transition from research and development to phased deployment of a Strategic Defense System (SDS). Meanwhile U.S. reinterpretation of the 1972 ABM Treaty claimed it did not prohibit development and testing of space-based missile defenses, including a system like the nuclear-powered X-ray laser. Recognizing, however, that live-fire testing of any strategic missile defense system posed nearly insurmountable challenges, the SDIO broke ground in March 1988 for a National Test Facility (NTF) at Falcon (later Schriever) Air Force Base (AFB) near Colorado Springs, Colorado. The NTF became the hub of the SDI National Test Bed, a distributed network of computers at numerous government, corporate, and academic sites across the United States that performed complicated modeling, simulation, and actual hardware-in-the-loop testing of various missile defense systems.

The first phase of SDS deployment included six systems: the Boost Surveillance and Tracking System, the Ground-based Surveillance and Tracking System, the Space-based Surveillance and Tracking System, the Exoatmospehric Reentry Vehicle Interceptor Subsystem, the Space-based Interceptor System, and the Battle Management Command, Control, and Communications System. Before implementation of the first phase occurred, however, the Cold War between the United States and Soviet Union ended, significantly diminishing the threat of attack by intercontinental ballistic missiles. Furthermore the high cost and technical difficulties associated with SDI had caused politically powerful Senator Sam Nunn of Georgia to propose that development be refocused toward a more limited system for protection against accidental or unauthorized launches.

By January 1991 President George H. W. Bush had restructured SDI, calling it Global Protection Against Limited Strikes (GPALS). Three components characterized GPALS: theater missile defense (TMD) capabilities, a national missile defense (NMD) system, and a space-based, boost-phase interceptor system known as Brilliant Pebbles. President William Clinton's administration canceled Brilliant Pebbles in 1993 and drastically cut TMD and NMD funding. In May 1993 Secretary of Defense Les Aspin renamed the SDIO the Ballistic Missile Defense Organization (BMDO) and reoriented its priorities toward TMD development. Those actions essentially terminated SDI as Reagan originally envisioned it, but the concepts and technological progress continued into the twenty-first century.

Deployment of an NMD system became the goal of some congressional Republicans during the late 1990s. The Commission to Assess the Ballistic Missile Threat to the United States, established under the 1998 Defense Authorization Act and chaired by Donald Rumsfeld, warned in July 1998 that the ballistic missile threat against the United States remained real and credible. In May 2001 President George W. Bush told a National Defense University audience that an NMD system would allow the United States to move away from massive retaliation and MAD as a strategy. Secretary of Defense Rumsfeld redesignated BMDO as the Missile Defense Agency in January 2002 and charged it with deploying elements of the NMD system as soon as practicable, while continuing to develop and test additional technologies. During 2004, DoD began placing in underground silos at Fort Greely, Alaska, and Vandenberg AFB, California, the first interceptors for the Ground-based Midcourse Defense segment of the NMD system.

Robert A. Kilgo

See also: Antiballistic Missile (ABM) Treaty

Bibliography

Craig Eisendrath, *Phantom Defense: America's Pursuit of the Star Wars Illusion* (2001).

Frances Fitzgerald, *Way Out There in the Blue: Reagan, Star Wars, and the End of the Cold War* (2000).

Peter L. Hays et al., *Spacepower for a New Millennium: Space and U.S. National Security* (2000).

U.S. Department of Defense. http://www.defense.gov/.

DOCTRINE AND WARFARE

Doctrine and warfare are intimately related, the former being suggested procedures for best handling military problems that tend to recur during the latter. Doctrine embodies generalizations drawn from concepts derived from inferences based on observations of military activities during war, exercises, and peacetime operations. It affects armed forces' force structure, organization, equipment, training, education, and leadership at all levels of command. Based as it is on the accumulated, distilled experience of warriors in combat or military exercises, doctrine is rooted in history but must be sufficiently flexible to function in present circumstances and to evolve as the nature of warfare or specific threats to a nation change. Introduction of the longbow, repeating rifle, machine gun, tank, and airplane all fostered significant changes in the nature of warfare and, consequently, in military doctrine. The perceived success of Allied airpower in winning World War II, for example, led to doctrinal principles associated with gaining control of the airspace and achieving air superiority.

When the Soviet Union and the United States both developed intercontinental ballistic missiles with thermonuclear warheads during the late 1950s, Cold War doctrine focused strategically on nuclear deterrence. That concept involved creating a perception in the opponent's mind that the cost of military aggression would outweigh any possible gain. The doctrine of nuclear deterrence evolved from simply massive retaliation in the early 1950s to assured destruction in the 1960s. Because U.S. and Soviet political leaders both endorsed the idea of assured destruction, the doctrine soon became known as Mutual Assured Destruction (MAD). A U.S. modification of MAD occurred in July 1980 when President Jimmy Carter adopted "countervailing strategy," which shifted the focus of planned, initial nuclear strikes from destroying population centers to killing Soviet leaders and attacking military targets, hopefully to induce a Soviet surrender before both countries experienced total destruction.

Military leaders have long recognized that changes in technology, tactics, and combat techniques affect the formulation of doctrine and vice versa. With respect to U.S. space systems, for example, technological "push" of doctrine occurred for several decades as satellites for various applications—reconnaissance and surveillance, early warning, meteorology, communications, nuclear detection, and navigation—improved the ability of U.S. military leaders to command and control forces worldwide and reduced the ability of potential enemies to effect a surprise attack. Satellites significantly enhanced so-called "situational awareness" and, thereby, as illustrated in the First Gulf War of 1991, not only allowed more efficient, effective coordination of forces in battle, but extended the battle area deep into enemy territory. Thereafter U.S. operations in Bosnia, Kosovo, and elsewhere increased doctrinal emphasis on minimization of civilian casualties, collateral damage to nonmilitary targets, and losses due to friendly fire. In that instance, the technological "pull" of doctrine led to installation of Global Positioning System (GPS) receivers on precision-guided munitions; equipping U.S. ground, sea, and air units with GPS receivers permitted so-called "blue force tracking."

When the attacks of 11 September 2001 indicated that international terrorism had replaced Russian nuclear forces as the primary, specific threat, U.S. military doctrine

BERNARD A. SCHRIEVER
(1910–2005)

(Courtesy U.S. Air Force)

From his appointment in 1954 as commander of the U.S. Air Force Western Development Division until his retirement as head of Air Force Systems Command in 1966, General Bernard Schriever led U.S. efforts to develop long-range, liquid- and solid-propellant ballistic missiles—Atlas, Titan, Thor, and Minuteman—and the first military satellites for applications ranging from reconnaissance and early warning to meteorology and communications. Under his leadership, the long-range missiles evolved to become reliable launch vehicles for human and robotic spaceflight from the late 1950s into the twenty-first century. Schriever was instrumental in implementing new management processes used for military high-technology acquisition.

Rick W. Sturdevant

shifted fundamentally from Cold War deterrence to preemption. President George Bush announced that the United States no longer would allow rogue states like Iraq, Iran, and North Korea to acquire weapons of mass destruction and, furthermore, would not wait to absorb an attack before striking against a growing terrorist threat. The consequences of this evolving doctrine became apparent in 2003 during the Second Gulf War, when space systems contributed more significantly than ever to the flexibility, precision, and lethality of a preemptive U.S. invasion of Iraq. That conflict also revealed the lack of an approved doctrine for joint space operations and the need, consequently, for better integration and coordination of theater-level space support.

A growing recognition that space systems had become essential to the success of U.S. military operations of all kinds tended to accelerate both doctrinal change and the "pull" that doctrine had on technological change. Although General Bernard Schriever, as early as February 1957, had emphasized the critical importance of achieving space control or space superiority, not until the twenty-first century did it become a key feature of U.S. military doctrine. That, in turn, drove pursuit of defensive and offensive counter-space technologies, plus efforts to develop new systems to achieve better situational awareness in space.

Meanwhile the superiority of U.S. military space capabilities caused other nations, like Russia and China, to adjust their military doctrines by reviving the concept of nuclear deterrence in the first case and maintaining a nuclear-counterattack capability in the second case. With regard to space doctrine, specifically, all Russian documents associated with military doctrine since 1992 identified protection of Russian space systems, including ground stations, as essential to national security. Although some experts, as recently as 2004, doubted whether China had an official military space doctrine, Chinese publications suggested that gaining space dominance was essential to securing information dominance, the latter being critical to military success. Both Russia and China publicly opposed the deployment of weapons in space, because they believed it would lead to a costly, destabilizing arms race in space and undermine global security.

Rick W. Sturdevant

See also: China, Russia (Formerly the Soviet Union), Russian Space Forces, United States Air Force

Bibliography

Federation of American Scientists. http://www.fas.org/.

Robert Frank Futrell, *Ideas, Concepts, Doctrine: Basic Thinking in the United States Air Force*, 2 vols. (1989).

Mark E. Harter, "Ten Propositions Regarding Space Power: The Dawn of a Space Force," *Air and Space Power Journal* 20, no. 2 (Summer 2006): 64–78.

I. B. Holley Jr., *Technology and Military Doctrine: Essays on a Challenging Relationship* (2004).

David E. Lupton, *On Space Warfare: A Space Power Doctrine* (1988).

First Gulf War

First Gulf War, also known as the Persian Gulf War or Operation Desert Storm, occurred from 16 January to 28 February 1991. In response to an August 1990 Iraqi invasion of Kuwait, a United Nations coalition of forces, led by the United States, liberated Kuwait and drove Iraqi President Saddam Hussein's troops back toward Baghdad. Because this conflict marked the first extensive use of space-based systems in actual combat, U.S. Air Force Chief of Staff General Merrill A. McPeak later called it "the first space war." Coalition forces relied on the full range of military and civil space capabilities, from communications, navigation, and meteorology to surveillance and early warning. As their understanding of space capabilities grew, battlefield commanders became increasingly dependent on space assets as their eyes, ears, and voice for command and control. Some observers, noting that Iraq made little use of satellites compared to the United States and its allies, argued this was the first information war. They reasoned that the swift defeat of the Iraqis, with minimal coalition casualties, resulted from an overwhelming information differential. Knowledge obtained from, or communicated through, coalition space systems became power.

Satellite communications (SATCOM) proved essential during the war and Operation Desert Shield, which preceded it. More than 90 percent of all military traffic—secure voice, data, facsimile, and teletype messages—to and from the operational theater went via satellite links. The lack of a viable communications infrastructure in many parts of Saudi Arabia and Kuwait increased the coalition's dependence on SATCOM for intra-theater command and control. At the height of operations, the system handled 700,000 telephone calls and 152,000 messages per day. Coalition forces had access to at least 15 individual U.S., British, and French military communications satellites whose

An Aerospace Audiovisual Service crew of videographers and photographers set up an Inmarsat satellite transmitter on the sand during Operation DESERT STORM in Saudi Arabia. (Courtesy U.S. Department of Defense)

terrestrial footprints covered the Persian Gulf. Using one Defense Satellite Communications System (DSCS) 2 and two DSCS 3 satellites, augmented by two British Skynet satellites, U.S. Central Command had a total superhigh frequency capacity of 115 Mbps for wartime operations. Fleet Satellite Communications/Air Force Satellite Communications and other space systems provided very high frequency and ultrahigh frequency (UHF) tactical links for air, ground, and naval forces. When coalition forces required more channel capacity than military systems could provide, they turned to such civil satellite systems as Intelsat, Inmarsat, and Arabsat. In addition, a 150-pound experimental Multiple Access Communications Satellite, launched by the Defense Research Projects Agency in 1990, demonstrated store-and-forward techniques in the UHF band by picking up logistical information over Saudi Arabia, dumping it over the United States, and vice versa.

Commercial television coverage of the war contributed substantially to increased SATCOM traffic. According to Intelsat, full-time channel usage increased from 2 to 22, with short-term reservations rising to more than 400 channels on one occasion in January. Peter Arnett and Cable News Network became familiar names as Arnett reported the news from Iraq as it was happening during the First Gulf War. Television viewers around the world, including the combatant forces, gained a unique perspective on the war.

Coalition forces benefited immensely from a near monopoly on space-based surveillance applications—photographic and radar reconnaissance, ocean surveillance, electronic eavesdropping, weather monitoring, and early warning. Imagery from civilian U.S. Landsat and French SPOT (Satellite Pour l'Observation de La Terre) multispectral imaging satellites proved useful in rapid production of up-to-date, spatially accurate maps and preflight video briefings for military pilots. Keyhole sensors did not provide a large field of view but were useful for highly detailed mapping of specific areas inside Iraq such as presidential palaces or military complexes. Lacrosse offered day-and-night radar imaging, and White Cloud ocean surveillance satellites monitored Iraqi combat aircraft movements. The Defense Meteorological Satellite Program, augmented by the U.S. National Oceanic and Atmospheric Administration's Geostationary Operational Environmental Satellites and Polar Orbiting Environmental Satellites in addition to European Meteosat spacecraft, contributed to the coalition forces' awareness of harsh, rapidly changing weather or environmental conditions, including desert sandstorms and smoke from burning oilfields. Timely, accurate data received directly from those satellites allowed mission planners to shift targets, types of aircraft, and types of weaponry quickly depending on weather conditions. Defense Support Program satellites detected Iraqi Scud missile launches, thereby alerting the crews of antiballistic missile Patriot batteries. Altogether, coalition forces relied on more than 30 military, civil, and commercial surveillance satellites.

The U.S. NAVSTAR Global Positioning System (GPS), although not fully operational during Desert Storm, supplied unprecedented navigational accuracy to coalition forces. With only 16 satellites in orbit at the time, GPS nonetheless enabled large numbers of troops to maneuver confidently across featureless desert terrain, spotters to mark minefields and pinpoint enemy targets with remarkable precision, and naval launchers to improve midcourse guidance of standoff missiles. Insufficient quantities of military-model GPS receivers on the eve of Desert Storm led to extensive use of portable, commercial models. According to one estimate, coalition troops used as many as 12,000 personal receivers.

Because space systems only supported and enhanced air, ground, and naval operations during the First Gulf War, some experts judged the sobriquet "first space war" inaccurate. Despite such quibbling, nearly all admitted that space finally had manifested itself as an important element of military power. As a consequence, the United States and other powerful nations began to fully integrate space-based capabilities into the planning and conduct of wartime operations.

Eric Thorson

See also: Military Communications, Navigation, Reconnaissance

Bibliography

Peter Anson and Dennis Cummings, "The First Space War: The Contribution of Satellites to the Gulf War," in *The First Information War*, ed. Alan D. Campen (1992).

Bruce D. Berkowitz, *The New Face of War: How War Will Be Fought in the 21st Century* (2003).

Michael Russell Rip and James M. Hasik, *The Precision Revolution: GPS and the Future of Aerial Warfare* (2002).

Military Space Doctrine

Military space doctrine refers to beliefs about how nations and their armed forces might best derive strategic utility from the use of outer space. Broadly construed, military doctrine consists of historically derived, officially taught precepts that guide the conduct of military operations. Although a large number of plans, programs, and perspectives constitute the foundation for U.S. military space doctrine, a few specific documents have been particularly important: Air Force Manual 1-6, *Military Space Operations* (September 1982); Department of Defense (DoD) Space Policy (March 1987); Air Force Space Policy (December 1988); DoD Space Policy (July 1999); the Space Commission Report (January 2001); Air Force Doctrine Document 2-2, *Space Operations* (November 2001); and Joint Publication 3-14, *Joint Doctrine for Space Operations* (August 2002), the first approved joint space doctrine. The United States came to depend militarily on space more than any other nation and, consequently, developed its military space doctrine to a greater degree.

U.S. aviation experts frequently spearheaded thinking about military uses of space. As early as 1945, Henry "Hap" Arnold's visionary "Third Report to the Secretary of War" and Theodore von Kármán's *Toward New Horizons* study viewed space as an arena for the natural extension of core Army Air Force doctrine and spaceflight as a potential means of "flying" higher, farther, and faster to conduct long-range strategic-attack missions. The RAND Corporation's first report, *Preliminary Design of an Experimental World-Circling Spaceship*, in 1946, presciently laid out the engineering challenges and conceptual utility for nearly every type of military space system subsequently built.

Even before the launch of the world's first artificial satellite, President Dwight Eisenhower's administration developed a comprehensive space policy that remained instrumental to structuring the international legal regime and military doctrine for space into the twenty-first century. This policy publicly emphasized using space for "peaceful purposes" such as the International Geophysical Year scientific satellite program. Although hidden, the most important part of the policy was to use those early civil programs for legitimizing overflight of the Soviet Union, thereby allowing space-based reconnaissance to enhance U.S. national security. The May 1955 National Security Council document 5520, "Satellite Program," codified Eisenhower's approach. De-emphasis of overt military uses of space through the peaceful purposes policy and Eisenhower's misgivings about the U.S. military ability to develop and operate spy satellites led to establishment in 1961 of the highly classified National Reconnaissance Office (NRO), whose existence was officially secret until 1992.

Eisenhower's policy conflicted with the predominant military view that saw potential defense-related utility in a wide range of space operations beyond reconnaissance. By 1958 the U.S. Air Force (USAF) had coined the word "aerospace," and USAF doctrine asserted air and space were a seamless medium under that service's operational purview for purposes of power projection. Initially, the USAF employed the aerospace concept quite successfully to advocate for space missions and, by the beginning of President John F. Kennedy's administration, had emerged as the lead service for space. The doctrine framed by the aerospace concept was thwarted repeatedly in its early years, however, both secretly via creation of the NRO and publicly through a string of canceled USAF attempts to develop systems like the Dyna-Soar spaceplane and the Manned Orbiting Laboratory. Moreover, neither the other services nor the DoD ever accepted the USAF definition of aerospace. Their objections gained greater doctrinal strength throughout time, as indicated by disappearance of the word in the DoD 1999 space policy statement and its de-emphasis in both the 2001 Space Commission Report and the 2002 Joint Publication 3-14.

Given its controversial and somewhat obscure early history, it was hardly a surprise that the aerospace concept failed to provide a firm foundation for developing space doctrine or compelling answers to questions about what the military services should do in space, how they should do it, or why. It certainly did not help that USAF doctrine manuals into the 1980s simply substituted the word "aerospace" for "air" and inappropriately ascribed airpower attributes like speed, range, and flexibility to space forces.

By the early 1980s, numerous critics inside and outside the USAF challenged the aerospace concept along with the development of spacepower theory and resulting doctrine. They built instead on Dennis Drew's model: the idea that doctrine should grow from the soil of history, develop a sturdy trunk of fundamental doctrine, branch out into doctrine for specific environments, and only then attempt to sprout organizational doctrine analogous to leaves. That model furnished a way to analyze comprehensively the aerospace concept and the first official USAF space doctrine: Air Force Manual (AFM) 1-6, *Military Space Doctrine*, released in 1982. Critics pointed out that the USAF had no doctrinal foundation for its aerospace concept. It was attempting to produce leaves on a nonexistent branch, because it had not developed environmental doctrine before promulgating the organizational doctrine in AFM 1-6. Too often, it was force-fitting space doctrine into the mold of air doctrine rather than developing doctrinal characteristics more appropriate for the space environment, such as emplacement, pervasiveness, and timeliness.

Other doctrinal and organizational developments during the 1980s–90s built on a more solid foundation. The USAF acknowledged the operational importance of space by establishing a Space Command in 1982, and DoD followed suit with a unified (multi-service) United States Space Command in 1985. Following those important organizational developments were noteworthy doctrinal statements: the 1987 DoD Space Policy and the 1988 Air Force Space Policy. The former promulgated four basic military space mission areas (Space Support, Force Enhancement, Space Control, and Force Application), while the latter advocated more specific mission responsibilities the USAF should perform under the 1987 DoD typology and, avoiding use of the "aerospace" term, stridently asserted spacepower would become as decisive in future combat as airpower then was.

Merrill McPeak, USAF chief of staff, emphasized the growing maturity and importance of space assets in enhancing the combat effectiveness of coalition forces during the First Gulf War in 1991 by labeling that conflict "the first space war" and, in June 1992, by adding the words "air and space" to the USAF mission statement. According to Thomas Moorman, McPeak's vice chief of staff, that change in wording formally legitimized USAF space operations and placed them conceptually on equal footing with air operations. In its November 1996 Global Engagement vision statement, the USAF moved still further from the aerospace concept by asserting a metamorphosis from an "air" force to an "air and space" force on an evolutionary path toward a "space and air" force.

Although these developments matured space doctrine significantly, critics did not perceive them as rapid or pervasive enough to encapsulate current knowledge about how space operations contribute to winning wars. This led Congress in 1999 to create the Commission to Assess United States National Security Space Management and Organization, generally known as the Space Commission. Chaired by Donald Rumsfeld, the commission included senior retired government officials and military space officers. Its January 2001 report, which represented the most important study ever completed on national security space, contained 13 major recommendations. Those included recognizing space as a top national-security priority, making significant organizational changes, better aligning NRO and DoD space activities, developing the means to deter and defend against hostile actions in and from space, and investing in science, technology, and professional development to ensure U.S. space leadership. Nearly all of the Commission's recommendations were implemented and, building on Joint Publication 3-14, the USAF released a *Counterspace Operations* doctrine (Air Force Doctrine Directive 2-2.1) in August 2004 that explicitly stated a concept of U.S. military operations to deceive, disrupt, deny, degrade, or destroy an adversary's space capabilities. Such subsequent developments, however, as the October 2002 merger of the United States Space Command into the United States Strategic Command and the July 2005 decision to make the NRO director and the under secretary of the Air Force (the DoD Executive Agent for Space) separate positions seemed contrary to what the commission sought.

While Russia, China, and other nations seldom release specific statements about their military space doctrines, one can infer much from their general statements, policies, programs, and actions throughout time. During the early 1960s Soviet military doctrine focused on achieving overwhelming offensive superiority in all environments, including outer space, and in using space to support a combined-arms approach to combat operations. Although Soviet public expressions after Moscow's accession to the 1967 Outer Space Treaty shifted toward viewing space as a sanctuary, forever free from weapons, the growth of Soviet military space capabilities indicated a continuation of the earlier doctrine. The Soviet Union sought to achieve and maintain military superiority in outer space sufficient to deny its use to adversaries and to assure maximum space-based support for Soviet offensive and defensive combat operations on land, at sea, in the air, and in outer space. Following the Soviet Union's collapse in 1991, Russia continued to realize that success in modern warfare depended increasingly on force enhancement through space-based capabilities. Financially strapped but seeking to secure its access to space, Russia became more vocal in its advocacy of space as sanctuary and actively argued for a multilateral treaty prohibiting the deployment of weapons in space.

Assessments of U.S. space capabilities and perceptions of U.S. military efforts after the 1990s to dominate space apparently influenced the military space doctrines of other nations, most notably China. Like the Russians, the Chinese believed that U.S. weapons in space would be detrimental to their security. China officially joined Russia in calling for a ban on such weapons. Conversely Chinese efforts from the 1980s onward to develop an antisatellite system suggested a doctrine of ensuring the survivability of its own space assets and negating U.S. capabilities in the event of hostilities. By the early twenty-first century, the United States, Russia, China, France, the United Kingdom, Japan, India, Israel, and practically every other nation with satellite access acknowledged in their respective doctrines a reliance, if not outright dependence, on space-based systems to enhance the combat capabilities of ground, air, and sea forces.

Peter L. Hays

See also: United States Air Force, United States Department of Defense

Bibliography

Steven J. Lambakis, *On the Edge of Earth: The Future of American Space Power* (2001).

Benjamin S. Lambeth, *Mastering the Ultimate High Ground: Next Steps in the Military Uses of Space* (2003).

Matthew Mowthorpe, *The Militarization and Weaponization of Space* (2004).

Michael E. O'Hanlon, *Neither Star Wars nor Sanctuary: Constraining the Military Uses of Space* (2004).

Space Security Index. http://www.spacesecurity.org/.

Operation Desert Storm. *See* First Gulf War.

Operation Iraqi Freedom. *See* Second Gulf War.

Persian Gulf War. *See* First Gulf War.

Second Gulf War

Second Gulf War, also called Operation Iraqi Freedom (OIF), commenced on 18 March 2003. The United States, the United Kingdom, and their coalition partners sought to eliminate the threat posed by Iraqi President Saddam Hussein's presumed weapons of mass destruction and his support of Al Qaeda's global terrorist activities. If Operation Desert Storm in 1991 marked the first extensive use of space-based assets in a war, OIF confirmed that the United States had become reliant on space systems to defeat opponents. More than 50 military satellites, augmented by civil and commercial satellites, supported OIF. Although U.S. President George Bush announced the end of major combat on 1 May 2003, a widespread insurgency with daily acts of terrorism continued to plague Iraq.

The fully operational Global Positioning System (GPS), with 27 satellites in orbit, contributed spectacularly to OIF. Coalition aircraft dropped more than 5,500 GPS-guided Joint Direct Attack Munitions with devastating accuracy, irrespective of time or weather conditions. Through a capability called GPS-Enhanced Theater Support, coalition forces received precise position, velocity, navigation, and timing information. Perhaps the most valuable GPS application was the Space-based Blue Force Tracking System, which helped coalition ground and air forces avoid incidents of fratricide. When the Iraqis attempted to block GPS signals with six Russian-made jamming devices, U.S. air strikes destroyed all of them.

Satellite communications (SATCOM) played an even greater role than in the First Gulf War. A comparison of the two conflicts revealed that even as the troop level shrank by four-fifths, demand for bandwidth quadrupled. Among the reasons for the latter was extensive SATCOM use for Predator and Global Hawk Unmanned Aerial Vehicle operations. Although military satellites supplied 75 percent of the SATCOM for the First Gulf War and the United States subsequently added Milstar to its capabilities, they furnished less than 32 percent during the Second Gulf War. In addition to relying on Intelsat and Inmarsat commercial services in OIF, coalition forces used the Iridium satellite system for mobile, tactical communications.

Space-based imagery, including that of weather patterns, contributed immeasurably to the success of OIF. Two years earlier, in February 2000, the Space Shuttle *Endeavour* had used an instrument specially modified for the Shuttle Radar Topography Mission to produce a three-dimensional map of targets in Iraq. As in the First Gulf War, National Reconnaissance Office satellites used daylight, low-light, and infrared cameras, in addition to radar, to yield classified surveys of activity on the ground. Unclassified high-resolution commercial images from DigitalGlobe's *Quickbird* and Space Imaging's *Ikonos* proved valuable because the United States could share them immediately with others. Keyhole Corporation's EarthViewer product overlaid *Quickbird* and *Ikonos* imagery on a *Landsat 7* base map, which gave both coalition forces and television news reporters real-time, three-dimensional models of current events. Various National Oceanic and Atmospheric Administration satellites and the European *Meteosat 5* supplied weather data for OIF. The coalition also received near real-time warning of sandstorms, rain, and other weather or environmental conditions from Moderate Imaging Spectroradiometer instruments aboard the NASA Earth Observing System *Terra* and *Aqua* Sun-synchronous research satellites via Satellite Focus, a secure Internet application developed by the Naval Research Laboratory.

Eric Thorson

See also: Global Positioning System, Military Communications, Reconnaissance

Bibliography
Bruce Berkowitz, *The New Face of War: How War Will Be Fought in the 21st Century* (2003).
Global Security. http://www.globalsecurity.org/.
Alexander H. Levis, ed., *The Limitless Sky: Air Force Science and Technology Contributions to the Nation* (2004).

EARLY WARNING

Early warning, the process of detecting, analyzing, and alerting designated authorities of an impending enemy attack in time for those authorities to take preventive, defensive, or counteroffensive actions. In modern military situations, the detection process usually involves the use of sophisticated sensor technologies, such as aerial or space-based reconnaissance and surveillance, long-range radar, and electronic eavesdropping. Analysis depends on complex computer systems, like those of the North American

Aerospace Defense Command inside Cheyenne Mountain near Colorado Springs, Colorado. Distribution of the alert involves equally complicated, secure communications networks that rely on fiber optics, microwave relays, and satellites.

Examples of both effective and failed early warning abound in history. Perhaps the most familiar example of effective warning in U.S. history is Paul Revere's ride in April 1775, and the best-known failure is the Japanese attack on Pearl Harbor in December 1941. From balloons during the U.S. Civil War in the 1860s to camera- or radar-equipped airplanes in the twentieth century, aerial platforms steadily enhanced early-warning capabilities. By the end of World War II, nations on both sides—the United States, Canada, the United Kingdom, the Soviet Union, Germany, and Japan—had ground-based radar systems for early warning of enemy air attacks. During the 1950s, the Distant Early Warning (DEW) Line of radar sites across northern Alaska and Canada, along with the White Alice Communications System that linked it to military command centers farther south, provided early warning of Soviet air attacks across the polar region.

As the Cold War evolved, the focus of both U.S. and Soviet early-warning radar systems expanded to include attacks by intercontinental or submarine-launched ballistic missiles and threats from Earth-orbiting satellites. Three BMEWS (Ballistic Missile Early Warning System) radar sites completed in the early 1960s gave the United States, Canada, and the United Kingdom early warning of an attack by intercontinental ballistic missiles launched from the Soviet Union. In the early 1980s, the U.S. Air Force (USAF) added four Pave PAWS (Phased Array Warning System) radar sites for early warning of nuclear missiles launched from Soviet submarines. For detecting, tracking, and identifying potentially threatening satellites, the U.S. military deployed a number of Baker-Nunn optical cameras at various locations worldwide from 1960 into the 1990s. Nine radar sites built during 1958–64 across the southern United States comprised the NAVSPASUR (Navy Space Surveillance) System for tracking objects in low-to-medium orbits; several Air Force space surveillance radars, like the powerful AN/FPS-85 at Eglin Air Force Base, Florida, began operating later in the 1960s. In the 1980s, the Air Force began using GEODSS (Ground-Based Electro-Optical Deep Space Surveillance) technology to augment and ultimately replace the Baker-Nunn cameras.

In 1961, the Soviet Union began fielding its early-warning system for missile attacks with a modified version of the Dnestr radar, originally designed for the IS (satellite destroyer) antisatellite system. Two Dnestr-M radar sites in the northern Soviet Union and a command center near Moscow became operational in February 1971. During the 1970s, new early-warning stations equipped with the Dnepr radar, an upgraded version of the Dnestr-M, covered western and southern approaches to the Soviet Union. A new generation of Daryal-type phased-array radars began augmenting and replacing the Dnestr-M and Dnepr radars during the mid-1980s. For space surveillance, the Soviet Union maintained two Dnestr radar stations from the 1970s onward, and Russia began trial operations with the Okno optical deep-space surveillance system in December 1999.

Both the United States and the Soviet Union pursued space-based detection of enemy missile launches, because that provided earlier warning than what long-range radars gave. The USAF Missile Defense Alarm System (MIDAS) program of the 1960s experimented with infrared-detecting satellites to spot launches of long-range missiles, and that led in the 1970s to the operational Defense Support Program (DSP). Although the USAF sought to replace DSP with a new Space-Based Infrared System in the 1990s, the service continued to rely solely on geostationary DSP satellites into the early twenty-first century. Based on the U.S. experience in the First Gulf War, the tactical warning provided by DSP satellites became increasingly important, even as those same satellites continued to provide strategic warning.

Although the Soviet Union did not begin its research for a space-based early-warning system until the late 1960s, essentially a decade behind the United States, it performed experimental flights during the mid-1970s. The space tier of the Soviet early-warning system, which consisted of Oko satellites in both highly elliptical and geosynchronous orbits, became fully operational by the early 1980s.

Rick W. Sturdevant

See also: Ballistic Missiles

Bibliography

Edwin Leigh Armistead, *AWACS and Hawkeyes: The Complete History of Airborne Early Warning* (2002).

Geoffrey Forden, "Reducing a Common Danger: Improving Russia's Early-Warning System," *Policy Analysis* 399 (3 May 2001), on Cato Institute website.

Global Security. http://www.globalsecurity.org/.

Pavel Podvig, ed., *Russian Strategic Nuclear Forces* (2001).

Ballistic Missile Early Warning System

Ballistic Missile Early Warning System (BMEWS), a network of radar and communications systems designed to detect and track intercontinental ballistic missiles launched from Russia toward North America.

Conceived by the United States Air Force in 1955, approved in late 1957, and built with highest national priority in the early 1960s, BMEWS consisted of three sites: Thule Air Base, Greenland; Clear Air Force Station (AFS), Alaska; and Fylingdales, England. Two types of L-band radars were selected for BMEWS: the General Electric AN/FPS-50 search radar, a stationary parabolic reflector 165 ft high and 400 ft wide; and the RCA (Radio Corporation of America) AN/FPS-49 mechanical tracking set, which included a dish antenna nearly 80 ft in diameter that could swing a full 360-degree circle and swivel vertically 90 degrees from horizontal. In the mid-1960s, Clear AFS received an improved version of the AN/FPS-49 tracking radar, the AN/FPS-92. The coverage areas of the three radar facilities overlapped to provide 99 percent assurance of detecting a missile attack and providing 15–37 minutes warning. BMEWS also could detect orbiting satellites and forwarded both missile warning and space surveillance data via dual pathways to processing centers deep inside the Cheyenne Mountain Complex near Colorado Springs, Colorado.

Beginning with the dual-faced AN/FPS-120 at Thule in June 1987, each of the BMEWS locations switched operations from the antiquated search and mechanical tracking radars to Raytheon ultrahigh frequency solid-state phased array radar (SSPAR) systems. A three-faced AN/FPS-126 SSPAR became operational at Fylingdales in October 1992, followed by a dual-faced AN/FPS-120 at Clear AFS in February 2002. This upgrade, which allowed multiple, simultaneous missile warning and space tracking, greatly increased the sites' capabilities while cutting operational and maintenance expenses almost in half.

Brad M. Evans

See also: North American Aerospace Defense Command, Space Surveillance

Bibliography

Global Security. http://www.globalsecurity.org/.

Eloise Paananen and Kenneth H. Drummond, *Sky Rangers: Satellite Tracking around the World* (1965).

Melvin L. Stone and Gerald P. Banner, "Radars for the Detection and Tracking of Ballistic Missiles, Satellites, and Planets," *Lincoln Laboratory Journal* 12, no. 2 (2000): 217–44.

Andrew Wilson, ed., *Jane's Space Directory 1993–1994* (1993).

Defense Support Program

Defense Support Program (DSP), the central component in the U.S. space-based early warning system. In geostationary orbit approximately 22,000 miles above the equator, DSP satellites detect missile and space launches and nuclear detonations. Spinning at 6 rpm, DSP satellites use infrared sensors to detect heat from missile and booster plumes against Earth's background. Operated by Air Force Space Command, DSP satellites make a surprise missile attack nearly impossible.

The historical roots of DSP are in the WS-117L Advanced Reconnaissance System Subsystem G, which became the Missile Defense Alarm System (MIDAS) program during the early 1960s. Program 461, the final element of MIDAS, proved the principle of space-based surveillance and contributed significantly to the design of DSP infrared sensors. Meanwhile, in 1963, the United States Air Force (USAF) undertook Program 266 to devise an improved infrared early warning system with satellites in geostationary, rather than low Earth, orbit. The designation changed to Program 949 in 1967 when the government awarded TRW and Aerojet development contracts for the spacecraft and sensor respectively. A security breach in 1969 resulted in further designation changes to Program 647, then DSP, when the effort achieved full functional capability. The first launch of a DSP satellite, called IMEWS (Integrated Missile Early Warning Satellite), occurred on 6 November 1970. All DSP satellites, except one, were launched from Cape Canaveral, Florida. Space Shuttle mission Space Transportation System-44D launched a DSP satellite on 24 November 1991.

Numerous improvements enabled DSP to keep pace with evolving missile threats. The original IMEWS satellite weighed 2,000 lb and had 400 watts of power, 2,000 infrared detectors, and a design life of 1.25 years. During the life of the program, the design underwent numerous improvements to enhance reliability, capability, and survivability. The weight grew to 5,250 lb and power to 1,275 watts, with detectors

The Air Force Space Command-operated Defense Support Program (DSP) satellites are a key part of North America's early warning systems. In their 22,300-mile geosynchronous orbits, DSP satellites help protect the United States and its allies by detecting missile launches, space launches, and nuclear detonations. (Courtesy Northrop Grumman)

increasing to 6,000 and the design life to five years. Sensor design improvements included improved resolution and above-the-horizon scanning capability for full hemispheric coverage. Increased onboard signal processing refined clutter rejection. Despite such changes, the USAF began studying alternatives in the early 1990s to replace DSP. By 1994, the service had selected the concept of a Space-Based Infrared System (SBIRS) as successor to DSP.

After DSP satellites (designed for strategic purposes) detected every Scud tactical missile launch in 1991 during the First Gulf War, this successful demonstration convinced the USAF to develop an Attack and Launch Early Reporting to Theater (ALERT) processing capability to provide DSP information more rapidly to battlefield commanders. At the same time, to ensure more efficient operations and prepare for SBIRS implementation, the USAF automated DSP overseas ground stations and upgraded the Continental United States (CONUS) station at Buckley Air Force Base, Colorado, to become the central processing facility for all DSP data.

David C. Arnold

See also: First Gulf War, North American Aerospace Defense Command, United States Air Force

Bibliography
Desmond Ball, *A Base for Debate: The US Satellite Station at Nurrungar* (1988).
Jeffrey T. Richelson, *America's Space Sentinels: DSP Satellites and National Security* (2001).
Fred Simmons and Jim Cresswell, "IR Eyes High in the Sky: The Defense Support Program," *Crosslink* 1, no. 2 (Summer 2000): 18–25.

Integrated Missile Early Warning Satellite. *See* Defense Support Program.

Missile Defense Alarm System

Missile Defense Alarm System (MIDAS), a developmental program to detect ballistic missile launches using space-based infrared sensors. It originated as Subsystem G in the Advanced Reconnaissance System or WS-117L proposal that Lockheed Missiles and Space Division submitted to the U.S. Air Force (USAF) in March 1956. The program designation changed to MIDAS after the USAF briefly relinquished oversight to the Advanced Research Projects Agency from March 1958 to November 1959. Design plans called for placing an infrared-sensing telescope in a rotating turret mounted in the nose of an Agena spacecraft. Engineers hoped, eventually, to place eight operational satellites in 2,000-mile polar orbits for constant monitoring of launches from the Soviet Union.

The USAF strongly supported MIDAS, because it would nearly double the warning time the Ballistic Missile Early Warning System provided for intercontinental ballistic missile launches. Many scientists were skeptical, however, that the system would successfully detect missile launches against Earth's background radiation. After six failures, beginning with *MIDAS 1* on 26 February 1960, the seventh flight, launched on 9 May 1963, operated long enough to detect nine missile launches. Meanwhile, the USAF had renamed MIDAS as Program 461. In 1966 a Research Test Series (RTS) with significantly improved MIDAS-like spacecraft and sensors resulted in successful 325- and 372-day flights, during which the satellites detected 139 rocket launches and identified four Soviet launch sites. Although MIDAS never achieved operational status, technical and scientific information derived from it contributed significantly to the Defense Support Program.

Eric Thorson

See also: Lockheed Corporation, United States Air Force

Bibliography
R. Cargill Hall, "Missile Defense Alarm: The Genesis of Space-based Infrared Early Warning," *Quest* 7, no. 1 (Spring 1999): 5–17.
Nicholas W. Watkins, "The MIDAS Project: Part I Strategic and Technical Origins and Political Evolution 1955–1963," *Journal of the British Interplanetary Society* 50 (June 1997): 215–24.

Missile Warning. *See* Early Warning.

Project M. *See* Oko.

Oko

Oko (Eye) or US-KS, a Soviet space-based early warning system introduced by the Soviet Union in 1972, declared operational in 1976, and expanded to a full, nine-satellite constellation by 1980. Originally led by designer Anatoli Savin of TsNII (Central Scientific-Research Institute) Kometa and scientist Mikhail Miroshnikov of the Vavilov State Optics Institute, Oko sometimes was called Project M (for Miroshnikov). Because of the usual placement in Molniya-type, highly elliptical orbit (HEO), each Oko satellite reached apogee twice daily over northern Europe, from where its infrared-sensing telescope could spot the flame from an ascending U.S. intercontinental ballistic missile against the blackness of space with minimal risk of false alarms due to reflected sunlight. Although the Russians had launched more than 80 Oko satellites from Plesetsk Cosmodrome by 2006, serious financial difficulties and the premature failure of several deployed satellites led to system deterioration after 1996, which left coverage gaps in Russia's ballistic missile early warning network. A further setback occurred in May 2001 when fire destroyed the Oko command-and-control center near Moscow, resulting in the loss of all launch-detection capability for four months.

To reduce the number of satellites needed for 24-hour surveillance, to achieve global coverage, and to improve infrared-detector sensitivity, the Soviet Union tested a new geosynchronous (GEO) early-warning spacecraft—Prognoz or US-KMO—in 1975. However, the first three operational GEO satellites, launched during 1984–87, were of the Oko type. Although the Soviet Union launched the first operational US-KMO satellite (*Kosmos 2133*) in 1990, and Russia launched several more from Baikonur Cosmodrome during the next 15 years, each succumbed to technical failure in a relatively short time. At the beginning of 2005 Russia had no operational US-KMO satellites and relied exclusively on Oko—two in HEO and one in GEO.

Rick W. Sturdevant

See also: Russia (Formerly the Soviet Union)

Bibliography
Geoffrey Forden, Pavel Podvig, and Theodore A. Postol, "False Alarm, Nuclear Danger," *IEEE Spectrum* 37, no. 3 (March 2000): 31–39.
Brian Harvey, *Russia in Space: The Failed Frontier?* (2001).
Pavel Podvig, "History and the Current Status of the Russian Early-Warning System," *Science and Global Security* 10, no. 1 (2002): 21–60.

Space-Based Infrared System

Space-Based Infrared System (SBIRS), an early warning and tracking system that emerged from the Strategic Defense Initiative (SDI) missile defense program that U.S. President Ronald Reagan initiated in March 1983. The first SDI system architecture in 1987 included two kinds of space-based infrared (IR) sensor systems. First, a Boost Surveillance and Tracking System (BSTS) would detect launches of hostile ballistic missiles, warn defenders of the attack, and establish initial tracks on the attacking missiles. Second, cued by data from BSTS, a Space Surveillance and

Tracking System (SSTS) then would develop detailed tracks on objects in the attacking wave. Cued by those tracks, other defensive systems that included radars and interceptors would home in on designated targets using their own onboard IR sensors. In addition to its SDI functions the BSTS would fulfill requirements specified in a 1981 Secretary of Defense master plan for tactical warning and attack assessment by augmenting and, ultimately, replacing the Defense Support Program (DSP) satellite constellation.

As the U.S. missile defense architecture evolved, advancing technologies forced program officials to reevaluate how best to meet requirements for sensor data. The Brilliant Pebbles (BP) interceptor concept, which entered the SDI architecture in 1990, represented a key development. Considering BP's increased IR-sensing capability, planners eliminated BSTS from the architecture. Because other agencies still had requirements for an early-warning or space-surveillance capability that BSTS would have satisfied, that sensor program transferred from the SDI Organization to the U.S. Air Force (USAF) and became known as the Follow-on Early Warning System (FEWS).

Shortly after gaining the FEWS program, the USAF received an alternative proposal calling for an improved DSP satellite, DSP 2, to handle early warning and surveillance. The competition between FEWS and DSP 2, tempered by the experience of providing Scud launch detection and warning to military and civilian populations during the First Gulf War in early 1991, generated a synthesis called the Alert Locate and Report Missiles (ALARM) system. ALARM would use an improved ground-station capability known as ALERT (Attack and Launch Early Reporting to Theater) to warn theater commanders of missile attacks and cue missile defenses within the theater.

While ALARM was evolving out of BSTS, the SSTS system underwent transformation into Brilliant Eyes (BE), a new space-tracking system consisting of numerous, smaller, less-expensive satellites. Before long, BE became known as the Space and Missile Tracking System (SMTS).

By mid-1994 the situation regarding future IR sensors bordered on chaos. To resolve the problem, the Department of Defense (DoD) launched a summer study to develop a set of overarching space-based IR requirements and to propose an approach for satisfying those requirements. The study group recommended consolidation of all space-based IR sensor programs under a single architecture and acquisition management plan, which the DoD did in early 1995. Known as SBIRS (pronounced "sibbers") the new architecture absorbed both ALARM and SMTS. The USAF became responsible for managing the entire SBIRS acquisition program, which included two major components: SBIRS-High and SBIRS-Low.

As the replacement for the DSP system, SBIRS-High would provide the surveillance capability associated with missile attack warning and attack assessment in addition to detection of nuclear detonations. It would consist of two satellites in highly elliptical orbit, four others in geosynchronous orbit (GEO), and a consolidated ground processing station. The SBIRS-High satellites would have short- and mid-wave IR sensors, in addition to other IR equipment for surveillance to ground level. These improved sensors would enable SBIRS-High satellites to image dimmer targets and develop more accurate data on missile launches and projected impact points than was possible with DSP satellites. Moreover, ground controllers would have the flexibility to point the sensors at areas of particular interest and to relook at multiple areas of interest when necessary.

In November 1996 Lockheed Martin became the SBIRS-High prime contractor. Operational certification of the first SBIRS-High increment, the consolidated master control station at Buckley Air Force Base, Colorado, occurred in December 2001. The first SBIRS payload entered highly elliptical orbit in June 2006. By December 2007 plans called for launching the first SBIRS-High GEO satellite in 2009.

SBIRS-Low, the progeny of BE and SMTS, would consist of 24 satellites in low Earth orbit (LEO). Sensors on those satellites would operate in four wavebands: short-wave IR, medium-wave IR, long-wave IR, and the visible portion of the electromagnetic spectrum. The LEO constellation would provide tracking data on longer-range theater missiles and on the warheads of strategic ballistic missiles. It also would complement SBIRS-High in such missions as battle-space characterization, space surveillance, and technical intelligence.

Despite its comprehensiveness the SBIRS concept failed to bring stability to the U.S. program for acquiring new space-based IR sensors. Given its critical missile-defense role and slippage in the USAF schedule for launching the first SBIRS-Low satellites, the program transferred to the DoD's Missile Defense Agency (MDA) in October 2001. The latter promptly restructured it to reduce programmatic risk and enhance the program's integration with the broader missile defense architecture. The transformed SBIRS-Low concept became the Space Tracking and Surveillance System (STSS).

The MDA planned to develop concomitantly a series of STSS research-and-development (R&D) LEO satellites and ground equipment for detection and tracking of ballistic missiles during the midcourse portion of their flight. Using information gained from the R&D satellites, MDA anticipated the incremental development of satellites with increasingly sophisticated capabilities in an evolving testbed. Data from STSS satellites would help in distinguishing decoys from warheads and in cueing missile defense systems to engage attacking missiles at the earliest possible moment.

Northrop Grumman received the prime contract for STSS in November 2002. By December 2007, the contractor expected to deliver the first two R&D satellites to MDA for launch in 2008. Planners expected the ultimate size of the operational STSS satellite constellation to vary from as few as nine to as many as 30 satellites, depending on the success of technological initiatives explored in the R&D satellites.

Rick W. Sturdevant

See also: Antiballistic Missile Systems

Bibliography
Federation of American Scientists Space Policy Project. http://www.fas.org/spp/.
Jay A. Moody, "Achieving Affordable Operational Requirements on the Space-Based Infrared System (SBIRS): A Model for Warfighter and Acquisition Success?" research paper for Air Command and Staff College, Report No. 97-0548 (1997).
Richard J. Newman, "Space Watch, High and Low," *Air Force Magazine* 84, no. 7 (July 2001): 35–38.

Space Surveillance and Tracking System. *See* Space-Based Infrared System.

ELECTRONIC INTELLIGENCE

Electronic intelligence (ELINT), interception and analysis of electronic emanations such as radio and television transmissions, telephone communications, radar signals, and guided missile telemetry. This requires special antennae or receivers of varying sizes and capabilities to collect the signals for analysis. Eavesdropping on radio communications developed with the birth of radio communications in the early twentieth century. Measuring the strength, range, azimuth, and other capabilities of radar began in 1940 when the German Luftwaffe attacked England in the Battle of Britain. The Royal Navy used intercepted radio messages to hunt and destroy the German battleship *Bismarck*. The United States cracked Japanese diplomatic and military codes before World War II, a fact scrupulously kept secret, leading to the decisive defeat of the Imperial Navy in the Battle of Midway. Collection of electronic intelligence continuously improved and expanded during the Cold War. Naval vessels such as the USS *Pueblo* and *Liberty* bristled with antennae, while land sites in Norway, Iran, and elsewhere pointed huge antenna dishes toward the Soviet Union. U.S. Air Force (USAF) and Navy planes such as RB-29s, RB-47s, P2V Neptunes, and P3 Orions made incessant peripheral flights around the Iron Curtain to tap into communications and measure ground radar capability by a technique called ferreting.

Space was a natural environment for electronic intelligence collection because satellites could use the altitude advantage to cover vast areas deep inside the communist bloc while staying beyond the reach of enemy fighters and surface-to-air missiles. One impetus for adapting electronic intelligence collection to space was the Soviet space program. Caught by surprise when *Sputnik* started the space age on 4 October 1957, the U.S. Central Intelligence Agency (CIA) had no intention of being surprised again or, almost as embarrassing, of being dependent on information from the Soviet Union. Furthermore, a great deal of important military information, such as ballistic missile telemetry, military communications, and radar capability could be collected in quantity from satellites flying roughly parallel to the equator in low Earth orbit (LEO). In July 1960, the U.S. Naval Research Laboratory's experimental *Galactic Radiation and Background (GRAB 1)* satellite with scientific and ELINT packages began collecting data on Soviet air defense radars. The first of 16 U.S. ELINT satellites known as "heavy ferrets" was launched in 1961 to monitor Soviet radio communications to learn whether an attack or all-out war was imminent. Improved models, some flying as low as 50 miles, others at about 1,000 miles, replaced the original ferrets. In 1971, another type of ELINT satellite, code-named Jumpseat, began flying eccentric orbits with a low perigee and an apogee so high they "hung" over the northern Soviet Union for as long as 12 hours. They listened to electronic traffic and eavesdropped on the Soviet Union's Molniya electronic intelligence satellites.

The USAF operated the next major U.S. ELINT program called Canyon. In 1968, the first satellite in the series went into a near-geosynchronous orbit. At that distance—almost 22,300 miles above the equator—it could listen selectively to the entire Soviet landmass. Canyon was, in turn, succeeded by a CIA program called Rhyolite-Aquacade. The first of those super-secret satellites was launched in 1971, but they

This 1982 work shows the Kosmos 389 satellite, which was launched in December 1970 and performed electronic intelligence (ELINT) missions. Kosmos 389 was the first in a series of "ferret" satellites that pinpointed sources of radar and radio emissions to identify air defense sites and command and control centers. Transmitted to ground stations, the data was used for Soviet targeting and war planning. (Courtesy Brian W. McMullin/Defense Intelligence Agency)

did not stay secret for long. Two men who worked for TRW, maker of the satellites, sold thousands of Rhyolite documents to the Soviet intelligence service (KGB) before being arrested. Robert Lindsey's book *The Falcon and the Snowman* (1979) noted the revelation during their trial that Rhyolite not only monitored Soviet ballistic missile telemetry but could eavesdrop on 11,000 telephone calls simultaneously. A series of improved versions named Orion, Chalet-Vortex, Magnum, Mercury, and Mentor succeeded Rhyolite. Bought and operated by the National Reconnaissance Office, the satellites supply downloaded intercepts to the National Security Agency at Fort George Meade, Maryland.

Soviet (now Russian) ELINT satellites all have flown under the ubiquitous name of Kosmos, though observers generally know their particular names. The first photo-reconnaissance satellites, Zenit (Zenith), which were first orbited in April 1962, were also the first electronic intelligence collectors. A series of improved spacecraft successively called Tselina O, D, R, and 2 replaced them. The first Tselina O (or *Kosmos 189*) was successfully launched in October 1967, followed by scores of others, sometimes seven per year. In addition to Russia and the United States, the People's Republic of China and France operate ELINT satellites, and Israel possibly uses them.

William E. Burrows

See also: National Reconnaissance Office

Bibliography
James Bamford, *The Puzzle Palace: A Report on America's Most Secret Agency* (1982).
William E. Burrows, *Deep Black: Space Espionage and National Security* (1986).
Jeffrey T. Richelson, *The U.S. Intelligence Community* (1999).

Acquacade. *See* United States Signals Intelligence.

Canyon. *See* United States Signals Intelligence.

Chalet. *See* United States Signals Intelligence.

Crystal. *See* KH-11 Kennan.

GRAB

GRAB (Galactic Radiation and Background) experiment, the world's first reconnaissance satellite placed in orbit, and the first space-based electronic intelligence (ELINT) system. Officially declassified in June 1998, GRAB collected information on Soviet air defense radar signals unobtainable by U.S. aircraft patrolling the borders of the Soviet Union.

The Naval Research Laboratory (NRL) proposed an ELINT satellite system in the spring of 1958, and the Office of Naval Intelligence sought political support for what it dubbed "Project Tattletale." Department of Defense (DoD), Central Intelligence Agency, and other endorsements led to President Dwight Eisenhower's approval on 24 August 1959 for full development of the highly classified GRAB system. It included the satellite and an overseas network of ground sites to receive data from the satellite and to record it on magnetic tape for distribution through the NRL to such organizations as the National Security Agency (NSA) and Strategic Air Command (SAC). GRAB used the NRL Space Science Division's SolRad (Solar Radiation) satellite program as a cover mission to develop the satellite.

On 5 May 1960, just four days after the Soviet Union's downing of a U-2 spy plane prompted Eisenhower to cease further aerial reconnaissance over Soviet territory, the president approved the first GRAB mission. A Thor Able-Star rocket launched from Cape Canaveral, Florida, carried *GRAB 1* atop the *Transit 2A* navigation satellite on 22 June 1960. In addition to the classified ELINT package, *GRAB 1* carried instrumentation to measure solar radiation, which allowed the DoD to acknowledge publicly the piggyback launch as *SolRad 1* without disclosing GRAB. It had a useful life of 90 days. Only one of four further attempts to launch a GRAB payload succeeded, that being on 29 June 1961. *GRAB 2* operated successfully for 14 months. Despite the limited success in launching GRAB missions, data from GRAB enhanced the effectiveness of SAC war planning and allowed the NSA to discover operational Soviet radar capable of supporting an antiballistic missile system. The Director of Naval Intelligence transferred control of GRAB to the National Reconnaissance Office on 14 June 1962. GRAB operations ended two months later as the NRO replaced GRAB with the POPPY system, which had its first launch on 13 December 1962.

Robert Kilgo

See also: SolRad, Transit, United States Navy

Bibliography

Dwayne A. Day, "Listening from Above: The First Signals Intelligence Satellite," *Spaceflight*, August 1999.

Federation of American Scientists Space Policy Project: Project Tattletale. http://www.fas.org/spp/military/program/sigint/grab.htm.

Robert A. McDonald, with Sharon K. Moreno, "Grab and Poppy: America's Early ELINT Satellites," National Reconnaissance Office (2005).

Naval Research Laboratory. http://www.nrl.navy.mil/accomplishments/solar-lunar-studies/solrad-grab/.

David Van Keuren, "Cold War Science in Black and White: US Intelligence Gathering and Its Scientific Cover at the Naval Research Laboratory, 1948–62," *Social Studies of Science* 31, no. 2 (2001).

Magnum. *See* United States Signals Intelligence.

Parcae. *See* White Cloud.

Rhyolite. *See* United States Signals Intelligence.

Russian Signals Intelligence Satellites

Russian signals intelligence satellites performed two basic functions: listening to foreign radio traffic and conversations, and monitoring electronic emissions from foreign sources to locate and classify types of equipment and facilities. Tselina (virgin land)-type satellites were essentially the "ears" that allowed the Soviet government to hear evidence of potentially hostile military activities, especially on land. An Electronic Intelligence (ELINT) Ocean Reconnaissance Satellite (EORSAT) system provided similar evidence with respect to surface vessels, but was relatively ineffective against submarines.

The Soviet Union began exploring space-based interception of electronic signals when designer Mikhail Zaslavskiy's TsNII 108 (later TsNIRTI—Central Scientific Research Radiotechnical Institute) developed a so-called Kust (bush) payload for installation on the first Zenit photoreconnaissance satellites. Based on initial studies and development of two experimental spacecraft by the OKB-586 (Special Design Bureau)—later KB Yuzhnoye—authorities decided in 1964 to produce an operational Tselina ELINT system consisting of two varieties of satellites—Tselina O for low-sensitivity, general observations, and Tselina D for high-sensitivity, detailed observations. TsNII-108 built the payloads, and OKB-586 handled overall spacecraft design. The first Tselina O satellite launched from Plesetsk Cosmodrome in October 1967, and the complete Tselina O system apparently earned official, operational certification in March 1972. A total of 40 Tselina O launches occurred, the last one in March 1982.

Meanwhile, the first of the much heavier, more complex Tselina D satellites launched from Plesetsk in December 1970. The Tselina D system became fully operational in December 1976, and the O and D versions of the satellites operated simultaneously until 1984. On 18 March 1980, in one of the worst launch disasters, a catastrophic explosion and fire on the pad killed 48 people and destroyed a Tselina D spacecraft atop a Vostok-2M rocket. More than 70 Tselina D launches occurred before the final one in December 1992, and the last Tselina D satellite apparently stopped operating in 1994. A variant known as Tselina R, with an overall weight equal to Tselina D but a payload weight only 56 percent of the Tselina D, was launched into the same type orbit as Tselina D four times between December 1986 and April 1993. The O, D, and R varieties of Tselina each had a lifespan of six months.

Design work on a Tselina 2 ELINT satellite began as early as March 1973. As the Soviet military's exploitation of space-based ELINT improved, the sensitivity and lifespan requirements of the Tselina 2 increased. The addition of new capabilities nearly doubled the overall weight and more than tripled the payload weight compared to Tselina D, but a Tselina 2 had twice the lifespan of the Tselina D. The first Tselina 2 prototype launched from Baikonur Cosmodrome in September 1984. Declared operational in December 1988, the full Tselina 2 system included a constellation of four satellites in high, circular orbits. Financial problems prevented Russia from maintaining such a constellation after the mid-1990s, and by January 2004 a single on-orbit Tselina 2 remained operational. A successful launch in June 2004 finally gave the Russians a second on-orbit Tselina 2 satellite.

Rick W. Sturdevant

See also: Russia (Formerly the Soviet Union), Russian Naval Reconnaissance Satellites

Bibliography
Brian Harvey, *Russia in Space: The Failed Frontier?* (2001).
Russian Space Web. http://www.russianspaceweb.com/.

Spook Bird. *See* United States Signals Intelligence.

Tselina. *See* Russian Signals Intelligence Satellites.

United States Signals Intelligence

United States Signals Intelligence (SIGINT), interception and analysis of electronic emissions from foreign sources, usually involving human communication but not excluding radio signals, such as those produced by radars. Human messages, either spoken words or written text, transmitted electronically can be intercepted to yield communications intelligence. Nonhuman electronic emissions from radars or other types of equipment can yield electronic intelligence. The United States tends to combine those two forms of intelligence under the category of SIGINT. With SIGINT, political and military leaders can assess threats posed by nations or terrorists.

The first major use of SIGINT by the United States came during World War I, when the Army Radio Intelligence Section used special equipment to eavesdrop on German conversations that revealed enemy tactics and strategy. The American Expeditionary Forces were able to exploit this information obtained through the interception and deciphering of radio signals to provide tactical planning for military operations. Technological advances after the war allowed the United States to monitor Japanese diplomatic communications leading up to World War II. During the war itself, based on intercepted communications from the Japanese fleet, the U.S. Navy (USN) positioned itself for a crucial victory at Midway. Teaming with the British, the USN read coded German naval transmissions using an Enigma cipher machine, thereby countering German submarines in the Atlantic. Translation of intercepted, low-level codes and ciphers also permitted Allied forces to move rapidly across France after the Normandy invasion.

In 1949 a three-way competition among Army, Navy, and Air Force cryptological units prompted the Secretary of Defense to establish the Armed Forces Security Agency (AFSA) to eliminate interservice rivalries. When the Korean War exposed deficiencies in the AFSA arrangement, President Harry Truman commissioned a study that led to formation of the National Security Agency (NSA) in November 1952. Its existence remaining secret until 1957, the NSA intercepted, decoded, and disseminated SIGINT under direct authority of the Secretary of Defense. During the 1950s, reconnaissance aircraft provided the primary means for obtaining SIGINT.

From the 1960s onward, various types of SIGINT satellites greatly enhanced NSA collection capabilities, but official information about those space-based systems remained highly classified and publicly unavailable. Unofficial sources claim the existence of several kinds of SIGINT satellites in geosynchronous or highly elliptical, Molniya-type orbits. They also refer to the evolution of geosynchronous SIGINT satellites from the Spook Bird/Canyon experimental satellites of the late 1960s, to Rhyolite in the 1970s, Chalet/Vortex and Magnum in the 1980s, and Acquacade from the mid-1980s into the 1990s. Some unofficial sources claim that Jumpseat satellites in highly elliptical orbits collected SIGINT data from the extreme northern portion of the Soviet Union. Undoubtedly SIGINT satellites permitted NSA to supply extremely valuable information to U.S. officials during the Cold War generally and in such specific instances as the Cuban Missile Crisis, the Vietnam War, and the two Persian Gulf Wars.

Rick W. Sturdevant

See also: National Reconnaissance Office

Bibliography

Desmond Ball, *Pine Gap: Australia and the U.S. Geostationary Signals Intelligence Satellite Program* (1988).

James Bamford, *Body of Secrets: Anatomy of the Ultra-Secret National Security Agency* (2001).
Global Security. http://www.globalsecurity.org/.

Vortex. *See* United States Signals Intelligence.

White Cloud

White Cloud, codenamed Parcae, a U.S. Navy satellite system that detected, located, and tracked ships by picking up their radar emissions. Each satellite, in reality an orbiting cluster of three subsatellites that continuously measured their distance apart and communicated via millimeter-wave radio links, could determine precisely the bearing of a seaborne emitter through interferometry, a technique based on calculation of the time difference between arrival of signals. The complete White Cloud surveillance system, which included four satellite clusters with orbital planes 60–120 degrees apart (at the equator) and several ground stations (regional reporting centers), bore the code name Classic Wizard but officially was designated the Naval Ocean Surveillance System (NOSS) in the mid-1970s.

Conceived by the Naval Research Laboratory, reportedly in response to a 1970 request from the Chief of Naval Operations, White Cloud succeeded two earlier Naval Research Laboratory–designed satellite systems—Galactic Radiation and Background (GRAB), and Poppy. Essentially three GRABs with interlinks, the first experimental White Cloud satellite was launched in December 1971. At the time, Navy planners envisioned complementing White Cloud with Clipper Bow, a satellite system that used active radar and electronic intelligence sensors, but high costs prompted cancellation of the latter program in fiscal year 1980. White Cloud became the key space-based ocean surveillance system, with the first three clusters orbited during 1976–80 and five improved versions following during 1983–87. A second generation, with updated reconnaissance and data communications equipment, launched during 1990–96. The first-generation system apparently monitored radars operating at 0.5–4 GHz; the second-generation satellites raised the upper limit to 10 GHz and used a different intracluster link, because the original one interfered with radio astronomy. By the 1990s the Navy and National Reconnaissance Office employed White Cloud to locate vessels at sea and fixed emitters on solid ground.

Norman Friedman

See also: National Reconnaissance Office, United States Navy

Bibliography

Norman Friedman, *Seapower and Space: From the Dawn of the Missile Age to Net-Centric Warfare* (2000).
Global Security. http://www.globalsecurity.org/.

MILITARY COMMUNICATIONS

Military communications involve transmission of strategic and tactical information necessary for successful military operations and overall cohesion. That information, which includes routine orders, directives for maneuvering forces, and intelligence, could be exchanged in oral, written, or pictorial form. Couriers delivered the information long ago, because that was considered the most secure method of relaying messages, but smoke or semaphore signals also sufficed. By the 1860s, armies could use the telegraph and, by World War I, the telephone. During World War II, radio became the dominant medium for military communications. Many naval units, especially submarines, depended on radio to relay their position and receive orders. Although land-based units had other options, radio remained the most reliable means to communicate among geographically dispersed units.

Radio, however, presented certain disadvantages. Atmospheric storms or changes in ionospheric conditions due to solar weather could disrupt radio traffic for significant periods. Using powerful transmitters, an enemy could jam critical transmissions, and radio stations presented inviting targets for attack by various means. While microwave afforded more directional control over signals and less vulnerability to adverse weather conditions, it required relay towers every 20 to 30 miles and exhibited the same vulnerabilities as radio to enemy attack. Enemy listening posts also listened in on radio traffic.

British Interplanetary Society member and Royal Air Force electronics officer Arthur C. Clarke published an article in the October 1945 issue of *Wireless World* magazine describing the potential of "extra-terrestrial relays" for communications worldwide, and the first RAND report in May 1946 acknowledged the utility of a "world-circling spaceship" for military communications. Not until the launch of *SCORE* (*Signal Communication by Orbiting Relay Equipment*) in December 1958 did the communications satellite become a reality, and not until October 1960 did the United States launch the Army's experimental *Courier 1B*, the world's first military communications satellite. During the 1960s, the U.S. Department of Defense (DoD) used Echo passive communications satellites and Syncom active communications satellites managed by NASA, along with Lincoln Experimental Satellites managed by the Massachusetts Institute of Technology (MIT) Lincoln Laboratory, to advance the development of ground, air, and seaborne terminal equipment.

The first wartime use of satellite communications occurred during the Vietnam conflict. In January 1965 NASA transferred control of its two Syncom satellites to the DoD. The U.S. military used them extensively during the next two years for Southeast Asia operations and augmented their capacity with leased commercial satellite circuits. When the U.S. Air Force launched the first seven Initial Defense Communications Satellite Program (IDCSP) satellites in June 1966, the DoD relied less on Syncom. Having established the practice in Vietnam, the DoD continued into the twenty-first century to use leased satellite channels for routine administrative and logistical communications, thereby reserving dedicated military systems for more sensitive command-and-control communications.

Cooperation among the United States, United Kingdom, and other North Atlantic Treaty Organization (NATO) members led to broader experimentation with IDCSP capabilities during the late 1960s. Separate Skynet and NATO communications satellite systems soon appeared. Although nearly identical to the IDCSP design, the first Skynet and NATO satellites differed from IDCSP in that they were larger, more powerful, occupied geostationary orbits, and possessed a command system for orbital station-keeping.

When the Soviet Union first contemplated satellites for relaying communications remains uncertain, but its General Staff was briefed on possible uses of artificial satellites in 1956. Although the Soviet Union showed an interest in highly elliptical orbits as early as 1960 and began designing a communications satellite for such an orbit in October 1961, the first successful Molniya launch did not occur until April 1965. Strela, the first Soviet dedicated military communications satellite system began with an August 1964 launch of three experimental satellites with store-and-dump capability. The Soviet Union began using geosynchronous orbits for military communications in the mid-1970s and, a decade later, introduced Potok (code-named Geizer) satellites for that purpose.

The late 1970s brought military communications satellites for maritime use. The U.S. Navy Fleet Satellite Communications (FLTSATCOM) system, which included an Air Force Satellite Communications system package, provided ultrahigh-frequency and extremely high–frequency capability to ships and aircraft. Previously, the mobility of such platforms, combined with power requirements and antenna sizes, had precluded naval or aerial use of strategic communications satellite systems. The Soviet Raduga in late 1975, joined in 1978 by its civilian equivalent, the Gorizont series, supported the rival superpower's naval communications to some degree.

During the 1980s–90s military communications satellites such as those in the U.S. Defense Satellite Communications System became larger, more powerful, and able to resist jamming attempts. Newer systems like the Ultrahigh-Frequency Follow-On replaced older ones like FLTSATCOM. In the case of the U.S. Milstar system, engineers designed the satellites to more likely survive a nuclear conflict. Furthermore, as the phenomenal growth in demand for satellite communications capacity outstripped the availability on dedicated military systems, U.S. forces relied increasingly on leased commercial satellite channels. Military operations, especially the First Gulf War, revealed the military importance of satellite communications, which furnished the primary means for exchanging information in near-real time among widely dispersed forces and senior leaders in Washington, DC.

The demand for more bandwidth grew exponentially with each new conflict, largely fueled by the need to transmit video imagery for a variety of purposes. Services once thought exotic, such as secure Internet access and real-time video feeds from Unmanned Aerial Vehicles (UAVs) became commonplace. By the early twenty-first century, "pilots" sitting at consoles in the United States were using satellite communications to maneuver UAVs over Afghanistan and Iraq. Through the application of direct-broadcast satellite technology, the U.S. military began constructing a Global Broadcast System to overcome bandwidth problems and, thereby, get correct information to users at the right time and place. This revolutionized the way the military used information to improve significantly its understanding of the battlefield situation and its application of destructive force as required.

In April 2007 DoD announced Internet Routing In Space (IRIS), a three-year, collaborative industry-government project to demonstrate Internet Protocol routing and dynamic bandwidth resource allocation capabilities from a geostationary, commercial communications satellite. By using fewer hops and fewer frequencies per message, IRIS promised to increase satellite capacity and to reduce transmission times between remote terminals, thereby further revolutionizing military communications.

Although some experts argued that military forces should depend more on cheaper, more survivable fiber-optic links and less on satellites, reliance on space-based communications continued to expand among the world's defense establishments at the beginning of the twenty-first century. Nations like China, Israel, Australia, and France launched their own dedicated military communications satellites, and others like the United Kingdom and Italy undertook development programs. Japan's defense agency planned to use high-capacity commercial satellite service to maintain contact with its troops participating in United Nations peacekeeping efforts from East Asia to the Middle East, and many other countries depended on commercial satellite service to meet some of their military communications requirements. That included the U.S. DoD, whose 2001 contract with Iridium Satellite LLC for global satellite telephone service to 20,000 users constituted 40 percent of the company's total operation.

Morland Gonsoulin

See also: Commercial Space Communications, First Gulf War, Syncom

Bibliography

Andrew J. Butrica, ed., *Beyond the Ionosphere: Fifty Years of Satellite Communications* (1997).
Brian Harvey, *Russia in Space: The Failed Frontier?* (2001).
Donald H. Martin, *Communication Satellites*, 4th ed. (2000).
Harry L. Van Trees et al., "Military Satellite Communications: From Concept to Reality," in *The Limitless Sky: Air Force Science and Technology Contributions to the Nation*, ed. Alexander H. Levis (2004), 175–209.
Steven J. Zaloga, *The Kremlin's Nuclear Sword: The Rise and Fall of Russia's Strategic Nuclear Forces, 1945–2000* (2002).

Advent

Advent, the world's first major geosynchronous satellite project. A communications satellite development effort originally undertaken by the Advanced Research Projects Agency in 1958 and dubbed Notus, it became Advent in February 1960 after the Department of Defense assigned it to the U.S. Army. Plans called for a three-axis-stabilized, solar-powered satellite weighing more than 1,000 lb with a secure command system and four repeaters, each capable of handling 12 one-way voice links or a single-spread-spectrum voice link. Technologically ambitious for its time, Advent suffered cost overruns, setbacks in the development of its command system, and excessive satellite-to-booster weight ratios. Delays in development of the Centaur upper-stage launch vehicle, which was essential for placing the heavy Advent satellite in the proper orbit, compounded the project's difficulties. Furthermore, the triode tube on which Advent's design relied already was obsolete. This host of problems compelled Secretary of Defense Robert McNamara to cancel the Advent program on 23 May 1962 without a single satellite having been launched. Throughout the next three decades, however, practically all of Advent's design characteristics became commonplace in the world of satellite communications.

James C. Mesco

See also: Centaur, Defense Advanced Research Projects Agency

Bibliography

U.S. Congress House Committee on Science and Astronautics, *Project Advent: Military Communications Satellite Program* (1962).
David J. Whalen, *The Origins of Satellite Communications, 1945–1965* (2002).

Air Force Satellite Communications—AFSATCOM. *See* Fleet Satellite Communications System (FLTSATCOM).

Altair. *See* Soviet and Russian Military Communications Satellites.

Courier

Courier, the world's first high-capacity communications satellite series. Proposed by the U.S. Army Signal Corps in September 1958 and built by Philco's Western Development Laboratories in Palo Alto, California, under supervision of the Army Signal Research and Development Laboratory at Fort Monmouth, New Jersey, each Courier satellite carried one analog and four digital tape recorders and could relay up to 100,000 words of teletype data per minute at ultrahigh frequency. Although operating primarily in a store-and-dump mode, the experimental, spherically shaped, 500 lb satellite could actively receive and transmit communications in real time. After a launch-vehicle malfunction prevented *Courier 1A* from achieving orbit in August 1960, a Thor Able-Star rocket successfully boosted *Courier 1B* into low Earth orbit on 4 October. On its second orbit, the satellite successfully relayed a message from President Dwight Eisenhower to the United Nations between ground stations in New Jersey and Puerto Rico. *Courier 1B*, the first communications satellite to use solar cells for recharging nickel-cadmium storage batteries, operated successfully for 17 days before its payload stopped responding to commands from the ground.

James C. Mesco

See also: Space Systems/Loral, United States Army

Bibliography

Max L. Marshall, *The Story of the U.S. Army Signal Corps* (1965).
Donald H. Martin, *Communication Satellites*, 4th ed. (2000).

Defense Satellite Communications System

Defense Satellite Communications System (DSCS), follow-on to the Initial Defense Communications Satellite Program (IDCSP) or IDSCS. Because IDSCS performance demonstrated that satellite communications could satisfy certain military requirements, the U.S. Department of Defense decided in June 1968 under Program 777 to develop a more advanced DSCS 2 satellite. In March 1969 the U.S. Air Force (USAF) awarded TRW Systems Group a DSCS 2 production contract. Unlike the IDSCS, the DSCS 2 had a command subsystem, attitude control and station-keeping features, a dual-spin configuration, and multiple channels with multiple-access capability, and it went into geostationary orbit. Each cylindrical DSCS 2, 9 ft in diameter and 6 ft high, weighed 1,350 lb. With four superhigh frequency (SHF) channels, each DSCS 2 could handle 1,300 two-way voice circuits or digital data at approximately 100 Mbps. Although originally designed to operate five years, subsequent modifications extended the lifetime of later DSCS 2 satellites to as much as 20 years.

Although the USAF performed on-orbit control of DSCS satellites for station-keeping purposes, the Defense Information Systems Agency (formerly Defense Communications Agency) determined the constellation's overall orbital configuration and assigned channel capacity to users. The USAF launched the first DSCS 2 pair in November 1971 and the last DSCS 2 in October 1982. Out of 16 DSCS 2 satellites, 12 successfully reached geostationary orbit. Operators retired the last operational DSCS 2 in October 1998.

USAF planning for DSCS 3 began in 1973, and General Electric received the development-and-manufacturing contract in February 1976. The USAF paired

Artistic rendition of a Defense Satellite Communications System (DSCS) Block 2 satellite in orbit above Earth, 1981 (Courtesy U.S. Department of Defense)

the first DSCS 3 for launch with the last DSCS 2 in October 1982. A series of corporate mergers during the 1990s resulted in Lockheed Martin Space Systems becoming the satellite manufacturer through launch of the 14th, final DSCS 3 in August 2003. Three-axis stabilized and weighing nearly 2,600 lb with a rectangular body measuring 6.8 × 6.3 × 6.4 ft, from which extended two solar arrays spanning nearly 38 ft, a DSCS 3 provided greater capacity and more flexibility than a DSCS 2, especially for mobile-terminal users. With six SHF channels, single- and multiple-beam antennas, and anti-jamming features, each DSCS 3 satellite could handle the equivalent of 2,600 phone conversations simultaneously.

The DSCS 3 satellites consistently exceeded their 10-year life expectancy, with the first one still operating for more than two decades. A service life enhancement program (SLEP) in the late 1990s improved by 200 percent the overall communications capacity of the last four DSCS 3 satellites and afforded tactical users a 700 percent increase in capacity for certain situations. Nonetheless skyrocketing demand for additional bandwidth to support unpiloted aerial vehicles, digital imagery, and other new combat systems outstripped what the five-satellite DSCS constellation could supply. By 1997, USAF officials were discussing a "DSCS replacement" system and, in January 2001, awarded The Boeing Company a Wideband Gapfiller Satellite (WGS) contract. The gapfiller program, with each satellite having 10 times the SHF capacity of a DSCS, was scheduled to begin augmenting DSCS as early as 2006 and, eventually, to replace it.

James C. Mesco

See also: Lockheed Martin Corporation, TRW Corporation, United States Air Force

Bibliography

Donald H. Martin, *Communication Satellites*, 4th ed. (2000).

David N. Spires and Rick W. Sturdevant, "From Advent to Milstar: The United States Air Force and the Challenges of Military Satellite Communications," in *Beyond the Ionosphere: Fifty Years of Satellite Communication*, ed. Andrew J. Butrica (1997).

Harry L. Van Trees et al., "Military Satellite Communications: From Concept to Reality," in *The Limitless Sky: Air Force Science and Technology Contributions to the Nation*, ed. Alexander H. Levis et al. (2004), 175–209.

Fleet Satellite Communications System

Fleet Satellite Communications system (FLTSATCOM), a military satellite constellation operated by the U.S. Navy for communication among surface ships, submarines, naval aircraft, and ground terminals. Air Force Satellite Communications (AFSATCOM) relied heavily on FLTSATCOM satellites for communication between the National Command Authority (president and secretary of defense) and U.S. strategic nuclear forces. The AFSATCOM portion also supported the White House Communications Agency, Air Intelligence Agency, reconnaissance aircraft, and special operations.

The Navy began developing FLTSATCOM in 1971 as the first truly operational satellite communications system for tactical users. Designed and manufactured by TRW, the first FLTSATCOM satellite was launched from Cape Canaveral, Florida, into geosynchronous orbit in February 1978. The system, which became fully operational in January 1981, operated primarily in the ultrahigh frequency (UHF) range but also used superhigh frequency for fleet broadcast uplink and, later, extremely high frequency on selected satellites to communicate with Milstar ground terminals.

In June 1996 the U.S. Air Force's 3rd Space Operations Squadron at Schriever Air Force Base, Colorado, turned over FLTSATCOM operations to the Naval Satellite Operations Center at Point Mugu, California. Although the last FLTSATCOM satellite was launched in 1989 and UHF Follow-On satellites began replacing the system in 1993, two FLTSATCOM satellites remained operational at the beginning of 2005. The AFSATCOM portion of FLTSATCOM that disseminated Emergency Action Messages authorizing use of nuclear weapons shifted to Milstar in the late 1990s.

Robert Kilgo

See also: United States Navy

Bibliography
Air University Space Primer (2003).
Donald H. Martin, *Communication Satellites*, 4th ed. (2000).

Gelios. *See* Soviet and Russian Military Communications Satellites.

Geyzer. *See* Soviet and Russian Military Communications Satellites.

Globus. *See* Soviet and Russian Military Communications Satellites.

Gran. *See* Soviet and Russian Military Communications Satellites.

Initial Defense Communications Satellite Program

Initial Defense Communications Satellite Program (IDCSP), the first operational military satellite communications system. This U.S. Air Force (USAF) program was authorized in 1964 as an experimental first step toward high-volume, strategic voice and data communications between large-antenna, fixed or transportable ground stations and large ship-borne equipment, but it was quickly pressed into active service. The satellites, built by Philco, were 24-face polyhedrons, 36 in across, and weighing 100 lb. Launched into sub-synchronous orbits in clusters of up to eight, IDCSP satellites used spin stabilization, fixed antennas, and body-mounted solar cells. Each satellite could handle up to five commercial-quality or 11 tactical-quality voice calls, 1,550 teletype messages, or digital imagery transmissions.

The USAF orbited 26 IDCSP satellites from 16 June 1966 to 13 June 1968. By the latter date, IDCSP had been renamed the Initial Defense Satellite Communications System (IDSCS). Although designed to last three years or less, IDSCS satellites averaged six years before failure. Two IDSCS terminals in Vietnam were in place by July 1967 and became best known for sending imagery to intelligence analysts in Washington, DC. The USAF began replacing IDSCS with DSCS 2 satellites in 1971, but three of the former remained operational as late as 1976. The system is sometimes referred to retroactively as DSCS 1.

Matt Bille

See also: Space Systems/Loral

Bibliography
Donald H. Martin, *Communication Satellites*, 4th ed. (2000).
David N. Spires and Rick W. Sturdevant, "From Advent to Milstar: The United States Air Force and the Challenges of Military Satellite Communications," in *Beyond the Ionosphere: Fifty Years of Satellite Communication*, ed. Andrew J. Butrica (1997).

Korund. *See* Soviet and Russian Military Communications Satellites.

Lincoln Experimental Satellites

Lincoln Experimental Satellites (LES), developed by the Massachusetts Institute of Technology Lincoln Laboratory to improve active communications satellite capabilities for the U.S. military.

Four technological objectives initially characterized the program: high-efficiency, solid-state transmitters; electronically despun antennas; smaller, mobile ground terminals; and techniques for on-orbit control at synchronous altitude. After a launch vehicle failure in February 1965 left *LES 1* tumbling uselessly in an improper orbit, its *LES 2* twin achieved the planned orbit in May. A superhigh-frequency (SHF), solid-state repeater and eight-horn, electronically switched antenna for omni-directional reception proved reliable for use with large Earth terminals, but testing also indicated an ultrahigh-frequency (UHF) downlink would accommodate smaller, less-complicated ground antennas. The next two LES satellites went into wrong orbits due to a launcher malfunction. However, *LES 3* generated a UHF signal that permitted compilation of multipath propagation data (signals that bounce off the terrain in many directions and hence arrive at an antenna from several paths) over a wide variety of terrains, which proved useful in designing airborne terminals. Similarly, *LES 4* measured spatial and temporal variations in radiation over a wide range of altitudes, which aided the design of future spacecraft. *LES 5* and 6, launched in July 1967 and September 1968 respectively, employed the military UHF band to relay signals from near-synchronous or geostationary orbit to airborne, seaborne, and fixed and mobile ground terminals. Insufficient funding prevented completion of *LES 7*, which was supposed to demonstrate a multiple-beam antenna for SHF reception. After their launch in March 1976, *LES 8* and *9* communicated directly with each other via extremely high frequency (EHF) cross-links. Subsequent procurement of military communications satellites included the technologies of onboard signal processing, cross-linking, and EHF transmission and reception tested on *LES 8* and *9*.

James C. Mesco

See also: Commercial Space Communications

Bibliography
Donald H. Martin, *Communication Satellites*, 4th ed. (2000).
William W. Ward and Franklin W. Floyd, "Thirty Years of Space Communications Research and Development at Lincoln Laboratory," in *Beyond the Ionosphere: Fifty Years of Satellite Communication*, ed. Andrew J. Butrica (1997).

Luch. *See* Soviet and Russian Military Communications Satellites.

Milstar

Milstar, a highly robust, flexible communications satellite system able to operate through all levels of conflict, including nuclear war, and to provide worldwide connectivity to many types of terminals on land, at sea, or in the air. Operated by the U.S. Air Force (USAF), Milstar provided high-priority, secure communications for all branches of the U.S. military after the mid-1990s.

After congressional opposition squelched plans for separate strategic and tactical systems in the late 1970s, planners defined a single "military strategic and tactical relay" concept in April 1981. The USAF contracted with Martin Marietta in 1983 for Milstar satellite development. A Titan IV-B rocket launched from Cape Canaveral, Florida, sent the first of two Milstar Block 1 satellites toward geosynchronous orbit on

Artist's rendering of Milstar, a strategic satellite system launched in 1994. (Courtesy U.S. Air Force)

7 February 1994. Even before engineers completed its on-orbit checkout, U.S. military forces used *Milstar 1* during September 1994 for Operation Uphold Democracy in Haiti. A post–Cold War Department of Defense decision in 1993 to replace Milstar with a lighter, cheaper satellite series limited the number of Milstar Block 2 satellites to four. After the first Block 2 satellite entered an unusable orbit in April 1999, the USAF successfully launched the remaining three, the last on 8 April 2003.

Milstar operated in the extremely high frequency range with spread-spectrum waveforms, which rendered its signals highly resistant to jamming and allowed smaller, highly mobile user terminals. It also provided ultrahigh-frequency and superhigh-frequency capabilities for compatibility with older terminal equipment. Crosslink antennas on each satellite minimized dependence on ground relay stations. The Block 1 satellites carried a low data rate payload built by TRW, which transmitted at a rate of 75 to 2400 bps suitable for nuclear war. Block 2 added a medium data rate payload supplied by Boeing, which could transmit at speeds up to 1.544 Mbps that were more suitable for conventional warfare.

Robert Kilgo

See also: TRW Corporation, United States Air Force

Bibliography

Harry L. Van Trees et al., "Military Satellite Communications: From Concept to Reality," in *The Limitless Sky: Air Force Science and Technology Contributions to the Nation*, ed. Alexander H. Levis et al. (2004): 175–209.

Air Force Link website: Fact Sheets, Milstar Satellite Communications System.

NATO Communications Satellites

NATO (North Atlantic Treaty Organization) communications satellites provided voice and data links among NATO members. During 1967–70, NATO gained experience by using Initial Defense Communications Satellite Program (IDCSP) satellites and operating two IDCSP ground stations. Meanwhile, in April 1968, the U.S. Air Force (USAF) contracted with Philco-Ford Corporation to build two NATO satellites similar to the first British Skynet, which used a variant of the IDCSP design. Those satellites, *NATO 2A* and *2B*, were launched into geostationary orbit in 1970 and 1971. They operated at higher powers and wider bandwidths than their predecessors, which allowed the use of smaller terminals. The USAF awarded Philco-Ford (later Ford Aerospace and Communications Corporation) a $27 million contract in March 1973 to develop a much larger, significantly higher-capacity NATO 3 satellite with both wide-beam and narrow-beam capabilities to support hundreds of users simultaneously with voice and facsimile services in the ultrahigh, superhigh, and extremely high frequency bands. During 1976–84, the United States launched four NATO 3 satellites, some of which remained operational into the 1990s. *NATO 4A* and *NATO 4B*, essentially modified versions of the first-generation Skynet 4, were built by Matra Marconi Space for the United Kingdom (UK) Ministry of Defence and launched in 1991 and 1993 to augment and eventually replace the NATO 3 series.

The NATO 4 satellites added spot-beam, Earth-coverage, and even more channel capacity to existing capabilities. In May 2004 NATO selected a consortium of France, Italy, and the United Kingdom to replace NATO 4. The consortium would employ French Syracuse, Italian Sicral, and UK Skynet satellites to satisfy superhigh- and ultrahigh-frequency requirements through 2019. On 1 February 2005 Paradigm Secure Communications, Ltd., established the first link (between a European ground station and frontline operational forces) under that new arrangement.

James C. Mesco

See also: Space Systems/Loral, Matra Marconi Space

Bibliography
Donald H. Martin, *Communication Satellites*, 4th ed. (2000).
SpaceRef. http://www.spaceref.com/.

Potok. *See* Soviet and Russian Military Communications Satellites.

Project Needles. *See* West Ford.

Raduga. *See* Soviet and Russian Military Communications Satellites.

Satellite Data System

Satellite Data System (SDS), spacecraft that supported the U.S. intelligence community and the U.S. Air Force in a variety of ways. The first SDS was launched in June 1976. That satellite and its many successors operated in highly elliptical orbit, dropping from an altitude of 24,000 miles over part of the Northern Hemisphere to approximately 200–240 miles over the Southern Hemisphere. Normally, the United States had two SDS in orbit at any given time.

As its primary function, SDS relayed imagery from U.S. electro-optical imagery satellites, making that data available to interpreters in near-real time. The SDS also hosted transponders for the Air Force Satellite Communications System, which handled command and control of U.S. strategic forces. Furthermore, the SDS relayed communications between the main satellite control facility in California and satellite tracking stations around the world. Beginning with the fourth SDS, they carried equipment to detect atmospheric nuclear explosions.

Jeffrey T. Richelson

See also: Nuclear Detection, Reconnaissance

Bibliography
Dwayne A. Day, "Relay in the Sky: The Satellite Data System," *Space Chronicle, Journal of the British Interplanetary Society* 59, Supplement 1 (2006): 56-62.
Federation of American Scientists. http://www.fas.org/.
Jeffrey T. Richelson, "The Satellite Data System," *Journal of the British Interplanetary Society* 37 (1984): 226–28.

SCORE, Signal Communication by Orbiting Relay Equipment

SCORE, Signal Communication by Orbiting Relay Equipment, the world's first communications satellite, launched from Cape Canaveral in December 1958. The experimental project demonstrated that a 9,000 lb Atlas intercontinental ballistic missile could achieve Earth orbit and that a space-based communications repeater would work.

In June 1958 the Advanced Research Projects Agency directed the U.S. Army Signal Research and Development Laboratory at Fort Monmouth, New Jersey, to build a communications payload weighing less than 150 lb for launch in a pod atop an Atlas B missile supplied by the U.S. Air Force. The resulting fully redundant system included two tape recorders, each with a four-minute storage capacity, inside the pod and four antennas (two for transmission and two for reception) mounted flush with the stainless-steel surface of the missile. Before launch, technicians loaded onto both *SCORE* recorders President Dwight Eisenhower's message wishing everyone around the globe peace and goodwill. On 19 December the satellite responded to commands from the ground and listeners worldwide heard for the first time a human voice from outer space.

The *SCORE* batteries failed after 12 days in orbit but not before the communications package successfully handled 78 voice and teletype messages, some real-time and others store-and-forward, among ground stations in Arizona, California, Georgia, and Texas.

Morland Gonsoulin

See also: Advanced Research Projects Agency

Bibliography

H. McD. Brown, "A Signal Corps Space Odyssey Part II: SCORE and Beyond," *Army Communicator* 7, no. 1 (Winter 1982): 58–68.

Deane Davis, "The Talking Satellite: A Reminiscence of Project SCORE," *Journal of the British Interplanetary Society* 52, no. 7/8 (July–August 1999): 239–59.

Donald H. Martin, et al., *Communications Satellites*, 5th ed. (2007).

Skynet

Skynet, the British military satellite communications system, originated in late 1966 when the United States agreed to produce for the United Kingdom (UK) satellites compatible with the U.S. Initial Defense Communications Satellite Program (IDCSP). In March 1967 the U.S. Air Force awarded Philco-Ford Corporation an IDCSP/A (for augmentation) production contract. By the time the first of these 285 lb, spin-stabilized satellites launched in November 1969, its designation had changed to *Skynet 1A*. The launch of its *Skynet 1B* twin in August 1970 proved less than successful when an apogee-motor failure prevented the satellite from achieving geostationary orbit.

Marconi Space Systems developed the similarly designed but heavier, more powerful, more reliable Skynet 2 series in the UK with U.S. assistance. Although a launch

Skynet Satellites

Satellite Name	Manufacturer	Launch Date	Comments
1A	Philco-Ford	22 November 1969	Operated 36 months
1B	Philco-Ford	19 August 1970	Apogee motor failed, leaving satellite in synchronous transfer orbit
2A	Marconi Space Systems	19 January 1974	Launch vehicle guidance failed; satellite decayed 25 January 1974
2B	Marconi Space Systems	23 November 1974	Operated until 1987
3			Never developed
4A	British Aerospace Dynamics	1 January 1990	Still active 2007
4B	British Aerospace Dynamics	11 December 1988	Decommissioned 1998
4C	British Aerospace Dynamics	30 August 1990	Still active 2007
4D	Matra Marconi Space	10 January 1998	Replaced *4B*; still active 2007
4E	Matra Marconi Space	26 February 1999	Still active 2007
4F	Astrium	7 February 2001	Still active 2007
5A	EADS Astrium	11 March 2007	Launched with *INSAT 4B*
5B	EADS Astrium	14 November 2007	Launched with *Star One C1*
5C	EADS Astrium	12 June 2008	Launched with *Turksat 3A*

malfunction in January 1974 doomed *Skynet 2A*, its *Skynet 2B* sibling achieved geostationary orbit in November 1974 and remained operational until 1987. Skynet 3 development ceased prematurely in favor of the 1,700 lb, three-axis-stabilized, nuclear-hardened, jam-resistant Skynet 4 Stage 1 satellites, three of which were launched between December 1988 and August 1990. Manufactured by British Aerospace Dynamics, these were the first solely British-built Skynet satellites. Three similar but more powerful Skynet 4 Stage 2 satellites, built by Matra Marconi Space and Astrium, became operational during 1998–2001.

Meanwhile, the Ministry of Defence (MOD) began defining requirements for the Skynet 5 system. Paradigm Secure Communications, a subsidiary of the European Aeronautic Defence and Space Company (EADS), signed a conceptually new Private Finance Initiative contract with MOD in October 2003 to meet those requirements. A subcontractor, EADS Astrium, delivered the first Skynet 5 satellite, which would be four times more powerful than the Skynet 4 Stage 2, to Kourou, French Guiana, where it was launched in March 2007.

James C. Mesco

See also: European Aeronautic Defence and Space Company (EADS), Space Systems/Loral, Matra Marconi Space, United Kingdom

Bibliography
Global Security. http://www.globalsecurity.org/.
Donald H. Martin, *Communications Satellites*, 4th ed. (2000).
Paradigm Secure Communications. http://www.paradigmsecure.com/.

Soviet and Russian Military Communications Satellites

Soviet and Russian military communications satellites have been launched into highly elliptical, semisynchronous orbits and medium Earth orbits since 1970 and into geostationary orbit since 1975. Although the Soviet Union launched the first Molniya 1 satellite into highly elliptical orbit in 1965, it introduced a military version with the secret designation Korund in 1970. Beginning in 1975 the Korund provided satellite communications for all branches of the Soviet military. Korund M, an improved model also designated Molniya 1T, has been used since 1983. In December 2006 Russia launched the first satellite in its Meridian series to replace the old Molniya 1 series. Meridian satellites provided communications links between seagoing vessels or ice-reconnaissance aircraft and coastal stations in the area of the North Sea route and across northern Siberia to the Russian Far East.

Flying in medium Earth orbits, Strela store-dump satellites, which probably were announced as part of the Kosmos series, serve global military and intelligence users. After several experimental launches in the 1960s, the first two generations of Strela satellites became operational in the 1970s. Strela 1M satellites were launched eight at a time between 1970 and 1992. The heavier Strela 2M satellites were launched singly between 1970 and 1994. Strela 3 satellites entered service in 1985, with most launches carrying six satellites at a time and later just two. In addition, since 1967, the Soviet Union regularly launched Tsiklon and Parus military navigation satellites, also announced officially as part of the Kosmos series, which provided communications links for naval forces.

Between 1975 and 1999, the Soviet Union and Russia operated geostationary satellites officially announced as Raduga but secretly named Gran. Each Raduga satellite had two three-channel transponders, at least one of which served the armed forces. In 1989 the Soviet Union began launching the more capable Raduga 1 model, secretly labeled Globus 1, with a longer lifetime. Beginning in 1982 the Soviet Union operated a type of geostationary Kosmos satellite, secretly designated Geyzer or Potok, for relaying data from photographic and electronic reconnaissance satellites. Between 1985 and 1995, the Soviet Union and Russia launched another type of geostationary data-relay satellite, referred to as Luch or Altair, for naval support.

Bart Hendrickx

See also: Molniya

Soviet and Russian Military Communications Satellites (Part 1 of 2)

Program Name	Announced Name	Launch Site (Launch Vehicle)	Orbital Parameters	Dates of First and Last Launch	Comments
Korund	Molniya 1	Baikonur/ Plesetsk (Molniya-M)	63° 1,500 × 39,000 km	1970–??	General military communications
Korund-M	Molniya 1/ Molniya 1T	Baikonur/ Plesetsk (Molniya-M)	63° 1,500 × 39,000 km	2 Apr 1983–18 Feb 2004	General military communications
Strela 1	Kosmos	Baikonur/ Kapustin Yar (Kosmos/ Kosmos-1)	various	18 Aug 1964–18 Sep 1965	Experimental store-dump satellites; launched in pairs, triplets, and quintuplets
Strela 1M	Kosmos	Plesetsk (Kosmos-3M)	74° 1,500 km	25 Apr 1970–3 Jun 1992	Operational store-dump satellites; launched in octuplets
Strela 2	Kosmos	Baikonur (Kosmos-1/ Kosmos-3)	56° 500 km	28 Dec 1965–27 Aug 1968	Experimental store-dump satellites
Strela 2M	Kosmos	Plesetsk (Kosmos-3M)	74° 800 km	27 Jun 1970–20 Dec 1994	Operational store-dump satellites
Strela 3	Kosmos	Plesetsk (Tsiklon-3/ Kosmos-3M)	82° 1,400 km	15 Jan 1985–23 Sep 2004	Operational store-dump satellites; usually launched in sextuplets
Tsiklon	Kosmos	Plesetsk (Kosmos-3M)	74°–83° 1,000 km	15 May 1967–27 July 1978	Naval communications; prime mission navigation

Soviet and Russian Military Communications Satellites (Part 2 of 2)

Program Name	Announced Name	Launch Site (Launch Vehicle)	Orbital Parameters	Dates of First and Last Launch	Comments
Parus/Tsiklon-B	Kosmos	Plesetsk (Kosmos-3M)	83° 1,000 km	26 Dec 1974–20 Jan 2005	Naval communications; prime mission navigation
Gran	Raduga	Baikonur(Proton)	geosynchronous	22 Dec 1975–5 Jul 1999	General military communications
Globus	Raduga 1	Baikonur(Proton)	geosynchronous	22 Jun 1989–27 Mar 2004	General military communications
Geyzer/Potok	Kosmos	Baikonur(Proton)	geosynchronous	17 May 1982–4 Jul 2000	Data relay for military satellites
Altair/Luch	Kosmos/Luch	Baikonur(Proton)	geosynchronous	25 Oct 1985–16 Dec 1994	Naval communications; prime mission Buran/Mir data relay
Gelios/Luch 1	Luch 1	Baikonur(Proton)	geosynchronous	11 Oct 1995	Naval communications; prime mission *Mir* data relay
Meridian	Meridian 1	Plesetsk(Soyuz-2/Fregat)	63° 1,000 × 39,800 km	24 Dec 2006	Military and government communications, especially between seagoing ships or ice-reconnaissance planes and coastal stations across northern Russia

Bibliography
V. Favorskiy, *"Voenno-kosmicheskie sily, kniga 1"* (1997).
Nicholas L. Johnson and David M. Rodvold, *Europe and Asia in Space, 1993–1994* (1995).
Donald H. Martin, *Communication Satellites*, 4th ed. (2000).

Strela. *See* Soviet and Russian Military Communications Satellites.

Tactical Communications Satellite

Tactical Communications Satellite (TACSAT), an experimental system, tested the feasibility of geostationary-satellite support to U.S. military users with a wide variety of mobile terminals as small as one foot in diameter. In January 1967 the U.S. Air Force selected Hughes Aircraft Company to design and build the *TACSAT 1* satellite. Measuring 25 ft high × 9 ft in diameter and weighing 1,600 lb, *TACSAT 1* entered Earth orbit in February 1969. Interim TACSAT operating capability using *TACSAT 1* and the sixth *Lincoln Experimental Satellite* (*LES 6*), both of which relied on ultrahigh and superhigh frequency bands, occurred in July 1970. *TACSAT 1* was the first satellite to employ "gyrostat" spin-stabilization techniques, developed by Hughes because the huge antenna structure atop the satellite's cylindrical body precluded conventional techniques. Operated by the Air Force Communications Service, *TACSAT 1* aided Apollo-capsule recovery operations by connecting aircrews with carrier and ground stations. By the time *TACSAT 1* experienced a crippling attitude-control failure in December 1972, its successful performance had convinced the U.S. Navy to begin acquiring a fully operational, tactical Fleet Satellite Communications capability.

James C. Mesco

See also: Hughes Aircraft Company

Bibliography
Boeing Company. http://www.boeing.com/.
Donald H. Martin, *Communication Satellites*, 4th ed. (2000).

Ultrahigh Frequency Follow-On System

Ultrahigh Frequency Follow-On (UFO) System, geosynchronous military communications satellites, each with 39-channel capacity, began replacing Fleet Satellite Communications (FLTSATCOM) and Hughes-built Leasat spacecraft in 1993. Controlled by the Naval Satellite Operations Center at Point Mugu, California, the UFO network provided global voice and data services to the U.S. Navy and various other tactical users with fixed or mobile terminals. Designed to be compatible with existing terminal equipment, trim cost, decrease power consumption, and increase security, the UFO satellites were customized versions of the popular Hughes Space and Communications Company (after 2000, Boeing Satellite Systems) 601 model. Because UFO satellites used primarily commercial off-the-shelf components for bus and payload modules, the builder could add capabilities for superhigh frequency (SHF), extremely high frequency (EHF), and Ka-band communications without lengthening production cycles or requiring additional engineering work.

The U.S. Navy awarded Hughes the nearly $2 billion UFO contract in July 1988. An Atlas booster malfunction in March 1993 sent the first UFO satellite into an unusable orbit, but the next nine launches between September 1993 and November 1999 succeeded. Boeing received authorization in January 2001 to build an eleventh UFO satellite, which achieved geostationary orbit in December 2003.

Weighing approximately 3,000 lb and designed to operate for 14 years, each UFO satellite had 39 UHF channels compared to 22 on Fleet Satellite Communications. In addition, the first two UFO satellite blocks (F1–3 and 4–7) had an SHF subsystem. The second, third, and fourth blocks (F4–7, 8–10, and 11) also had an enhanced EHF package compatible with Milstar ground terminals. Only the third block (F8–10) carried a Ka-band Global Broadcast Service (GBS) payload, provided under a March 1996 contract modification. After 1999 the three UFO satellites with high-capacity GBS revolutionized the U.S. military's use of intelligence data, especially the increased ability to use television and video imagery in near–real time to dominate the battlefield.

Morland Gonsoulin

See also: Hughes Aircraft Company, United States Navy

Bibliography
Boeing Company. http://www.boeing.com/.
Donald H. Martin, *Communication Satellites*, 4th ed. (2000).
SPAWAR. http://enterprise.spawar.navy.mil/.

West Ford

West Ford, also known as Project Needles, an experiment that involved placing approximately 480 million hair-like copper filaments in low Earth orbit to serve as a passive communications medium. Motivation for the experiment came after scientists discovered that a high-altitude thermonuclear bomb test in August 1958 had destroyed a portion of the ionosphere used by U.S. military forces as a reflector for high-frequency radio signals. During an Army study soon thereafter, Walter Morrow from the Massachusetts Institute of Technology Lincoln Laboratory and Harold Meyer from TRW Corporation considered the problem and proposed the creation of an artificial ionosphere using belts of extremely fine wires, which they called resonant scatterers, in two circular orbits—one polar and one equatorial.

Encouraged by the U.S. military's need for secure, reliable, and survivable communications in a thermonuclear environment, Lincoln Laboratory proposed the West Ford project to demonstrate Morrow's and Meyer's concept. The filaments, each with a length of 0.7 in and a diameter of 0.0007 in, acted as tiny dipole antennas to reflect microwave signals transmitted from ground systems at 8 GHz. Engineers designed an 88 lb dispenser carrying 43 lb of dipoles to ride atop an Atlas-Agena B launched from

Vandenberg Air Force Base, California. After an initial failure on 21 October 1961 when the dispenser failed to open, a successful launch and dispersion occurred on 9 May 1963. Throughout the course of four months, the tiny wires spread on average 1,300 ft apart to form a belt measuring slightly more than 9×18 miles in a nearly circular, nearly polar orbit at an altitude of approximately 2,266 miles.

Before the belt became fully extended, technicians successfully transmitted voice and data signals at 20 kbps in both directions between Camp Parks near Pleasanton, California, and Millstone Hill near Westford, Massachusetts. Once the belt's density decreased, however, the data rate declined to a disappointing 100 bps. West Ford experimentation ceased in 1965 as the dipoles began to fall from orbit. Much to the satisfaction of optical and radio astronomers who originally feared the West Ford project might interfere with their research, most of the dipoles had reentered the atmosphere by early 1966.

Morland Gonsoulin

See also: High-Altitude Nuclear Tests

Bibliography

Donald H. Martin, *Communication Satellites*, 4th ed. (2000).

William W. Ward and Franklin W. Floyd, "Thirty Years of Space Communications Research and Development at Lincoln Laboratory," in *Beyond the Ionosphere: Fifty Years of Satellite Communication*, ed. Andrew J. Butrica (1997), 79–93.

Wideband Gapfiller Satellite. *See* Defense Satellite Communications System.

Wideband Global System

Wideband Global System (WGS), originally called the Wideband Gapfiller System when the U.S. Air Force awarded an initial development contract to Boeing Satellite Systems Inc. in January 2001, was planned to augment Defense Satellite Communications System (DSCS) 3 and Global Broadcast Service (GBS) capabilities and, potentially, fill a gap chronologically between the loss of DSCS 3 and GBS and a future Advanced Wideband System that became part of the projected Transformational Satellite Communications System. The U.S. government eventually agreed to purchase five WGS satellites, and the Australian Ministry of Defence agreed in October 2007 to pay for a sixth WGS satellite. On 10 October 2007 an Atlas V booster launched the first WGS satellite from Cape Canaveral, Florida, into geostationary orbit.

That single *WGS 1* satellite, with an on-orbit weight of 7,600 lb, had as much communications capability as all the DSCS 3 satellites combined and an operational life expectancy of 14 years compared to 10 for a DSCS 3. Featuring both X-band and Ka-band links for reception and transmission, the WGS satellite offered users with different types of terminals greater communications flexibility than DSCS 3. An uplink in one band, for example, could be connected to downlinks in either or both bands. Through digital processing, the 39 channels on a WGS satellite could be divided into 1,875 independently routable sub-channels. The *WGS 1* satellite effectively doubled

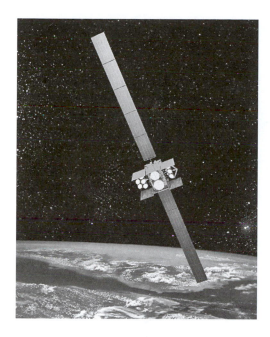

The Wideband Global SATCOM (WGS) satellite is the successor to the Defense Satellite Communications System Block III (DSCS-III) satellite. One WGS satellite has about 12 times the bandwidth of a DSCS-III satellite. (Courtesy U.S. Strategic Command)

the bandwidth or communications capacity that existing U.S. military satellites provided to the armed services.

Rick W. Sturdevant

See also: Boeing Company

Bibliography

Boeing Satellite Systems. http://www.boeing.com/defense-space/space/bss/.
Donald H. Martin et al., *Communication Satellites*, 5th ed. (2007).

MILITARY EXPERIMENTAL SATELLITES

Military experimental satellites demonstrate the feasibility of cutting-edge technologies, test new systems and subsystems for risk mitigation, provide early capabilities pending operational constellations, improve operational designs, measure various aspects of the space environment, and investigate problems associated with human spaceflight. While details about other spacefaring nations' military experimental satellites are scanty, information about U.S. military experimental satellites is more accessible. These include the Missile Defense Alarm System, *GRAB* (*Galactic Radiation and Background*), *Courier*, Lincoln Experimental Satellites, *SCORE* (*Signal Communication by Orbiting Relay Equipment*), *Tactical Communications Satellite* (*TacSat*), and Samos.

The United States launched many military experimental satellites, plus numerous experimental payloads riding piggyback on other spacecraft, as part of the Department

Artist's rendition of the TacSat 2 microsatellite. (Courtesy U.S. Air Force)

of Defense (DoD) Space Test Program (STP). Created in May 1965 and officially chartered on 15 July 1966 as an activity under the executive management of the U.S. Air Force (USAF), the STP recorded its first launch on 29 June 1967. Two extremely small satellites on that mission, the Army's *Sequential Correlation of Range 9* and the Navy's *Aurora 1*, improved geodetic-survey accuracy worldwide and obtained ultraviolet background-radiation data for surveillance satellites. Originally many of the experimental payloads rode a Space Test Experiments Platform into orbit. Later a medium-class platform, like the nearly 5,500 lb *Advanced Research and Global Observation Satellite* that launched in February 1999, carried nine different experimental payloads. More than 85 Space Shuttle flights, beginning with Space Transportation System (STS)-4 in 1982, also carried STP experiments. By November 2007 the STP had flown 467 experiments on 187 spaceflights.

Virtually every operational application for military satellites benefited from STP or other experimental satellite projects. For space-based positioning, navigation, and timing, the Naval Research Laboratory's *Timation 3*, launched in July 1974 and subsequently renamed *Navigation Technology Satellite 2*, contributed directly to the Global Positioning System by successfully demonstrating the use of high-accuracy atomic clocks in space. Between October 1985 and May 1999, several small DoD communications satellites, known as "lightsats," tested operationally the concept of developing and launching spacecraft quickly for direct support of military forces in the field; two of those lightsats, designated Multiple Access Communications Satellites, launched in May 1990 and supported Marines in Iraq, National Science Foundation activities in Antarctica, and Army special forces in Thailand. For a year after its launch in December 2006, the USAF experimental *TacSat 2* microsatellite demonstrated numerous advanced capabilities: web-based payload tasking, collection of tactically relevant imagery and signals intelligence, information distribution from tactical systems used by aircraft and unpiloted aerial vehicles, real-time signal geo-location and identification of emitters using satellite and aircraft platforms, and more. Earlier the Ram Burn Observations experiment and the Shuttle Ionospheric Modification with

Pulsed Localized Exhaust experiment, both of which flew on more than one Space Shuttle mission between December 2001 and July 2002, supplied data for the Space-Based Infrared System. In January 2003, partly as a risk-reduction effort for the National Polar Orbiting Environmental Satellite System, the Navy's *Coriolis* satellite imaged disturbances in the solar wind and measured wind speed and direction on the ocean surface.

To improve space situational awareness (SSA), the U.S. military launched several experimental satellites in the early twenty-first century to demonstrate autonomous rendezvous-and-proximity operations with other robotic spacecraft. Beginning with *Experimental Satellite System* (*XSS*) *10* in January 2003 and *XSS 11* in April 2005, the USAF proved it could perform on-orbit inspection and other sophisticated SSA missions at fairly low cost. In 2007 the Defense Advanced Research Projects Agency and NASA participated jointly in the highly successful Orbital Express mission that involved deployment of two experimental satellites—a surrogate next-generation serviceable satellite (*NextSat*) and a prototype servicing satellite (*Autonomous Space Transfer and Robotic Orbiter*)—to validate, during a three-month period, the feasibility of robotic, autonomous on-orbit refueling and reconfiguration of satellites.

Rick W. Sturdevant

See also: Courier; Defense Advanced Research Projects Agency; GRAB; Lincoln Experimental Satellites; Missile Defense Alarm System; *SCORE, Signal Communication by Orbiting Relay Equipment*; Tactical Communications Satellite

Bibliography

Defense Advanced Research Projects Agency. http://www.darpa.mil/.

Donald H. Martin et al., *Communications Satellites*, 5th ed. (2007).

SMC History Office, *Historical Overview of the Space and Missile Systems Center, 1954–2003* (2003).

MILITARY GEODETIC SATELLITES

Military geodetic satellites enabled exact determination of points on Earth's surface and more precise measurement of Earth's size, shape, and gravity field, information required for guiding missiles accurately over intercontinental distances or for tracking satellites in orbit. Consequently both the United States and the Soviet Union emphasized the importance of geodesy during the Cold War. In 1960 the U.S. Department of Defense published the first in a series of increasingly improved World Geodetic Systems and announced a joint Army, Navy, NASA, and Air Force (ANNA) project to build a geodetic satellite. On 31 October 1962, that effort resulted in the successful launch of the world's first dedicated geodetic satellite, *ANNA 1B*, manufactured by Johns Hopkins University Applied Physics Laboratory (APL). Instrumentation onboard *ANNA 1B* included a U.S. Army radio-ranging package called Sequential Collation of Range (SECOR). During 1964–69 the Army sent 13 SECOR satellites built by

Cubic Corporation into space and used them to determine previously uncertain positions of Pacific islands by bringing them within the same geodetic global grid. Experimentation with SECOR, sometimes called the Electronic and Geodetic Ranging Satellite (EGRS), contributed technologically to the U.S. Navy's Timation system and, ultimately, to the Global Positioning System. Also in the 1960s, APL used U.S. Navy Transit navigation satellites and a network of 13 ground stations to collect Doppler data for a more precise model of Earth's gravity field.

The Soviet Union Defense Ministry flight-tested Sfera geodetic satellites for military topographical research during 1968–72 before launching fully operational Sferas during 1973–80 to provide data for improving the accuracy of long-range missiles. Low-orbiting, second-generation Geo IK or Musson satellites, which the Soviet Union/Russia launched during 1981–94, employed high-intensity flashing lights, a Doppler system, radio transponder, radar altimeter, and laser reflectors to help the Soviet armed forces develop a global geodetic data base and characterize Earth's gravitational field. In 1989 two passive Etalon satellites covered completely with laser reflectors went into higher orbits to allow the Soviet military to fully characterize Earth's gravitational field at the altitude and inclination planned for the Global Navigation Satellite System constellation. During the 1990s students at Mozhaisky Military University of Space Engineering in St. Petersburg, Russia, designed the *Zeya* satellite for geodesy and navigation tests from Sun-synchronous orbit. Like the Sfera, Geo IK, and Etalon satellites, *Zeya* was manufactured under Chief Designer Mikhail Reshetnev's leadership, but it failed to operate and reentered Earth's atmosphere in October 1999. Although designed primarily for photo-reconnaissance, Zenit 4MT/Orion and Zenit 8/Oblik satellites built under Dmitri Koslov's leadership and launched during 1971–82 and 1984–94 respectively, contributed to geodesy.

Meanwhile the U.S. military services built on their work in the 1960s, beginning with the April 1970 launch of the Army's *TOPO 1* topographic-geodetic satellite, manufactured by Cubic Corporation, into Sun-synchronous orbit. In March 1985, *Geosat*, manufactured by APL for the Navy, carried a radar altimeter (to measure sea-surface height to within 5 cm) and two tape recorders that, beginning in November 1986 under the Exact Repeat Mission, collected precise altimeter readings of the same ocean locations every 17 days for more than three years. *Geosat* became the satellite to capture sea-level deviations associated with progression of the 1987 El Niño phenomenon in the equatorial Pacific. After the highly successful, five-year Geosat mission, the Navy contracted with Ball Aerospace and Technologies Corporation to develop a series of Geosat Follow-On (GFO) radar-altimeter satellites for continuous observation from a Geosat Exact Repeat Orbit to enhance the effectiveness of Navy weapon and sensor systems. Launched in February 1998, accepted operationally in November 2000, and controlled by Naval Network and Space Operations Command's Naval Satellite Operations Center, *GFO 1* had completed 129 exact-repeat cycles by the end of 2006.

Rick W. Sturdevant

See also: Johns Hopkins University Applied Physics Laboratory, Space Geodesy

Bibliography

Global Security. http://www.globalsecurity.org/.

W. Fred Boone, *NASA Office of Defense Affairs: The First Five Years* (1970).

Navy Geosat Follow-On (GFO) Altimeter Mission. http://ibis.grdl.noaa.gov/SAT/gfo/bmpcoe/default.htm.

Steve M. Yionoulis, "The Transit Satellite Geodesy Program," *Johns Hopkins APL Technical Digest* 19, no. 1 (1998): 36–42.

MILITARY INSTITUTIONS

Military institutions enhance national security by providing the organizational infrastructure needed to research, develop, and acquire space-related systems; to launch and operate satellites; and to process data relayed from space. Beginning in the late 1940s, some U.S., Soviet, and other military leaders envisioned using powerful rocket-powered weapon systems to enable the evolutionary migration to space of reconnaissance, surveillance, and other functions traditionally performed from land, sea, or air platforms. In the United States, the Department of Defense (DoD) and its individual services—Army, Navy, and Air Force—relied on various government, corporate, academic, and nonprofit institutional arrangements in their quest to enter space. The same was true, to one degree or another, for other aspiring space powers like the Soviet Union, France, and the United Kingdom. As the reality of spaceflight approached in the late 1950s, military leaders, who assumed they would run all space programs, were disappointed by the creation of civilian institutions to oversee many scientific and prestigious, human spaceflight activities. In the United States, for example, the National Aeronautics and Space Administration (NASA) absorbed most of the Army's and Navy's space-related infrastructure, and the Air Force was directed to support NASA missions. Even space-based reconnaissance missions, although heavily supported with U.S. military resources, were managed by civilians in the National Reconnaissance Office, which oversaw collection, processing, and distribution of highly sensitive data to appropriate military and civil users.

Historical parallels and divergences exist in how national military institutions approached the research and development (R&D), acquisition, and operation of space systems. As with earlier R&D activities, the U.S. Army relied primarily on its arsenal system and the U.S. Navy on a design bureau. By contrast, the U.S. Air Force (USAF) relied more on corporate contracts managed originally through its Western Development Division, which underwent various name changes to become the Space and Missile Systems Center (SMC) in 1992. The Soviet Union (later Russia) and China more clearly separated the R&D and manufacture of space systems from military oversight by placing those processes in government-owned design bureaus and production plants, respectively. Military space systems emerged from requirements defined by the Ministry of Defense and from competition among design bureaus for contracts to satisfy those requirements. In all instances, however, nations predictably gave operational responsibility for military space systems to military institutions like the Russian Space Forces and the USAF. Even India's air force, in April 2005, announced it had begun establishing a strategic aerospace command to operate military space systems supporting its three military services and the civilian Indian Space Research Organisation.

In both Russia and the United States, military institutions controlled rocket testing and space-launch facilities. Nonetheless, at White Sands Missile Range, Cape Canaveral Air Force Station, and Vandenberg Air Force Base, U.S. private industry played a significant role in building and launching the rockets that carried military, civil, and commercial satellites into orbit. Similarly, in the Soviet Union, the Space Forces controlled the launch centers at Baikonur, Kapustin Yar, Plesetsk, and Svobony. Until the collapse of the Soviet Union in 1991, the Soviet military controlled all civilian and military launches. When it came to on-orbit control of early U.S. reconnaissance satellites, Lockheed employees and USAF officers sat side by side at the Air Force Satellite Control Facility in Sunnyvale, California. From the dawn of the space age onward, seldom did military institutions effectively fulfill space-related missions without civil or commercial assistance; seldom did civil or commercial institutions succeed in space-related ventures without support from military institutions. By the beginning of the twenty-first century, most countries with military space assets exhibited institutional patterns that increasingly centralized military space operations but simultaneously linked military, civil, and commercial space users in complex webs of interdependency.

Rick W. Sturdevant

See also: China, Russia (Formerly the Soviet Union), United States

Bibliography
National Space Studies Center. http://space.au.af.mil/.
Asif A. Siddiqi, *Challenge to Apollo: The Soviet Union and the Space Race, 1945–1974* (2000).
David N. Spires, *Beyond Horizons: A Half Century of Air Force Space Leadership* (1997).

Advanced Research Projects Agency. *See* Defense Advanced Research Projects Agency.

The Aerospace Corporation

The Aerospace Corporation, established as a public, nonprofit corporation in 1960 to ensure the United States Air Force (USAF) would have access to the technical expertise required to meet current and future space-related challenges. The system development and acquisition approach that Brigadier General Bernard A. Schriever had instituted in 1954 for the intercontinental ballistic missile (ICBM) program had relied on Ramo-Wooldridge Corporation for systems engineering. That arrangement, which placed the corporation in a privileged position, generated criticism from other aerospace firms and Congress. Even after October 1958, when it merged with Thompson Products to become Thompson-Ramo-Wooldridge (TRW) and its Space Technology Laboratory (STL) became an independent subsidiary, the criticism persisted.

The conflict-of-interest charges and congressional scrutiny compelled the USAF to seek an alternative way to perform the functions previously handled by TRW and STL. In June 1960 a USAF-sponsored organizing committee established the Aerospace

Corporation in Inglewood, California, adjacent to the USAF Ballistic Missile Division headquarters. By year's end, the new, nonprofit corporation had hired more than 1,700 employees and had accepted responsibility for systems engineering and technical direction of 12 major USAF programs.

In August 1960 Ivan A. Getting, formerly Raytheon's vice president for research and engineering, became The Aerospace Corporation's first president and remained in that position until his retirement in 1977. During that time and beyond, The Aerospace Corporation provided systems engineering and technical direction for the design, testing, evaluation, and initial operation of virtually every new USAF space and missile program. Those included the Defense Satellite Communications System, the Defense Support Program, the Defense Meteorological Satellite Program, and the Global Positioning System, in addition to various antisatellite and antiballistic missile programs. Corporation engineers also contributed extensively to programs for new space launch vehicles, Earth observation spacecraft, and ICBMs.

By the early twenty-first century, The Aerospace Corporation's primary customer remained Air Force Space Command's Space and Missile Systems Center, the Ballistic Missile Division's direct descendant. The corporation also performed a wide variety of engineering-related services for other entities, including the National Reconnaissance Office, NASA, National Oceanic and Atmospheric Administration, foreign governments, and international agencies. With main offices in Inglewood, Chantilly, Virginia, and Colorado Springs, Colorado, the corporation remained active in such cutting-edge areas as strategic awareness, missile defense, space-based surveillance, space-based radar, homeland security, and national law enforcement.

David N. Spires

See also: Systems Engineering, TRW Corporation

Bibliography
The Aerospace Corporation, *The Aerospace Corporation: Its Work, 1960–1980* (1980).
The Aerospace Corporation. http://www.aero.org/.

Air Force Space Command. *See* United States Air Force.

Army Ballistic Missile Agency. *See* Marshall Space Flight Center.

Army Space Command. *See* United States Army.

Army Space and Strategic Defense Command. *See* United States Army.

Baikonur. *See* Russian Space Launch Centers.

Ballistic Missile Defense Organization. *See* United States Department of Defense.

Blue Cube. *See* United States Air Force Satellite Control Facility.

Camp Cooke. *See* Vandenberg Air Force Base.

Cape Canaveral Air Force Station

Cape Canaveral Air Force Station (AFS), the East Coast launch area for unpiloted U.S. Department of Defense (DoD), NASA, and commercial space launches. President Harry Truman signed legislation in May 1949 to establish a Joint Long Range Proving Ground (JLRPG) for guided missiles on the inactive, World War II–era Banana River Naval Air Station along Florida's eastern shore. Responsibility for operation and maintenance of the JLRPG (renamed Patrick Air Force Base in August 1950), which included jurisdiction over the launch area at nearby Cape Canaveral and downrange facilities on Grand Bahama Island, went to the U.S. Air Force (USAF).

Cape Canaveral's proximity to the Atlantic Ocean meant the U.S. military could safely test fire over water, thereby minimizing the danger that existed at White Sands Proving Ground, New Mexico, of an errant missile impacting a populated area. The inaugural launch at Cape Canaveral on 24 July 1950 was the U.S. Army's two-stage Bumper 8. Throughout the next seven years, the U.S. military services test-launched from Cape Canaveral AFS a variety of other missiles: Lark, Matador, Snark, BOMARC, Redstone, X-17, Navaho, Jupiter, Bull Goose, Polaris, Thor, and Atlas.

After the Soviet Union launched *Sputnik*, the world's first artificial satellite, Cape Canaveral became the premier location for U.S. space launches. From 1958 onward, hundreds of different space missions, from Earth-circling satellites belonging to the

The Bumper V-2 was the first missile launched at Cape Canaveral on 24 July 1950. (Courtesy NASA/Kennedy Space Center)

DoD, NASA, foreign nations, or commercial enterprises, to lunar probes and inter-planetary spacecraft, launched from Cape Canaveral AFS. All of NASA's Mercury and Gemini crewed spaceflights originated there, as did such later robotic craft as *Mars Pathfinder*, *Stardust*, and *Deep Space 1*. Although NASA, in the mid-1960s, established Kennedy Space Center adjacent to Cape Canaveral AFS for launching Apollo and, later, Space Shuttle crews, it relied heavily on military support for range safety and downrange tracking. Meanwhile, the station continued testing missiles: Titan, Minuteman, Hound Dog, Pershing, Poseidon, and Trident.

In 1989, creation of the Spaceport Florida Authority—renamed Florida Space Authority (FSA)—by that state's legislature led to millions of dollars in state-financed improvements to launch infrastructure at Cape Canaveral AFS. By 2004, FSA had secured a real-property license from the USAF to operate and maintain Launch Complex 47 at Cape Canaveral AFS. Furthermore, FSA spent $300 million to refurbish Launch Complex 41, the previous Titan III pad, for Atlas V launches. Another $24 million from FSA financed construction of a Delta IV Horizontal Integration Facility. By 2005, Cape Canaveral AFS included more than 1,500 facilities spread over 16,000 acres and supported a workforce of 10,000 people.

Kevin M. Brady

See also: Kennedy Space Center

Bibliography
William Barnaby Faherty, *Florida's Space Coast: The Impact of NASA on the Sunshine State* (2002).
Florida Space Authority. http://www.spaceflorida.gov/.
Patrick Air Force Base. http://www.patrick.af.mil/.

Cheyenne Mountain. *See* North American Aerospace Defense Command.

Consolidated Space Operations Center. *See* United States Air Force Satellite Control Facility.

Cosmodromes. *See* Russian Space Launch Centers.

Defense Advanced Research Projects Agency

Defense Advanced Research Projects Agency (DARPA), established as ARPA on 7 February 1958 in direct response to the launch of *Sputnik* by the Soviet Union. It cen-tralized all U.S. Department of Defense (DoD) space research and development (R&D) activities, thereby elevating the priority of space technology while simultaneously limit-ing interservice rivalries and wasteful duplication of effort. ARPA assigned specific space projects to the individual services, with the U.S. Air Force (USAF) receiving more than 80 percent of ARPA funds. Until the formation of NASA in October 1958, ARPA served as the U.S. government's focal point for administering all space activities.

After 1960, when ARPA transferred civil space programs to NASA and military space programs to the individual services, it concentrated on R&D for ballistic missile defense, nuclear detection, and counterinsurgency capabilities. Renamed DARPA in March 1972, the organization's R&D focus became directed energy, information processing, and tactical technologies. It developed time sharing and the ARPANET, progenitor of the Internet, and made great strides with artificial intelligence and virtual reality. In the late 1970s DARPA projects included infrared sensing for space-based surveillance, high-energy laser technology for space-based missile defense, advanced computing applications, stealth technology, cruise missiles, very large-scale integrated (VLSI) electronic circuitry, and graphic-design software.

During the 1980s DARPA continued to advance information processing through the Strategic Computing Program and undertook hypersonic R&D for the National Aero-Space Plane project. DARPA supported the development of new, small-satellite launchers by issuing the first launch-services contracts for Orbital Sciences' Pegasus and Taurus in 1988–89. It also examined new concepts for small, lightweight satellites and cosponsored with Phillips Laboratory's *DARPASAT*, a 203 kg satellite built by Ball Aerospace and launched on a Taurus in March 1994. Although the designation changed back to ARPA in February 1993, then again to DARPA in March 1996, the organization preserved its fundamental character as a flexible, forward-looking, bureaucratically unconstrained R&D group able to study ideas or approaches others might find too risky or outlandish. By 2005 DARPA had eight program offices spending almost $3 billion on more than 200 projects ranging from robotics, biological defense, and information exploitation to advanced computing, weapon-delivery platforms, and novel technologies for space control.

Rick W. Sturdevant

See also: Antiballistic Missile Systems, National Aero-Space Plane (NASP)

Bibliography
Barber Associates, *The Advanced Research Projects Agency, 1958–1974* (1975).
Defense Advanced Research Projects Agency. http://www.darpa.mil/.
Frank L. Fernandez, "DARPA's Role in Radical Innovation," *Johns Hopkins APL Technical Digest* 20, no. 3 (1999): 250–52.

Johns Hopkins University Applied Physics Laboratory

Johns Hopkins University Applied Physics Laboratory (APL), a not-for-profit center for scientific research and engineering established at Silver Spring, Maryland, in March 1942 to develop a way for U.S. Navy vessels to defend themselves more effectively against enemy aircraft. The project resulted in development of a proximity fuse that dramatically increased the effectiveness of antiaircraft shells and other artillery. After the launch of *Sputnik* in October 1957, APL engineers, analyzing its beeping radio signal, observed the Doppler shift and, from that, formulated a satellite-based navigation concept. With Advanced Research Projects Agency and, later, naval funding, APL researchers pursued that concept and developed the Transit navigation satellite. Three months after the first Transit launch in September 1959, APL created its space department to improve Transit technology and to conduct other space-related scientific and technological studies. By 1964 APL had designed and built 15 navigation satellites and eight related research satellites, achieving along the way many firsts that became basic tools or standards for the design and development of space systems.

By the mid-1960s APL began developing application satellites for NASA. These included three geodetic Earth-orbiting satellites and a magnetic field satellite, which measured Earth's gravitational and magnetic fields, and the small astronomy series satellites, which identified X-ray sources in the universe. It also built instruments flown on various NASA scientific and planetary spacecraft. APL also developed U.S. military research satellites, including the Navy geodetic satellite that supplied the first continuous, comprehensive set of oceanographic altimeter readings.

When President Ronald Reagan's Strategic Defense Initiative began in the 1980s, APL devised various military research satellites, including the *Midcourse Space Experiment*, a technology demonstrator that identified and tracked ballistic missile signatures. As the Cold War ended, however, APL's focus shifted from approximately 70 percent defense-related to 80 percent NASA programs by the mid-1990s. APL designed and built science instruments flown on various NASA spacecraft, including *Galileo*, *Ulysses*, upper atmospheric research satellites, *Geotail*, *Cassini*, and the *Mars Reconnaissance Orbiter*. For its noteworthy success with the *Near Earth Asteroid Rendezvous* spacecraft in the late 1990s, APL received three more NASA planetary missions: the *Comet Nucleus Tour*, which launched in August 2002; *MESSENGER* (*Mercury Surface, Space Environment, Geochemistry, and Ranging*), which launched in August 2004; and the New Horizons Pluto–Kuiper Belt mission, which launched in January 2006. By then, APL had built more than 60 spacecraft and designed 175 space instruments, making it one of the world's foremost research and development institutions for cutting-edge science and technology in both the civilian and the national-security space arenas.

Incigul Polat-Erdogan

See also: Asteroids, Comets, Mercury, Planetary Science, Pluto and Kuiper Belt Objects, Space Geodesy, Transit

Bibliography
Helen Gavaghan, *Something New under the Sun: Satellites and the Beginning of the Space Age* (1998).
Johns Hopkins University Applied Physics Laboratory. http://www.jhuapl.edu/.
William K. Klingaman, *APL—Fifty Years of Service to the Nation: A History of the Johns Hopkins University Applied Physics Laboratory* (1993).

Kapustin Yar. *See* Russian Space Launch Centers.

Kosmicheskie Voiska. *See* Russian Space Forces.

Missile Defense Agency. *See* United States Department of Defense.

MITRE Corporation

MITRE Corporation, a private, nonprofit company chartered in July 1958 to provide engineering and technical services to the U.S. government. At Secretary of the Air

Force James Douglas's request, the Massachusetts Institute of Technology's Lincoln Laboratory created MITRE specifically to perform systems engineering for the SAGE—the Semi-Automatic Ground Environment—continental air defense system. Thereafter, the corporation became known internationally for its technical expertise and innovative contributions to the design and integration of sophisticated electronic command and control systems. By 2005 MITRE operated Federally Funded Research and Development Centers for the Department of Defense, the Federal Aviation Administration (FAA), and the Internal Revenue Service.

Building on its SAGE experience and subsequent work on command-and-control systems at the NORAD Cheyenne Mountain Complex near Colorado Springs, Colorado, MITRE expanded its role beginning in the late 1970s to encompass space-related systems. Focusing on jam-resistant, secure, survivable communications for the U.S. military, MITRE supported the Defense Satellite Communications System and helped plan for Milstar. To assist the Strategic Defense Initiative during the 1980s, MITRE established a technical evaluation facility for definition of command, control, and communications architectures and for simulation of key processes. As systems architect and integrator for the U.S. Army's Task Force 21 in the 1990s, MITRE linked more than 70 disparate battlefield systems, including commercially available, precision, lightweight Global Positioning System (GPS) receivers.

MITRE also assisted the FAA and NASA. It developed air traffic control systems in the 1960s and, during the 1990s, began working on integration of GPS into the National Airspace System. MITRE supported acquisition and analysis of computer systems for mission control, simulation, training, and administrative applications at Johnson Space Center in Houston, Texas, and contributed to such NASA programs as the Space Shuttle, *International Space Station*, and Tracking and Data Relay Satellite System.

Rick W. Sturdevant

See also: Military Communications

Bibliography

Davis Dyer and Michael Aaron Davis, *Architects of Information Advantage: The MITRE Corporation since 1958* (1998).
Stephen B. Johnson, *The United States Air Force and the Culture of Innovation, 1945–1965* (2002).

National Geospatial-Intelligence Agency

National Geospatial-Intelligence Agency (NGA), created on 24 November 2003 when President George W. Bush signed the 2004 Defense Authorization Act, which included a provision changing the name of the National Imagery and Mapping Agency (NIMA) to NGA. This change marked the latest step in the evolution of an intelligence and defense-support agency established by Congress on 1 October 1996.

Converging trends in technology, congressional interest in economy, and the agreement of the Central Intelligence Agency (CIA) and the Department of Defense, which were influenced by lessons from Operations Desert Shield/Desert Storm (First Gulf War), came together in 1996 to establish NIMA. The Defense Mapping Agency, the

National Photographic Interpretation Center (NPIC), the Central Imagery Office, and the Defense Dissemination Program Office were merged into NIMA. Congress also brought selected imagery-related parts of the CIA, Defense Airborne Reconnaissance Office, Defense Intelligence Agency, and National Reconnaissance Office into NIMA. NPIC had been formed in 1961 by joining photo interpreters and related staff from the U.S. armed forces with those of the CIA under the latter's leadership. The Defense Mapping Agency was created in 1972 after the Vietnam War from the mapping, charting, and geodesy functions of the Army, Navy, and Air Force. The establishment of NIMA reflected a government-wide trend to consolidate functions for improvement of national security and for more efficient use of technology with resulting dollar savings.

The years 1996–2003 saw these organizations learning to work together. During that time, NIMA created animated renditions of imagery and geospatial data, allowing users to visualize inaccessible terrain. This enabled NIMA to assist in resolving international disputes, such as those between Peru and Ecuador or between Israel and Lebanon. NIMA also provided maps and visualizations that gave the Dayton Peace Accord diplomats from the Balkans graphic views of the boundary locations they were debating. In February 2000 NIMA and NASA teamed to create the Space Shuttle *Endeavour*'s Shuttle Radar Topography Mission (SRTM). That single flight provided the most detailed measurements of Earth's elevation ever gathered. In addition, NIMA contributed to homeland defense and disaster relief efforts, supported U.S. armed forces overseas, and developed safer airway charts. After the terrorist attacks of 11 September 2001 on the World Trade Center and Pentagon, NIMA's mission was redefined to include contributions to homeland security and helping safeguard prominent events in the United States or overseas, while assisting the armed force's work in Afghanistan and Iraq.

The agency's name change in 2003 reflected both a new product NGA was developing and the growing unity of its parts. Geospatial intelligence, that new product, combined traditional geospatial data, imagery, and other resources to present digital representations of world locations, along with naturally occurring and human activities. Congress designated the NGA as the funding agency to support U.S. private remote sensing corporations, with imaging satellite contracts to DigitalGlobe in 2003 and to Orbimage in 2004.

Martin K. Gordon

See also: Private Remote Sensing Corporations, Reconnaissance

Bibliography
Federation of American Scientists. http://www.fas.org/.
National Geospatial-Intelligence Agency. https://www1.nga.mil/Pages/Default.aspx.

National Imagery and Mapping Agency. *See* National Geospatial-Intelligence Agency.

National Reconnaissance Office

National Reconnaissance Office (NRO) designs, builds, and operates U.S. reconnaissance satellites. Directed by the Under Secretary of the U.S. Air Force (USAF), the

NRO distributes its products to the Central Intelligence Agency (CIA), the Department of Defense (DoD), and an expanding list of customers. Intelligence analysts use those products to warn of potential trouble spots around the world, plan military operations, and monitor the environment. The NRO is staffed by DoD and CIA personnel.

In 1958, President Dwight D. Eisenhower directed the CIA to develop a recoverable reconnaissance satellite, because technological problems had delayed the Air Force WS-117L satellite program. The CIA awarded Lockheed Corporation a contract to develop a recoverable Corona satellite.

Based on President Eisenhower's orders, Secretary of the Air Force Dudley Sharp established the Office of Missile and Satellite Systems, headed by the Under Secretary of the USAF, on 31 August 1960. Within a year, however, pressure within President John F. Kennedy's administration for a more formalized relationship between the USAF and the CIA prompted Under Secretary of the Air Force Joseph V. Charyk, on 24 July 1961, to recommend establishment of a National Reconnaissance Program (NRP). By early September, discussions resulted in creation of the NRP and establishment of the NRO, managed jointly by the Under Secretary of the USAF and the CIA deputy director of plans. Not until May 1962 did the DoD and the CIA agree on a single NRO director, Secretary Charyk.

Initially, Charyk organized the NRO into four programs, each having responsibility for system design and development, coordination with contractors, and operations. Program A included all USAF satellite intelligence programs (for example, Samos) managed by the USAF Special Projects Office in what became the Air Force Space and Missile Systems Center at Los Angeles Air Force Base, California. Program B encompassed CIA satellite programs, which were the responsibility of the CIA deputy director for Science and Technology. Program C comprised the Navy component (for example, GRAB). Program D covered aerial surveillance programs (for example, U-2) before 1969. The USAF Strategic Air Command subsequently controlled those aircraft until its disestablishment in 1992, when control of the remaining U-2s reverted briefly to the NRO before devolving to the Defense Airborne Reconnaissance Office in late 1993.

The William Clinton administration ordered the declassification of the organization's existence and its name in September 1992, followed by the location of its headquarters in Chantilly, Virginia, in 1994. Early the following year thousands of images from the Corona, Argon, and Lanyard satellite systems during the 1960s and early 1970s became publicly available. For the first time, in December 1996, the NRO gave advance notice of a reconnaissance satellite launch.

Declassification, the end of the Cold War, and growth in commercial remote sensing during the 1990s compelled the NRO to reassess its role. In August 1996 a panel led by retired Navy Admiral David E. Jeremiah examined a dozen issues related to "Defining the Future of the NRO for the 21st Century" and recommended organizational, procedural, and other changes to improve performance. Meanwhile, revelations of financial mismanagement within the NRO resulted in the appointment of a new director who instituted quarterly execution reviews and other steps to restore confidence among congressional overseers. Reports by the National Commission for the Review of the NRO in August 2000 and the Commission to Assess United States

National Security Space Management and Organization in January 2001 promoted integration of more NRO and USAF activities based on the "best practices" of each.

David C. Arnold

See also: Corona, GRAB, KH-9 Hexagon, KH-11 Kennan, Lacrosse, Reconnaissance, Samos

Bibliography

William E. Burrows, *Deep Black: Space Espionage and National Security* (1987).

R. Cargill Hall, "The NRO in the 21st Century: Ensuring Global Information Supremacy," *Quest* 11, no. 3 (Fall 2004): 4–10.

Clayton Laurie, *Congress and the National Reconnaissance Office* (2001).

Robert Perry, *Management of the National Reconnaissance Program, 1960–1965* (1969).

National Security Agency

National Security Agency (NSA), the largest and most secretive U.S. intelligence organization. Its headquarters is situated at Fort Meade, halfway between Washington, DC, and Baltimore, Maryland. At the beginning of 2006, the NSA was building a new warning hub and data warehouse at Buckley Air Force Base (AFB) near Denver, Colorado, and was planning to move a small portion of its workforce from Fort Meade to Buckley AFB.

Created in 1952 by President Harry Truman, the NSA eavesdrops on worldwide communications, attempts to break the encryption systems of other nations, and develops encryption systems for U.S. government communications. To eavesdrop on communications signals, many with limited range, the agency first has to capture them. Traditionally, that involves using land-based listening posts and specially designed aircraft, surface ships, and submarines with sensitive antennas and receiving equipment. During the 1950s, for example, U.S. military aircraft regularly flew along the periphery of the Soviet Union to collect both communications intelligence (Comint), such as telephone conversations and messages, and electronic intelligence (Elint), such as radar signals and telemetry. Once radar signals were intercepted and analyzed, Elint specialists could find ways to jam or spoof them, thereby enabling bombers to penetrate hostile borders undetected during wartime.

Flights along, and occasionally across, the Soviet borders were extremely hazardous, with nearly 50 aircraft attacked and more than 100 crewmembers killed throughout the years. In May 1960 the shootdown of Francis Gary Powers's U-2 effectively ended aerial surveillance over the Soviet Union. Fortunately during the late 1950s NSA and Naval Research Laboratory scientists secretly had begun exploring ways to eavesdrop from space and, on 5 June 1960, President Dwight Eisenhower approved launching the first U.S. spy satellite. Known as *GRAB—Galactic Radiation and Background*, a scientific designation to cover up its true mission—the soccer ball–size satellite gave the NSA, for the first time, the ability to map the entire Soviet radar system. In August 1962 a top-secret agreement concerning NSA participation in the National Reconnaissance Office (NRO) specified that the latter would develop satellites for

---

SAMUEL C. PHILLIPS
(1921–1990)

As director of the Minuteman intercontinental ballistic missile development program (1959–63), U.S. Air Force General Samuel Phillips's innovative management techniques ensured its success. Assigned to direct NASA's Apollo program (1964–69), he successfully applied the same management methods. After heading the Space and Missile Systems Organization, National Security Agency, and Air Force Systems Command, he retired from active duty in 1975 to manage TRW Corporation's Defense Systems Group. Phillips served both as a management consultant to NASA after the 1986 *Challenger* disaster and as a member of the National Research Council's Committee for the Human Exploration of Space.

Rick W. Sturdevant

---

signals intelligence (SIGINT), with NSA advising on their operation and the desired format of material to be collected.

During the next four decades the NSA relied increasingly on cutting-edge technology to conduct SIGINT from space on countries around the world. Microwave signals, which carry a wide variety of communications from telephone calls to Internet communications to computer data transfers, became a key target. Because microwaves travel in a straight line, and Earth is curved, those signals continue on into space at the speed of light and can be intercepted by NSA satellites in geostationary orbit. They also collect signals transmitted by commercial ground stations to communications satellites like those operated by Intelsat.

SIGINT satellites retransmit the data they collect to NSA listening posts around the world. Using giant parabolic dishes, those listening posts also eavesdrop on the downlinks from dozens of on-orbit governmental and commercial satellites. After a brief analysis, the listening posts forward the raw material to NSA headquarters at Fort Meade, where cryptolinguists, cryptoanalysts, and other specialists translate and decipher the data. The small percentage of information NSA finds useful is then turned into reports, which it sends to the White House, the Central Intelligence Agency, the Pentagon, and other federal agencies.

James Bamford

See also: GRAB, United States Signals Intelligence

Bibliography
James Bamford, *Body of Secrets: Anatomy of the Ultrasecret NSA, From the Cold War to the Dawn of a New Century* (2001).
———, *The Puzzle Palace: A Report on NSA, America's Most Secret Intelligence Agency* (1982).
National Security Agency. http://www.nsa.gov/.

National Security Space Architect. *See* United States Department of Defense.

Naval Center for Space Technology. *See* United States Navy.

Naval Research Laboratory. *See* United States Navy.

Naval Space Command. *See* United States Navy.

North American Aerospace Defense Command

North American Aerospace Defense Command (NORAD), a binational military organization activated by the United States and Canada in September 1957 and, by formal agreement on 12 May 1958, headquartered in Colorado Springs, Colorado. Originally named North American Air Defense Command, it was established to centralize operational control of all North American air-defense forces under a single commander appointed by the president of the United States and the prime minister of Canada. With the emergence of the intercontinental ballistic missile (ICBM) and space systems in the 1960s, NORAD assumed broader responsibilities for early warning and space surveillance. The former responsibility involved detection of both ICBM and submarine ballistic missile launches; the latter involved detection, tracking, and identification of satellites (active and inactive), rocket bodies, and other debris. Renewal of the NORAD agreement in May 1981 reflected this by officially changing "Air" to "Aerospace" in the organization's name. From the 1960s into the twenty-first century, Canadian political opposition precluded ballistic missile defense becoming part of the NORAD mission.

NORAD blast doors, Cheyenne Mountain, Colorado. (Courtesy U.S. Air Force)

Beginning in 1965, the NORAD Combat Operations Center, hosted by the U.S. Air Force and located in Cheyenne Mountain, southwest of Colorado Springs, served as the collection-and-processing facility for data from radar, optical, satellite, and other sensor systems worldwide. Altogether these constituted the Integrated Tactical Warning and Attack Assessment (ITW/AA) "system of systems," which was responsible for detection, validation, and warning of an attack against North America by aircraft, missiles, or space vehicles. A sophisticated communications switching system sorted and routed incoming data to separate air, missile, and space centers for analysis, after which an assessment of the threat was dispatched to the appropriate U.S. and Canadian civilian leaders and military authorities. The space center in Cheyenne Mountain also maintained the world's most complete catalog of artificial objects in space and warned spacecraft operators, including crews onboard the Space Shuttle and the *International Space Station*, if orbital maneuvers were required to avoid an impending collision.

During the First Gulf War in 1991, the NORAD Missile Warning Center warned coalition forces of Iraqi Scud missile launches, which alerted Patriot antimissile batteries to intercept the Scuds. With the Soviet Union's reorganization and the end of the Cold War in the early 1990s, NORAD expanded its traditional air-defense mission to include counter-narcotics operations, such as tracking small-engine aircraft smuggling drugs into countries in North America. NORAD also placed greater emphasis on monitoring commercial space activities. After the attacks of 11 September 2001 in the United States, NORAD became the initial focal point to handle the threat of hijacked commercial airliners being used against civilian targets in either the United States or Canada. Furthermore, in early 2006, newly appointed Canadian Minister of National Defence Gordon O'Connor announced that renewal of the NORAD agreement, due to expire in May, would expand NORAD's mission responsibilities to include maritime surveillance.

James C. Mesco

See also: Canada, Early Warning, First Gulf War, Space Debris, United States

Bibliography
Richard G. Chapman Jr., *Legacy of Peace: Mountain with a Mission, NORAD's Cheyenne Mountain Combat Operations Center, the Cold War Years: 1946–1989* (1996).
Ann Denholm Crosby, *Dilemmas in Defence Decision Making: Constructing Canada's Role in NORAD, 1958–96* (1998).
North American Aerospace Defense Command. http://www.norad.mil/.
Kenneth Schaffel, *The Emerging Shield: The Air Force and the Evolution of Continental Air Defense, 1945–1960* (1991).

Pacific Missile Range. *See* Vandenberg Air Force Base.

Plesetsk. *See* Russian Space Launch Centers.

RAND Corporation

RAND Corporation, a nonprofit institution dedicated to improvement of policy and decision making through research and analysis of major public welfare and national security challenges, including those related to space and missile systems. It was the first of the so-called "think tanks" that influenced U.S. military and social policies during and after the Cold War.

At the end of World War II, General of the Army Air Forces Henry H. "Hap" Arnold was concerned about the impact of scientific and technical developments on future warfare. Believing advanced research and development efforts would greatly benefit military planning and operational capabilities, he worked with representatives from the military, industry, and academia to initiate Project RAND (a contraction of research and development) in December 1945 under contract to Douglas Aircraft Company of Santa Monica, California. Project RAND spun off and became an independent, nonprofit corporation on 14 May 1948.

RAND released its first study, titled *Preliminary Design of an Experimental World-Circling Spaceship*, on 2 May 1946. It focused on the engineering feasibility of putting an artificial satellite in Earth orbit for reconnaissance, meteorology, communications, and other applications. Subsequent RAND reports examined rocket propulsion, orbital mechanics, and space policy. A *Project FEED BACK Summary Report* in 1954 represented the culmination of several RAND studies that encouraged the Air Force to proceed with development of a reconnaissance satellite.

In large measure, RAND's work during 1946–54 laid the foundation for America's space program. Over the next half century, in addition to continuing its space-related studies, the corporation contributed significantly to the development of digital computing, artificial intelligence, systems analysis, socioeconomic policy making, public health programs, and counterterrorism efforts.

Clayton K. S. Chun

See also: Ballistic Missiles, Corona

Bibliography

Martin J. Collins, *Cold War Laboratory: RAND, the Air Force, and the American State, 1945–1950* (2002).

Merton E. Davies and William R. Harris, *RAND's Role in the Evolution of Balloon and Satellite Observation Systems and Related U.S. Space Technology* (1988).

RAND, *Project Air Force 50th, 1946–1996* (1996).

RAND Corporation. http://www.rand.org/.

Russian Space Forces

Russian Space Forces, "Kosmicheskie Voiska" (KV) in Russian, also known as "space troops," a branch of the Russian Armed Forces responsible for the procurement, testing, launching, and control of military and civilian spacecraft; tracking space objects; and strategic aerospace defense.

The Space Forces originated in the mid-1950s from the Engineering Brigades of the Supreme Command Reserves, specialized artillery units for testing the first Soviet intercontinental ballistic missile, R-7. On 4 October 1957 a military crew under Colonel Alexander Nosov opened the space era by launching an R-7 rocket with the first artificial satellite, *Sputnik*.

GHERMAN STEPANOVICH TITOV
(1935–2000)

After becoming a fighter pilot in 1957 Gherman Titov underwent cosmonaut training in 1960. He flew the world's first daylong space mission in *Vostok 2* during 6–7 August 1961, becoming the second Soviet cosmonaut. In the mid-1960s he headed the team training to pilot the Spiral military spaceplane. For almost two decades from 1972 until his retirement in 1991, Titov served in the Russian Space Forces, rising to the rank of the colonel general and first deputy commander. He directed qualification testing for such Soviet Union military and civilian space systems as the Bor space glider, Almaz, Soyuz T, Tselina, Yantar, and the Zenit launch vehicle.

Peter A. Gorin

(Courtesy AP/Wide World Photos)

The first specialized military space unit, the Third Directorate (procurement of space systems), was created in September 1960 under the Chief Directorate of Rocket Armaments (GURVO) of the Strategic Missile Forces (RVSN). In October 1964 the new Central Directorate of Space Systems (TsUKOS) combined all space-related military units inside the Strategic Missile Forces. The Space Forces subsequently changed names and affiliations several times, beginning in March 1970 when TsUKOS was reorganized into the Chief Directorate of the Space Systems (GUKOS). In November 1981 GUKOS was removed from the Strategic Missile Forces and subordinated directly to the Ministry of Defense. GUKOS underwent further reorganization in November 1986 to become the Directorate of the Space Systems Commander (UNKS) and again in August 1992, when it became a separate service known as the Military Space Forces (VKS). From 1997 to 2001 the Space Forces lost its independence and went back under the Strategic Missile Forces command. Perceiving that reorganization as a mistake because it hampered development of military space capabilities, the Russian president reestablished the Space Forces (KV) as a separate service under the Ministry of Defense on 24 March 2001.

Existence of the Space Forces remained secret in the Soviet Union until officially acknowledged in 1991. Until 1992 the Space Forces controlled all military and civilian space activities. It operated three space launch centers: Baikonur (Tyura-Tam), Kapustin Yar, and Plesetsk, in addition to the military mission control center (GNIITs KS) with its network of ground- and sea-based receiving stations. Beginning that year, the newly created Russian Space Agency shared responsibilities for civil missions. In 1997 a new military space launch center, Svobodny, became operational. A larger portion of the Baikonur facilities was transferred to the Russian Space Agency's jurisdiction in 1998.

When the National Air Defense service disbanded in 2000, the Space Forces acquired additional functions and infrastructure for strategic antimissile and aerospace defense.

Peter A. Gorin

See also: Russia (Formerly the Soviet Union), Russian Federal Space Agency

Bibliography
Russia's Arms Catalog, vol. 6, *Missiles and Space Technology* (1998).

Russian Space Launch Centers

Russian space launch centers, called "cosmodromes," located deep inland, thus having their launch azimuths (which determine a satellite's orbital inclination to the equatorial plane) limited by so-called "fall fields"—large uninhabited areas where spent rocket stages can impact without harming people or property. All cosmodromes are essentially military bases restricted to visitors and with no precise maps publicly available. Russia's four cosmodromes are Kapustin Yar, Baikonur, Plesetsk, and Svobodny.

Kapustin Yar, the unofficial name for the Central State Multi-Purpose Test Range (GTsMP-4), lies between Volgograd and Astrakhan in the steppes of southwestern Russia. Founded in 1946 as a missile proving ground, the range saw tests of more than 200 different weapons systems, including live nuclear munitions. Suborbital space missions began in 1949, using R-1, R-2, R-5, and R-11 rockets for geophysical and

GIOVE-A atop the Soyuz launcher on pad six at Baikonur Cosmodrome in Kazakhstan. (Courtesy European Space Agency)

biological experiments. The first successful orbital mission occurred in 1962 with a Kosmos rocket based on the R-12 missile launched from the underground silo. During 1962–87, 84 space missions originated from Kapustin Yar, and an 85th mission occurred in 1999. They included mostly small Kosmos military and civilian satellites, the experimental rocket planes, and Interkosmos international projects. In 2007 only one space launch complex for the Kosmos 3M rocket was still operational. Space launches from Kapustin Yar are limited by its narrow launch azimuth range of 48–49 degrees, thus it primarily serves as a multipurpose proving ground for the Strategic Missile Forces.

Baikonur is located in a semidesert area of the Republic of Kazakhstan in Central Asia, with launch azimuths ranging from 56 to 66 degrees. Officially designated as the Scientific Research Test Range (NIIP-5), since its establishment in 1955 it was casually called Tyura-Tam after the local railroad station. The Soviet press introduced the name Baikonur in 1961 for the purpose of misinformation, since there was a small town of Baikonur 350 km northeast of the cosmodrome. Leninsk, the cosmodrome's administrative center, was formally renamed Baikonur in 1995. A majority of the Soviet intercontinental ballistic missiles (ICBMs) were tested at Baikonur. With the launch of *Sputnik* in 1957, it became the world's first spaceport. At its peak, Baikonur accommodated 52 different launch systems and 34 technical complexes, making it the largest Soviet space center. More than 1,100 space missions, including all piloted spacecraft, all space stations, and all lunar or planetary probes originated from Baikonur. By 2007, however, only 12 launch complexes for five rocket types remained operational. Those included double launch pads for each of the Dnepr, Soyuz, Strela, and Zenit space launchers, in addition to quadruple launch pads for the Proton-K rocket. The enormous facilities for the Energia superbooster and a Buran reusable space plane dominate the center of the cosmodrome. Some of them were built in the 1960s for the N-1 lunar rocket but later reconfigured for a new mission. Unfortunately, following termination of the Energia/Buran program in 1993, those unique structures slowly decayed. After dissolution of the Soviet Union, the independent Republic of Kazakhstan negotiated a long-term lease of Baikonur to Russia. In 2005 Russia and Kazakhstan announced joint plans to use the new Angara space launch system at Baikonur. The Russian Space Forces continue to control the cosmodrome, but the Russian Federal Space Agency has operated 80 percent of the facilities since 1998.

Plesetsk, Europe's only spaceport, lies in the dense forests of northwestern Russia near the town of Plesetsk in the Arkhangelsk Region. Founded in 1957 as the first Soviet ICBM site and reorganized into a test range in 1964, Plesetsk continuously served as a strategic missile base, proving ground, and cosmodrome. Approximately 1,900 space missions, mostly military, have launched from this site since 1966, making it the busiest Russian cosmodrome. Renamed the State Test Cosmodrome 1 in 1994, Plesetsk became the Russian Space Forces' primary launch center. Although Plesetsk has a wider range of the launch azimuths, 62–83 degrees, than Baikonur, its greater distance from the equator requires substantially more propellant for identical payloads. By 2007 Plesetsk had seven operational launch complexes for five rocket types, which include two pads for Soyuz and two for Kosmos 3M, in addition to single pads for Molniya, Rokot, and Start space launchers. An ongoing conversion of the two unfinished Zenit pads for the new Angara launch system promised to give this cosmodrome the capability to handle practically any kind of space mission, including piloted ones.

Svobodny, situated in far eastern Siberia 40 km northwest of the town of Svobodny, Amursk Region, was founded in 1996 as the State Test Cosmodrome 2 on the site of a decommissioned ICBM base. Space launches from Svobodny commenced in March 1997 using lightweight, mobile Start-1 boosters. The decision to build the new low-latitude cosmodrome was prompted by the early intentions of the Russian Space Forces to leave Baikonur. Svobodny offered the possibility of launch azimuths ranging from 51 to 62 degrees and 90 to 104 degrees. However, the original plans to convert five existing missile silos for Strela space launchers and build double launch pads for the Angara launch system languished due to funding problems. In 2005 the Russian government announced its intention to close Svobodny by the year 2009.

In 2007 the Russian government announced a decision to build in the Amursk Region in 2008–18 an entirely new cosmodrome, called Vostochny (eastern), with a capability to launch piloted missions.

Peter A. Gorin

See also: Launch Facilities, Russian Federal Space Agency, Russian Launch Vehicles

Bibliography
Russian Center for Space Infrastructure Operations (TsENKI). http://faculty.fordham.edu/siddiqi/sws/rsl/russian_space_links.html.

Space and Naval Warfare Systems Command. *See* United States Navy.

Strategic Defense Initiative Organization. *See* United States Department of Defense.

Svobodny. *See* Russian Space Launch Centers.

United States Air Force

United States Air Force (USAF), the U.S. Department of Defense (DoD) executive agent for space and, therefore, the military department primarily responsible for research, development, acquisition, and operation of defense-related space systems.

Even before it became a service separate from the Army on 18 September 1947, the nascent USAF had expressed its interest in space. In 1946 the Army Air Forces sponsored Project RAND's initial report, "Preliminary Design of an Experimental World-Circling Spaceship," which predicted an artificial Earth-circling satellite could be developed and launched within five years. Realistically the USAF had little interest in pursuing satellite development during an era of frugal budgets and an air-atomic defense strategy that focused on bombers rather than missile and space systems. Seeking to forestall Navy and Army claims, however, USAF leaders asserted in 1948 that their service had exclusive responsibility for military activity in space. From that point forward the USAF role in space proceeded along two broad tracks. One involved periodic efforts to convince national leaders to designate the USAF officially responsible for military space activities, including satellite launches and operations, deployment of weapons, and sending humans into orbit or beyond. The other centered on institutionalizing space throughout the armed forces by transferring space-related activities from the research and development (R&D) realm to the operational. Advocates believed that "normalizing" and "operationalizing" space would help the USAF achieve leadership of the nation's military space program.

The USAF renewed its interest in satellites during the early 1950s when the Dwight Eisenhower presidential administration acted to defend the nation from surprise attack. By 1956 that effort included USAF development programs for Atlas and Titan intercontinental ballistic missiles, the Thor intermediate-range ballistic missile, and a military reconnaissance satellite. The last program led to the first early warning, infrared satellite, to a defense meteorological satellite, and to reconnaissance satellites operated by the National Reconnaissance Office (NRO). At the same time, the Eisenhower administration promoted a "freedom of space" policy to allow unrestricted passage of satellites, civil or military, over any nation.

Following the Sputnik satellite launches in late 1957, the USAF coined the term "aerospace" to justify its claim to space leadership. When the Eisenhower administration preferred a civilian-led effort under the newly created NASA, albeit dependent on military support, the USAF campaigned unsuccessfully for designation as the DoD "executive agent" for military space. By the end of Eisenhower's presidency, the USAF had failed to achieve the space leadership position claimed by its most ardent representatives. Soon thereafter, USAF space officers found their prerogatives subordinated to Pentagon review agencies, and the service lost its reconnaissance satellite program to the newly established NRO. Despite those setbacks, the USAF had compiled an impressive list of achievements that included acquiring several types of long-range ballistic missiles that also served as space boosters, providing the infrastructure to support both military and civil space launches, and gaining management responsibility for both ground- and space-based early warning systems in addition to the ground-based space surveillance network. USAF leaders also thwarted two Army and Navy attempts to create a multiservice, unified command for military space activities. Responsible for nearly 80 percent of the military space budget, the USAF found itself the leading service for military space.

In the spring of 1961 Secretary of Defense Robert McNamara designated the USAF the service for military space R&D, which meant the Army and Navy had to submit their requirements and acquire their space-related capabilities through the USAF. In response the USAF agreed to establish Air Force Systems Command (AFSC), thereby providing an organizational focus for acquiring space and missile systems. The early days of the John Kennedy presidential administration fueled high hopes among USAF space leaders, but NASA soon emerged as the dominant organization for human spaceflight. Efforts to make military-crewed spaceflight the focal point of a space-oriented service ended in 1969 with cancellation of the USAF Manned Orbiting Laboratory. With the larger USAF space agenda unfulfilled, AFSC assumed most of the service's operational responsibility for space, thereby setting the stage for a future contest between other operational elements and the R&D community.

During the 1970s proliferation and maturation of satellite systems, along with the prospect of Space Shuttle operations, reinvigorated the USAF space program. Traditionally the USAF and DoD assigned operational management of space systems to the command or agency with the greatest need for a particular space-based capability. Systems like the Space Shuttle or Global Positioning System, however, had multiple capabilities that blurred functional lines. This compelled a reassessment of the importance of space capabilities, both in terms of application and organization, for operational commanders. The growing debate focused on whether the R&D or operational community should launch and control space systems. Gradually the question of whether the USAF should establish a new, major operational command for space came into focus. By the early 1980s the Ronald Reagan presidential administration's interest in an expanded military space program, one relying on the Space Shuttle, antisatellite weaponry, and ballistic missile defense, provided an important boost for organizational changes already underway within the USAF. On 1 September 1982, after acceding to Army and Navy demands for creation of a unified space command in the near future, the USAF created an operational Air Force Space Command (AFSPC).

During its first decade, AFSPC systematically acquired responsibility for early warning and space surveillance systems, on-orbit control of military satellites, and space launch. Along the way, it worked to establish effective relations with the unified command, U.S. Space Command created in 1985 by the Joints Chief of Staff, and its other service components. It spearheaded USAF efforts to develop and procure expendable launch vehicles following the Shuttle *Challenger* tragedy in 1986. When the First Gulf War occurred in 1991, AFSPC space systems that had been acquired for strategic purposes during the Cold War provided critical theater-level support to the forces of the United States and its coalition partners.

The experience of the First Gulf War caused USAF leaders to assert with renewed vigor their vision for the nation's military space program and the role the USAF should play. Providing space support to military forces in tactical, rather than strategic, situations received extraordinary emphasis. Because space systems had become so vital to victory on the battlefield, space control—the task of protecting friendly assets and denying enemies uses of their assets—became a central focus of the USAF and DoD. Finally in 2003 DoD approved the recommendation of a high-level, congressionally mandated space commission to designate the USAF as DoD Executive Agent for Space. What USAF space enthusiasts had sought for a half century happened. The USAF faced the long-standing challenge of according space the same priority it gave to aviation when it came to funding, organizing, training, and equipping forces.

David N. Spires

See also: Antisatellite Systems, Ballistic Missiles, Doctrine and Warfare, Early Warning, United States

Bibliography
Air War College, Space Operations and Resources. http://space.au.af.mil/.

R. Cargill Hall and Jacob Neufeld, eds., *The U.S. Air Force in Space: 1945 to the Twenty-first Century* (1998).
David N. Spires, *Beyond Horizons: A Half Century of Air Force Space Leadership* (1997).
———, *Orbital Futures: Selected Documents in Air Force Space History*, 2 vols. (2004).

United States Air Force Satellite Control Facility

United States Air Force Satellite Control Facility (AFSCF), the U.S. Air Force (USAF) organization that tracks military satellites, receives and processes telemetry transmitted by them, and sends commands to them. Operated by Air Research and Development Command's 6594th Test Wing, it began supporting the Corona program in early 1959. The AFSCF consists of control nodes, a central scheduling facility, remote tracking sites, and communications links connecting them. By June 1960 satellite controllers had moved from an interim center in Palo Alto, California, to a permanent facility in nearby Sunnyvale. Originally called the Satellite Test Annex, the center was renamed Sunnyvale Air Force Station in 1971 and Onizuka Air Force Station in 1986 to honor deceased *Challenger* astronaut Ellison Onizuka. The Sunnyvale complex functioned as a central command and control node for nine tracking stations built around the globe in 1959–61. The primary contractors for the initial systems were Lockheed and Philco. They worked under the leadership of Air Force Systems Command engineers. Operational responsibility for the AFSCF transferred to the newly established Air Force Systems Command in 1961 and to Air Force Space Command in 1987. Periodically the USAF closed some early tracking stations and opened others. During the 1980s the USAF constructed the Consolidated Space Operations Center near Colorado Springs, Colorado, in preparation for military Space Shuttle operations. It became the primary command and control node for military satellites.

The AFSCF hardware and software underwent many upgrades, beginning with one that removed mainframe computers from each tracking station and centralized command and control computing in Sunnyvale. During the late 1980s an Automated Remote Tracking Station (ARTS) program introduced equipment based on personal computers at old stations, thereby reducing station personnel. The USAF built new ARTS-equipped tracking stations in Colorado Springs and on Diego Garcia in the Indian Ocean. New systems increased reliability, trimmed maintenance costs, and allowed the network to support a steadily increasing workload. By 2004 the satellite control network performed more than 70,000 contacts annually for more than 100 military and civilian spacecraft.

David C. Arnold

See also: Command and Control, Corona, Military Communications

Bibliography
David C. Arnold, *Spying from Space: Constructing America's Satellite Command and Control Systems* (2004).
David N. Spires, *Beyond Horizons: A Half Century of Air Force Space Leadership* (1997).

United States Army

The United States Army was heavily involved since the end of World War II in the development and use of space-based systems and missile defense. Most notably under Operation Paperclip in January 1946, the Army brought Wernher von Braun's team of German rocket scientists along with engineering documents and V-2 hardware to the United States. During the next 15 years, first at Fort Bliss, Texas, and later under the auspices of the Army Ballistic Missile Agency (ABMA) at Redstone Arsenal, Alabama, von Braun's team designed and built increasingly powerful rockets, including the Jupiter-C that launched *Explorer 1*, the first U.S. satellite in January 1958. Over the vigorous opposition of Major General John B. Medaris, who had commanded the ABMA during 1955–58 before becoming responsible for all Army Ordnance rocket- and space-related programs, von Braun's team transferred to NASA in 1960.

During the 1940s–50s Army Ordnance funded research and development of rocket technology at the California Institute of Technology's Jet Propulsion Laboratory (JPL) in Pasadena, California. JPL produced a series of increasingly powerful rockets that culminated with the Corporal, the first U.S. operational, surface-to-surface guided missile, and the WAC (Without Attitude Control) Corporal, the first U.S. high-altitude test rocket. Reentry technology also captured JPL's attention under the Bumper WAC project. Before the end of 1957 the Army directed JPL to begin evaluating specific satellite designs, which led to preparation of *Explorer 1*. In December 1958 contractual oversight of JPL's space-related work transferred from the Army to NASA.

The Tiros program, which also transferred to NASA in April 1959, was directed initially by the Army Signal Research and Development Laboratory (SRDL) at Fort Monmouth, New Jersey. When *Tiros 1*, the world's first meteorological satellite, was launched in April 1960 under NASA auspices, SRDL remained responsible for the payload and for operation of a Tiros ground station at Fort Monmouth's Camp Evans Project Diana site, from whence the Army had bounced radar signals off the Moon in 1946.

In 1958 SRDL undertook development of the communications package for Project SCORE (Signal Communication by Orbiting Relay Equipment) and began working on the Courier and Advent programs. The service created a new field agency, the U.S. Army Satellite Communications Agency, to integrate satellite communications ground terminals into the Army Signal Corps. The Army's role as integrator expanded in the late 1960s when it became responsible for fielding, operating, and maintaining the ground-terminal segment of the Initial Defense Satellite Communications System. Technological evolution and greater use of the Defense Satellite Communications System (DSCS) during the 1970s–80s increased the importance of the Army's DSCS command and control responsibilities.

The desire to formulate a clear policy with respect to Army use of space systems, fostered establishment of a field element in 1984 as a precursor to a group, then an agency, that led ultimately to activation of an Army Space Command (ARSPACE) at Colorado Springs, Colorado, in 1988. As a component of the United States Space Command, ARSPACE ensured integration of Army requirements into planning for space support during military operations worldwide. Meanwhile, activation of an

Army Space Institute in 1986 and a Space Demonstration Program in 1987 aimed to introduce service members to the use of space systems in strategic and tactical situations. Those organizational changes, combined with experience from the First Gulf War in 1991, resulted in a deployable Contingency Operations–Space concept in 1994 that evolved into the Army Space Support Team.

Beginning with the Nike program in the 1950s, ballistic missile defense (BMD) became a space-related responsibility of the Army. During the 1960s the Army worked on components of an advanced antiballistic missile (ABM) system called Sentinel, renamed Safeguard in 1969 and fielded at the Stanley R. Mickelsen Safeguard Complex in North Dakota during the early 1970s. Although that complex operated only a brief time, the service's interest in BMD planning led to establishment of the U.S. Army Strategic Defense Command (USASDC) in 1985. To provide an overall focal point for space and missile defense matters, the service established the U.S. Army Space and Strategic Defense Command (USASSDC) in 1992. The USASSDC, which included ARSPACE and elements of the USASDC, became the U.S. Army Space and Missile Defense Command (SMDC) in October 1997. An important BMD step occurred in October 2003 when SMDC and the Colorado Army National Guard activated the 100th Missile Defense Brigade (Ground Missile Defense), the first organization of its kind, to operate the initial part of an integrated, national BMD system. An operational Ground-based Midcourse Defense (GMD) capability drew nigh with activation of the Alaska Army National Guard Missile Defense Space Battalion in January 2004, followed by installation of the first ballistic missile interceptor at Fort Greely, Alaska, in July 2004.

Morland Gonsoulin

See also: Antiballistic Missile Systems; Jet Propulsion Laboratory; Military Communications; von Braun, Wernher

Bibliography
John B. Medaris, *Countdown for Decision* (1960).
Redstone Arsenal. www.redstone.army.mil.
U.S. Army Space and Missile Defense Command. www.smdc.army.mil.
James Walker and James T. Hooper, *Space Warriors: The Army Space Support Team* (2003).
James Walker et al., *Seize the High Ground: The U.S. Army in Space and Missile Defense* (2003).

United States Department of Defense

United States Department of Defense (DoD), established by an act of Congress in 1947 as the National Military Establishment and renamed as the DoD in 1949 with authority over the military departments of the Army, Navy, and Air Force (USAF). Over time, various "combat support" organizations such as the National Security Agency, Defense Intelligence Agency, Defense Information Systems Agency, and National Geospatial Intelligence Agency also came under DoD purview to perform functions beyond the scope of the particular services. The DoD initially assigned responsibility for space issues to the Research and Development (R&D) Board's Committee on Guided Missiles. That body supported satellite capability studies by RAND,

The Pentagon in Arlington, Virginia, houses the Department of Defense. Built during World War II, it is the world's largest office building. Approximately 25,000 people work there. (Courtesy Library of Congress)

the Air Force–funded research corporation that had issued a pathbreaking report in 1946 on the feasibility of an Earth-circling spaceship. By the early 1950s all three services within DoD pursued separate space and missile programs. The various services' claims to primacy regarding military space missions, coupled with the larger question of who should oversee obviously nonmilitary space activities, continued into the late 1950s. From that time into the twenty-first century, acquisition of space systems and operational use of those systems became like two sides of a periodically re-minted coin in terms of changing policies, processes, and organizations.

In response to the Dwight Eisenhower presidential administration's concerns about possible surprise attacks and the nation's ability to respond, the USAF rushed, after 1954, to develop intermediate-range and intercontinental ballistic missiles together with a Weapon System 117-L space-based reconnaissance system. Simultaneously the Army pursued development of a Project Orbiter satellite in addition to a military communications satellite. The Navy, besides working toward an electronic-surveillance satellite, received primary responsibility for Project Vanguard, an effort sanctioned by the National Security Council to launch a civilian scientific satellite as part of the International Geophysical Year.

Following the Soviet Union's launch of Sputnik satellites in October and November 1957, Eisenhower's administration responded to public pressure and answered the Soviet space challenge by creating a comprehensive national program consisting of military, civilian, and national security elements. In February 1958 Secretary of Defense (SecDef) Neil McElroy created the Advanced Research Projects Agency (ARPA) to centralize all DoD space R&D activities. This accomplished two important administration objectives, namely to elevate the priority that the military placed on developing space technology, and to limit interservice rivalries that wasted resources through duplication of effort. Later that year, however, ARPA lost much of its control of the national space program to the newly established, civilian National Aeronautics and Space Administration (NASA). At the same time Eisenhower's administration embarked covertly on a space-based reconnaissance program for national security that, by 1961, transferred from the USAF to the newly established, secret National Reconnaissance Office (NRO).

Meanwhile, in August 1958, President Eisenhower signed the Defense Reorganization Act. That legislation created the position of Director for Defense Research and Engineering (DDR&E), which, as the sixth-highest-ranking official in DoD, reported directly to the SecDef. The following year DDR&E assumed ARPA's space-related R&D responsibilities, thereby becoming the SecDef's principal adviser on all new military projects related to satellites and launch vehicles. In March 1961 the John Kennedy presidential administration enhanced DDR&E's authority over military space programs through DoD Directive 5160.32, "Development of Space Systems." It granted DDR&E authority to establish guidelines for the conduct of space activities by the individual services but also consolidated the acquisition of military space systems under the Air Force.

Although the allocation of space-related responsibilities and the creation of organizations to handle them stabilized somewhat by the early 1960s, the question of how best to organize and plan for the acquisition of military space systems recurred often during the next 40 years and into the twenty-first century. With respect to the budgeting process, Secretary of Defense Robert McNamara, in 1962, introduced the Planning, Programming and Budgeting System (PPBS), which enabled him, on a continuous basis, to use cost-effectiveness studies or systems analysis to assess alternatives and allocate DoD resources according to his priorities. The influence of the PPBS varied depending on who served as SecDef but, throughout the next four decades, to a significant degree, affected the acquisition, operation, and maintenance of all military systems, including launch vehicles and satellites. Not until 2004 did DoD implement major changes, including a two-year budget cycle, under the Planning, Programming, Budgeting, and Execution (PPBE) process, which shifted the focus from straight financial discipline to program performance and results. The PPBE came at a time when the problem of large-scale acquisition programs running over budget and beyond schedule was rampant and required fixing.

Throughout the 1960s and much of the 1970s DDR&E played the central role in overseeing military space and missile programs. Although renamed the Office of the Under Secretary of Defense for Research and Engineering in 1977, DDR&E reemerged as a subordinate office within the Office of the Under Secretary of Defense for Acquisition and Technology with passage of the Military Retirement Reform Act of 1986. In addition, the Assistant Secretary of Defense for Command, Control, Communications, and Intelligence or ASD (C3I), an office created by the National Defense Authorization Act for fiscal year (FY) 1984, coordinated with the Under Secretary of Defense for Acquisition on DoD space architectures and technology programs.

During the early 1990s budgetary concerns and mounting congressional pressure prompted DoD to devise a more centralized, cost-effective acquisition strategy for space systems. In 1994 a comprehensive review of DoD space management practices resulted in three organizational changes designed to improve coordination and integration of military space activities. First, in December 1994, the SecDef established the position of Deputy Under Secretary of Defense for Space (DUSD/Space) to serve in the Office of the Secretary of Defense as the key element for space policy and liaison with Congress and interagency groups. The DUSD/Space also oversaw space programs, such as launch and support, reconnaissance and surveillance, tactical warning and attack assessment, and the Global Positioning System. Yet the ASD (C3I) remained responsible for all space C3I activities. A second initiative in March 1995 produced the DoD Space Architect (DoD SA) to consolidate development of architectures for space systems, thereby ending so-called "stovepipe" projects that began, progressed, and concluded with little or no regard for how they might interface with or affect other capabilities. That office became the SecDef's primary designer and overseer for space launch and tactical intelligence systems. Together with DUSD/Space, the DoD SA developed and maintained a comprehensive space system master plan. A third major organizational shift occurred in December 1995, when SecDef William Perry and Director of Central Intelligence John Deutch established the Joint Space Management Board to integrate policy, requirements, architectures, and acquisition for defense and intelligence space programs.

Those changes notwithstanding, the William Clinton presidential administration remained apprehensive about DoD's inefficiency, high costs, and tardiness in pursuing commercial alternatives to military space capabilities. It issued two Defense Reform Initiative Directives (DRIDs) in 1998, which enhanced the role of the ASD (C3I). The DRIDs abolished both the DUSD/Space and the DoD SA, replacing the latter with the National Security Space Architect (NSSA). Designed to focus directly on war fighters' needs, the NSSA reported directly to the ASD (C3I), which also assumed the space policy, space systems and architectures, space acquisition and management, and space integration responsibilities formerly handled by the DUSD/Space. The ASD (C3I) worked closely with the Under Secretary of Defense for Policy to ensure the integration of space policy decisions with broader national security policy decisions. By 2005 the Deputy Assistant Secretary of Defense for Command, Control, Communications, Intelligence, Surveillance, Reconnaissance (C3ISR), within the ASD (C3I), handled space policy in addition to development and integration of space control and space support programs. The 1998 directives replaced the Joint Space Management Board with the National Security Space Senior Steering Group (NSS-SSG). With the ADS (C3I) as one of three cochairs, the NSS-SSG sought widespread participation by civilian and military agencies in addressing defense and intelligence space management and integration issues.

Meanwhile in the early 1980s DoD renewed its attention to the operational side of the space coin. Maturation of most military space systems beyond the R&D stage to fully operational status, combined with plans for military use of the soon-to-be-completed Space Shuttle, led the USAF to establish a major command for space operations in September 1982. The Navy and Army followed suit in October 1983 and April 1988, respectively. Each of those organizations provided operational space capabilities to the

United States Space Command (USSPACECOM), a joint-forces organization established in September 1985. Disestablishment of USSPACECOM in October 2002 resulted in the transfer of its responsibilities to United States Strategic Command.

Also during the 1980s President Ronald Reagan announced a new R&D initiative for missile defense popularly known as Star Wars. To manage this complex effort involving thousands of military, civilian, and contractor personnel, DoD created the Strategic Defense Initiative Organization (SDIO) in March 1984. Its name changed to the Ballistic Missile Defense Organization in May 1993 after President William Clinton's administration reoriented the R&D toward theater missile defense. In January 2002 Secretary of Defense Donald Rumsfeld issued new guidance on execution of the missile defense program, with an eye toward deployment of an operational system, and renamed BMDO the Missile Defense Agency (MDA).

As a result of the vital role space systems played during 1991 in the First Gulf War and subsequent advances in the technological support such systems provided to military forces, DoD undertook far-reaching initiatives to improve the handling of its space responsibilities. Further impetus came from the report of a high-level commission chartered by the National Defense Authorization Act for FY 2000 to assess weaknesses in the national security space program. The commission's recommendations, implemented by the George W. Bush presidential administration, included designating the Secretary of the Air Force as DoD executive agent for space and creating an Under Secretary of Defense for Space, Intelligence, and Information to provide senior-level advocacy for space within DoD. In May 2004 DoD acted to unify its space efforts further by creating the National Security Space Office, which combined three DoD space organizations: the Transformational Communications Office, the NSSA, and the NRO's National Security Space Integration directorate. During the first decade of the twenty-first century DoD encouraged all three services to educate a cadre of professionals, officers and civilians, whose careers would focus on the acquisition and operation of military space systems.

David N. Spires

See also: National Aeronautics and Space Administration, Strategic Defense Initiative

Bibliography
Air War College, Space Operations and Resources. http://space.au.af.mil/.
Joshua Boehm and Craig Baker, "A History of United States National Security Space Management and Organization," Federation of American Scientists. http://www.fas.org/spp/eprint/article03.html.
Report of the Commission to Assess United States National Security Space Management and Organization (2001).
United States Department of Defense. www.defense.gov.

United States Navy

United States Navy, historically committed to the development and use of long-range rockets and space systems to enhance U.S. sea power. In November 1945 the newly established Navy Committee for Evaluating the Feasibility of Space Rocketry within the Bureau of Aeronautics recommended giving satellite development high priority. Two months later the Naval Research Laboratory (NRL) began a development program for the large, high-altitude Viking rocket. Beginning in October 1946 the NRL used V-2 rockets, which U.S. forces had captured in Germany at the end of World War II, to launch scientific payloads from White Sands Proving Grounds, New Mexico, into the upper atmosphere. As part of the International Geophysical Year, the Navy became the first U.S. service to attempt a satellite launch. Its Vanguard booster exploded seconds after launch on 6 December 1957, leaving the Army to orbit the first U.S. satellite on 31 January 1958.

Although the Navy's Vanguard subsequently launched several early U.S. satellites, that program, including its Minitrack satellite tracking system, was transferred to NASA in 1958. Navy interest in developing space boosters effectively ceased in 1961, when the Air Force gained responsibility for all U.S. military satellite launches. The service continued, however, to develop offensive and defensive missiles. These included submarine-launched, nuclear Polaris and later Trident ballistic missiles for strategic applications. By 2005 the Navy sought to develop and deploy a Standard Missile 3 (SM 3) on Aegis-class cruisers and destroyers for purposes of ballistic missile defense.

For the U.S. Navy and other navies around the world, space systems solved what had seemed to be entirely intractable problems. Most obvious was the need for long-distance communications with moving forces. Satellites were more reliable than bouncing high-frequency (HF) radio signals off the ionosphere, could operate at far higher frequencies, and offered much greater data rates. In addition, satellite radio seemed to overcome an important defect of terrestrial HF, which was that anyone intercepting the latter could locate the transmitter. Satellites also offered precise navigation, because their positions could be set and monitored. This possibility was first realized with Transit satellites and later with the Global Positioning System (GPS). Given precise positioning, the Navy could field an accurate sea-based, nuclear ballistic missile system. Furthermore, because satellites afforded truly global reconnaissance coverage, the Navy developed the White Cloud system specifically to locate enemy naval forces. During the late 1970s, under a Department of Defense program called Tactical Employment of National Capabilities or TENCAP, it combined data obtained from other organizations, particularly the National Security Agency, with White Cloud information and distributed the product to the fleet anywhere in the world. Conversely, only by using satellites could the Soviet Union reliably locate U.S. naval forces on the world's oceans. The U.S. Navy, therefore, became intensely interested in monitoring Soviet space activity. For many years the Navy operated the Naval Space Surveillance or NAVSPASUR radar "fence" across the southern United States.

The NRL, particularly its Naval Center for Space Technology (NCST) after 1986, spearheaded development of naval-related space hardware. In June 1960 the first NRL Solar Radiation (SOLRAD) satellite went into orbit carrying a highly classified Galactic Radiation Background (GRAB) experiment, the world's first operational electronic-intelligence payload. Based on a concept formulated by its scientist Roger L. Easton in April 1964, the NRL initiated the TIMATION space-based navigation project that contributed technologically to the GPS program's ultimate success. Living Plume Shield experiments in the early 1980s led to an evolutionary NRL program that profoundly influenced development of tactical communications equipment. In January 1994, the NCST *Clementine* spacecraft launched for lunar orbit and subsequently rendezvoused with an asteroid.

The Naval Electronics Laboratory's early work on satellite systems for ocean surveillance resulted in it becoming, in 1966, the core of the newly established Naval Electronic Systems Command. Recognizing the vital role of space systems in its operations, the Navy renamed that command the Space and Naval Warfare Systems Command (SPAWAR) in May 1985. SPAWAR acquired, fielded, and supported space-related communications and intelligence systems. Meanwhile, in October 1983 the Navy established Naval Space Command at Dahlgren, Virginia, to conduct its space operations.

For tactical use, the Navy began developing the Fleet Satellite Communications (FLTSATCOM) system, which operated at ultrahigh frequency (UHF), in the 1970s. During the First Gulf War in 1991, however, limited satellite capacity necessitated hand delivery of critical, complex orders to Navy aircraft carriers. Consequently the Navy drastically increased the communications capacity on board its vessels by adding extremely high frequency (EHF) and superhigh frequency (SHF) capabilities via the Ultrahigh Frequency Follow-On satellite system that began replacing FLTSATCOM in 1993. By 2005 vessels carried a half dozen different satellite antennas, which competed with other equipment for limited deck space and compelled the Navy to consider placing a series of broadband (multi-satellite) antennas together atop the superstructure of its next-generation combatant ship.

Norman Friedman

See also: Electronic Intelligence, Global Positioning System, Military Communications, V-2 Experiments, Vanguard

Bibliography

David H. DeVorkin, *Science with a Vengeance: How the Military Created the U.S. Space Sciences after World War II* (1992).

Naval Center for Space Technology. http://www.nsstc.org/.

Naval Historical Center. www.history.navy.mil.

Norman Friedman, *Seapower and Space: From the Dawn of the Missile Age to Net-Centric Warfare* (2000).

United States Space Command. *See* United States Department of Defense.

Vandenberg Air Force Base

Vandenberg Air Force Base (VAFB), the only military base in the United States from which government and commercial satellites are launched into polar orbit. This base near Lompoc, California, is also the only location from which intercontinental ballistic missiles (ICBMs) are launched over the Pacific Ocean to test the accuracy of warhead guidance, using targets at Kwajalein Atoll in the Marshall Islands. Located approximately 150 miles northwest of Los Angeles, California, VAFB was named in October 1958 to honor General Hoyt S. Vandenberg, second chief of staff of the United States Air Force (USAF). Although initially assigned to Air Research and Development Command, host responsibilities for the base transferred to Strategic Air Command in 1958 and subsequently to Air Force Space Command in 1991.

The Army activated the base, originally called Camp Cooke, in October 1941 as a training center for armored and infantry troops. Deactivated between February 1946 and August 1950, the installation reopened to train armored and infantry forces during the Korean War, only to close again in February 1953. Transformation of Camp Cooke to the nation's first space and ballistic missile operational and training base began in 1957, when the Army transferred the northern three-fourths to the USAF. Shortly thereafter the Navy received Point Arguello, the remaining portion of the original camp, as a launch site for its Pacific Missile Range. The USAF had selected VAFB as perfect for launching America's Corona polar-orbiting reconnaissance satellites, because rockets launched there did not go over land until they reached the South Pole, thus ensuring safety and security. Virtually every type of U.S. space booster and ICBM has been launched from pads or underground silos at VAFB. In September 1996 the California Spaceport at Space Launch Complex 6 on VAFB became the first federally approved commercial launch facility, offering an alternative to government subsidized operations.

David C. Arnold

See also: Atlas, Corona, Delta, Titan

Bibliography
David N. Spires, *Beyond Horizons: A Half Century of Air Force Space Leadership* (1997).
VAFB. www.vandenberg.af.mil.

White Sands Missile Range

White Sands Missile Range, a 4,000-square-mile area in south-central New Mexico that the U.S. Army designated in July 1945 as White Sands Proving Grounds to support weapon development and testing programs. Taking its name from nearby White Sands National Monument, the range supported Trinity, the world's first nuclear explosion, followed in September 1945 by its first missile firing, a U.S. Navy Tiny Tim. In October 1945 White Sands accommodated the launch of a WAC-Corporal, an upper-atmospheric research rocket developed by the Jet Propulsion Laboratory at the California Institute of Technology in Pasadena, California. During 1946–52, the range witnessed 73 V-2 rocket launches, many carrying scientific instruments to the edge of space. With U.S. Air Force permission, Convair tested three MX-774 missiles (precursors of the Atlas ICBM) at White Sands in 1948, and the Navy launched several Viking high-altitude research rockets there during 1949–55.

More than 38,000 missile firings had occurred at White Sands by 1990, but the range also had other space-related uses. NASA tested both the Little Joe II Apollo escape system and Apollo spacecraft engines there during the 1960s. In the early 1970s NASA selected the Northrup Strip at White Sands as an alternate landing site for the Space Shuttle, and the *Columbia* orbiter used the site in 1982. A High Energy Laser Systems Test Facility, operated by the Navy, opened at White Sands in 1985. It became the location for successfully aiming a Mid-Infrared Advanced Chemical Laser beam at an orbiting satellite in 1997 and for the successful interception and destruction of a Katyusha rocket using the Army's Tactical High Energy Laser/Advanced Concept Technology Demonstrator in 2000.

Joel W. Powell

See also: Sounding Rockets, V-2 Experiments

Bibliography

Joel W. Powell and Keith J. Scala, "Historic White Sands Missile Range," *Journal of the British Interplanetary Society* 47 (1994): 82–98.
White Sands Missile Range. www.wsmr.army.mil/wsmr.asp.

NAVIGATION

Navigation, the process of determining location and direction. For thousands of years, military units navigated by referring to local landmarks or celestial bodies. The first of these options required visible features, either natural or artificial (beacons or buoys) and either maps or individuals familiar with the terrain. Celestial navigation was a more technical process best performed by someone specially trained, and it generally worked only on clear nights. Navigation between points determined celestially or from maps involved the notoriously inaccurate interpolation of one's position based on assumed velocity, direction, and time—a process known as dead reckoning. Consequently, mariners often drifted off course, and military commanders found themselves fighting battles without accurate knowledge of the location of their own (or enemy) forces through most of recorded history.

Throughout the centuries, the invention of several instruments greatly improved navigation. Most significant were the magnetic compass, cross-staff, astrolabe, and sextant. During the eighteenth century, chronometers (clocks) built by John Harrison finally gave navigators a tool for determining longitude with reasonable accuracy. Soon after World War I, radio navigation systems allowed users to locate themselves by reference to the known location of one or more transmitting stations. Since the signal strength and positional accuracy of all electronic navigation systems declined with distance, anyone navigating beyond the range of those systems would be reduced to using more primitive methods.

The need for an accurate, globally available alternative to existing navigational capabilities became acute when the U.S. Navy developed the Polaris submarine-launched ballistic missile system in the late 1950s. The accuracy of any ballistic missile depended on knowing the position of its launch site. Absent some form of global navigation aid, a Polaris submarine crew faced the prospect of having to intermittently surface for celestial-position fixes, which rendered the vessel more vulnerable to discovery and possible attack by the enemy. To address this problem, the Navy sponsored two development programs: Transit and Timation. Transit, based on the Doppler effect, became the first operational space-based navigation system in 1964. Although adequate for the Navy's needs, Transit had two limitations that prevented it from achieving widespread use. It was slow, requiring a lengthy observation time to determine position, and it provided location only in two dimensions, latitude and longitude. A contemporary Soviet Union system, Tsiklon, operated on the same principle as Transit, with the same limitations. Tsiklon's derivatives included the Tsiklon M Parus system, and an all-service military system called Kristal, both deployed in the 1970s. While the U.S. Merchant Marine used the military Transit system, the Soviet merchant marine relied on a separate, semi-civilian satellite navigation system called Tsikada that became fully operational in 1986. The Soviet Union also developed a fully civilian navigational system called Nadezhda, deployed in the late 1980s.

The U.S. military, especially the Air Force (USAF), needed a space-based, radio navigation system for a wider range of uses. The Navy's experimental Timation system incorporated highly accurate atomic clocks on a pair of satellites. Meanwhile, the USAF conceptualized System 621B, which would provide location in three dimensions—latitude, longitude, and altitude. These competed with an Army concept called SECOR (Sequential Correlation of Range). In 1968 the U.S. Department of Defense (DoD) established a Navigation Satellite Executive Group to coordinate satellite navigation projects. By 1973 the services reached a compromise that combined elements of the Navy and Air Force concepts. It became the NAVSTAR Global Positioning System (GPS).

Deployment of NAVSTAR GPS began in 1974, initially using two modified Timation satellites. Ten Block 1 GPS satellites followed during 1978–85, and 24 Block 2 satellites later in the century.

As the constellation of GPS satellites became operational, receiver equipment was developed and tested for users. It benefited fully from the revolution in miniature electronics wrought by the development of integrated circuits. By 2000 civilian users could purchase pocket-sized GPS receivers for as little as $100. Military receivers, although slightly larger and more expensive, still were inexpensive enough to allow their use on board all ships and aircraft and in most ground combat units.

The consequences of this revolution in navigational precision were first evident in the 1991 First Gulf War, where units and even individuals knew their positions to an accuracy of a few feet. Combined with precision weapons and space-based communications, GPS allowed U.S. forces to maneuver and fire with unprecedented accuracy, thereby contributing to the extremely rapid and efficient execution of Operation Desert Storm. A decade later, U.S. Special Forces and CIA operators fighting the Taliban in Afghanistan used GPS receivers and satellite communications to order strikes by aircraft carrying GPS-guided weapons. Unlike earlier laser-guided weapons, these could operate in any weather and did not require illumination of the target. The GPS-guided weapons were so accurate that, in some cases, no explosive filling was necessary, because the kinetic energy of a concrete-filled bomb case was sufficient to achieve desired results. Although highly accurate GPS-guided weapons could minimize collateral damage, their accuracy could be a double-edged sword if the wrong coordinates were specified.

The nonmilitary uses of GPS soon became apparent. In 1983 when Soviet pilots shot down Korean Airlines Flight 007 after it inadvertently entered Soviet airspace, U.S. President Ronald Reagan announced that international civil operators would have access to GPS signals. Reagan directed the Department of Transportation, in 1987, to collaborate with the DoD on civil use of GPS. At the 10th Air Navigation Conference of the International Civil Aviation Organization in 1991, the United States formally offered to make GPS available to all nations. Civil operators had access to the unencrypted Standard Positioning Service when the GPS system became fully operational in 1995. Military users could access a more accurate, encrypted signal until 1 May 2000, when encryption was ended by presidential executive order, thereby giving all users equal access to the more accurate GPS signals.

To enhance the position accuracy obtained from GPS, several nations undertook space-based augmentation programs to broadcast correction signals from non-GPS satellites. The United States began developing a Wide Area Augmentation System for precision flight approaches. Europe pursued deployment of the European Geostationary Navigation Overlay System (EGNOS), and Japan invested considerably in a Multi-transport Satellite Augmentation System (MSAS) for navigation in Asia and the Pacific region. The Indian Space Research Organisation, in cooperation with Raytheon Corporation, planned to demonstrate the technology for its own Gagan (GPS-aided GEO Augmented Navigation) system to bridge the gap between EGNOS and MSAS.

The obvious value of GPS for military and civil purposes caused other nations to ponder the negative implications of a U.S. monopoly over this capability. The Soviet Union had begun developing its own Global Navigation Satellite System, similar to GPS, in the 1970s and declared it operational in 1993. In 2002, the European Space Agency in collaboration with the European Union agreed to develop a civilian-controlled global satellite navigation system called Galileo. During 2000–2003, China launched three Beidou (Big Dipper) geostationary navigational satellites for its military forces' exclusive regional use.

John D. Ruley and Rick W. Sturdevant

See also: Ballistic Missiles

Bibliography
Norman Friedman, *Seapower and Space: From the Dawn of the Missile Age to Net-Centric Warfare* (2000).
William H. Guier and George C. Weiffenbach, "Genesis of Satellite Navigation," *Johns Hopkins APL Technical Digest* 18, no. 2 (1997): 178–81.
Bradford W. Parkinson et al., "A History of Satellite Navigation," *Navigation* 42, no. 1 (1995): 109–64.
Dava Sobel, *Longitude: The True Story of a Lone Genius Who Solved the Greatest Scientific Problem of His Time* (1997).

Department of Defense Gravity Experiment

Department of Defense Gravity Experiment (DODGE), a U.S. Navy satellite designed and built by the Space Department, Applied Physics Laboratory (APL),

Johns Hopkins University to explore gravity-gradient stabilization at near-synchronous altitudes. After pioneering the Transit navigation satellite in the late 1950s, APL shifted to satellite-based geodetic research and applications, particularly experiments to stabilize spacecraft by affecting their gravity gradients. *DODGE*, launched on 1 July 1967, belonged to the latter category. It carried magnetic-damping devices and 10 booms that radio commands could extend or retract independently along three different axes. In addition to becoming the first satellite to stabilize its motion along the yaw axis using a reaction wheel, *DODGE* provided the measurements of Earth's magnetic field at near-synchronous altitudes that mathematicians needed to calculate fundamental constants for use in controlling future high-altitude spacecraft.

Perhaps the most significant accomplishment of *DODGE* was a feat entirely unanticipated by its designers. The satellite used two onboard television cameras to check its alignment with Earth during the various gravity-gradient exercises. Clyde Holliday, an APL scientist who 20 years earlier had captured the first photographs of Earth from space using cameras mounted in captured German V-2 rockets fired from White Sands Proving Ground, proposed scanning Earth systematically using a succession of red, green, and blue filters covering the *DODGE* camera lens. Transmitted from space, the resulting pictures were combined on the ground into the first, full-disc color image of the planet. That historically significant picture subsequently appeared on the cover of the September–October 1967 issue of the *Johns Hopkins Journal*. Before ceasing operation early in 1971, *DODGE* supplied thousands of color and black-and-white pictures of Earth.

John Cloud

See also: Johns Hopkins University Applied Physics Laboratory, United States Navy

Bibliography
B. B. Holland, "Doppler Tracking of Near-Synchronous Satellites," *Journal of Spacecraft and Rockets* 6, no. 4 (1969): 360–65.

JHU/APL, *History of the Space Department: The Applied Physics Laboratory, 1958–1978*, SDO-5278 (January 1979).

Global Navigation Satellite System

Global Navigation Satellite System (GLONASS—also called Uragan), the Russian counterpart to the U.S. Global Positioning System, operated in the early twenty-first century by the Coordinational Scientific Information Center of Russia's Ministry of Defense. The Soviet military stated a requirement in 1970 for a single, highly accurate satellite radio navigation system for its forces. A 1976 decree established GLONASS to meet that need. Plans called for 24 active satellites, 8 in each of three orbital planes, at an altitude of approximately 19,100 km and transmitting coded signals in two frequency bands to allow Soviet forces to determine precisely and instantaneously via ranging measurements their time, position, and velocity. The first GLONASS satellite launch occurred in October 1982, and Russian President Boris Yeltsin declared a status of initial operational capability in September 1993. No sooner had a full constellation of active GLONASS satellites been achieved in early 1996 than financial shortages prevented the Russians from maintaining it. By October 2005, the GLONASS constellation contained only 14 operational satellites.

In the late 1990s, Russia began exploring civil use of GLONASS and the possibility of integrating it with the European Union's proposed Galileo system. Introduction of GLONASS M satellites, first launched in December 2001, brought significant improvements in signal quality, offered a second civil code, doubled navigational accuracy, and more than tripled design life to seven years. To further improve GLONASS capabilities after 2005, the K series would add a third civilian frequency, provide a payload for search and rescue, cut the satellite's mass in half, and increase design life to 10 years.

James C. Mesco

See also: Commercial Navigation, Space Navigation Policy

Bibliography

Brian Harvey, *Russia in Space: The Failed Frontier?* (2001).
Nicholas L. Johnson and David M. Rodvold, *Europe and Asia in Space 1993–1994* (1995).
Space and Tech. www.spaceandtech.com.

Global Positioning System

Global Positioning System (GPS), the first space-based navigational system that enabled users to determine precisely their location in three dimensions and time within billionths of a second. The complete NAVSTAR (Navigation Satellite Time and Ranging) GPS operational constellation, operated by the U.S. Air Force (USAF), contained 24 satellites, plus several spares, orbiting Earth every 12 hours along six orbital planes at 18,000 km altitude. Consequently, users anywhere on Earth could receive radio signals from five to eight GPS satellites, a minimum of three being needed to triangulate a user's position. Originally developed for strategic military purposes, the operational GPS quickly proved essential for conducting theater warfare. Sales of GPS equipment to civil, commercial, and individual users, however, grew from $807 million in 1994 to $6 billion in 2000, far surpassing those to the U.S. military. Millions of nonmilitary users worldwide freely accessed GPS for many different applications.

In 1973, the Navy and Air Force formed a joint NAVSTAR GPS program office, which merged capabilities from several ongoing efforts: the operational Transit system, the Naval Research Laboratory's experimental Timation program, and USAF Program 621B. During 1978–85, the USAF successfully launched 10 GPS Block 1 satellites as operational prototypes, followed by the first fully operational Block 2 version in February 1989. Significant gaps existed in global coverage until December 1993 when the USAF first achieved a full 24-satellite constellation, and not until April 1995 did GPS officially achieve full operational status.

GPS originally had two broadcast frequencies—L1 and L2. The L1 channel provided Standard Positioning Service (SPS) through a feature called Selective Availability (SA) that intentionally degraded the signal. Users with SPS obtained a

Illustration of NAVSTAR Global Positioning System (GPS) Block 2A satellite in medium-Earth orbit, 1989. (Courtesy U.S. Department of Defense)

relatively coarse positioning accuracy of 100 m horizontally and 156 m vertically and time within 340 nanoseconds. After Korean Airlines Flight 007 strayed off course in September 1983 and Soviet fighter-interceptors shot it down, the United States announced SPS would be universally available at no cost to everyone. Both the L1 and L2 channels furnished Precise Positioning Service (PPS) to users with proper access codes. The encrypted PPS signal initially delivered positioning accuracy of 22 m horizontally and 27.7 m vertically and time within 200 nanoseconds. As U.S. military operations after the mid-1990s demonstrated, further refinement of PPS capability produced even greater accuracy. Meanwhile, mushrooming civilian and commercial dependence on GPS caused President William Clinton to direct in May 2000 that SA be discontinued to give the civil sector the same level of GPS accuracy as the military.

Even as it planned for a more sophisticated Block 3, the USAF launched several variants of Block 2 satellites during 1989–2005. Block 2A (Advanced), built by Rockwell and launched during 1990–97, increased the duration of the satellites' ability to operate autonomously without ground-control contact from 14 to 180 days but with degraded accuracy. Block 2R (Replenishment) satellites, built by General Electric AstroSpace (later Lockheed Martin) and launched during 1997–2005, improved autonomous operational capability through intersatellite links and extended design life from 7 to 10 years. The first of eight Block 2R-M (Modernized) satellites, built by Lockheed Martin, launched in September 2005 to provide increased signal strength, two additional signal frequencies (one military and one civil), and better antijamming capability. Later, Block 2F satellites supplied by Boeing would add a third nonmilitary signal, L5, intended primarily for civil aviation.

By 2006 the U.S. military used GPS for tracking friendly or enemy land, sea, and air forces globally during peacetime and war. The system allowed pilots to launch precision-guided munitions from safer distances with greater assurance of hitting their intended targets, enabled rescuers to locate downed pilots more quickly, and permitted operators to fly unpiloted aerial vehicles from control centers halfway around the world. Timing signals provided by GPS ensured synchronization of communications and computing systems for early warning, air defense, space surveillance, satellite control, and more.

Roger Handberg

See also: Commercial Navigation, Economics of Space Navigation, Space Navigation Policy

Bibliography
Andrews Space and Technology. http://www.andrews-space.com/.
NavtechGPS. www.navtechgps.com.
Bradford W. Parkinson et al., *Global Positioning System: Theory and Applications*, vol. 1 (1996).
Michael Russell Rip and James M. Hasik, *The Precision Revolution: GPS and the Future of Aerial Warfare* (2002).

Navstar. *See* Global Positioning System.

Nova. *See* Transit.

Oscar. *See* Transit.

Parus. *See* Tsiklon.

Program 621B. *See* Global Positioning System.

Timation. *See* Global Positioning System.

Transit

Transit, the world's first navigational satellite system, was conceived by scientists at the Johns Hopkins University Applied Physics Laboratory (APL) shortly after the launch of *Sputnik* in October 1957. After they determined the orbit of *Sputnik* by analyzing the Doppler shift of its radio signals, the APL scientists suggested they could use the Doppler effect from a satellite in a precisely known orbit to compute the position of a receiver on Earth. The U.S. Navy enthusiastically sponsored further work on the concept, because it promised accurate, all-weather, global navigation for its ships, especially Polaris submarines carrying nuclear-tipped ballistic missiles. In September 1958 the Advanced Research Projects Agency formally initiated the Transit development program, which it transferred to the Navy one year later. Under Richard Kershner's leadership, the APL began designing and building the satellites.

Operationally Transit spanned nearly four decades from the first successful launch—*Transit 1B*—in April 1960 to the system's replacement by the Global Positioning System and deactivation in December 1996. It achieved initial operational capability in 1964 and full capability in 1968. The Transit satellite constellation normally consisted of six polar-orbiting satellites (three active and three spare) at an altitude of approximately 600 nautical miles supported by three ground stations. Each satellite provided two-dimensional positioning accuracy within 600 ft for slow-moving military ships, but could do little to assist fast-moving airplanes or cruise missiles. In 1967 Vice President Hubert Humphrey announced the availability of Transit signals to tens of thousands of commercial vessels from all nations, which greatly improved the safety and efficiency of shipping operations worldwide.

Transit included two satellite versions: the 110 lb Oscar and the 350 lb Nova. Beginning with *Oscar 18* in the late 1960s the Navy awarded Radio Corporation of America (RCA) a satellite-production contract. RCA received another contract in 1977 to build several Novas. The Transit Improvement Program in 1969 helped provide radiation-hardened satellites, and a dual-launch method called Stacked Oscars on Scout established a temporarily oversize constellation of 12 satellites in 1988. On average Transit satellites had an operational lifetime of roughly 14 years, with a couple lasting more than 20 years.

The legacy of Transit included scientific discoveries, engineering innovations, the first commercial uses of space navigation, and products beyond the navigational realm. Besides the first electronic-memory computer in space, Transit significantly advanced the fields of geodesy and space science. Moreover, Transit technology spurred development of such biomedical devices as a rechargeable cardiac pacemaker, an automatic implantable cardiac defibrillator, and a programmable implantable medication system.

James C. Mesco

See also: Johns Hopkins University Applied Physics Laboratory, Radio Corporation of America (RCA), United States Navy

Bibliography

Robert J. Danchik and Pryor L. Lee, "The Navy Navigation Satellite System (TRANSIT)," *Johns Hopkins APL Technical Digest* 11, no. 1/2 (1990): 97–101.

Federation of American Scientists. www.fas.org.

Vincent L. Pisacane, ed., "The Legacy of Transit," *Johns Hopkins APL Technical Digest* 19, no. 1 (January–March 1998): 1–71.

Tsikada. *See* Tsiklon.

Tsiklon

Tsiklon (cyclone), the first Soviet Union military navigation and secure communications system, was employed continuously into the twenty-first century by the Russian

Tsiklon-Type Satellites (as of December 2007)

Satellite	Years of Launches	Number of Launches
Tsiklon	1967–1978	29
Parus (Tsiklon B)	1976–2007 (in service)	98
Tsikada	1976–2004 (in service)	31

Navy, especially its nuclear missile submarines. Based on a constellation of six satellites in near-circular 1,000 km orbits with an average inclination of 83 degrees and a spacing of 30 degrees between the satellites' orbital planes, the system provided global coverage. Like the U.S. Transit satellite navigation system, Tsiklon used Doppler shift to achieve a geopositioning accuracy of better than 100 m.

Conceived in 1962 and first launched for testing in 1967, the Tsiklon's development was initially hampered by high geopositioning errors. Debugging allowed the system to enter experimental service in 1971. The Soviet Navy commissioned an improved operational version, Tsiklon B, in 1976. Based on a Parus (sail) satellite first launched in 1974, the Tsiklon B completely replaced the original Tsiklon version in 1978.

Along with the Tsiklon B, another version of that navigation system, called Tsikada (cicada), was introduced for military and civilian naval use in 1976. Unlike Tsiklon B, the Tsikada constellation had only four satellites similar to Parus in mass and shape but with different subsystems. Tsikada satellites were not used for military communications. Since 1982 10 of the Tsikada satellites carried distress-signal relay equipment for the COSPAS-SARSAT international rescue system, which the Soviet Union, United States, Canada, and France had formed in 1977. Seven of those 10 satellites were publicly announced as Nadezhda (hope) without reference to their Tsikada origin.

Launched from Plesetsk on Kosmos 3M rockets, the spin-stabilized Tsiklon-type satellites had a cylindrical body about 2 m in diameter and length. Covered by solar cells and thermal radiators, they weighed approximately 900 kg and had an average on-orbit lifespan of three years. OKB-10 (Experimental Design Bureau) from Zheleznogorsk, later known as the Applied Mechanics Scientific-Production Association (NPO PM), originally designed both Tsiklon and Parus satellites, but further system development transferred to Poliyot Corporation in Omsk in the 1970s. By 2007 Russia had launched 157 Tsiklon-type satellites total (including failures)—almost all of them under the Kosmos cover name—and continued to use the Tsiklon B and Tsikada constellations along with its more advanced Global Navigation Satellite System (GLONASS).

Peter A. Gorin

See also: Kosmos, Russian Space Forces, Russian Space Launch Centers, Russian Launch Vehicles

Bibliography
Nicholas L. Johnson and David M. Rodvold, *Europe and Asia in Space 1993–1994* (1995).
Russia's Arms Catalog, vol. 6, Missiles and Space Technology, 1996–1997 (1998).

Uragan. *See* Global Navigation Satellite System.

PLANETARY DEFENSE

Planetary defense, a term coined by U.S. Air Force (USAF) officer Lindley Johnson in 1993, broadly defined as protecting Earth from natural objects, such as asteroids and comets, coming from outer space. The first close-up images of the Moon, as seen through Earth-based telescopes, showed round scars of all sizes on the lunar surface and generated considerable debate about whether volcanoes or large rocks from space made them. The question was resolved in favor of the latter when robotic spacecraft began exploring the solar system in the early 1960s. A 12-year Grand Tour of the four major outer planets by *Voyager 2*, launched in 1977, revealed that every solid body the spacecraft passed was covered with impact craters. As for Earth, astronomer Edmond Halley suggested in 1694 that comet impacts might have caused terrestrial catastrophes, and mining engineer Daniel Barringer suggested in 1902 an impact origin for Meteor Crater in Arizona. Geologist Eugene Shoemaker conclusively proved in 1960 that an object from space had collided with Earth to form that particular crater. Thirty years later, identification of a giant, hidden crater at Chicxulub on the Yucatán Peninsula reinforced the theory that a large object struck Earth 65 million years ago and, possibly, sent the dinosaurs into extinction.

The first step in planetary defense involved understanding the nature of the threat from a Near Earth Object (NEO)—asteroids or comets whose orbits caused them to come alarmingly close to Earth. In July 1981 NASA sponsored a three-day workshop on "Collision of Asteroids and Comets with Earth: Physical and Human Consequences."

An illustration of a massive asteroid crashing into Earth. (Courtesy Don Davis/NASA)

That same year, University of Arizona scientists Thomas Gehrels and Robert McMillan began Spacewatch, a program to detect and collect data on NEOs. Not until the 1990s, however, did the threat posed by Earth-orbit-crossing asteroids gain widespread political attention. Adopting a term coined by Arthur C. Clarke in his 1973 science-fiction novel *Rendezvous with Rama*, NASA delivered to the U.S. Congress in January 1992 a "Spaceguard Survey Report" calling for international collaboration to locate and catalogue kilometer-size or larger Earth-crossing asteroids and comets. At the same time, a NEO Interception Workshop investigated methods for destroying or altering the orbit of any object posing a threat and concluded that using nuclear detonations to alter an object's orbit afforded the best chance for success.

Observation of the collision of comet Shoemaker-Levy 9 with Jupiter in July 1994 confirmed in many experts' minds that humans should take more seriously the probability of such an event on Earth. Consequently, the International Astronomical Union (IAU) created the Spaceguard Foundation in 1995; Lawrence Livermore National Laboratory in California hosted the first international technical meeting on planetary defense in May 1995; the United Nations Office of Outer Space Affairs, in 1996, stated the need for an international network of telescopes for NEO searching and tracking; Spaceguard UK was established in 1997 to promote British NEO activities; and the National Research Council of the U.S. National Academy of Sciences gave its highest priority to NEO efforts in 1998. Release of the films *Armageddon* and *Deep Impact* during 1998 fueled popular awareness of the threat posed by NEOs, and newspapers warned of future threats from recently discovered asteroids. The following summer, a conference titled "International Monitoring Programs for Asteroid and Comet Threat (IMPACT)" in Italy resulted in the Torino scale—similar to the Richter scale for earthquakes—for categorizing the risk and potential severity of an impact by newly discovered objects. At a subsequent meeting in 2001, NEO researchers agreed to experiment with a more complex Palermo scale that added time of possible impact to the Torino scale factors of impact energy and event probability.

A Spaceguard survey to identify and track potentially threatening asteroids and comets gained momentum throughout the 1990s, with a goal of cataloging by 2009 at least 90 percent of all such objects greater than 1 km in diameter. Directed by astronomer Donald Yeomans of the NASA Jet Propulsion Laboratory (JPL), the survey depended on a variety of advanced telescopic systems or projects funded by NASA, the USAF, academic institutions, or space agencies in other countries. For example, the Near Earth Asteroid Tracking (NEAT) project involved mounting a JPL detector on USAF Ground-based Electro-Optical Deep Space Surveillance (GEODSS) and other USAF camera systems atop Haleakala on Maui, Hawaii. The Lincoln Near Earth Asteroid Research (LINEAR) project, funded by the USAF and NASA, used a pair of GEODSS telescopes at Lincoln Laboratory's Experimental Test Site on the White Sands Missile Range in Socorro, New Mexico. Other contributors included Spacewatch, the Lowell Observatory Near Earth Object Search (LONEOS), and University of Arizona–Australian National University (ANU) collaboration using the U.S. Catalina Schmidt telescope and the ANU Uppsala Schmidt telescope. Russia, Germany, the United Kingdom, and others established Spaceguard centers. Australia's space weather agency, Ionospheric Prediction Service Radio and Space Services, cooperated with USAF scientists in 2001 to undertake Project Wormwood at Learmonth Solar Observatory.

By the end of January 2006 the highly automated JPL Sentry System for collision monitoring reported 96 NEOs that potentially could impact Earth during the next 100 years. Compared to the significant progress in NEO detection and tracking, plans for interception or deflection lagged far behind.

William E. Burrows

See also: Asteroids, Comets, Space Surveillance

Bibliography

James C. Mesco, "Watch the Skies," *Quest* 6, no. 4 (Winter 1998): 35–40.

NASA Near-Earth Object Program. neo.jpl.nasa.gov.

Project Wormwood. http://www.ips.gov.au/IPSHosted/neo/.

Spaceguard UK. www.spaceguarduk.com.

RECONNAISSANCE AND SURVEILLANCE

Reconnaissance and surveillance, observing an area of Earth to determine what is there. Although surveillance and reconnaissance are similar in purpose, the main difference between them lies in duration and specification. Surveillance aims to collect information continuously, whereas reconnaissance operations focus on more specific localities or targets. The first military space mission was reconnaissance. There are several ways of reconnoitering from space, the primary methods being visual and radar reconnaissance. Visual reconnaissance is accomplished by launching human observers or cameras that return images to Earth. Although the photographic images can be in color, analysts generally prefer black-and-white pictures because they show greater detail. The biggest problem with visual reconnaissance is that it is useless when a target is covered by clouds. Radar can penetrate cloud cover but returns lower-quality images. Because radar reconnaissance is more technologically challenging and generally provides poorer imagery, only a few countries operate dedicated radar satellites.

By 2005 several countries possessed military satellite reconnaissance systems. These included the United States, Russia, France (in cooperation with Germany, Italy, and Spain), Japan, China, and Israel. In addition, Canada, India, and Brazil operated civilian remote sensing satellites that have limited military uses. Since the late 1990s, several private companies have offered to sell satellite imagery of increasingly high quality. Virtually any nation can buy detailed pictures of any desired location.

The United States was the first country to consider using satellites for reconnaissance. In 1946 the U.S. Air Force (USAF) asked the RAND Corporation in Santa

Monica, California, to study the feasibility of launching satellites for military purposes, reconnaissance being high on the list. RAND conducted an extensive study of reconnaissance satellites and their capabilities in 1954, which proposed an atomic-powered satellite carrying a television camera. When the USAF actually began a reconnaissance satellite program, the television camera proved impractical and a "film-scanning" system was chosen instead. That system would snap a photo, develop the film onboard the satellite, scan the image, and transmit it to Earth. Engineers substituted solar panels for the atomic power supply. Still, USAF funding for the program remained minimal until after *Sputnik* in October 1957.

Among the many challenges confronting the early U.S. reconnaissance satellite program, the most pressing was time. It would take several years for the film-scanning system to become operational. Consequently, in 1957, RAND engineers Merton Davies and Amrom Katz proposed that the USAF develop an interim reconnaissance satellite that would take pictures and return the exposed film to Earth in a recoverable capsule. Such a system could photograph far more area than the film-scanning system and, moreover, could achieve operational status sooner. President Dwight D. Eisenhower directed that this program, Corona, be secret and managed by the Central Intelligence Agency (CIA), with the USAF providing support.

With *Sputnik* signaling the advent of the space age, satellite reconnaissance received much greater attention in the United States, which suspected the Soviet Union was fielding many intercontinental ballistic missiles with thermonuclear warheads. This prompted the USAF to fund a Samos satellite program. In addition to film-scanning satellites, Samos soon included a film-return satellite and signals intelligence satellites known as "ferrets." After a string of failures in 1959 and early 1960, a Corona satellite successfully returned the first exposed film to Earth in August 1960. Despite their graininess and relative lack of detail, the pictures provided a wealth of information about military sites in the Soviet Union. Corona satellites helped demonstrate that the Soviet Union did not have more ICBMs than the United States. While Corona became increasingly successful and the quality of its images significantly improved, Samos suffered numerous failures and its film-scanning technology proved too limited to be useful. By 1960 the USAF canceled the film-scanning versions and began developing a different satellite to replace Corona, but technical problems also prevented it from ever becoming operational. Meanwhile in the late 1960s the CIA began developing an improved film-return satellite, KH-9 Hexagon, to replace Corona.

Despite its success Corona could only photograph objects on the ground about 6–9 ft long. Because this was not good enough to make detailed observations and measurements, the USAF began another satellite program called Gambit. Like Corona, Gambit returned its film to Earth, but could photograph much smaller objects. While Corona operations ceased in 1972, Gambit continued until 1985. Some of the last Gambit satellites could reputedly photograph objects as small as a baseball.

The next major advance in reconnaissance satellite technology occurred in 1976, when the United States launched a satellite known as KH-11 Kennan, later renamed Crystal. Employing technology similar to a common digital camera, Kennan transmitted images directly to the ground. Although the images still were black and white,

U.S. and Soviet Union/Russian Military Reconnaissance and Surveillance Satellites

Country	System	Image Type	Delivery Method	Operational Period
United States	Corona	Film	Reentry capsule	1960–1972
	Samos	Film	Digital readout	1960–1962
	KH-5 Argon	Film	Reentry capsule	1961–1964
	KH-6 Lanyard	Film	Reentry capsule	1963
	KH-7/8 Gambit	Film	Reentry capsule	1963–1984
	KH-9 Hexagon	Film	Reentry capsule	1971–1985
	Parcae	N/A	Transmitted	1976–1990
	KH-11 Kennan	Electro-optical	Transmitted	1976–1995
	Lacrosse	Radar	Transmitted	1988–
	Advanced KH-11	Electro-optical	Transmitted	1990–
	NOSS-2	N/A	Transmitted	1990–
	NOSS-3	N/A	Transmitted	2001–
Soviet Union/ Russia	Zenit	Film	Reentry capsule	1961–1994
	Teslina	N/A	Transmitted	1967–
	RORSAT	N/A	Transmitted	1970–1988
	Yantar	Film	Reentry capsule	1974–
	EORSAT	N/A	Transmitted	1975–
	Terilen	Electro-optical	Transmitted	1982–

probably revealed only those objects larger than a softball, and took several minutes to transmit, this was much faster than the days or weeks of waiting with the film-return method. Along with the increase in speed came a change in the way national leaders used information obtained from reconnaissance satellites. Whereas satellite imagery originally had been useful only in the preparation of long-range plans and studies, its increased timeliness now made it useful in crisis situations that required nearly immediate, high-level decisions. Later versions of Kennan probably were in use in the early twenty-first century. Because they still could not see through clouds, intelligence agencies turned to other technologies such as radar, used in the Lacrosse system.

The Soviet Union developed systems similar to the United States, usually trailing about three to seven years behind. Its first reconnaissance satellite, Zenit, was similar to the first spacecraft used to launch a human into orbit. Unlike Corona, Zenit returned both the film and the camera to the ground in a large capsule. The Soviet Union later developed a higher-resolution system called Yantar. Not until the 1980s did the Soviet Union have a satellite, Terilen, capable of transmitting images to the ground in "real time." The Russians continued to use modified versions of Zenit and Yantar in the early twenty-first century, although economic problems limited the number of satellites launched.

Ocean surveillance from space became a high priority for both the Soviet Union and the United States. The Soviet Radar Ocean Reconnaissance Satellite (RORSAT) program commenced operations in 1970, followed by the Electronic Intelligence (ELINT) Ocean Reconnaissance Satellite (EORSAT) program in 1974. As the designations imply, the nuclear-powered RORSAT used large radar antennas to bounce signals off the ocean and, thereby, locate ships; the EORSAT located naval forces by detecting and triangulating on their radio and radar emissions. Meanwhile, in 1971, the U.S. Navy began launching Parcae ocean surveillance satellites that functioned much like EORSAT. The U.S. Navy also contemplated, but never launched, an active, space-based radar system called Clipper Bow that resembled RORSAT in capability.

Both the Soviet Union and the United States explored the possibility of using humans on space stations equipped with large cameras to conduct reconnaissance. The United States abandoned its program, the Manned Orbiting Laboratory (MOL), in 1969 before launching any missions. The Soviet Union launched several Almaz space stations with synthetic aperture radar for reconnaissance, but eventually abandoned the idea because robotic systems were more effective than humans at taking photographs from space.

Certainly U.S. and Soviet satellite reconnaissance capabilities contributed immeasurably toward ensuring the Cold War did not escalate into nuclear war between the superpowers. Because each could assess the changing military posture of other, neither could realistically expect to achieve the element of surprise normally associated with a first strike. Furthermore both knew they could use information obtained from satellite reconnaissance to ensure parity in offensive nuclear forces. Similarly both used satellite reconnaissance to monitor compliance with strategic arms limitation or reduction agreements.

In the early twenty-first century, commercial satellites, like *Ikonos* and *QuickBird* operated by American companies, supplied imagery of virtually any place on Earth for a fee. Commercial satellites generally revealed terrestrial objects 2–3 ft long, which proved useful for many military and civilian purposes. The products of both military and commercial reconnaissance satellites were still not the animated films depicted in many spy movies. Sometimes referred to as "the poor man's reconnaissance," commercial imagery dramatically increased the power of a military force by allowing users to know what adversaries were doing.

Dwayne A. Day

See also: Commercial Remote Sensing, National Reconnaissance Office

Bibliography
Dwayne A. Day et al., eds., *Eye in the Sky: The Story of the Corona Spy Satellites* (1999).
Brian Harvey, *Russia in Space: The Failed Frontier?* (2001).
Jeffrey T. Richelson, *America's Secret Eyes in Space: The U.S. Keyhole Satellite Program* (1990).
L. Parker Temple, *Shades of Gray: National Security and the Evolution of Space Reconnaissance* (2005).

Arkon. *See* Russian Imaging Reconnaissance Satellites.

Baker-Nunn Camera System. *See* Space Surveillance.

Big Bird. *See* KH-9 Hexagon.

Blue Gemini

Blue Gemini is a name associated with various U.S. Air Force (USAF) plans to use Gemini space capsules for military purposes separate from NASA. In the autumn of 1961, Secretary of the Air Force Eugene Zuckert and Secretary of Defense Robert McNamara speculated on adapting Gemini or Dyna-Soar technology to transport USAF astronauts and equipment to an orbital test station for defense-related experiments. The USAF began planning in June 1962 for a Military Orbital Development System (MODS) that would rely on Gemini spacecraft as ferry vehicles.

To train astronauts for projected MODS missions, one proposal envisioned USAF pilots training on six NASA Gemini missions. Another allowed for independent USAF flights on capsules purchased from NASA. One idea, touted by Secretary McNamara, involved transferring Project Gemini management from NASA to the USAF. While expanded military involvement held the prospect of more funding for Gemini, NASA feared a USAF takeover might jeopardize its schedule for placing humans on the Moon. Conversely, senior USAF officers perceived Blue Gemini as a threat to the troubled Dyna-Soar project. Blue Gemini fizzled in January 1963, when the two sides agreed simply to USAF experiments aboard NASA Gemini flights.

Nevertheless, prospects for USAF Gemini flights increased in December 1963 when Secretary McNamara canceled Dyna-Soar and announced the Manned Orbiting Laboratory (MOL) program. A modified Gemini capsule—Gemini B—would ferry USAF astronauts to and from the MOL. Although NASA successfully flight-tested a robotic Gemini-B in November 1966, MOL's cancellation in June 1969 ended USAF Gemini plans.

James C. Mesco

See also: Dyna-Soar, Gemini, United States Air Force

Bibliography
Dwayne A. Day, "The Blue Gemini Blues," *Spaceflight* 49, no. 6 (June 2007): 226–34.
Mark Erickson, *Into the Unknown Together: The DoD, NASA, and Early Spaceflight* (2005).
Barton C. Hacker and James M. Greenwood, *On the Shoulders of Titans: A History of Project Gemini* (1977).

Corona

Corona, the first U.S. photoreconnaissance satellite system. During the series of 145 launches between August 1960 and May 1972, Corona satellites photographed vast portions of Earth's surface. Exposed film was returned to Earth in a reentry capsule, which descended by parachute for recovery by specially equipped U.S. Air Force (USAF) cargo planes. The United States used the imagery to track military targets and operations in communist-bloc nations and to understand Sino-Soviet strategic capabilities. Because the primary mission of the photoreconnaissance satellite system remained highly classified until the 1990s, government officials publicly presented the launch series as Discoverer—a cover involving biomedical experiments with mice and monkeys.

> ## CLARENCE L. BATTLE
> ## (1915–2002)
>
> Colonel Clarence "Lee" Battle accepted assignment to the U.S. Air Force Western Development Division in 1954. He became chief of the systems engineering division for Program WS-117L, the service's first satellite program. During 1958–63, he directed the Corona reconnaissance satellite development office, where he emphasized certain management principles, such as selecting a small group of good people, demanding high-quality performance with rigorous analysis and correction of failures, focusing on mission accomplishment, and avoiding busy work. Those principles, which fostered success in the Corona program, became known throughout the National Reconnaissance Office as "Battle's Laws."
>
> *Rick W. Sturdevant*

In 1958, President Dwight D. Eisenhower directed the Central Intelligence Agency (CIA) to develop a film-return reconnaissance satellite, because the USAF program WS-117L, which was to use a television camera and radio transmission, had fallen significantly behind schedule. Lockheed Corporation's satellite proposal basically rephrased earlier RAND Corporation work. After the CIA awarded Lockheed the contract, James W. Plummer established Lockheed's Advanced Projects Facility to spearhead the project. The contractor for the Corona camera was Itek, and Eastman Kodak provided the film. Using a proven Thor intermediate-range ballistic missile (IRBM) as a first stage and a Lockheed upper stage, the Agena, which was manufactured specifically for the Corona program, the Lockheed team went from go-ahead to first launch in nine months. USAF Colonel Charles Murphy, working for the CIA, determined that rather than taking pictures of particular things, the satellite should photograph wide areas, and interpreters could examine the images for details of specific sites. Using this search method, photo interpreters discovered that many targets were not where U.S. nuclear targeting planners thought.

The USAF first attempted to launch a Corona satellite on a Thor/Agena booster from Vandenberg Air Force Base, California, in January 1959. While still on the launch pad, the Agena upper stage malfunctioned when small, solid rockets prematurely fired. Engineers called this launch attempt *Discoverer 0*. The following month, *Discoverer 1*, carrying only a light engineering payload, became the world's first polar-orbiting satellite. In April 1959, *Discoverer 2* became the first satellite stabilized in orbit in all three axes, the first maneuvered on command from Earth, and the first to eject a reentry vehicle on command and send it back to Earth. The first successful capsule recovery occurred on the *Discoverer 13* mission in August 1960 with splashdown in the Pacific. A few days later, the *Discoverer 14* mission ended with the first midair recovery of a reentry vehicle from space when a USAF C-119 Flying Boxcar snagged the capsule over the Pacific. *Discoverer 14*, the first Corona mission to return photographs, provided more imagery of the Soviet Union than the 24 previous U-2 flights combined. Throughout the next decade, technological improvements enabled Corona missions to last longer and return more detailed imagery. Resolution improved from approximately 25 ft initially to 6 ft. Corona produced the first mapping of Earth from space and the first stereo-optical data from space.

As much as any system during the 1960s and early 1970s, Corona ensured U.S. national security and helped prevent a nuclear war between the world's superpowers. Among the significant contributions of the Corona program, the returned pictures proved that a "Missile Gap" did not exist but that a Chinese nuclear program did, and it revealed the existence of a Soviet program to compete against the United States in the race for the Moon. By providing the "national technical means," Corona helped verify nuclear arms control agreements between the United States and Soviet Union during the Cold War.

In February 1995, the National Reconnaissance Office, which was formed shortly after the *Discoverer 14* mission to manage Corona, declassified the program and 800,000 images. Transferred to the National Archives and Records Administration, those photographs added more than a decade of valuable information to the scientific database on environmental changes.

David C. Arnold

See also: Agena, Lockheed Corporation, National Reconnaissance Office, RAND Corporation, Delta, United States Air Force Satellite Control Facility

Bibliography

Dwayne A. Day et al., *Eye in the Sky: The Story of the Corona Spy Satellites* (1999).
Jonathan E. Lewis, *Spy Capitalism: ITEK and the CIA* (2002).
National Reconnaissance Office. www.nro.gov.
Curtis L. Peebles, *The Corona Project: America's First Spy Satellites* (1997).

Defense Meteorological Satellite Program

Defense Meteorological Satellite Program (DMSP), a U.S. Air Force (USAF) system that employed two or three Sun-synchronous satellites in polar, low Earth orbits to collect visible and infrared images of cloud cover, measure atmospheric moisture and temperature at various altitudes, and monitor plasma characteristics in the space environment.

During the early Corona missions in 1960–61, recovery of film canisters containing pictures of clouds substantiated RAND Corporation's warning from the early 1950s that successful satellite photoreconnaissance would depend on accurate, timely meteorological forecasts over the Eurasian landmass and elsewhere. Meanwhile, NASA had launched the first Radio Corporation of America (RCA)–built, experimental Tiros satellite in April 1960 and, with the Department of Commerce (DoC), had gained responsibility for developing a single National Operational Meteorological Satellite System to satisfy all civil and military requirements. Senior civilians and USAF officers in the Department of Defense (DoD), the newly created National Reconnaissance Office (NRO), and Strategic Air Command (SAC) doubted NASA's ability to deliver before 1963 a weather satellite capable of keeping its spin axis

An illustration of a Defense Meteorological Satellite Program (DMSP) Block 5 satellite. (Courtesy U.S. Air Force)

perpendicular to its orbital plane and, thereby, efficiently scanning Earth's surface. Engineers and high-level managers at RCA harbored similar doubts and, in November 1960, formally proposed to the USAF a separate, highly classified, Tiros-derived satellite system.

Joseph Charyk, under secretary of the USAF and head of the NRO, gained approval in July 1961 for a "minimum" program with a deadline of 10 months for launching the first satellite on a NASA Scout booster. Under its first director, Lieutenant Colonel Thomas Haig, who insisted that he handpick a small cadre of USAF officers and that his team retain responsibility for system engineering, the high-risk program proceeded under a succession of names—Program II, P-35, 698BH, 417, Defense Systems Applications Program, and ultimately DMSP. The USAF team conceived how to keep the satellite's spin axis perpendicular to its orbital plane by using a magnetic torque device, then used fixed-price, fixed-delivery contracts to ensure swift implementation by RCA and its subcontractors. Although the first DMSP Block 1 launch attempt in May 1962 failed, a second in August succeeded. Its imaging of weather in the Caribbean permitted effective aerial reconnaissance over Cuba during the October 1962 missile crisis. When Scout launchers proved unreliable in four of five attempts, Haig switched to Thor rockets and successfully placed four DMSP Block 1 satellites in orbit during 1964.

Meanwhile, SAC established a DMSP control center one floor below Global Weather Central at Offutt Air Force Base (AFB), Nebraska, with dedicated ground stations at Fairchild AFB, Washington, and Loring AFB, Maine. In June 1964, DMSP data on tropical storms began flowing to the U.S. Weather Bureau, because Tiros data had proven inadequate. By mid-1965 a specially designed DMSP satellite provided daily coverage over Southeast Asia to a tactical station at Tan Son Nhut Air Base near Saigon, Vietnam, thereby allowing cancellation of aerial weather reconnaissance and more precise planning of Vietnam War operations. Those impressive results prompted DoD officials to pursue DMSP as a highly classified, permanent system. In March 1973, Under Secretary of the Air Force and NRO Director John McLucas publicly announced DMSP's existence.

In the 1970s the RCA (subsequently General Electric and later Lockheed Martin) Astro Space–manufactured DMSP Block 5 spacecraft abandoned Tiros-derived technology. It used a three-axis-stabilized platform and a Westinghouse Operational Linescan System. This provided a nearly constant stream of higher-resolution visual and infrared imagery day and night. The National Oceanic and Atmospheric Administration (NOAA), at considerable cost savings, adopted a variant of the DMSP Block 5D military spacecraft for its Tiros-N civil series. When President Jimmy Carter directed in November 1979 that civil and military meteorological satellite programs would remain separate until the next block change, USAF officials refrained from designating a DMSP Block 6. Nonetheless, convergence of military and civil polar-orbiting meteorological satellite technologies periodically led to calls for merging management of the programs under NOAA.

Despite DMSP's crucial support to coalition combat air operations during the First Gulf War in 1991, post–Cold War budget constraints increased political support for the military-civilian merger concept. In May 1994 President William Clinton directed DoD and DoC to combine polar-orbiting weather capabilities. Transfer of DMSP from the USAF to NOAA effective 1 October 1998 marked the first time a U.S. civilian agency gained operational control of a military satellite system. Under the new arrangement, the USAF continued to acquire and launch DMSP satellites and to maintain backup operational capability. With its launch of updated satellites in December 1999 and October 2003, the USAF completed a DMSP Block 5D-3 constellation. Looking toward the future, an integrated program office staffed by NOAA, USAF, and NASA representatives sought to acquire the next-generation National Polar Orbiting Environmental Satellite System.

James C. Mesco

See also: National Polar Orbiting Environmental Satellite System, Tiros

Bibliography
Dwayne A. Day, "The Clouds Above, the Earth Below," *Spaceflight* 47, no. 8 (August 2005): 302–11.
R. Cargill Hall, "A History of the Military Polar Orbiting Meteorological Satellite Program," *Quest* 9, no. 2 (2002): 4–25.
National Geophysical Data Center. www.ngdc.noaa.gov.

Discoverer. *See* Corona.

Electronic Intelligence Ocean Reconnaissance Satellite. *See* Russian Naval Reconnaissance Satellites.

Gemini-B. *See* Blue Gemini.

Ground-Based Electro-Optical Deep Space Surveillance System. *See* Space Surveillance.

Helios

Helios, Europe's first space-based military reconnaissance system. Its name, derived from the mythological Greek Sun god, connoted the satellite's operation in Sun-synchronous orbit. At a summit meeting with Germany in 1984, France proposed a joint military reconnaissance satellite program. Although Germany decided against participating, Italy and Spain contributed to the largely French-financed $2 billion project. Construction began during February 1993 on *Helios 1A*, which resembled the *SPOT 4* (Satellite Pour l'Observation de la Terre) civil observation satellite, at the Matra Marconi Space facility in Toulouse, France. Weighing 2.5 tons and designed to operate five years, *Helios 1A* launched in July 1995 on an Ariane 4 rocket from Kourou, French Guiana, to provide 1 m resolution from an orbital altitude of approximately 700 km. The launch of *Helios 1B* occurred in early December 1999.

In April 1994 France had initiated development of an improved Helios version. Financed by France with small contributions from Belgium, Spain, and Greece and built by European Aeronautic and Defence and Space Company (EADS) Astrium, Helios 2 satellites carried two sensors: a medium-resolution instrument with a wide field of view in the visible and low-infrared spectrum, and an extremely high-resolution, infrared instrument. Compared to the dozens of images produced daily by the Helios 1 series, Helios 2 satellites delivered about 100 images of higher quality, in terms of resolution, contrast, and electronic noise. *Helios 2A* achieved orbit in December 2004, and *Helios 2B* achieved orbit in December 2009.

James C. Mesco

See also: Europe; European Aeronautic and Defence and Space Company (EADS); France; Matra Marconi Space; SPOT, Satellite Pour l'Observation de la Terre

Bibliography
Federation of American Scientists. www.fas.org.
Space.com. www.space.com.

Hexagon. *See* KH-9 Hexagon.

Kennan. *See* KH-11 Kennan.

KH-9 Hexagon

KH-9 Hexagon, successor to the U.S. Corona photoreconnaissance satellite. Like its predecessor, the KH-9 photographed large geographic areas, thereby allowing imagery analysts to search for new targets and count known ones, such as bombers at Soviet Union airfields and submarines in port. While other satellites with higher resolution occasionally focused on specific targets for technical assessment, the KH-9, operated by the National Reconnaissance Office, became the main intelligence workhorse from 1971 to 1986.

Comparable in size to a school bus, the Lockheed-built satellite had at its rear solar panels and a propulsion system for maintaining orbit. A camera system developed by Perkin-Elmer occupied the center portion, with four reentry vehicles at the front. Two powerful cameras exposed long strips of film 9 in wide that were stored in capsules for return to Earth, where C-130 aircraft captured them in midair. Camera resolution was probably about 1 m, approximately the same as commercial imagery satellites of the early twenty-first century. A dozen KH-9 missions carried mapping cameras and a fifth reentry vehicle. Several people involved with the program described the KH-9 as the most complicated mechanical system ever sent into space.

Because of the satellite's weight, launching a KH-9 required the largest space booster then in the U.S. military's inventory, the Titan III. Because of its size, U.S. Air Force crews that launched the KH-9 nicknamed it "Big Bird." Although early KH-9s remained in orbit for only 30–45 days, later ones stayed up as long as 275 days. A rocket explosion resulted in the premature loss of one KH-9, but 19 others successfully achieved orbit. In addition to their primary payload, most carried small subsatellites that were deployed in orbit to monitor foreign radar signals. By all accounts, the KH-9 was a successful system despite the complexity of its hardware.

Dwayne A. Day

See also: Lockheed Corporation, National Reconnaissance Office

Bibliography
Dwayne A. Day, "Pushing Iron: On-orbit Support for Heavy Intelligence Satellites," *Spaceflight* 46, no. 7 (July 2004): 289–93.
Jeffrey T. Richelson, *America's Secret Eyes in Space: The U.S. Keyhole Spy Satellite Program* (1990).

KH-11 Kennan

KH-11 Kennan, a series of high-resolution, electro-optical, photoreconnaissance satellites advocated by the Central Intelligence Agency's Directorate of Science and Technology, developed by TRW Corporation, and operated by the National Reconnaissance Office. Originally dubbed Program 1010 and first launched on 19 December 1976, the KH-11 used a photo-diode system to convert light into electrical impulses for immediate transmission of images to ground stations. After the first two flights, however, Kennan satellites employed a true charge-coupled device (CCD) array. Earlier imaging satellites like Corona relied on cameras, lenses, and film-return technology, but Kennan and its successors used telescopic instruments with mirrors and onboard digital processing. The KH-11 measured 64 ft in length and 10 ft in diameter, weighed 30,000 lb, and used a 92 in mirror to focus light on its CCD sensors. Launched from Vandenberg Air Force Base, California, into Sun-synchronous, polar orbits, Kennan satellites supposedly could view objects on the ground as small as 6 in. The last of nine Kennan satellites went into orbit on 6 November 1988, and at least one remained in operation through October 1995. Experts believe that KH-11s supplied the first evidence of the Iraqi invasion of Kuwait in 1990. Around that time, KH-12 "Improved Crystal" imagery satellites probably began to replace the Kennan. Considering the size, shape, and contractors involved, some observers suggest that NASA derived the *Hubble Space Telescope* from the KH-11.

Eric Thorson

See also: *Hubble Space Telescope*, National Reconnaissance Office, TRW Corporation

Bibliography
William E. Burrows, *Deep Black: Space Espionage and National Security* (1986).
Jeffrey T. Richelson, *America's Secret Eyes in Space: The U.S. Keyhole Spy Satellite Program* (1990).

Lacrosse

Lacrosse, an imaging reconnaissance satellite that used radar to penetrate darkness, smoke, and cloud cover to see objects on the ground and at sea. The United States deployed its first radar satellite, known as *Quill*, in December 1964. Little is publicly known about *Quill* other than that a Thor-Agena rocket placed it in orbit, and the satellite recorded its data on magnetic tape that was returned to Earth in a recoverable capsule developed for the Corona program. Although the *Quill* satellite worked, its limited utility precluded a second launch.

The U.S. Air Force began lobbying for a new imaging radar satellite during the 1970s. In January 1982 the successful test of a prototype resulted in a controversial decision to develop an operational system called Lacrosse. Built by Martin Marietta (later Lockheed Martin) and operated by the National Reconnaissance Office, Lacrosse relied on synthetic aperture radar technology to beam microwave energy to Earth and detect return signals reflected to space. The first Lacrosse satellite was launched aboard a Space Shuttle in December 1988, but others went into orbit on Titan 4 boosters.

A major weakness of Lacrosse, later named Onyx, was that it apparently could see only objects larger than approximately 3 ft long. As of April 2005 only five Lacrosse/Onyx satellites had been launched. A replacement originally was planned as part of the Future Imagery Architecture (FIA) class of satellites. In 2005, however, the radar component of the FIA apparently was being merged with the Space Based Radar program into a new system, renamed Space Radar and scheduled for operation after 2010.

Dwayne A. Day

See also: National Reconnaissance Office

Bibliography
Dwayne A. Day, "Early American Ferret and Radar Satellites," *Spaceflight* 43, no. 7 (July 2001): 288–93.

Manned Orbiting Laboratory

Manned Orbiting Laboratory (MOL), the second major United States Air Force (USAF) human spaceflight program. On 10 December 1963, the same day Defense Secretary Robert S. McNamara canceled the Dyna-Soar program, he authorized the USAF to develop a space station—the MOL. Plans called for two military astronauts in a Gemini-B or "Blue Gemini" capsule attached to the MOL—essentially a pressurized Titan IIIC upper-stage fuel tank—to enter a polar, sub-synchronous orbit, perform experiments for a month in the MOL, then abandon it and return to Earth aboard the Gemini-B.

Despite USAF studies delineating specific military requirements for various types of space-based, human reconnaissance and satellite inspection, the Department of Defense (DoD) waged a battle from the beginning to prove the utility of crewed military space operations. Meanwhile USAF planners added experiments P-14 (a sensitive radar antenna) and P-15 (high-resolution optics), thereby transforming MOL specifically into a crewed reconnaissance platform rather than a generic platform for determining the usefulness of military astronauts. That change in emphasis gave MOL a new lease on life, because politicians envisioned MOL's capacity to unite human observation and

Early concept drawing of the Manned Orbiting Laboratory, 1960. (Courtesy NASA)

reconnaissance satellite technology as enhancing U.S. knowledge about Soviet nuclear forces. Not to be outdone, the Soviet Union responded to MOL development by initiating its Almaz program in 1965.

Following the success of several DoD crewed reconnaissance experiments on two NASA Gemini missions, President Lyndon B. Johnson announced approval on 25 August 1965 of $1.5 billion for MOL development. Despite increasing USAF-NASA collaboration on MOL and launch of a mock-up on a 33-minute suborbital flight in November 1966, significant project delays and cost increases haunted the endeavor after early 1967. Other circumstances affected MOL's continuation: congressional critics thought MOL duplicated efforts in NASA's Apollo Applications (Skylab) program; McNamara opposed USAF efforts to integrate any crewed space program into national defense policy; and the Vietnam War siphoned funds from USAF and NASA space programs. Consequently on 10 June 1969, President Richard M. Nixon canceled MOL in favor of a new generation of automated reconnaissance satellites, reputedly the KH-9 Hexagon. The MOL space-suit development and dietary contracts transferred to NASA, as did seven of the 14 remaining MOL astronauts.

Roy F. Houchin

See also: Almaz, Dyna-Soar, *Skylab*, Space Stations

Bibliography

Roy F. Houchin, "Interagency Rivalry: NASA, the Air Force, and MOL," *Quest* 4, no. 4 (Winter 1995): 36–39.

Donald Pealer, "A History of the Manned Orbiting Laboratory (MOL) Part 1," *Quest* 4, no. 3 (Fall 1995): 4–16.

———, "Manned Orbiting Laboratory (MOL) Part 2," *Quest* 4, no. 4 (Winter 1995): 28–35.

———, "Manned Orbiting Laboratory (MOL) Part 3," *Quest* 5, no. 2 (Summer 1996): 16–23.

Military Orbital Development System. *See* Blue Gemini and Manned Orbiting Laboratory.

Minitrack. *See* Space Surveillance.

Nuclear Detection

Nuclear detection, the capability to watch for nuclear detonations by potential adversaries and to determine their location and yield, concerned the U.S. military since it first used atomic bombs against Japan late in World War II. The U.S. Air Force (USAF) first deployed a modified RB-29 bomber in 1947 to monitor Soviet Union efforts to develop a nuclear bomb. On 3 September 1949, a specially equipped RB-29 detected debris from the first Soviet atomic bomb detonation five days earlier. During 1950–53 the U.S. military had begun installation of seismic stations around the globe to augment airborne sampling. By 1958 the U.S. Army Signal Corps had established 11 stations worldwide to monitor acoustic signals in the atmosphere, and the

U.S. Navy had equipment on ships and land bases to collect radioactive particles falling from clouds.

In 1959, the Advanced Research Projects Agency and USAF undertook Project Vela to develop other ways for monitoring nuclear detonations in near-real time. The project included three separate detection systems: Vela Uniform to detect underground nuclear tests by measuring seismic signals, Vela Sierra to detect atmospheric nuclear tests, and Vela Hotel to detect nuclear tests in space. The United States used Project Vela detection systems to monitor compliance with the 1963 Treaty Banning Nuclear Weapon Tests in the Atmosphere, in Outer Space and Under Water, often shortened to the Limited Test Ban Treaty.

The USAF launched 12 Vela satellites—six Vela Hotel and six Advanced Vela—in pairs during 1963–70. Manufactured by TRW Corporation, all the Vela satellites orbited at altitudes between 63,000 and 70,000 miles to ensure they were above the Van Allen radiation belts. Launched during 1963–65, Vela Hotel satellites carried X-ray, neutron, and gamma-ray detectors. Analysis of data collected by *Vela 5* led Ray Klebasabel, a scientist at Los Alamos National Laboratory, to discover serendipitously during the late 1960s that gamma-ray bursts occurred as a natural, cosmic phenomenon. Advanced Vela satellites, launched during 1967–70, carried silicon photodiode sensors called bhangmeters that added the capability to detect atmospheric nuclear explosions. The Vela mission continued until 1984, when *Vela 9* ceased operations after 15 years.

Meanwhile, during the 1970s, Defense Support Program (DSP) early-warning satellites had begun to provide nuclear detection as a secondary function. Together with sensors on Global Positioning System satellites in the early 1980s, this became known as the Integrated Operational Nuclear Detection System (IONDS). By the late 1980s, IONDS became the U.S. Nuclear Detonation Detection System. At the beginning of the twenty-first century, anticipating the end of the DSP satellite series and its replacement in geostationary orbit by Space-Based Infrared System (SBIRS) satellites, military planners began developing high-capability, lightweight Space and Atmospheric Burst Reporting System sensors for SBIRS.

Robert A. Kilgo

See also: Defense Support Program, TRW Corporation

Bibliography
Air University Center for Space Studies. space.au.af.mil.
Robert S. Norris and William M. Arkin, "Soviet Nuclear Testing, August 29, 1949–October 24, 1990," *Bulletin of the Atomic Scientists* 54, no. 3 (May–June 1998): 69–71.

Ofeq. *See* Israel.

Onyx. *See* Lacrosse.

Orlets. *See* Russian Imaging Reconnaissance Satellites.

Quill. *See* Lacrosse.

Radar Ocean Reconnaissance Satellite. *See* Russian Naval Reconnaissance Satellites.

Russian Imaging Reconnaissance Satellites

Russian imaging reconnaissance satellites, nearly 800 launched since 1962, were categorized by U.S. and Western European analysts in terms of generations (first, second, third). The Soviet Union and Russia, instead, applied series names (Zenit, Yantar, Terilen/Neman, Orlets, and Arkon). Development of Zenit 2, the first generation of Soviet photoreconnaissance satellites, began under Chief Designer Sergei Korolev in 1956–57, with the first launch in April 1962 and regular use until 1970. A Zenit 2 carried four cameras that could look straight down or obliquely at Earth's surface, and each camera could take up to 1,500 black-and-white photographs with 7 m resolution over broad areas. After approximately eight days, a spherical module containing the reusable cameras and exposed film returned to Earth via parachute.

Even as the Soviet Union endeavored to improve its space-based, robotic imaging capabilities during the mid-1960s, a branch of Korolev's OKB-1 (Experimental Design Bureau) headed by Dmitri Kozlov and Vladimir Chelomey's OKB-52 competing bureau sought to develop military, crewed space stations for reconnaissance. A number of Soyuz-based plans, originally based on modernization of the one-person Vostok spacecraft, never came to fruition. Chelomey's Almaz reconnaissance platform (publicly announced as Salyut), when used with Korolev's Soyuz spacecraft, allowed military cosmonauts to experiment with cameras and ejection of film-return capsules on two flights in the mid-1970s. While the experiments showed the value of looking at the same location in many different spectral bands, Soviet defense officials found human reconnaissance from space too expensive and, consequently, decided to rely on robotic satellites.

The second generation of Soviet photoreconnaissance satellites included the Zenit 4 and Zenit 2M (Gektor) series for high-resolution and broad-area imagery, respectively. With a normal flight duration equivalent to Zenit 2, the Zenit 4 carried two cameras and augmented Zenit 2 imagery from November 1963 to August 1970. Carrying a more advanced camera system than Zenit 2 and operating up to16 days, the Zenit 2M series first launched in March 1968 and ran through March 1979.

A third generation of Zenit film-return satellites with still more advanced optics extended from 1968 to 1994. The Zenit 4M (Rotor), 4MK (Germes), and 4MKM (Gerakl) performed high-resolution photography during 1968–74, 1969–77, and 1977–80, respectively. A medium-resolution 4MKT (Fram) and a topographic-mapping 4MT (Orion) operated during 1974–85 and 1971–82, respectively. The Soviet military used the last two Zenit varieties, the broad-area 6U (Argon) and the geodetic/cartographic 8 (Oblik) during 1976–84 and 1984–94, respectively.

Yantar, the fourth and fifth generations of Soviet military imaging satellites and the first with designs substantially different from Zenit, debuted in 1974. The high-resolution, maneuverable Yantar 2K (Feniks) model, which the Soviet Union first

DMITRI ILICH KOZLOV
(1919–2009)

A lead designer of the nuclear-tipped R-5 and R-7 ballistic missiles at OKB-1 (Experimental Design Bureau) during the 1950s, Dmitri Kozlov headed the former OKB-1 Branch 3 (later known as TsSKB Progress) from 1961 to 2003, with responsibility for designing all derivatives of the R-7 for space launch. In the 1960s–90s, Kozlov initiated development of a unique military piloted reconnaissance spacecraft, Zvezda, and directed the design of Zenit and Yantar military photoreconnaissance satellites. During the 1970s–80s, he ventured into development of Soviet Union civil imaging and scientific satellites, including Resurs-F, Bion, and Foton.

Peter A. Gorin

launched successfully in December 1974, carried two detachable, successively releasable film canisters in addition to its main camera module that returned at the completion of the satellite's 30-day mission. After a 28th successful launch in June 1983, the Yantar 2K was replaced by an almost identical twin model, the Yantar 4K1 (Oktant), which had been tested on orbit in April 1979 and first launched operationally in June 1982. By the end of 1983, however, the Oktant yielded to the 4K2 (Kobalt). First launched in August 1981, a Kobalt satellite carried as many as 20 film-return capsules and an operational life of approximately two months. By 1998, the operational life of the Kobalt had doubled and by the 82nd Kobalt launch in September 2004 reportedly had tripled. In addition to the high-resolution Yantar variants, the Soviet military began launching the Yantar 1KFT (Kometa/Siluet) in February 1981 for high-precision topographic surveys. Russians continued launching Kometa satellites, with the 21st entering low Earth orbit in September 2005.

With the initial launch of a Yantar 4KS1 (Terilen) at the end of December 1982, the fifth generation of Soviet imaging reconnaissance satellites entered the scene, being accepted operationally in 1985. The first in a series of electro-optical satellites, Terilen allowed digital transmission of visual and infrared imagery, with 2-m resolution, to Earth via Potok (Geizer) communications-relay satellites. Although a more capable 4KS2 model had been planned, the Soviet Union canceled it in June 1983 in favor of a modernized 4KS1 that would become Yantar 4KS1M (Neman) and replace Terilen in 1991. Neman extended the satellite's operational lifetime from Terilen's maximum of 207 days to 238–259 days. In 1991 the cash-strapped Russian military extended Neman missions beyond one year.

Sixth- and seventh-generation, high-resolution, broad-area photoreconnaissance satellites, which employed a Yantar-like bus with film-return photographic technology, comprised a series called Orlets. Seven Orlets 1 (Don) satellites, each weighing 6,500 kg with a non-returnable camera module and carrying eight film-return capsules, orbited between July 1989 and August 2003 with an average service life of two to three months. In August 1994 after the launch of several experimental mockups in 1986–87, the Soviet Union launched an Orlets 2 (Yenisey) version that weighed nearly twice as much as the Don, operated three times longer, and carried 22 film-return capsules.

In June 1997, Russia launched its first Araks (Arkon) military imaging reconnaissance satellite. A high-altitude, electro-optical platform, the Arkon could transmit digital images in real time or store them for later downlink. The satellite could capture up to 15 frames of the same area per pass with a resolution of 2–10 m and with a choice of eight bands ranging from optical to infrared. A second Arkon, launched in July 2002, operated more than one year but short of its design life, which some experts believed was three years.

Observers expected Russia's Monitor-E remote sensing satellites, developed primarily for civil purposes and first launched in August 2005, would have military applications. Weighing only 750 kg, these lightweight satellites could be manufactured and launched for considerably less cost than the older, heavier series but could provide equivalent or better performance. Engineers anticipated improving resolution of the color or black-and-white imagery from 8 m to less than 1 m and hoped to extend the operational lifetime to five years.

Rick W. Sturdevant

See also: Almaz, Russia (Formerly the Soviet Union)

Bibliography
Phillip S. Clark, "Russian Fifth Generation Photoreconnaissance Satellites," *Journal of the British Interplanetary Society* 52, no. 4 (April 1999): 133–50.
Peter A. Gorin, "Black 'Amber': Russian Yantar-Class Optical Reconnaissance Satellites," *Journal of the British Interplanetary Society* 51, no. 8 (August 1998): 309–20.
Brian Harvey, *Russia in Space: The Failed Frontier?* (2001).
Konstantin Lantratov, "Soyuz-Based Manned Reconnaissance Spacecraft," *Quest* 6, no. 1 (Spring 1998): 5–21.

Russian Naval Reconnaissance Satellites

Russian naval reconnaissance satellites, two types—Radar Ocean Reconnaissance Satellite (RORSAT) and Electronic Intelligence (ELINT) Ocean Reconnaissance Satellite (EORSAT)—developed by the Soviet Union during the Cold War to locate and identify potentially hostile naval vessels. During the late 1950s Vladimir Chelomey's OKB-52 (Experimental Design Bureau) developed a new cruise missile capable of striking enemy ships beyond the range of Soviet shipboard radars. To fully exploit that missile's capability, the Soviet navy worked in 1959–60 with OKB-52 and Chief Designer Aleksandr Raspeltin's KB-1 bureau to define requirements for radar and ELINT ocean reconnaissance satellites that could pinpoint enemy ships at greater distances. Official authorization to proceed with the space-based system came in two government decrees during 1960–61. Dubbed MKRTs (Naval Space Reconnaissance and Targeting System), it included two subsystems: US-A (Upravlayemyi Sputnik–Aktivniy or Controlled Satellite–Active) and US-P (Upravlayemyi Sputnik–Passivniy

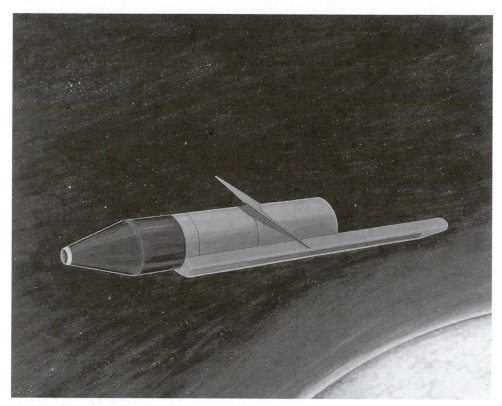

Illustration of a Soviet Radar Ocean Reconnaissance Satellite (RORSAT). (Courtesy Defense Intelligence Agency)

or Controlled Satellite–Passive), which U.S. analysts named RORSAT and EORSAT, respectively.

RORSAT utilized nuclear energy to satisfy the radar's demand for large, sustained amounts of power. After several launches in the mid-1960s of RORSAT prototypes with chemical batteries instead of nuclear reactors, the Soviet Union began testing reactor-equipped US-A satellites in October 1970. Initially, immediate on-orbit failure or an operational lifespan of a few days characterized RORSAT performance. Finally, a RORSAT launched in December 1973 became the first of its kind to operate for its full design life. After 44 days, its nuclear reactor separated and boosted itself to a high, storage orbit to avoid radioactive contamination of Earth's surface when the rest of the spacecraft deorbited and reentered Earth's atmosphere. Not until May 1974 did the first pair of US-A satellites enter Earth orbit only two days apart to pass successively over the same area, as prescribed in operational plans.

Although the Soviet government officially declared RORSAT operational in October 1975, design problems and equipment failures continued to frustrate Soviet engineers and diminish the system's reliability. Several accidents involving explosion or deorbiting of the nuclear reactor embarrassed Soviet officials and threatened the terrestrial environment. One especially serious incident occurred when the control system on

Kosmos 954, a RORSAT launched in September 1977, failed at the end of the satellite's operational life to boost the reactor into a higher orbit before the spacecraft naturally deorbited. On 24 January 1978 radioactive debris from *Kosmos 954* spread over a large area in northern Canada, sparked international protests, and necessitated a major cleanup effort. The incident caused a two-year hiatus in RORSAT launches while Anatoliy Savin's engineers at TsNII (Central Scientific-Research Institute) Kometa redesigned the safety system to minimize the recurrence of such an accident. Apparent problems with the improved safety system on *Kosmos 1900* in September 1988, however, generated international pressure that led to the program's cancellation.

Meanwhile, EORSAT or US-P development had taken longer, the initial launch of an apparently inert mockup not occurring until December 1974. Actual on-orbit testing probably began with the second EORSAT launch in October 1975. After termination of the RORSAT program in 1988, the Soviet navy increased its reliance on EORSAT and established a four-satellite constellation of the latter type for the first time, briefly expanding it to six satellites in 1990. On average each EORSAT operated more than one year and occasionally as long as two years. During the late 1980s the Soviet Union began introducing US-PM and US-PU, two modernized versions of EORSAT. The technical differences between those newer versions remained unclear to analysts outside Russia, but those two became the mainstay for Russian naval reconnaissance since at least 1993. In May 2004 one such version launched from Baikonur Cosmodrome to replace another, which had disintegrated on orbit a few months earlier.

Rick W. Sturdevant

See also: Electronic Intelligence, United States Navy

Bibliography

Brian Harvey, *Russia in Space: The Failed Frontier?* (2001).
Leo Heaps, *Operation Morning Light: Terror in Our Skies—The True Story of* Cosmos 954 (1978).
Fritz Muse, "RORSATS: The Veiled Threat," *Journal of the British Interplanetary Society* 56, Supplement 1 (2003): 42–49.
Asif A. Siddiqi, "Staring at the Sea: The Soviet RORSAT and EORSAT Programmes," *Journal of the British Interplanetary Society* 52, no. 11/12 (November–December 1999): 397–416.

Samos

Samos, a U.S. Air Force (USAF) satellite program, still partially classified in 2007, which concentrated initially on radio relay of scanned photographs from Earth orbit but later used by the National Reconnaissance Office to test film-recovery techniques and electronic-intelligence payloads. The program began in 1956 as part of Weapon System 117L, a broadly conceived effort to develop a reconnaissance satellite. It became Sentry when the USAF divided WS-117L into three separate projects in 1958 and, for unknown reasons, was renamed Samos in August 1959. Lockheed became the Samos prime contractor. For photoreconnaissance, Samos satellites carried one of four payloads: E-1 or E-2 frame-readout cameras and E-5 or E-6 recoverable panoramic cameras, all of which differed in focal length, ground resolution, and coverage. Two of the first three Samos missions, all of which carried readout cameras, suffered launch failures; only *Samos 2*, launched on 31 January 1961, achieved orbit and delivered poor results. Another eight Samos missions during 1961–62 carried film-return capsules, with disappointing results, which led swiftly to the program's cancellation.

The legacy of Samos, beyond its dismal performance, remains somewhat mysterious. Following the failed recovery of the *Samos 6* capsule in March 1962, the remaining four E-5 cameras went to a warehouse for later use on KH-6 Lanyard mapping missions. Before the "Blue Gemini" project, Lockheed even designed a "life cell" within the Samos E-5 capsule to explore the military role of human spaceflight. Finally, the Samos E-2 film-readout system became the basis for NASA's Lunar Orbiter camera, which returned imagery of the Moon's surface on five flights during 1966–67, thereby aiding selection of landing sites for Project Apollo.

James C. Mesco

See also: Lockheed Corporation, Lunar Orbiter

Bibliography

Dwayne A. Day, "From above the Iron Curtain to around the Moon," *Spaceflight* 47, no. 2 (February 2005): 66–71.

———, "The Samos E-5 Recoverable Satellite," *Spaceflight* 44, no. 10 (October 2002): 424–31; 45, no. 2 (February 2003): 71–79; and 45, no. 9 (September 2003): 380–89.

SAR-Lupe

SAR-Lupe (Synthetic Aperture Radar and Lupe, German for magnifying glass), Germany's first reconnaissance satellite system, used X-band radar technology to acquire extremely high-resolution images day or night in all types of weather. The Federal Office of Defense Technology and Procurement contracted in 2001 with a consortium of European space companies, led by OHB-System AG of Bremen, for five identical, 770 kg satellites and a ground station at Gelsdorf near Bonn. Operating at an altitude of 500 km in three near-polar orbital planes, the satellites could provide Earth-coverage between 80° N and 80° S latitude and could deliver more than 30 images daily in stripmap or spotlight modes. A unique ground test of SAR-Lupe capability occurred in October 2004 when the satellite, mounted inside a radome, produced a high-quality image of the *International Space Station*.

Originally, plans called for launching *SAR-Lupe 1* in mid-2005, but concerns about the satellite's security system delayed the schedule. After completing extensive tests at the Industrieanlagen-Betriebsgesellschaft mbH (IABG) Ottobrunn center near Munich, *SAR-Lupe 1* was transported to Plesetsk and launched successfully onboard a Russian Cosmos 3M rocket on 19 December 2006. The remaining four SAR-Lupe satellites were launched at six-month intervals between July 2007 and July 2008. Meanwhile, on 30 July 2002, Germany and France had signed an agreement to operate

SAR-Lupe and Helios satellites cooperatively within a European Union reconnaissance system that other member states could join.

Rick W. Sturdevant

See also: Germany

Bibliography

OHB-System. www.ohb-system.de.

Chris Pocock, "Space Radar: Germany's SAR-Lupe Constellation Puts Europe Ahead," *C⁴ISR Journal* 5, no. 10 (November/December 2006): 34–37.

Space Detection and Tracking System. *See* Space Surveillance.

Space Surveillance

Space surveillance, the close or continued observation by any means—radar, optical, passive receiver—of artificial or naturally occurring objects, generally in Earth orbit, in order to accrue information or take action when warranted. Several nations, especially the United States and the Soviet Union and Russia, have conducted space surveillance since 1957. Tracking and analysis of satellite orbits provides scientific data on upper-atmosphere density and winds and Earth's gravitational field. That information is useful in predicting satellite decays to avoid them being mistaken for incoming missiles or to warn populated areas of potential impact. As the quantity of active satellites and artificial orbital debris increased over time, tracking became critical to avoid potentially catastrophic collisions in space. Surveillance sensors also collect information about the size, shape, structure, composition, operation, and purpose of foreign satellites that might pose a threat to national security.

The U.S. Space Surveillance Network (SSN) originated in preparation for the International Geophysical Year in 1957, when both the United States and the Soviet Union stated they would launch satellites. Minitrack, a system designed by the Naval Research Laboratory to track its Vanguard satellite, consisted of a worldwide network of radio receivers that enabled analysts to calculate the position of the satellite through triangulation. Concurrently the Smithsonian Astrophysical Observatory established 12 sites worldwide from which large, specially designed Baker-Nunn cameras could photograph orbiting satellites. On 17 October 1957 a Baker-Nunn camera in Pasadena, California, filmed the orbiting rocket body of *Sputnik*. Meanwhile, on 5 October, the Massachusetts Institute of Technology Lincoln Laboratory's Millstone Hill radar at Westford, Massachusetts, had become operational and subsequently proved valuable for identifying small satellites at long range. Within three years, the U.S. Air Force (USAF) began acquiring Baker-Nunn cameras for military space surveillance. In early November 1957 under Project Harvest Moon (later SPACETRACK), Air Force Cambridge Research Center at Bedford, Massachusetts, began operating a primitive filter center to collect all available orbital data on *Sputnik* and predict the satellite's behavior. On 19 December 1958 under Project Shepherd, that filter center became the Interim National Space Surveillance and Control Center until replaced by a new

Space Defense Center inside the Cheyenne Mountain Complex near Colorado Springs, Colorado, 1973. (Courtesy U.S. Air Force)

center at Hanscom Field in January 1960. By mid-1961 that center's space surveillance functions had transferred to USAF facilities in Colorado Springs, Colorado, where they remained in the early twenty-first century.

The Advanced Research Projects Agency (ARPA), created by the U.S. Department of Defense in February 1958, proved instrumental in promoting the integration of various planned and existing military tracking efforts. Under the 496L program in late 1959, ARPA authorized construction of a Navy Space Surveillance (NAVSPASUR) fence in addition to the FPS-85 phased array radar at Eglin Air Force Base (AFB), Florida, to increase U.S. tracking capability. ARPA also built the Space Detection and Tracking System (SPADATS), an advanced (for its time) computer system to process and integrate information from tracking sites around the world. ARPA assigned the USAF the task of using SPADATS to develop and maintain a catalog of all Earth-orbiting objects. Installed initially at Ent AFB in Colorado, SPADATS became operational on 1 July 1961. In late 1965 it moved into the nearby North American Air Defense Command (NORAD) Cheyenne Mountain Complex, where the system regularly received data from approximately 600 sensors of various kinds.

During the next four decades the SSN incorporated new radar, optical, and passive sensor systems and deactivated others around the globe. During the 1980s the USAF fielded PAVE PAWS solid-state phased array radars and the Ground-based

Electro-Optical Deep Space Surveillance (GEODSS) system. Several additional systems became operational in the 1990s: improvements at the Maui Optical Tracking and Identification Facility (MOTIF) in Hawaii; a Transportable Optical System (TOS); and the AN/FPS-129 HAVE STARE or Globus II X-band, mechanical radar at Vardo, Norway. By the late 1990s the SSN had its first space-based system—the Ballistic Missile Defense Organization's *Midcourse Space Experiment* (*MSX*) satellite, with its Space-Based Visible (SBV) sensor. The network catalog listed more than 26,000 artificial objects, many of which already had reentered the atmosphere, but the SSN continued to track nearly 9,000 objects, approximately 20 percent of which were functioning payloads or spacecraft. Under the right conditions, it could track objects as small as 1 cm at 1,000 km. Among other significant accomplishments throughout the years, the SSN ensured collision avoidance for crews aboard the Space Shuttle, *Mir*, and the *International Space Station*.

A network of large, ground-based radars formed the backbone of the Soviet Space Surveillance System (SSS). The Dnepr "Hen House" radar system developed in the 1960s was augmented by the Daryal-UM Large Phased Array Radar (LPAR) system in the 1980s. By the late 1990s, following the collapse of the Soviet Union, only nine radar facilities remained intermittently operational due to funding difficulties. To augment the radar information, Russia received data from optical and electro-optical sites in its homeland and in Kazakhstan, Tajikistan, Ukraine, Georgia, Armenia, and Turkmenistan.

Brad M. Evans

See also: Advanced Research Projects Agency, Ballistic Missile Early Warning System, International Geophysical Year, North American Aerospace Defense Command

Bibliography
Global Security. www.globalsecurity.org.
Michael J. Muolo, *Space Handbook: A War Fighter's Guide to Space* (1993).
Eloise Paananen and Kenneth H. Drummond, *Sky Rangers: Satellite Tracking around the World* (1965).
Andrew Wilson, ed., *Jane's Space Directory 1993–1994* (1993).

Terilen. *See* Russian Imaging Reconnaissance Satellites.

Vela Satellite. *See* Nuclear Detection.

Weapon System 117L. *See* Corona, Missile Defense Alarm System, and Samos.

Yantar. *See* Russian Imaging Reconnaissance Satellites.

Zenit. *See* Russian Imaging Reconnaissance Satellites.

Space and Society

Space and society have been intertwined since humans first began speculating about the nature of the night sky. Ideas about the cosmos necessarily grew from the knowledge and beliefs of individuals, who in turn drew from the societies in which they lived. In the earliest societies, people used the movement of stars to mark calendars and associated this movement with astrological or religious portents. Ancient Greeks were perhaps the first people to attempt natural as opposed to supernatural explanations for the world around them, and they developed physical and mathematical explanations for the heavens as well. Centuries later, medieval Europeans tied Greek physical explanations to a Christian cosmology, which both encouraged and inhibited deeper philosophical and natural inquiry. By the 1600s, empirical observations and theoretical advances in astronomy and physics helped to lead the Scientific Revolution. The new "mechanical philosophy" inspired changes to older institutions, such as governments, universities, and companies, and spurred the creation of learned societies to support the development of new sciences and technologies. These included scientific knowledge in fields such as ballistics, chemistry, and electromagnetism, and new or improved technologies, such as rockets, petrochemicals, and radios.

The impact of space on society, and vice versa, has been evident in media and popular culture. From the literary genre of science fiction, which dates to at least the mid-1800s with the publication of Mary Shelley's *Frankenstein*, to the speculative nonfiction of the twentieth century, there has been an evolution of popular visions of spaceflight. Writers such as Jules Verne extrapolated from then-current developments in science and technology to imagine the future, including stories about human travels into space. Verne's stories in particular were often cited as inspiration by young people, some of whom later made their mark in science and technology. Speculative nonfiction was the product of spaceflight visionaries and prophets who worked to turn their dreams into a reality and established a literary foundation for the space age. By the nineteenth and early twentieth centuries, Konstantin Tsiolkovsky in Russia, Robert Goddard in the United States, and Hermann Oberth in Germany began to develop the theoretical and experimental foundations of spaceflight that would eventually turn fiction into fact. This was the golden age of rocketry in which youthful groups of like-minded individuals began to form amateur societies, beginning in the 1920s, that provided the nucleus of what later became a major industrial effort in support of national interests following World War II. These societies were concentrated in the United States and Europe and included the German VfR (Verein für Raumschiffahrt or

Dr. Wernher Von Braun with Walt Disney during a visit to NASA's Marshall Space Flight Center in 1954. (Courtesy NASA/ Marshall Space Flight Center)

Society for Spaceship Travel), 1927; the American Interplanetary Society, 1930; and the British Interplanetary Society (BIS), 1933. There were similar groups in the Soviet Union, including the Society for the Study of Interplanetary Communications, 1924, and the Group for Investigation of Reactive Motion (GIRD), 1931. These groups created publications and held lectures to popularize rocketry and spaceflight and (with the exception of BIS) began to experiment with rockets.

Individuals like Goddard and amateur groups would have remained marginal but for the fact that rocket technologies had potential military applications. Military officials in several nations observed rocket tests and stunts with amusement and then with growing interest. Goddard began to receive private funding in 1929 for his rocket experiments, but the U.S. military did not initially take an interest in rocket work, despite interest from its military counterparts in other nations. The German Army recruited Wernher von Braun, who began his work on army missiles in December 1932. Over the next few years, the German Army either recruited other rocket enthusiasts or took steps to keep them out of rocketry. The Soviet Union brought GIRD into the government fold in 1933. Military interest also had negative effects. In both fascist Germany and the communist Soviet Union, this effectively marginalized the remaining amateur groups that could not compete with government funding and were actively suppressed in the interest of secrecy. In Western democracies, early amateur rocket and space groups either evolved into professional organizations supporting the newly forming rocket and space industry or went out of existence.

Both the Soviet Union and Germany began serious development of large-scale rocketry in the 1930s, but Soviet efforts were derailed in the Stalinist purges starting in 1937. Marshal Mikhail Tukhachevsky, in whose organization the Soviet rocketeers existed, was executed and all people and organizations associated with him were

JULES VERNE (1828–1905)

(Courtesy Library of Congress)

A nineteenth-century French author, Verne described many future discoveries and inventions, such as submarines, helicopters, guided missiles, space travel, and other technologies in his scientific adventure novels. In 1863 he wrote *Cinq Semaines en Ballon* (*Five Weeks in a Balloon*), which was his first book to appear in English (1869). He continued with *A Journey to the Centre of the Earth* (1872), *From the Earth to the Moon Direct, in 97 Hours 20 Minutes: And a Trip Round It* (1873), and other novels with a scientific basis and remarkably accurate forecasts of the scientific achievements of the twentieth century. Verne's visionary stories inspired early rocket pioneers, including Konstantin Tsiolkovsky, Robert H. Goddard, and Hermann Oberth.

Pablo de León

immediately under suspicion. Sergei Korolev, Valentin Glushko and other rocket experts were arrested and imprisoned, remaining imprisoned until late in World War II. German success in developing the V-2 intermediate range ballistic missile impressed military establishments around the world, and by the end of the war, the United States, the Soviet Union, the United Kingdom, and France were developing rockets. Each of these nations sought to enlist the postwar services of German rocket experts to assist domestic rocket efforts. The United States acquired the most prominent of these experts, including Wernher von Braun, Walter Dornberger, and other leading members of the V-2 rocket team.

Between the end of World War II in 1945 and the launch of *Sputnik* in 1957, long-range rocketry was a primary area of competition between the Western democracies and the Soviet Union, as it was clear that the integration of rockets and nuclear warheads would form a nearly unstoppable and devastating weapon. In the Soviet Union, development of long-range missiles became a priority immediately following World War II as a means to threaten the United States. The United States already had the B-29 bomber to deliver nuclear bombs, and rockets had to compete with aircraft as the primary means of delivering nuclear weapons. In both nations, scientific experiments frequently flew on early rockets, as military needs dictated the scientific need to understand the upper atmosphere and near-Earth space. These experiments also included some astronomy experiments, as scientists recognized that many electromagnetic wavelengths from space could not be seen from Earth's surface, but could be observed from space. Phenomena such as aurorae and the newly discovered ionosphere drew scientific attention and led to a proposal for a worldwide scientific collaboration, dubbed the

International Geophysical Year (IGY), to study these and other physical global issues in 1957–58 through measurements taken around the world.

Public interest in the prospects of human spaceflight were reflected in popular culture, ranging from widely circulating magazines, such as *Collier's*, to the mention of spaceflight in songs, to the array of extraterrestrial panoramas offered by space artists, such as Chesley Bonestell. The era following World War II also saw significant growth in the number of space-themed movies and books aimed at public entertainment (as opposed to education). Perhaps this reflected an increase in public interest in technology following the technical developments made during World War II. Prior to the V-2 rocket, many people thought of rocketry and space travel, to the extent they thought of it at all, as science fiction, the stuff of "Buck Rogers" pulp magazines. The V-2 changed perceptions, and in the United States, experts such as von Braun realized that gaining support for spaceflight in a democracy required popularization and education of the general public. In the 1950s he appeared on Walt Disney's *Disneyland* television show and published articles in a series of space articles in *Collier's* magazine in 1952–54. The series also presented a series of steps for space exploration that remained influential: Earth-orbiting space stations built using reusable shuttles, then using that station as a base for trips to the Moon and then to Mars. Science-fiction movies symbolized the movement of space ideas from marginal "cult" status in pulp magazines of the 1930s to mainstream entertainment for the masses, portrayed in ever more realistic-appearing forms. Movies such as *Destination Moon*, *The Day the Earth Stood Still*, and *The War of the Worlds*, attracted large audiences and sent the message to the general public that spaceflight was soon likely to become reality. This heightened level of public expectation in turn served to drive reality so that spaceflight was no longer confined to the silver screen or the pages of pulp magazines. Rather, among some there was a growing sense of impatience that spaceflight had not yet arrived. This was also reflected in professional ranks, with concerns by some members of established professional societies who were reluctant to closely associate with dreams of spaceflight for fear of damaging their credibility, as opposed to others who wished to take an active role to make spaceflight a reality. Such a philosophical schism influenced the formation of the American Astronautical Society in 1954.

The U.S. Dwight Eisenhower presidential administration decided to support IGY as a means to a national security end. U.S. officials were desperate to acquire information about the military capabilities and intentions of the Soviet Union and had secretly authorized high-altitude overflights of the Soviet Union to gather imagery. They realized that eventually the Soviet Union would confirm and protest the presence of these aircraft in their national airspace, and a reconnaissance satellite was the next logical step to provide continued monitoring of Soviet capabilities. To ensure that a satellite would legally be allowed to "fly" over the Soviet Union, Eisenhower wanted the first U.S. satellite to support a nonmilitary scientific mission. If the Soviet Union did not protest, then it would set a precedent for reconnaissance satellites. Two teams competed for the mission. Even though Wernher von Braun's Army team had a launch vehicle that was nearly ready to place a spacecraft in orbit, the Vanguard Navy team had a more sophisticated scientific proposal and was selected to make the initial

U.S. entry into orbit. This was of critical importance in establishing the proper tone for participation in IGY.

Despite predictions by the U.S. Army Air Force's Project RAND as early as 1946 that the launch of an Earth orbiting satellite could have international repercussions, the Western media response to the launch of *Sputnik* on 4 October 1957 caught military and political leaders in the United States and the Soviet Union by surprise. *Sputnik* was not a priority for Soviet leaders, who were focused on the development of intercontinental ballistic missiles (ICBM). A scientific satellite for IGY was a secondary concern. Only after the Western press began to publicize *Sputnik* as a major Soviet triumph did Soviet leaders realize that they had scored a propaganda coup. The Eisenhower administration was not surprised by the launch, and some were quietly pleased, as *Sputnik* established the legal precedent of spacecraft flying over foreign nations. Although the Eisenhower administration could say nothing about this, it nonetheless tried to project an air of confidence. To the media, *Sputnik* proved that the Soviet Union could launch a nuclear-tipped ICBM at the United States, and that it had pulled ahead in the development of critical technologies. After the November 1957 launch of *Sputnik 2* with Laika the dog aboard, the administration's opponents sensed a political opportunity and held congressional hearings, ostensibly to find out how the Soviet Union had beaten the United States into space and apparently gained the lead in ICBMs. International observers saw this as a major victory for the Soviet Union over the United States, broadening the appeal of the communist model over democracies.

In December 1957 the first U.S. attempt to launch a satellite by the Navy's Vanguard program failed spectacularly in the full glare of the media and in stark contrast to the shroud of secrecy behind which the Soviet Union managed its rocket program. This failure led the Eisenhower administration to allow von Braun's Army rocket team to launch the Jet Propulsion Laboratory *Explorer 1* satellite. Congressional legislation established the Defense Advanced Research Projects Agency to coordinate the military space program and NASA to run the civilian space program. Congress also passed the National Defense Education Act, which aimed to support science, engineering, and language programs to develop the technical personnel needed to compete with the Soviet Union. Since that time, and particularly in the United States, spaceflight has served as an educational catalyst. Through 1958 and 1959, the United States and the Soviet Union began their human spaceflight programs and a series of launches aimed at placing ever larger payloads further into space, with the Soviet Union successfully acquiring the first picture of the Moon's far side. The space race was a ready-made media event, with news networks assigning dedicated reporters to follow the successes and failures of the U.S. space program.

The next year saw another major U.S. setback as the Soviet Union successfully shot down a U-2 reconnaissance aircraft. This was a major intelligence and political incident, but one that the Eisenhower administration had been planning for since the mid-1950s. At that time, the U.S. Air Force (USAF) began a reconnaissance satellite program called WS-117L. The program developed into several components, one of which, Corona, was run by the Central Intelligence Agency (CIA) under the cover name Discoverer. After 12 consecutive failed test flights, the 13th flight in

August 1960 succeeded, and the next flight returned the first images of the Soviet Union. Contrary to military and political critics who believed the Soviet Union was ahead of the United States in ICBM deployment, the Corona images showed that the converse was true; the advantage was in favor of the United States. To run the Corona program, the Eisenhower administration secretly established the National Reconnaissance Office, a joint operation of the CIA and USAF.

During the 1960 U.S. presidential campaign, Richard Nixon, who was well aware of the Eisenhower administration's secret reconnaissance satellite program, kept that information secret. John Kennedy won the election based partly on the charge that the Eisenhower administration had allowed the Soviet Union to get ahead in the deployment of ICBMs, despite his receipt of intelligence briefings between July 1960 and Election Day that questioned this conclusion. The evidence that the missile gap was a myth was strong by fall 1960 and incontrovertible one year later.

Soon after assuming office, a failed CIA-supported attempt to overthrow Cuba's Fidel Castro by an invasion of anti-Castro Cuban exiles embarrassed the Kennedy administration. In April 1961 the Soviet Union placed the first man in orbit, Yuri Gagarin. President Kennedy searched for some way to counter these foreign policy failures; the space program offered one means to do so. After Alan Shepard's suborbital flight in May 1961, Kennedy challenged the United States to support a manned lunar landing as a way to beat the Soviet Union in space. Through the rest of the 1960s, the two superpowers escalated their competition in space, with robotic spacecraft to the Moon, Venus, and Mars, and efforts to achieve ever-more ambitious firsts in human spaceflight, with a manned Moon landing as the ultimate goal.

The human spaceflight programs were ready-made human event stories, with astronauts and cosmonauts riding inherently dangerous rockets to an exotic destination with fabulous views. While the Soviet Union exploited its successes, it kept a tight grip on the media and hid imperfections in a shroud of secrecy. By contrast, the U.S. human spaceflight program operated in the full glare of the media. *Life* magazine purchased the rights to the life stories of the Mercury astronauts, while Walter Cronkite gave glowing television coverage on the evening news. As a result of its allure, the human spaceflight program precipitated an abundance of memorabilia, from ephemera to full-scale replicas. One of the unanticipated outcomes of the lunar race was the view of Earth from space, images of which were an inspirational factor in the development of the environmental movement, which took off in the 1970s.

The United States ultimately won the race by landing a man on the Moon in July 1969, outspending the Soviet Union roughly $21 billion to $12 billion. While having more funding was important, just as crucial to the outcome was the United States's much better organization and leadership as compared to the Soviet Union. The U.S. effort was managed through a single organization, NASA, which contracted with major corporations throughout the country, drawing to the manned program many of the best and brightest engineers and scientists. By contrast, the Soviet effort was fragmented among several competing design bureaus. The secretive Soviet system encouraged backroom dealings and bureaucratic competition, while the Soviet Union's political leadership never gave the manned lunar program sufficient attention to enforce a centralized organization and plan. Its attempts to coordinate among the design bureaus only created compromises among the bureaus, which was no substitute for a single coherently planned and developed program. The Soviet human spaceflight program was reactionary, responding to U.S. plans by redirecting its design bureaus to meet and beat the announced U.S. goals. In the early 1960s this was achievable with the powerful R-7 rocket, able to orbit its relatively simple Vostok and Voskhod capsules. However, the much more sophisticated and powerful systems needed for successful orbital rendezvous capabilities (Soyuz) and lunar landings (N-1 launcher and lunar lander) proved too difficult to achieve on a reactionary basis. Soviet leaders kept the manned lunar program secret, and when they lost the race, they publicly claimed that the Soviet Union had never planned a manned lunar landing. Instead they stated that their goal had always been to perform more cost-effective robotic lunar landings, and that they had always aimed to put cosmonauts on Earth orbiting space stations. The truth only came out two decades later.

More quietly the United States and the Soviet Union used scientific and human spaceflight programs to strengthen alliances around the world. Although much attention was focused on the human spaceflight competition between the United States and the Soviet Union, many other nations shared an interest in the political, military, and commercial opportunities presented by a presence in space. Space programs emerged in a number of nations, such as Australia, Canada, India, Israel, Italy, and Japan, in addition to national consortia such as the European Space Agency (ESA), primarily for Western European nations, and Interkosmos, for many Soviet bloc nations.

The U.S. government negotiated agreements with many countries for NASA spaceflight communications and tracking stations at remote locations, in which local personnel learned to operate advanced communications systems. Developed nations took advantage of the U.S. offer to launch two science spacecraft for each nation for free, and after those two launches for the price of the launch. In these cases, engineers from advanced democracies learned firsthand from NASA how to build and operate satellites. China began to develop its ballistic missile programs with help from the Soviet Union in the late 1950s. Its leader, Tsien Hsue-shen, had learned rocketry in the United States, attending the Massachusetts Institute of Technology and becoming a professor at the California Institute of Technology before being deported back to China in the mid-1950s on the unproven accusations that he was a communist sympathizer. With knowledge gained from both superpowers at its beginning, China had to develop its space program further without help from either the United States or the Soviet Union. The Soviet Union and its communist allies created the Interkosmos program in 1970 as a means to involve communist nations in scientific space endeavors.

NASA's offer greatly assisted European nations individually and also as a group through the European Space Research Organisation (ESRO). West European nations formed ESRO to collaborate in the development and operation of science satellites. It was one of several mechanisms of west European integration, which the U.S. government supported as a means to counter Soviet power. European nations also created the European Space Vehicle Launcher Development Organisation (ELDO) to jointly build a European launcher, primarily to avoid dependence on the United States. The United States did not support ELDO directly, because U.S. policy was to encourage such dependence on the United States. ELDO did not succeed in developing its Europa launch vehicles, but it formed the basis for the Ariane launcher of the 1970s.

As the number of nations interested in space activities increased, so too did the need to develop international law and policy in this new realm. Many of these matters were taken up by the United Nations Committee on the Peaceful Uses of Outer Space, which coordinated the preparation of a number of agreements and treaties for possible adoption by member nations. In addition to the original Outer Space Treaty, these included the Agreement on the Rescue of Astronauts, the Anti-Ballistic Missile Treaty, the Convention on International Liability for Damage Caused by Space Objects, the Moon Treaty, the (partial) Nuclear Test Ban Treaty, and the Registration Convention.

The success of national space programs was aided by the activities of advocacy groups at the local, national, and international levels. When successful, they proved to be an effective extension of national space programs, particularly in the development of space policy (and the not-so-trivial matter of funding allocation). Among the advocacy groups in the United States were the National Space Society and the Planetary Society. There have also been multinational advocacy efforts, primarily aimed at promoting unified policy among spacefaring nations. These included the International Astronautical Federation (whose membership comprised individual space societies) and the International Academy of Astronautics (whose membership comprised individuals of particular note in the field of astronautics). Youth in particular was served by the formation of the Space Generation Advisory Council in the 1990s under the auspices of the United Nations.

During the 1960s the military competition between the United States and the Soviet Union continued. Both sides continued to build their nuclear arsenals, though the U.S. buildup was somewhat less frantic as it already had the lead. The Kennedy administration quietly slowed the U.S. ICBM buildup, since it was clear by fall 1961, based on Corona imagery, that the United States led the Soviet Union in ICBM deployment, in addition to a huge lead in nuclear warheads. After vocal objections to U.S. reconnaissance satellite overflights in the early 1960s, the Soviet leadership quietly stopped talking about them after 1963, when Soviet reconnaissance satellites began to take imagery of the United States. Soviet leaders realized that reconnaissance satellites were extremely valuable to them and to the United States. After experimentation with a piloted spaceplane called Dyna-Soar, the USAF began to develop the Manned Orbiting Laboratory for manned reconnaissance from space. The Richard Nixon presidential administration canceled it in 1969 to save money and also because it began to appear duplicative of NASA's proposed Skylab program. In response the Soviet Union created the Almaz military space station program, which continued into the 1970s.

Both nations tested nuclear weapons in space in the late 1950s and early 1960s, enough to realize that the electromagnetic pulse and residual radiation from nuclear explosions in space would make space unusable by either nation. As both gained significant benefits from space, they agreed to stop further nuclear tests in space. The Outer Space Treaty of 1967 formalized these agreements, banning "weapons of mass destruction" from space. By omission, the treaty implicitly accepted reconnaissance satellites and left open more testing on antisatellite systems, which both nations fielded in the 1960s.

Two of the most important commercial space developments of the 1960s were the formation of the International Telecommunications Satellite Organization (Intelsat) in 1964, and the U.S. creation of the privately owned Communications Satellite (Comsat) Corporation to manage U.S. interests in the international communications consortium. Established by an act of Congress in 1962 and incorporated in 1963, Comsat Corporation established a crucial foothold for private industry in space. Its role as the manager of U.S. interests in Intelsat, and as Intelsat's majority shareholder into the early 1970s, ensured that profitability would be a crucial factor in the goals and management of Intelsat. Even though Intelsat's primary initial shareholders were the United States, Canada, west European nations, and Japan, private interests were hardly guaranteed a role, because most nations (with the important exception of the United States) assigned government postal and communications organizations to represent and control national interests in the new consortium. Despite objections from the Soviet Union, which proposed that only nations could operate in space, the Outer Space Treaty of 1967 formally allowed private enterprises in space, but under supervision of their governments and with governments assuming liability for any damages. West European nations banded together to negotiate with the United States as a group and extracted an agreement that the initial Intelsat accord that gave Comsat Corporation controlling shares in the organization would be renegotiated in the early 1970s. The new definitive agreement assigned shares to nations based on their usage of the system, which by the late 1970s assured that the United States no longer controlled Intelsat. This had the concrete result that by the late 1970s U.S. companies had much less chance of winning contracts to build Intelsat comsats.

The rise of Intelsat presaged a growing emphasis on economic issues and a concurrent shift away from prestige as a primary motive for space activities. By 1966, NASA's budget was decreasing, the first year in a series of budget reductions that lasted into the mid-1970s. Public support for and interest in the Apollo program steadily decreased through the decade, with a short-lived increase from *Apollo 8*'s December 1968 circumlunar flight to *Apollo 11*'s July 1969 lunar landing. After *Apollo 11*, public support for NASA waned as the Moon landings seemed increasingly irrelevant in the United States in the midst of public unrest related to the Vietnam War and public angst associated with the emergence of the civil rights movement. NASA's proposal to continue an expansive human flight program with a human mission to Mars failed miserably, as both the Richard Nixon presidential administration and the U.S. Congress wanted to drastically slash spaceflight funding. Believing that the nation needed heroes, Nixon nonetheless wanted to maintain a U.S. human spaceflight program. In January 1972 he authorized a minimal human spaceflight program that also offered potential long-term cost benefits: the reusable Space Shuttle. However, the Nixon administration required that the Shuttle be developed for roughly $5 billion, much lower than NASA's desired development budget, and it also had to meet USAF needs to launch large reconnaissance satellites. The resulting large system was not fully reusable, which ultimately meant that the Shuttle would not meet its ostensible long-term goal of reducing the cost of placing payloads in orbit. This would only become apparent in the mid-1980s. In the meantime, in 1973 NASA placed its first space station, *Skylab*, in orbit, performing three long-duration missions into 1974. It also supported the collaborative Apollo-Soyuz Test Project (ASTP) in 1975, in which NASA launched its final Apollo capsule to rendezvous with the Soviet Soyuz in Earth orbit. However, ASTP represented little more than a minor display of political détente, a policy that the Nixon presidential administration was pursuing with the Soviet Union.

To cut U.S. direct costs, NASA offered Canada and west European nations a role in the Space Shuttle program. French leaders believed that this was a ruse to divert Europeans from building a European launcher that would reduce dependence on the United States. West Germany and other nations looked more favorably on the offer. Eventually the Europeans agreed to both build a European launcher, Ariane, and participate in the Shuttle program by building the *Spacelab* experiment module. With ELDO floundering, the Europeans formed ESA on the basis of ESRO, but unlike this scientific organization, ESA could also develop "applications" programs, such as comsats and launchers.

The development of Ariane succeeded, and the Europeans formed Arianespace, the first commercial space launch corporation, to market the new vehicle to the growing commercial comsat industry. With the Shuttle's flight rate much lower than hoped or advertised, Arianespace's marketing efforts succeeded, and Ariane obtained a significant share of the growing market to launch private and government comsats. The 1973 Intelsat agreement allowed for regional comsats, and several nations and companies subsequently purchased comsats from U.S. and later European satellite manufacturers. Nations such as Indonesia consisting of thousands of islands, and companies such as Pan American Satellite Corporation that marketed satellite communications to Latin America, found significant national benefits and commercial profits. They, along with several other nations and companies and the newly founded International Maritime Satellite Organisation (Inmarsat), led to the expansion of satellite communications in the 1970s–80s. After the loss of *Challenger* in 1986, the Ronald Reagan presidential administration removed NASA's Shuttle from competition for commercial payloads, instead promoting commercial expendable launchers, such as Delta and Atlas. It assigned the Department of Transportation responsibility for commercial launch licensing.

The Space Shuttle program also emphasized the scientific and research potential of microgravity. To take advantage of *Spacelab* and the Shuttle's ability to return experiments to Earth (and to provide justification for the Shuttle in the absence of a space station to build and supply), the U.S., Japanese, and European governments and some corporations sponsored microgravity research in materials and life sciences. This proved quite useful for understanding the behaviors of these materials and organisms in space and for future space exploration. However, the hoped-for breakthroughs in medicines and materials did not materialize. The Soviet Union and later Russia also performed microgravity experiments on Salyut and *Mir* space stations with similar results. Research on the *International Space Station* in the early twenty-first century has had equally mixed returns. As of early 2009 the practical applications of microgravity research to life on Earth have remained limited.

The 1970s was a low period in U.S. political and military influence, with defeat in Vietnam and increased communist expansion and influence in southeast Asia, Africa, and Latin America. By contrast, Soviet power was waxing, with continued massive growth in its nuclear arsenal. The two superpowers agreed to limit antiballistic missile (ABM) systems in the 1972 ABM Treaty, but further agreements were not forthcoming. The Soviet Union's occupation of Afghanistan in 1979 renewed Cold War fears and along with the U.S. hostage crisis in the new Iranian Islamic republic, Ronald Reagan was elected U.S. president in 1980.

The Reagan administration (1981–89) advocated development of defenses against ICBMs through the Strategic Defense Initiative (SDI), which if deployed would violate the ABM Treaty. Derisively called "Star Wars" by detractors, SDI was vigorously opposed by the Soviet Union, which was secretly attempting to develop a similar system of its own. Despite domestic and foreign political opposition and development difficulties, the U.S. program continued through various incarnations to become the Missile Defense Agency. Concerned about the potential range of North Korean missiles, in December 2001 the George W. Bush presidential administration announced that it would withdraw from the ABM Treaty and soon thereafter began to deploy an ABM system in Alaska.

The Soviet Union tenaciously pursued its space station programs in the 1970s–80s. These stations were publicly identified under the name Salyut, but were in fact civilian Long-Duration Orbital Stations or Almaz military stations. The military station experiments showed that robotic reconnaissance was more cost effective than reconnaissance by humans. Nonetheless, civilian stations continued to have political and scientific value, so they continued through *Salyut 7* and further to the Mir program, the first launch of which occurred in 1986. In 1976 the Soviet Union implemented a "guest cosmonaut program" as part of Interkosmos, in which nine guests cosmonauts flew to *Salyut 6* from 1978 to 1981.

As the space race peaked and was then consigned to history books, memorializing space and educating the public became growth industries. The Kennedy Space Center Visitor Complex opened in 1968, and the U.S. Space and Rocket Center in Huntsville, Alabama, opened in 1970. The Smithsonian Institution National Air and Space Museum in Washington, DC, opened in 1976, becoming the most popular of the Smithsonian's museums. The Memorial Museum of Cosmonautics opened in 1981 at the 20th anniversary of Gagarin's flight. While Young Cosmonaut Groups had existed in the Soviet Union since the 1960s, the first space camp for young people opened in the United States in Huntsville in 1982. Interdisciplinary space education came into being in the late 1980s, with the University of North Dakota's Department of Space Studies offering master's degrees, and the International Space University's later master's programs offering degrees in space studies and space management.

The demise of the Soviet Union in 1991 and the creation of the noncommunist Russia marked the official end of the Cold War and the further spread of space capabilities. Reconnaissance satellites, which had been among the most tightly held secrets of the Cold War, were commercialized with high-resolution images for sale by Space Imaging Corporation in 1999. In the 1990s the U.S. government debated the level of resolution it would allow these companies to provide, with competition from France,

Russia, India, Israel, and others pressuring the government to allow higher resolution than would otherwise have been the case. Russia entered the commercial space launch market, offering several launchers for hire at prices with which the Europeans and Americans could not compete, sometimes in conjunction with U.S. or European companies to assist with sales. The United States responded by negotiating for quotas and price floors on Russian and Chinese launchers to protect the U.S. launch industry. North Korea attempted, but failed, a satellite launch in 1998, while Iran successfully launched its first satellite in February 2009.

The U.S. Space Station Freedom program was saved by transforming it into a diplomatic tool to assist Russia's turn to democracy. Russia and the United States, along with Japan and west European nations, joined in 1993 to form the International Space Station program as an expanded (with the inclusion of Russia) evolution of the Space Station Freedom program, with the first component placed in orbit in 1998. China also began to offer commercial launch services through state-owned Great Wall Corporation. Human spaceflight capabilities emerged in China with the launch of the first taikonaut, Yang Liwei, in October 2003. Russia began to offer commercial flights on its Soyuz vehicle starting in 1990 with Tokyo Broadcasting System paying for a reporter to fly to *Mir*. The first private paying citizen, Dennis Tito, flew aboard Soyuz to the *International Space Station* in 2001, while Scaled Composites Corporation successfully launched the first private suborbital flight, *SpaceShipOne*, in 2004.

The 1990s and 2000s saw the dramatic growth of satellite navigation in both military and commercial uses. The United States began development of the Global Positioning System (GPS) in the mid-1970s, but only in the early 1990s was it approaching full capabilities. In 1991 the Gulf War demonstrated the capabilities of satellite navigation to aid troop movements and enabling the development of precision weaponry. It enabled precise troop movements in the featureless desert of southern Iraq and guided new smart bombs and missiles to within feet of their targets, greatly enhancing the effectiveness of U.S. weaponry. After the war, civilian uses grew rapidly, so that by the 2000s applications of GPS signals were a multibillion-dollar industry. Its success spurred the European Union to collaborate with ESA, and later other nations such as India and China, to build a global navigation system, called Galileo, that was not dependent on U.S. goodwill.

Commercial satellite communications continued its steady growth in the 1990s, and a large spike in comsat launches occurred in the late 1990s due to the deployment of new low-Earth-orbit satellite constellations. Led by Iridium, ICO Global Communications, Globalstar, and Orbcomm, these constellations aimed to provide Internet store-and-forward services and global telephone service. However, their business models failed to account for competition from ground-based cellular phone networks, which expanded rapidly in the 1990s and 2000s and provided mobile phone service at prices with which the satellite systems could not compete. All went bankrupt in the 2000s. To launch these and other projected satellite constellations into orbit, a number of privately funded launch vehicles began development in the late 1990s. However, these ventures collapsed along with the comsat constellations, leaving the launch market with traditional providers. In the meantime private competition forced international consortia, such as Intelsat and Inmarsat, to rethink their business models. Both were fully privatized by the early 2000s to allow access to the U.S. market because the United States would not allow them to compete with fully privatized vendors in the United States unless they were privatized.

The tragic loss of the Shuttle *Columbia* on its return from space in 2003 made replacement of the Shuttle an even more pressing issue. Along with the continuing cost overruns of the *International Space Station*, the loss of *Columbia* forced a fundamental reassessment of NASA's programs. In January 2004 the George W. Bush administration announced the Vision for Space Exploration, in which NASA would retire the Shuttle by 2010, replace it with new launch vehicles, return to the Moon for potential permanent habitation, and eventually send human expeditions to Mars. This led to the creation of the Constellation program, which was actively developing the Ares I launcher and Orion crew capsule in early 2009. NASA began preliminary discussions with other nations regarding their potential participation in this new initiative by matching national interests with program elements, similar to how Canada established a technical niche through the development of the Canadarm robotic system.

From the 1960s through the early 2000s, space science advanced rapidly, changing scientific understanding of the cosmos and humanity's place in it. Theories about the expanding universe, black holes, and the Big Bang were verified through ground-based and space-based measurements in the 1960s and 1970s. Space probes and rovers showed that while Mars was not the home of a dying civilization, as projected by Percival Lowell in the 1900s, it had once had flowing rivers and almost certainly had water just under its surface. The Voyager mission to the outer planets discovered that the solar system was far more diverse than scientists had imagined. From the 1990s well into the 2000s results from the *Hubble Space Telescope* provided scientists and the public alike with spectacular images from outer space. The discovery of ever-increasing numbers of planets outside the solar system showed that planets were common, providing tantalizing evidence that life may also be common. Observations of Earth from space provided critical data to understand humanity's impact on its home planet.

Michael L. Ciancone and Stephen B. Johnson

MILESTONES IN THE DEVELOPMENT OF SPACE AND SOCIETY

1865 Publication of *De la Terre à la Lune* (*From the Earth to the Moon*) by Jules Verne.

1898 Publication of *War of the Worlds* by H. G. Wells.

1902 Release of movie *Le Voyage dans la Lune* by Georges Melie.

1903 Russian scientist Konstantin Tsiolkovsky publishes the first article on rocketry and space exploration.

1913 Article in French *Journal de Physique* by Robert Esnault-Pelterie on the use of rockets for space exploration.

1916 Publication of *La Conquête de l'Espace* (*The Conquest of Space*) by Victor Coissac is the first book of nonfiction in any language on the use of rockets for human spaceflight.

1919 Publication of "A Method of Reaching Extreme Altitudes" by Robert Goddard.

1923 Publication of *Die Rakete zu den Planetenräumen* (*The Rocket into Planetary Space*) by Hermann Oberth

Publication of *Si puó giá tentare un viaggio dalla terra alla luna?* (*Is It Possible Yet to Attempt a Voyage from the Earth to the Moon?*) by Luigi Gussalli.

1926 16 March: U.S. scientist Robert Goddard launches world's first liquid-fuel rocket.

1927 5 July: German Society for Spaceship Travel (Verein für Raumschiffahrt, or VfR) founded.

1928 Publication of *Das Problem der Befahrung des Weltraums* (*The Problem of Space Navigation*) by Hermann Noordung (pseudonym for Herman Potočnik).

Publication of first of 10 volumes on interplanetary flight by Nikolai Rynin.

1929 7 January: Debut of Buck Rogers comic strip.

1930 Movie *Frau im Mond* (*Woman in the Moon*) directed by Fritz Lang.

Soviet Union Chief Designer Sergei Korolev cofounds Russian Group for Investigation of Reactive Motion (GIRD).

Publication of *L'Astronautique* by Robert Esnault-Pelterie.

4 April: American Interplanetary Society founded.

27 September: VfR establishes Raketenflugplatz (rocket airfield) near Berlin.

1931 *The Conquest of Space* by David Lasser is the first book of nonfiction in English on the use of rockets for human spaceflight.

Formation of public rocket clubs in the Soviet Union.

1932 October: German Army hires German scientist Wernher von Braun to help with rocket experiments.

1933 September: Soviet Union Army assumes control of GIRD and replaces it with RNII (Jet Propulsion Research Institute).

13 October: Formation of British Interplanetary Society.

1934 VfR dissolves.

1936 Publication of "Liquid-Propellant Rocket Development" by Robert Goddard.

1938 Soviet Chief Designers Sergei Korolev and Valentin Glushko falsely accused of subversion and imprisoned.

30 October: Radio broadcast of *War of the Worlds*.

1941 Formation of Reaction Motors, Inc.

1942 3 October: First successful launch of A-4 (later known as the V-2) rocket.

1944 Korolev and Glushko released from prison.

1 July: Jet Propulsion Laboratory (JPL) begins operation.

1 August: German engineer Eugen Sänger releases final report on rocket-boosted winged spaceplane.

1945 Publication of article in *Wireless World* by Arthur C. Clarke on the use of geostationary satellites for communications.

September: German scientists, including von Braun, arrive in United States under Operation Paperclip.

1946 Release of RAND report on "Preliminary Design of an Experimental World-Circling Spaceship."

May: Decree issued by the Soviet Union to found the postwar missile program

August: Korolev begins long-range ballistic missile development for Soviet Scientific Research Institute NII-88.

1947 Woomera Rocket Range established in Australia.

1948 11 June: United States Air Force (USAF) begins suborbital biological flights with monkeys as passengers.

1949 9 February: USAF forms Department of Space Medicine.

1950 Movie *Destination Moon* directed by George Pal.

Alexandre Ananoff organizes first International Astronautical Congress (Paris).

26 April: Korolev becomes Chief Designer of Special Design Bureau No. 1 (OKB-1) of NII-88.

1951 Formation of International Astronautical Federation (IAF).

12 October: Von Braun lectures at First Symposium on Spaceflight in New York.

1952 11 February: *Collier's* magazine publishes concepts for piloted spaceflight discussed at First Symposium on Spaceflight.

1953 Publication of *The Mars Project* by von Braun.

1954 22 January: American Astronautical Society (AAS) founded.

9 July: X-15 Project begins as joint venture among National Advisory Committee for Aeronautics (NACA), USAF, and U.S. Navy.

1955 Chinese engineer Tsien Hsue-Shen expelled from the United States.

2 June: Soviet Union founds Baikonur Cosmodrome.

December: USAF approves high-altitude human-occupied balloon flights under Project Manhigh.

1956 China begins national space program.

January: Soviet Union approves development of scientific satellite.

1957 4 October: *Sputnik* becomes first human-made object to reach Earth orbit.

3 November: *Sputnik 2* carries first animal, a dog, into Earth orbit.

1958 U.S. National Defense Education Act enacted.

1 February: United States launches its first satellite, *Explorer 1*, into Earth orbit.

2 July: Soviet government authorizes development of Raketoplan.

1 October: NASA begins operation; Project Mercury starts.

18 December: Launch of *SCORE* (*Signal Communication by Orbiting Relay Equipment*), the first satellite to relay communications.

1959 United Nations (UN) establishes Committee on the Peaceful Uses of Outer Space (COPUOS).

NASA awards *Life* magazine rights for exclusive access to Mercury astronauts.

2 January: Launch of Soviet spacecraft *Luna 1*, the first human-made object to reach escape velocity.

9 April: NASA announces its first group of astronauts.

28 May: U.S. Army Ballistic Missile Agency (ABMA) launches monkeys Able and Baker on suborbital spaceflight.

12 September: Launch of Soviet spacecraft *Luna 2*, the first probe to impact the Moon.

4 October: Launch of Soviet spacecraft *Luna 3*, which returns the first images of far side of the Moon.

1960   IAF establishes the International Academy of Astronautics.

International Institute of Space Law (IISL) founded by the IAA.

1 July: ABMA becomes Marshall Space Flight Center.

1961   12 April: Soviet launch of Yuri Gagarin, first human to fly in space, on *Vostok 1*.

5 May: U.S. launch of Alan Shepard Jr., first American to fly in space, on *Freedom 7*.

25 May: U.S. President John F. Kennedy declares United States should land a man on the Moon by end of decade.

10 July: U.S. launch of *Telstar*, the first active, direct relay communications satellite.

19 September: NASA establishes Manned Spacecraft Center (later Johnson Space Center).

19 December: CNES (French Space Agency) established.

1962   Thumba Equatorial Rocket Launching Station (TERLS) established by India.

20 February: U.S. launch of John Glenn Jr., first American to orbit Earth, on *Friendship 7*.

26 April: First launch (*Ariel 1*) of bilateral space program between the United Kingdom and the United States.

1 July: NASA establishes Launch Operations Center (later Kennedy Space Center).

11 July: NASA selects lunar orbit rendezvous (LOR) as mode for Apollo lunar landings.

31 August: U.S. Communications Satellite Act signed by President John F. Kennedy.

28 September: Launch of first Canadian satellite, *Alouette 1*.

1963   Merger of the American Rocket Society and the Institute of Aeronautical Sciences to form the American Institute of Aeronautics and Astronautics (AIAA).

16 June: Soviet launch of Valentina Tereshkova, first woman to fly in space, on *Vostok 6*.

19 July: X-15 Flight 90 is first rocket-powered aircraft to reach space.

26 July: U.S. launch of *Syncom 2*, the first geostationary satellite.

4 August: Partial Nuclear Test Ban Treaty opened for signature.

10 December: U.S. Department of Defense cancels Dyna-Soar (X-20) program; approves start of Manned Orbital Laboratory (MOL).

1964 29 February: European Space Vehicle Launcher Development Organisation (ELDO) established.

20 March: European Space Research Organisation (ESRO) established.

25 July: Intelsat established.

12 October: Soviet launch of first multi-crewed spaceflight (*Voskhod 1*).

15 December: Launch of first satellite built by a European country (*San Marco 1* by Italy).

1965 18 March: Soviet cosmonaut Alexei Leonov conducts first spacewalk (*Voskhod 2*).

26 November: Launch of first French satellite, *Asterix*.

1966 Television series *Star Trek* begins.

14 January: Sergei Korolev, founder of the Soviet space program, dies.

3 February: Soviet spacecraft *Luna 9* achieves first soft landing on the Moon.

16 March: First docking in space during *Gemini 8*.

1967 First space history symposium of the IAA.

27 January: UN Outer Space Treaty opened for signature.

27 January: U.S. crew of *Apollo 204* (*Apollo 1*) dies in launch pad fire.

23 April: Soviet cosmonaut Vladimir Komarov dies during *Soyuz 1* reentry.

26 November: Launch of first Australian satellite, *WRESAT*.

1968 Release of movie *2001: A Space Odyssey* directed by Stanley Kubrick.

Kennedy Space Center Visitor Complex opens.

27 March: Soviet cosmonaut Yuri Gagarin dies in training exercise.

9 April: First launch from French Centre Spatiale Guyanais (CSG) in Kourou, French Guiana.

22 April: UN Rescue and Return Agreement goes into effect.

24 December: U.S. crew of *Apollo 8* is first to orbit Moon.

1969 Indian Space Research Organisation (ISRO) created.

National Space Development Agency (NASDA) created.

21 February: First all-up test of the Soviet N1 rocket ends in failure.

20 July: American Neil Armstrong first human to walk on Moon during *Apollo 11*.

15 September: NASA Space Task Group issues report on post-Apollo space program.

1970 Soviet Union establishes Interkosmos program.

U.S. Space and Rocket Center opens in Huntsville, Alabama.

11 February: Launch of first Japanese satellite, *Ohsumi*.

24 April: Launch of first Chinese satellite, *Dong Fang Hong 1*.

17 November: Soviet spacecraft *Luna 17* delivers *Lunokhod 1* rover to surface of the Moon.

15 December: Soviet spacecraft *Venera 7* makes soft landing on Venus.

1971   19 April: Soviet launch of first space station (*Salyut 1*).

29 June: Soviet crew of *Soyuz 11* dies during reentry.

15 November: Intersputnik treaty deposited with UN.

1972   5 January: U.S. President Richard Nixon approves Space Shuttle program.

1 September: UN Convention on International Liability for Damage Caused by Space Objects goes into effect.

1973   14 August: ESRO signs memorandum of understanding with NASA to build *Spacelab*.

1974   22 May: TsKBEM merges with KB EnergoMash to form NPO Energia.

1975   15 April: European Space Agency (ESA) established.

17 July: Docking of Apollo and Soyuz spacecraft (Apollo-Soyuz Test Program).

August: Formation of L-5 Society.

1976   Publication of *The High Frontier* by Gerald K. O'Neill.

UN Registration Convention goes into effect.

17 February: Soviet Union cancels crewed lunar program and approves Energia-Buran Soviet shuttle.

1 July: U.S. National Air and Space Museum (NASM) opens.

20 July: U.S. spacecraft *Viking 1* lands on Mars.

1977   Release of movie *Star Wars* by George Lucas.

1978   24 January: The Soviet *Kosmos 954* naval reconnaissance satellite reenters the atmosphere, spreading nuclear material over western Canada.

2 March: Soviet launch of the first mission under the Interkosmos program, *Soyuz 28*.

18 May: First ESA group of astronauts enter service.

1979   11 July: U.S. spacecraft *Skylab* disintegrates after reentering the atmosphere.

16 July: Inmarsat founded.

1980   Formation of Planetary Society.

1981   12 April: First piloted mission of U.S. Space Shuttle.

1982   U.S. Space Camp opens in Huntsville, Alabama.

1983   18 June: U.S. launch of Sally Ride, first American woman to fly in space, on STS-7.

30 August: U.S. launch of Guion Bluford, first African American to fly in space, on STS-8.

1984   25 January: U.S. President Ronald Reagan approves space station project.

27 August: NASA begins Teacher in Space Project.

5 October: U.S. launch of Marc Garneau, first Canadian to fly in space, on STS-41-G.

1985   Glavkosmos organized.

Publication of *Cosmos* by Carl Sagan.

1986   U.S. National Commission on Space issues report on "Pioneering the Space Frontier."

Alcântara (Brazil) launch facility opened.

28 January: U.S. Space Shuttle *Challenger* suffers catastrophic failure during ascent.

13 March: *Giotto* probe (ESA) encounters Halley's Comet.

1987 Australian Space Office established.

Formation of National Space Society.

12 April: International Space University (ISU) founded.

1988 Italian Space Agency (ASI) created.

Space Camp–Florida opens in Titusville, Florida.

29 September: Launch of first Space Shuttle flight after *Challenger* disaster, STS-26.

15 November: Only flight of Soviet space shuttle *Buran*.

1989 1 March: Canadian Space Agency (CSA) begins operation.

20 July: President George H. W. Bush announces Space Exploration Initiative.

1990 Augustine report published.

German Agency for Space Issues (DARA) formed.

2 December: Soviet launch of Japanese journalist, Toyohiro Akiyama, first paying spaceflight passenger to *Mir* space station.

1991 31 July: United States and Soviet Union sign agreement to conduct joint space activities.

8 December: Soviet Union dissolves.

1992 Russian Space Agency (RSA) established.

31 July: U.S. launch of astronaut Franco Malerba, first Italian to fly in space, on STS-46.

5 October: NASA and Russian Space Agency approve Shuttle-Mir program.

1993 Gagarin Cosmonauts Training Center begins offering space camp programs in Russia.

April: Inter-Agency Space Debris Coordination Committee formed.

21 June: First flight of commercially developed *Spacehab* module.

2 September: United States and Russia agree to merge space station programs.

1994 Brazilian Space Agency established.

3 February: U.S. launch of Sergei Krikalev, first Russian to fly aboard a Space Shuttle, on STS-60.

1996 First year in which global commercial space revenues exceed government space expenditures.

18 May: Creation of the X-Prize.

29 June: First Space Shuttle docking with *Mir*.

1997 Publication of *The Case for Mars* by Robert Zubrin.

4 July: U.S. spacecraft *Mars Pathfinder* lands on Mars.

1998 Formation of Mars Society.

4 December: First ISS assembly flight links *Unity* node and *Zarya*.

1999   China Aerospace Science and Technology Corporation (CASC) created.

Formation of Space Generation Advisory Council.

2000   11 October: U.S. launch of 100th Space Shuttle flight.

2001   23 March: *Mir* disintegrates after reentering the atmosphere.

28 April: Soviet launch of U.S. entrepreneur Dennis Tito, first commercial space-flight passenger, on *Soyuz TM-32*.

2003   Japan merges ISAS, NAL, and NASDA to form the Japanese Aerospace Exploration Agency (JAXA).

1 February: U.S. Space Shuttle *Columbia* breaks apart during reentry.

15 October: China's first piloted spacecraft (*Shenzhou 5*) reaches orbit.

2004   Rosaviakosmos reorganized into Russian Federal Space Agency (Roskosmos).

14 January: U.S. President George W. Bush announces Vision for Space Exploration.

21 June: *SpaceShipOne* is first privately funded spaceflight.

4 October: Mojave Aerospace Ventures wins X-Prize.

2005   26 July: U.S. launch of first Space Shuttle flight after *Columbia* disaster, STS-114.

2006   7 November: ISRO endorses plan for human spaceflight program in India.

Michael L. Ciancone and Stephen B. Johnson

See also: Civilian and Commercial Space Applications, Human Spaceflight and Microgravity Applications, Military Applications

Bibliography

William E. Burrows, *This New Ocean: The Story of the First Space Age* (1999).

Steven J. Dick and Roger D. Launius, eds., *Societal Impact of Spaceflight* (2007).

John M. Logsdon, ed., *Exploring the Unknown: Selected Documents in the History of the U.S. Civil Space Program*, 7 vols. (1995–2009).

ECONOMICS OF SPACE

The economics of space has been an underappreciated aspect of space endeavors. Economics in general refers to the organization of the productive and distributive aspects of life. For space activities, microeconomics and political economy are the two subdisciplines that are most relevant, since microeconomics focuses on specific industrial sectors, and space activities have been tightly intertwined with governments.

Spaceflight began in the late 1950s as an almost exclusive government domain. In the Soviet Union, the government dominated all aspects of life, and the state controlled all aspects of spaceflight. The first spaceflights by the United States were also government run, but private industry had roles right from the start. For example, for *Explorer 1*, the first U.S. satellite, the program was managed by the U.S. Army, but the Jet Propulsion Laboratory, a branch of the California Institute of Technology, a private university, developed the satellite, and the launch vehicle was manufactured by Chrysler Corporation. Other satellite projects were either military, or in the case of NASA, government-run civilian activities. The dominance of governments was enshrined in the Outer Space Treaty of 1967, which directed that private activities in space must be supervised by their governments, and the governments assumed all liabilities.

In the 1960s the rivalry between the Soviet Union and the United States drove the vast majority of worldwide space funding to intelligence-gathering satellites, human spaceflight, and space science. NASA's Mercury, Gemini, and Apollo programs ensured a huge spike in funding for U.S. human spaceflight, while robotic voyages to the Moon and the planets spurred science funding. Similarly, the Soviet Union spent significant sums on its human spaceflight and science programs. The Apollo program alone cost between $21.8 and $25 billion in then-year dollars (dollars valued in the years in which they were spent), and the Soviet manned lunar landing program has been estimated at roughly half that amount. In the United States, spy satellites, whose costs even in 2009 remain classified, reputedly cost an amount similar to Apollo. Space science cost far less, but still several billion dollars in the 1960s. Space launch vehicles also required major funding in the 1950s–60s, as these enabled all other space applications.

Communications satellites drew far less funding in the 1960s, but were important for establishing an entry for private commercial ventures in space. After initial experimentation by NASA, the U.S. military, and American Telephone and Telegraph Corporation, commercial communications satellites proved to be profitable. To build and manage the first system of three geosynchronous satellites, in 1962 the U.S. Congress passed a law to form Comsat (Communications Satellite) Corporation, which acted as the negotiator for the United States with Europe and Japan to create Intelsat (International Telecommunications Satellite Organization), the consortium that controlled the satellite system. Intelsat grew slowly in the late 1960s–70s, when it was joined by Inmarsat (International Maritime Satellite Organisation), a similar international consortium for maritime communications. Purely private satellites (not controlled by governments or government consortia) began to be developed in the 1970s, but only in the 1980s–90s did they begin to rival the government-run programs in costs and revenues. The development of satellite television beamed directly to homes proved to be an extraordinarily profitable application, which by the early twenty-first century was the largest single subsector of satellite communications.

In the 1970s–80s, human-spaceflight funding decreased significantly from its peak in the 1960s. Space-science funding also shrank in the United States during this period, but by this time other nations, including Canada, western European nations, and Japan were spending significant sums on space science. Europe, along with China and India, spent large sums on launchers to develop their own indigenous launch capabilities so as not to depend on the United States or the Soviet Union, whom these new space powers assumed would not be willing to launch military or commercial payloads that could compete with them. The European Ariane launcher proved to be an effective competitor to the U.S. Space Shuttle and expendable launchers, particularly after the loss of the Shuttle *Challenger* in 1986.

The 1990s saw dramatic growth in the commercial aspects of space communications and navigation. A major spike in private funding for space communications occurred with the development of medium-Earth-orbit satellite communications constellations, such as Iridium. In the first decade of the twenty-first century, these became insolvent and the companies went bankrupt. However, the need to launch many satellites also drew a surge in private funding to build private launch vehicles. With the collapse of the satellite constellations, these efforts largely proved fruitless. The Global Positioning System (GPS), which remained a U.S. Air Force–funded and operated system, proved to be a major commercial success, with numerous private vendors worldwide creating commercial applications that used the GPS signals. By the early twenty-first century, several nations were developing their own regional navigational satellites, or joining the European Galileo consortium to develop a combined system. The demise of the Soviet Union, the rise of capitalist Russia, and the emergence of China led to new competition in space launches.

In the early twenty-first century, private funding and revenues exceeded government funding for space endeavors for the first time. These included the first space tourists, which promised to open space to more private individuals over time. Other uses of space, such as mining or collecting solar energy from space, remained speculative. By 2006 total expenditures on space activities were about $146 billion, with roughly $83 billion in the communications sector.

Stephen B. Johnson

See also: Commercial Space Applications, National Aeronautics and Space Administration

Bibliography
Roger Handberg, *International Space Commerce: Building from Scratch* (2006).
Stephen B. Johnson, "The Political Economy of Spaceflight," in *Societal Impact of Spaceflight*, ed. Steven J. Dick and Roger D. Launius (2007), 141–91.
John L. McLucas, *Space Commerce* (1991).

Economics of Human Spaceflight

Economics of human spaceflight has been dominated by national governments in their quest for international prestige and technological leadership. Soon after the launch of *Sputnik* by the Soviet Union in October 1957, the United States decided to place a man in space to promote the superiority of American democracy over Soviet communism. Project Mercury soon became a high priority and NASA's most expensive endeavor, and the Soviet Union responded with its Vostok program.

The R-7 launcher gave the Soviet Union a significant advantage in the early 1960s space race. Sergei Korolev's OKB-1 (Experimental Design Bureau-1) used it to launch the first man and the first woman into space (Yuri Gagarin on *Vostok 1* and Valentina Tereshkova on *Vostok 6* respectively), the first three-man crew (*Voskhod 1*), and the first spacewalk (Alexei Leonov on *Voskhod 2*). With Gemini, NASA surpassed Soviet efforts, and then Apollo gave the United States a victory in the early space race, powered by some $25 billion in government funding. By contrast the Soviet Union's estimated $13 billion, combined with bureaucratic infighting and competing lunar programs, was insufficient to overcome American efforts.

After Apollo, NASA's human flight funding slumped from a peak of just more than $3 billion in 1967 to just under $1 billion in 1974, in then-year dollars ($14.75 billion to $3.38 billion in 2006 dollars). These funding constraints were a major factor in the design and development of the Space Shuttle. In 1984 the U.S. Space Station Freedom (SSF) program and NASA's human spaceflight funding slowly grew to support it. As part of the Space Shuttle program, the European Space Agency (ESA) developed Spacelab, marking its entrée into human spaceflight. Unlike Apollo, both the Space Shuttle and SSF programs were motivated by economic and technological concerns; the Shuttle to provide low-cost access to space, and SSF to develop new technologies to improve the U.S. economy.

By contrast, Soviet resources remained relatively stable, as it refocused efforts to military and civilian space station programs (first Salyut/Almaz and then Mir) and began the Energia-Buran shuttle development program. Energia-Buran cost an estimated $4.5 billion (then-year dollars) from the late 1970s until its demise in the late 1980s. However, the economic support for space programs collapsed along with the Soviet Union itself in the late 1980s and early 1990s, and the new Russian space program transformed its Interkosmos guest cosmonaut program into a commercial program for paying customers. The first of these was a Japanese reporter, whose company reputedly paid $12 million (then-year dollars) for a ride to *Mir* in 1990. In 1993 several factors led to the reformulation of the SSF into the International Space Station (ISS) program. These included the Russian desire to keep its space programs alive, the U.S. motivation to support the fledgling capitalist nation and its fears that Russian space workers would work for rogue nations, and the U.S. hope that a program tied to critical foreign relations would ensure more congressional support. The United States paid Russia for a variety of services, including launching astronauts to *Mir* and building *ISS* hardware, while Russia funded some of its own modules and some logistics flights. When the Space Shuttle fleet was grounded after the *Columbia* disaster of February 2003, the United States paid Russia to transport astronauts to the *ISS*.

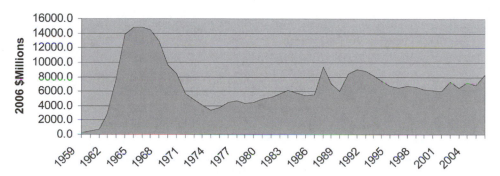

NASA's Human Spaceflight Budget, in 2006 millions of dollars.

In the 1990s China began its relatively low-budget space program (estimated at approximately $240 million per year in 2002) by purchasing and modifying Russian Soyuz capsules, and paying for its taikonaut (yuhangyuan) training at Star City, Russia. These efforts produced results with China's first successful piloted spaceflight launch in 2003. China's motives appeared similar to the initial motivations of the United States and the Soviet Union: to garner world prestige and to inspire their people, particularly young people, to study technical subjects.

The loss of the Space Shuttle *Columbia* in 2003 brought attention to the economic and technical flaws of the Space Shuttle and renewed debate about the future of the *ISS*, which depended on the Shuttle to complete its assembly. The Shuttle's operational costs were far higher than for expendable launchers, and the cost of the *ISS* grew far beyond the initial estimates of $8 billion for SSF to more than $25 billion by the early twenty-first century for the United States alone, without any obvious economic benefits in return. Along with ongoing concerns about NASA's long-term goals, these led U.S. President George W. Bush's administration to announce in January 2004 its Vision for Space Exploration, which directed NASA to replace the Space Shuttle, return humans to the Moon, and eventually go to Mars. Since NASA was not going to receive significant new funding for the program, the new Constellation program had to be funded by scaling back other programs in space science, technology research and development, aeronautics, and the eventual retirement of the Shuttle. In 2010 U.S. President Barack Obama proposed to cancel Constellation and shift some of the funding to commercial companies to launch humans and cargo into low Earth orbit, ostensibly in an effort to lower costs.

By the 1990s government financial constraints made the Russian human flight program "commercial" in outlook. From a Russian viewpoint, it mattered little whether its customers were governments or individuals. When negotiations to transfer *Mir* to private ownership failed, American businessman Dennis Tito became the first space tourist, paying a reputed $20 million for a flight to the *ISS* in 2001. Citing safety concerns, NASA and ESA fought to keep Tito off the *ISS*, while Russia was eager to capitalize on its ability to launch humans into space.

By 2007 the future of space tourism looked promising, with several more private citizens paying Russia to fly to *ISS* on Soyuz spacecraft, and Scaled Composites Company's successful suborbital flight of *SpaceShipOne* in June 2004, which reputedly cost $27 million to develop. Scaled Composites and others soon began development of suborbital vehicles for tourist flights. Bigelow Aerospace launched *Genesis 1*, an inflatable habitat, in July 2006 in an early structural test of a future orbiting hotel.

Stephen B. Johnson

See also: Human Spaceflight Programs, Politics of Prestige, Tourism

Bibliography
Stephen B. Johnson, "The Political Economy of Spaceflight," in *Critical Issues in the History of Spaceflight*, ed. Steven J. Dick and Roger D. Launius (2007).
NASA Historical Data Books, vols. 1–6 (1974–2000).

Economics of Space Communications

Economics of space communications provided the first major commercial success of satellites, when the *Early Bird* communications satellite garnered significant revenues after its launch in 1965. Prior to that time, the initial communications satellites were experimental, such as the *SCORE* (*Signal Communication by Orbiting Relay Equipment*) satellite in 1958, which used a tape recorder to store and forward U.S. President Dwight Eisenhower's Christmas holiday greeting to the world. *Courier 1B*, launched in 1960, was the world's first active repeater satellite. *Telstar*, funded and designed by American Telephone and Telegraph (AT&T) Company, launched in July 1962, was the first active, direct-relay communications satellite. In 1963 the United States launched *Syncom 2*, the first geostationary satellite. Its successor, *Syncom 3*, was used to broadcast the Olympic Games from Japan to the United States in October 1964 and later for military links to Vietnam.

In the 1960s the first application for satellite communications was in international telephony. Fixed-point telephones convey calls to an Earth station, where the calls are then relayed to a geostationary satellite, which then retransmits the signal to another Earth station. When placed over the Atlantic or Pacific oceans, geosynchronous communications satellites were cost-effective as compared to telephone cables laid across the bottom of these oceans. For example, in 1955 the Trans-Atlantic Cable TAT 1 cost $35 million, providing 36 telephone circuits, whereas in 1965, the *Early Bird* satellite cost $7 million for 240 circuits, earning nearly $1 million in revenues in its first six months of operation. The geostationary orbit is practical for communications because ground-based antennae, which must be directed toward the satellite, can operate effectively without the need for expensive satellite tracking motion systems.

International telephony via satellite was initially implemented through Intelsat, an international consortium set up for the purpose. It was managed by Communications Satellite Corporation (Comsat), which was established by the U.S. Congress in 1962 to develop and operate the international satellite communications system for the United States. Investors believed in the profit potential of Comsat, as they significantly oversubscribed the initial $200 million stock offering. Intelsat developed several generations of satellites, each providing greater performance for lower cost. Unwilling to participate in Intelsat, the Soviet Union created Intersputnik to handle the satellite communications needs of the Soviet bloc nations.

Large nations with scattered populations (such as the Soviet Union, Canada, and Indonesia) found satellite communications important for domestic political and economic development and used domestic systems, such as the Soviet Union's *Molniya* (1965, using a highly-inclined orbit), Canada's *Anik* (1972), and Indonesia's *Palapa* (1976). Military needs drove the development of military communications satellites, but the U.S. and Soviet military used civilian and commercial systems also. The 1972 final Intelsat agreements allowed for the creation of regional communications systems if they did not compete with the international system, leading to a series of European and Asian systems. The first private systems were Western Union's *Westar 1* (1974), and Radio Corporation of America (RCA) Americom's *Satcom 1* (1975). By the 1980s–90s, many nations had developed or purchased their own geosynchronous communications satellites for domestic communications and national television broadcasts, while a number of commercial companies broke into the market. These "fixed satellite services" exceeded $7 billion in revenues by 2003.

The recognition that television broadcasts over wide regions were greatly enhanced by using satellites spurred further growth, as private vendors such as DIRECTV, Echostar, and SES marketed directly to consumers, bringing hundreds of channels directly to consumers through "Very Small Aperture Terminals" (small home antennae), in what became known as "Direct Broadcast Satellite" services. By 2003 these services earned revenues of more than $28 billion. Expanding beyond television, satellite radio came into existence in the early 21st century, and by 2003 its revenues exceeded $153 million. Another new application was satellite Internet service, which remained a small but growing market in 2007, catering to rural and highly mobile customers, both civilian and military.

Mobile communications was spawned by the military and civilian needs, leading to experimental satellites such as *Tactical Satellite* (1969) and *Marots* (later *Marecs*—1981). Inmarsat was created in 1979 to manage the international maritime mobile communications system, this time with the Soviet Union and Eastern bloc nations participating.

The Little Low Earth Orbit (LEO) satellite service market, which uses constellations of satellites in LEO or medium Earth orbit, is made up mostly of applications whose messages do not require immediate delivery, but also includes global telephone service. The store-and-forward satellites store data for downlink to the appropriate gateway station destination as it rotates around Earth. Certain applications also require a real-time or near real-time mode of delivery. Such applications include the intranet or point-of-sale transactions market and certain segments of the supervisory control and data acquisition market. Real-time mode is possible only with full deployment of a constellation with a large number of satellites and the corresponding operation of gateways in the geographic areas targeted for service. Finally, as of 2006, two LEO constellations provided handheld telephony primarily to remote areas and international communications: Iridium and Globalstar. The Iridium system became operational in 1998 and had 66 active satellites in orbit in 2006. As of 2006 Globalstar operated with 48 active satellites. Both of these systems went bankrupt due to competition from ground-based cellular service, and in 2003 were operated under new ownership, earning roughly $600 million.

By the early twenty-first century, there existed a significant overcapacity of satellite communications capabilities in orbit. The migration of broadcasting from analog to digital signals made the satellite industry more efficient by allowing more content to be broadcast per transponder. In addition, the merger of major capacity lessors (for example, within the direct to home—DTH—satellite television sector) reduced the need for transponder capacity to broadcast programming. Even after a wave of consolidation in the fixed satellites services operators market, there remained about 30 operators worldwide with satellites in orbit as of 2006. Consolidation improved profitability as the consolidated companies procured satellites, insurance, and launches cheaper for their fleets in addition to rationalizing operating and marketing costs. More important, it enabled the consolidated companies to reallocate satellites among their fleets depending on changing market conditions.

The market for satellite communications expanded at a sustained rate during the last decades of the twentieth century, with annual revenues by 2006 estimated in the range of $50–60 billion. Governments, and particularly the military, continued to use commercial satellite capacity for their various communications needs, peaking during wartime.

Benoit Denis and Stephen B. Johnson

See also: Commercial Satellite Communications, Military Communications, Space Telecommunications Policy

Bibliography

Andrew J. Butrica, ed., *Beyond the Ionosphere: Fifty Years of Satellite Communication* (1997).
International Space Business Council, *State of the Space Industry 2004* (2004).
David J. Whalen, *The Origins of Satellite Communications, 1945–1965* (2002).

Economics of Space Navigation

Economics of space navigation grew significantly in importance at the end of the twentieth century, fueled by a plethora of civilian uses for military navigational satellite systems, particularly the Global Positioning System (GPS). Military applications were the original motivation for satellite navigation, including precise delivery of weapons to targets; more accurate location and direction of friendly forces on land, sea, and air; and synchronization of electronic systems through GPS-accurate clocks.

A U.S. Marine consulting a Global Positioning System (GPS) unit, 2005. (Courtesy U.S. Department of Defense)

Developed and deployed by the Navy in the late 1950s–60s at an annual cost of roughly $24 million per year, Transit was the first operational satellite-based navigation system used for location of ballistic missile submarines and surface ships. Transit comprised a total of seven low-altitude polar-orbiting satellites, whose signals were made available to civilian users in 1967. It was quickly adopted by a large number of commercial marine navigators and owners of small pleasure craft.

After Transit's proven success, three U.S. armed services began experimental programs to improve its capabilities. The Navy's Timation, the Army's Sequential Correlation of Range (SECOR), and the United States Air Force's (USAF) Program 621B provided technological elements of the tri-service GPS program started in 1973. GPS was the only fully functional global satellite navigation system as of 2007, requiring 24 satellites for full functionality, with four each in six different orbital planes. The USAF has kept up to three spares in orbit also, with annual research, development, and operating costs in the early twenty-first century on the order of $250–500 million.

The system has provided a Standard Positioning Service and an encrypted higher-precision signal. After the Soviet Union shot down a commercial airliner (Korean Air Flight 007) when it strayed into Soviet airspace in 1983, the U.S. administration, under President Ronald Reagan, made GPS signals available for civilian use. After that time the GPS signal was free for anyone to use, although the U.S. military reserved the right to make it unavailable in emergencies. In 2000 the U.S. administration, under President William Clinton, turned off the high-precision signal encryption, making this more accurate signal available to civilian and commercial users, and stated that U.S. policy was to keep the signal available at all times for civilian and commercial purposes.

Though designed for military applications, GPS was soon used for a host of civilian and commercial applications. This was driven by the development and miniaturization of GPS receivers, which between 1990 and 2006 declined in price by a factor greater than 15 while new functionalities were added to the marketed products. By 2006 the technology was considered accessible to virtually everyone in developed countries. GPS found its way into the mobile handsets market, nuclear detection systems, navigation, cars,

GPS Application Market Value by Application, 2003

Application	Value in Millions, USD
Maritime	450
Aviation	400
Location-based Services	400
Telecom	250
Element Management System	200
Telematics	200
Leisure	175
Surveying	150

Source: Frost & Sullivan

boats, planes, construction equipments, movie-making gear, farm machinery and other agricultural applications, robotic vehicles, laptop computers, and animal tracking, essentially becoming a universal utility. Between 1996 and 2000 GPS applications markets grew from annual revenues less than $2 billion to the range of $6–8 billion.

The Soviet Union, and later Russia, developed and operated navigation systems for military and civilian purposes, the best known of which is Global Navigation Satellite System (GLONASS), operated by the Ministry of Defense. Manufactured by Polyot, the GLONASS system was launched in 1982 and became fully operational in 1993. Like the U.S. GPS, the GLONASS system required 24 satellites for full functionality and provided two types of signals, one for military purposes and a degraded signal for commercial purposes. Because Russia was unable to keep the full system operating after the collapse of the Soviet Union in 1991, and due to its signal being less accurate than GPS, commercial uses of GLONASS typically combine GLONASS signals with GPS signals to improve accuracy.

To reduce dependence on the United States and to promote European industry, which by the early twenty-first century handled a significant portion of total worldwide sales of navigational receivers, the European Union and European Space Agency (ESA) agreed in March 2002 to introduce their alternative to GPS, called Galileo. This system, funded also by India and China, was designed to operate on a commercial basis with lower precision signals free to users, but higher precision signals available for a fee, depending on the precision needed. As a precursor to Galileo, ESA, the European Commission, and EUROCONTROL (which oversees air traffic control in Europe) developed the European Geostationary Navigation Overlay System (EGNOS) at a cost of roughly €300 million to supplement the GPS and GLONASS systems by reporting on the reliability and accuracy of the signals, allowing position to be determined to within 5 m. EGNOS, which operates by sending signals to three geostationary satellites and back to a network of ground stations, began operations in July 2005. The first Galileo satellite, *GIOVE A* (Galileo In-Orbit Validation Element) was launched on 28 December 2005, at a cost of roughly €150 million. ESA and Galileo Industries signed a €950 million contract the next month to begin building the full constellation.

Benoit Denis and Stephen B. Johnson

See also: Commercial Navigation, Navigation

Bibliography
European Space Agency, Galileo. http://www.esa.int/esaNA/galileo.html.
Paul B. Stares, *The Militarization of Space: U.S. Policy, 1945–1984* (1985).
U.S. Department of Commerce, *Trends in Space Commerce* (2001).

Economics of Space Reconnaissance and Remote Sensing

The economics of space reconnaissance and remote sensing was initially monopolized by military and intelligence organizations in the United States and the Soviet Union. Satellite Earth observation developed during the late 1950s and the 1960s as a continuation of aerial photography for the purposes of identifying military and industrial capabilities, such as nuclear production plants and test sites, military facilities and large weapons, and missile launch sites. New instruments, forming images in the infrared in addition to visible light, produced valuable information. Programs such as Corona and Samos in the United States, and Zenit in the Soviet Union, drove remote sensing technology development, as military and intelligence requirements led to the creation and expansion of both government and industry organizations, such as the National Reconnaissance Office, Itek Corporation, and Lockheed Company in the United States, and OKB-1 (Experimental Design Bureau) in the Soviet Union.

Even at the beginning of the twenty-first century, the massive expenditures on these systems in the 1950s–60s (and after) remained classified, but in the United States were purported to be on the scale of the human spaceflight program. Civilian analysts in the early twenty-first century estimated U.S. reconnaissance expenditures in the range of $6–8 billion per year. With reconnaissance satellites unable to remain on orbit for long, because they had to return their film to Earth, each superpower launched dozens of reconnaissance satellites in the 1960s and early 1970s. However, the reconnaissance-driven development of high-resolution digital imaging, sent back to Earth using radio waves instead of film, meant that later systems first deployed in the late 1970s and 1980s, such as the KH-11 in the United States and the Yantar Terilen in the Soviet Union, launched fewer satellites, though the ones launched were significantly more complex and expensive than their predecessors in the 1960s. The military focused initially on achieving the highest possible resolution, which by the 1970s reportedly reached roughly the 1 m resolution range. China's Fanhui Shi Weixing satellites, first launched in 1974, comprised that nation's first reconnaissance system.

By the 1990s, military remote sensing satellites were being commissioned, manufactured, and used by regions outside the United States and Russia. In 1995 Israel launched its first Ofeq satellite, as did France with its first Helios satellite. In 1996 Chile emerged with *FASat* (*Fuerza Aéra de Chile Satellite*), designed and developed by Surrey Satellite Technology of the United Kingdom. Japan launched the first of its Information Gathering Satellites in 2003, driven largely by concerns about North Korean military capabilities. While by 1999 it was more cost effective for these nations and others to purchase commercial satellite imagery for national intelligence (which they frequently did), they nonetheless desired control of their satellites and were willing to spend the hundreds of millions of dollars or more to develop them.

Along with reconnaissance satellites, military planners recognized the need for wide-angle, low-resolution imagery from weather satellites to aid tactical operations and to ensure that the new reconnaissance satellites did not take pictures of clouds. In the United States, the Defense Meteorological Satellite Program initially supported the Corona satellite reconnaissance program, but soon was utilized for tactical forecasts, while the Tiros satellites were developed by the U.S. Army and later NASA for civilian weather forecasts. The Soviet Union developed its weather satellites, known as Meteor, while other nations (including Japan, India, and China) and Europe ultimately developed their own systems, contributing to a worldwide network of satellites placed in geosynchronous and polar orbits. U.S. weather satellite expenditures in the early twenty-first century were about $300 million from the military and an equivalent figure from civilian sources. Worldwide, the demand for weather satellites has remained quite stable, as they have become essential utilities maintained by national governments for the good of their national economies and military capabilities.

With the launch of *Landsat 1* in 1972, civilian scientists began to assess land use. Developed by NASA and operated by the Department of Commerce, Landsat spawned a number of new uses by scientific users and government planners and analysts. These "commercial" uses prompted the U.S. government to privatize Landsat operations through the Earth Observation Satellite Company (EOSAT) in 1985. EOSAT's consequent difficulties, due to congressional restrictions and the large increase in prices to fund operations, led to a backlash against the privatization, and in 1992 the privatization policy was revoked, transferring control of future Landsat satellites to the National Oceanic and Atmospheric Administration (NOAA).

Landsat's success spurred several nations to develop their remote sensing satellites, under government or semiprivate control. France's Satellite Pour l'Observation de la Terre (SPOT) satellites, first launched in 1986, were developed by the French Space Agency (CNES) but were transferred after launch to SPOT Image, a private company partly owned by the French government, to operate on a commercial basis. European nations, through the European Space Agency, developed and operated the European Remote Sensing satellite series, first launched in 1991, and later *Envisat*. India made remote sensing one of its primary space objectives; the Indian Remote Sensing (IRS) satellite series, first launched in 1988, showcased India's burgeoning space capabilities. Starting in 1992, IRS imagery was marketed through the private corporation Antrix.

The 1992 congressional act that transferred future Landsats to NOAA created a regulatory framework for high-resolution commercial remote sensing satellites in the United States. Drawing from military and civilian technologies, commercial companies quickly formed. These included Orbimage, Space Imaging, which merged in 2005 to create GeoEye, and Earthwatch, which became DigitalGlobe in 2001. These companies launched their remote sensing satellites and marketed satellite imagery for a plethora of end users, including military and intelligence agencies, regional and local planning, scientific studies of global climate and land use changes, prospecting, and news media, among others. These companies competed with government civilian agencies, and their private marketing companies, and with some military organizations to sell imagery.

The market for reconnaissance and remote sensing products can be divided into several pieces, including the manufacture of the satellites, the provision of the raw images, and the "value-added" services that modify and incorporate imagery into products tailored for specific uses. Value-added services are the largest by revenues, followed by satellite manufacturing, with the raw imagery sales the smallest piece, remaining less than $1 billion annually in the early twenty-first century. By 2006 the value-added market, which included Geographic Information System (GIS) Software, GIS Data, and GIS Value-added Services Market exceeded $5 billion.

Raw imagery, the mapping data, is generally provided in panchromatic (black and white) or multispectral (different colors) form, with the resolution depending on the application. Weather satellites use low resolutions on the order of kilometers per pixel, whereas high-resolution imagery was available commercially at 1 m resolution by 1999. Military resolutions remain classified, but are reputedly on the order of 2–4 in. The trend was to provide higher resolutions, which created downward pressure on imagery pricing. This imagery was a key base layer for innumerable government and commercial data products.

Federal, state, and local governments remained the dominant users of remote sensing data into the early twenty-first century. With U.S. government policy requiring annual purchases of commercial imagery to keep U.S. private companies viable, these data suppliers built business plans around large-scale yearly purchases of commercial data in the hundreds of millions of dollars from the U.S. government alone. With greater awareness of threats to the environment, governments, nongovernmental organizations, and other end users also demanded ways to develop a better understanding of the effects of environmental processes. At the end of the twentieth century, environmental applications grew in importance and revenues, such as crop monitoring and crop health, water temperature monitoring, and smog level monitoring.

By the early 2000s mergers and acquisitions in remote sensing markets signaled an industry-wide desire to bring more capabilities in house. Many of these capabilities were outside the traditional definitions of remote sensing and GIS. In many cases information technology, defense electronics, communications, software, and positioning companies became subsidiaries or holders of companies heretofore solely known for remote sensing and GIS. These changes continued the transition of remote sensing and GIS software markets from the government-dominated domain into commercial and consumer markets.

Benoit Denis and Stephen B. Johnson

See also: Commercial Remote Sensing, Private Remote Sensing Corporations, Reconnaissance and Surveillance, Remote Sensing Value-added Sector

Bibliography
John C. Baker et al., eds., *Commercial Observation Satellites: At the Leading Age of Global Transparency* (2001).
Parker Temple, *Shades of Gray: National Security and the Evolution of Space Reconnaissance* (2005).
U.S. Department of Commerce, *Trends in Space Commerce* (2001).

Economics of Space Science

Economics of space science has been controlled by governments since inception of the space age, motivated by several factors, including practical military and civilian needs to understand the space environment, international political competition and cooperation, national technological development, and curiosity about the nature of the universe and of Earth. These motivations have combined to create relatively stable government funding of space science from the 1950s through the early twenty-first century. However the internal competition among scientists and research and development organizations to acquire that funding has often been fierce.

The Cold War competition between the United States and the Soviet Union was the primary motivation for the first phase of space science activities. From 1945 to 1957 the military funded space science as part of flight testing of ballistic missiles in the United States and Soviet Union, and also in support of the planned International Geophysical Year (IGY) of 1957–58. Funding of the IGY Vanguard program was secretly driven by the desire of U.S. President Dwight Eisenhower's administration to have a peaceful scientific satellite set a precedent for later space reconnaissance missions flying over the Soviet Union. Much of this early work focused on understanding the upper atmosphere and near Earth space, through which ballistic missiles had to travel. Following the formation of NASA in October 1958, the new U.S. civilian space agency allocated roughly 10–15 percent of its funding in its first decade to space science, much of it to compete with the Soviet Union in a robotic race to the planets. The Soviet Union provided major resources to engage in the race, with robotic missions to the Moon, Venus, and Mars.

Unlike the Soviet Union, the United States also supported classical astronomy with space observatories. While ground-based observatories were far less expensive, spacecraft such as the Orbiting Astronomical Observatories and the High Energy Astronomy Observatories provided access to electromagnetic wavelengths not visible on Earth's surface because they were absorbed by the atmosphere, such as ultraviolet and X rays. Other satellites such as the *Hubble Space Telescope* observed in the visible spectrum that could be seen from Earth's surface, but without atmospheric distortions. By the 1990s the development of adaptive optics made this less an advantage for space-based systems. Within NASA, planetary scientists and classical astronomers competed for influence, promoting planetary probes and observatory missions respectively. In the 1960s planetary probes, which were key to the robotic space race, predominated, but by the 1970s and 1980s observatory missions grew in importance, with NASA's funding of astronomy and physics programs overtaking its planetary programs by the late 1970s.

The United States used space science as a tool of national policy, to foster good relations with its allies. In 1959, it offered to launch, free of charge, science satellites from each nation. By the mid-1960s this was clarified to limit this to two free launches. The offer was extended to national groups, such as the European Space Research Organisation (ESRO). Japan and Western European nations took advantage of the offer. In Western Europe, ESRO was created in 1964 to provide a mechanism for European nations to pool their resources to build and launch expensive space science satellites. Its success was written into the charter of the European Space Agency (ESA), which was created in 1975. ESA Member States were required to support space science, technology programs, and basic infrastructure, with contributions determined as a percentage of their nation's gross national product. Space science projects were frequently used in bilateral and multilateral efforts, with different nations contributing scientific instruments, the satellite bus, and launches, with exchanges of the spacecraft's scientific data a major feature of the bargain. National governments typically then funded their own universities, corporations, and government laboratories to build their contributions to the project. The Soviet Union mirrored the U.S. strategy, offering cooperative relationships on scientific space missions with its communist bloc allies through the Interkosmos program.

Microgravity science was a relative latecomer in scientific efforts but became a major topic of space science by the 1970s–80s, usually tied to human spaceflight. The Spacelab program was a good example. Built by ESA as its contribution to the Space Shuttle program, *Spacelab* was a laboratory in which astronauts could conduct a variety of experiments, most of them related to microgravity phenomena. NASA and European nations did not want to fly an empty laboratory, so they placed significant funding into the creation of microgravity experiments, creating new scientific disciplines and organizations in the process. This was opposed by traditional astronomers and physical scientists, who believed it was a waste of money, at least in comparison to funding their fields. Similar funding battles occurred in the United States, where NASA directly funded microgravity science, which grew into the hundreds of millions of dollars per year with the International Space Station program in the 1990s.

While some private companies funded microgravity experiments, the promises of revolutionary materials and medicines proved elusive. The greatest uses of microgravity science were to assist with the space program itself, through better understanding of microgravity effects on solids, fluids, and organisms, including humans.

Government funding of space science generally flowed through major science organizations and space agencies, with advice on how to allocate those funds given by prominent scientists and scientific organizations. For example, in the United States the bulk of space science funding funneled through NASA, including funding for some ground-based observatories of direct benefit to the space agency's missions. The National Science Foundation (NSF) also funded space science, though the bulk of its funding went to ground-based initiatives. Both depended on the National Research Council (NRC) for advice, and the NRC in turn provided decennial assessments that helped guide both NASA and NSF astronomy programs.

Competition for space science funding has been a multisided affair, involving government laboratories, universities, nonprofit companies, and corporations. For scientists, whether working in government laboratories, universities, or nonprofit corporations, space science funding could make or break their careers, particularly because the missions were few in number but well funded if the scientist won the competition to build an instrument. For private companies, space science was usually a small but highly visible portion of their business, providing opportunities for them to advertise and recruit personnel, which was generally difficult to do when most of their products were highly classified. Government laboratories often had an insider's edge in the competition, but faced national policies that tried to direct funding to private companies and universities. For example, ESRO member nations worked hard to ensure funding of scientific instrument development was kept away from ESRO scientists and steered to organizations within their nations, whether in industry, academia, or government. Over time, however, the cost and complexity of these instruments has resulted in more funding to government and corporate scientists and engineers.

By 2007 global space science funding totaled roughly $6–7 billion annually.

Stephen B. Johnson

See also: Astrophysics and Planetary Science, European Space Agency, European Space Research Organisation, Japan, Microgravity Science, National Aeronautics and Space Administration, Universities

Bibliography
J. Krige and A. Russo, *A History of the European Space Agency*, 2 vols. (2000).
NASA Historical Data Books (series).
National Research Council, *The Decade of Discovery in Astronomy and Astrophysics* (1991).
State of the Space Industry (series).

Economics of Space Technology Research and Development

The economics of space technology research and development (R&D) has been an important element of space programs from their inception in the late 1950s. One of the major considerations of civilian space programs has been the stimulation of the economy from the investments in space technology. Space R&D has had long-term impacts on industrial development and on improvements in the quality of life. It has stimulated the education of a new generation of scientists and engineers specializing in space science and applications and initiated the development of new industries around the space infrastructure, including a number of telecommunications services that would otherwise be inefficient or impossible to deliver.

The importance of space has to be put in economic perspective. Space is not a particularly large segment of the economy. In 2004 the gross domestic product (GDP) of the major industrialized nations was more than $30 trillion. Worldwide, annual total expenditures on space activities were between $100–150 billion, or roughly one-half of 1 percent of the GDP.

Space technologies have received a large amount of attention, despite the rather small direct contribution to GDP. This is the result of four factors: the significant political and security impact of space, the long tradition of public interest in the unknown and in exploration, the perception that high-technology industries such as space spur a variety of technological advances and spin-offs, and the growing importance of the space infrastructure to the development of the rapidly growing information and high-technology economic sectors of the world.

The economic infrastructure from in-place satellites has included a wide range of services that is often not visible to the average consumer. Global navigation satellites, for example, provide location services for boaters and hikers and also provide crucial timing information for communications, pagers, and other services. Weather satellites provide visual television pictures of clouds for the news and economically important forecasts that help farmers, electric utilities, and other weather-dependent industries to optimize their long- and short-term business decisions. Remote sensing satellites show land use patterns, provide business and governments with valuable environmental information, and contribute to disaster relief through satellite images and analyses of natural disasters.

Examples of several major space research-driven technologies include propulsion, materials, electronics and avionics, communications, and instruments. The aerospace industry has required these technologies to explore and use the outer space environment. In addition the companies producing these products and services for the aerospace industry have developed remarkable consumer applications based on these innovations that have improved the quality of life and created jobs and business opportunities.

NASA and the European Space Agency (ESA), in addition to national space programs, have made consistent and concerted efforts to move the knowledge and innovations of their R&D investments to the commercial mainstream. Through information dissemination, publications, programs, and funding demonstration projects, they have encouraged small and large firms to commercialize these technologies. Taking technology from the R&D stage to the commercial stage is not a straightforward process. It often involves iterations, false starts, and loops back to research. Actual technology transfer may involve manufacturing of goods, provision of services, improvement of processes, and the improvement of the quality of life through social benefits that are not directly measurable by GDP.

Most space technology innovations directly improve the efficiency, safety, and costs of new space and defense equipment and services. However they also frequently impact other economic sectors. Only a few studies have attempted to measure the economic benefits from these (and other) NASA investments. One study in the mid-1970s selected four technological improvements (integrated circuits, cryogenics, gas turbine engines, and a computer program simulating large structures) and analyzed NASA's role in deploying these innovations into commercial applications. The study calculated that the aggregate impact from only those four technologies showed cumulative economic benefits larger than the entire NASA budget for 1975. Other studies focusing on particular industries or technologies have also found successful examples of space research having an impact on economic sectors unrelated to space activities. In the life sciences, for example, a 1997 study showed that a selection of firms that received NASA R&D funding modified these technologies and improved such diverse products as sunglasses, thermometers, instruments used in noninvasive medical diagnosis, and materials used in firefighting clothing. NASA funds stimulated many multiples of additional privately funded space R&D, and the space-related innovations were a direct link to new medical and consumer goods and services.

Estimates of the long-run productivity benefits and economic returns to space R&D have ranged from 3:1 to more than 14:1. Any government expenditure will create jobs and income, as long as the money is being spent. However, the productivity benefits associated with R&D are far more significant because the technological impacts on the economy continue to the indefinite future. Thus it is significant that all the studies of space benefits, whether on the U.S. economy or foreign economies, have shown large positive effects much greater than the space R&D funds expended. Often the innovations are a combination of knowledge from a variety of sources, but had it not been for the specific needs of the space programs, these improvements would either not have been developed or would have been developed at a much slower pace. Examples of these specific needs include requirements such as very lightweight strong materials, miniaturization, photovoltaic energy cells, and advanced optical sensors.

NASA, the Department of Defense (DoD), ESA, and national space agencies have line items in their budgets for research and technology development and transfer. Research is oriented toward advancing knowledge while development is aimed at working prototypes. In 2005 NASA's budget allocated approximately $5 billion to research and $10 billion to development. In the commercial sector R&D activities focus on efficient operations to develop new products to satisfy consumer markets and generate investor profits. Not only do R&D expenditures dedicated to space activities influence space technology, but in fact, advances in electronics, biotechnology, optics, chemicals, and many other sectors also contribute to space technology just as space technology has spun off innovations into many other fields.

To facilitate the process of taking new space technology and encouraging commercial uses, governments have established a number of special programs. They have included the Small Business Innovation Research program in which 2.5 percent of the major government R&D agency budget (roughly $1.3 billion in 2005) is earmarked for small businesses in the United States who compete for the funds to develop innovations that will be useful for agency missions in addition to having commercial value. Extra independent R&D funds were awarded primarily along with defense contracts to encourage contractors to further develop their technologies for commercial and government uses. These IR&D funds totaled about $2 billion out of a $70 billion DoD R&D budget. Many agencies created special technology transfer offices to collect and disseminate information about different innovations developed by R&D programs and to find private sector partners and users for those innovations. Each agency also created patent and licensing programs to obtain patents for government employees who developed new technology and to administer licensing programs that allowed private firms to use those technologies commercially, sometimes with royalty payments to the government.

Henry Hertzfeld

See also: Economics and Technology in Space Policy, Nations

Bibliography
David Baker, *Inventions from Outer Space: Everyday Uses for NASA Technology* (2000).
Henry Hertzfeld, "Space as an Investment in Economic Growth," in *Exploring the Unknown, Volume III: Using Space*, ed. John Logsdon et al. (1998), 385–400.
"Symposium on Technology Transfer in the Space Sector," *Journal of Technology Transfer* 27, no. 4 (2002). *Spinoff* (annual).

Economics of Space Transportation

The economics of space transportation is a crucial underlying factor in getting to and returning from space. Space transportation is a complicated, expensive, and risky business involving huge initial investments in hardware and technology, along with a supporting infrastructure for building launch vehicles, providing a safe launch site, and monitoring the vehicles in flight. It has been marked by dramatic successes in addition to some well-publicized catastrophic failures. Overall, launches into Earth orbits have been successful approximately 95 percent of the time.

Predictions during the 1960s–70s estimated that the cost of getting to space would decrease by orders of magnitude, from more than $10,000/lb to as low as $100/lb at the end of the twentieth century. The development of the Space Shuttle in the early 1970s was also heralded with the promise of cheaper access to space. With falling launch costs, space would open new markets and offer consumers new services. Unfortunately these predictions were not realized, and the cost of access to space was about the same in 2007 as in 1965. In the early twenty-first century the use of space remained limited, and both governments and new entrepreneurs in private companies have continued working toward getting there more efficiently.

The decade of the 1960s was characterized by rapid developments in launching capabilities, mainly in the ability to launch increasingly heavy payloads. The economics of launch vehicles in this era was fairly simple. Government mission objectives of the Cold War in both the United States and the Soviet Union trumped all normal economic constraints, and for a unique and brief period in history, access to space was a goal that overrode cost/benefit analyses, at least in easily measurable monetary terms. That period ended in the 1970s, and after that the development of new launch vehicles and other space programs was slowed by technological and budgetary constraints.

The development of the Space Shuttle during the 1970s and the first launch in 1981, however, promised a new era in space transportation. Although the Shuttle could return to Earth and depart again to space, it was not a truly reusable vehicle. It proved to be expensive to operate because the fleet never met its original mission model of flying more than 30 times per year, which would have spread the high development and operations costs over many flights. Even government and private research and development (R&D) experiments were difficult to manifest on the Shuttle because of launch delays and the dedicated use of the Shuttle to construct the *International Space Station*. NASA set a date of 2010 for the last Shuttle flight. As of 2007 launch technology appeared inadequate for the development of a reusable Single Stage to Orbit (SSTO) vehicle.

At the beginning of Shuttle operations and before the Shuttle *Challenger* accident in January 1986, NASA initiated a policy of putting all U.S. payloads (including commercial telecommunications satellites) on the Shuttle and phasing out new R&D investments for expendable launch vehicles (ELVs). This policy had two major long-term economic effects: it encouraged other nations to develop competitive commercial launch vehicles, and it put the United States behind in ELV technology. After *Challenger*, the Commercial Space Act of 1988 reformulated U.S. policy to stimulate commercial development of ELVs and to have more than one vehicle capable of putting payloads in space. The Shuttle and its major competitor during the 1980s, the European/French launcher Ariane, were heavily supported by their respective governments and had similar pricing policies for commercial payloads.

A true market price did not develop in this industry because of government support and the dual use (combined security and commercial uses) of launch vehicles.

In addition the development of a commercial launching business in historically non-market driven economies such as Russia and China in the 1990s added even more complicating factors to the international pricing of launch vehicles. The industry has been characterized by deeply discounted prices for the vehicles from nonmarket nations, politically instituted restrictions on the number of U.S. payloads that can be flown on some of these vehicles, and export controls on U.S. space technology. Although economic competition exists among launch vehicle providers (for example, they do bid on prices for individual launches), the actual pricing of a launch does not resemble the open competition for most goods and services.

Governments continue to regulate access to space because of its strategic importance. Commercial uses of space are encouraged, but within limits defined by a combination of complex rules. These regulations range from the ordinary and expected areas of public safety and financial responsibility (making sure that the commercial launch vehicles are adequately insured), administered by the Federal Aviation Administration of the Department of Transportation in the United States, to national security policies designed to ensure that sensitive technologies are not made generally available to all nations.

Since the mid-1980s the global aerospace industry has responded with improvements in ELV reliability and performance and also with proposed designs for innovative suborbital vehicles that may someday evolve into less expensive access to low Earth orbit space. From an economic standpoint two factors have led to a consistent oversupply of launch vehicles: first, the political aspect of launch vehicles that makes nations desire independent access to space at any cost and second, incorrect long-term market analyses predicting communications satellites and services that did not materialize.

However as the accompanying chart indicates, commercial launches were significantly less frequent in the 2000s than in the 1990s, averaging about 20 launches per year worldwide. Making the case for space transportation as a growing and profitable economic activity rests on two demand factors: robust government programs

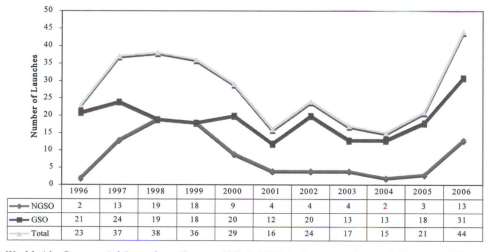

	1996	1997	1998	1999	2000	2001	2002	2003	2004	2005	2006
NGSO	2	13	19	18	9	4	4	4	2	3	13
GSO	21	24	19	18	20	12	20	13	13	18	31
Total	23	37	38	36	29	16	24	17	15	21	44

Worldwide Commercial Launches. (Source: U.S. DOT/FAA Commercial Launch Forecast, 2006)

purchasing private launches, and/or the development of the demand for new services in space that would increase the need to get there and back. The projected increase in launches for 2006 reflects improved economic conditions and a catching-up from the depressed levels in the early 2000s. In 2007, longer term projections were for relatively level year-to-year commercial launches averaging about 20 per year.

In 2004 the launch vehicle industry generated between $4–5 billion per year in revenues. Heavy-lift launch ELVs such as Atlas 5 (Lockheed-Martin Corporation), Delta 4 (Boeing Company), Ariane 5 (Arianespace), SeaLaunch, Proton (manufactured by Khrunichev and marketed by International Launch Services), will continue to be relatively expensive with costs averaging between $5,000–10,000/lb lifted into low Earth orbit. Some market price reductions may be possible, if demand increases and companies can produce more vehicles, but major price reductions will have to await new large R&D investments by governments and many years to develop, perfect, test, and make new vehicles operational.

Launching smaller payloads could result in lower costs per pound through the innovative work of entrepreneurial private companies. The successful X Prize flight in 2004 by Scaled Composites' vehicle, *SpaceShipOne*, demonstrated that reaching the edge of space in a reusable and relatively inexpensive vehicle was possible. However as of 2007 it was yet to be proven that these vehicles could operate safely and on schedule over time, make a profit, and be scaled to the size and energy requirements needed to place payloads into orbit. Other new companies, such as Space Exploration Technologies Corporation (SpaceX), were developing launchers capable of putting all but the heaviest payloads into orbit at less cost than current vehicles. As of 2010 only one of these entrepreneurial firms (SpaceX) had had a successful launch. None had yet demonstrated a reliable launch system, although several appeared closer to their objectives. The biggest economic challenge for launch vehicle manufacturers in the twenty-first century will be to prove that the success rate and cost of new vehicles will actually create the market opportunities that will open new uses of space and push space activities into the mainstream of commerce.

Henry Hertzfeld

See also: Arianespace; Boeing Company; Commercial Space Launch; Lockheed Martin Corporation; Russian Launch Vehicles; Space Shuttle; Space Transportation Policy

Bibliography
S. J. Isakowitz et al., *International Reference Guide to Space Launch Systems*, 4th ed. (2004).
John M. Logsdon, ed., *Exploring the Universe, Volume 4: Accessing Space* (1999).
U.S. Department of Transportation, Office of Commercial Space Transportation. http://www.faa.gov/about/office_org/headquarters_offices/ast/.

EDUCATION

Education has been a prominent support mechanism and beneficiary of space programs since the beginning of the space age. High levels of technical capability, which required equally high levels of scientific and technical education, were needed for spaceflight to become possible. College degrees in science or engineering were necessary for spaceflight's "rocket scientists" to succeed. However, public interest in human spaceflight has also been important by fostering general interest in scientific and technical topics and education.

In the public debate about the launches of *Sputnik* and *Sputnik 2* in October and November 1957, political, military, industrial, and educational leaders lamented the perceived lack of interest of U.S. students in technical subjects. Against this, they contrasted the much greater number of technical degrees given in the Soviet Union per capita. President Dwight Eisenhower signed the National Defense Education Act on 2 September 1958, authorizing just under $1 billion to fund science, mathematics, and foreign language teachers, programs, equipment, and students. It also highlighted the two-way connection between space programs and education: space programs needed technically educated workers, but technical education programs also needed space programs. Space projects captured the imagination and inspired greater student interest in these subjects, and these students then improved national technical capabilities, which in turn improved military and economic competitiveness. This logic was frequently used in other nations as a justification for spaceflight.

NASA has supported a variety of educational programs at universities and in elementary and secondary schools. In the 1960s NASA Administrator James Webb created the Sustaining University Program to maximize the benefits of space programs on Earth for research, education, and economic development. In 1989 NASA's university programs became the National Space Grant College and Fellowship Program, with funding to every state. NASA also spent significant resources developing and distributing primary and secondary school educational materials for teachers, by the 1990s often through the Space Grant program.

NASA and other space agencies and organizations also directly or indirectly supported the development of space museums and science centers, such as the U.S. Space and Rocket Center founded in 1968 near the NASA Marshall Space Flight Center in Huntsville, Alabama. This was also the site of the U.S. Space Camp, with its first students arriving in 1982. The Soviet Union's Young Cosmonauts program began in the 1960s for the same reasons. Museums focused on space exploration were very successful, exemplified by the Smithsonian Institution's National Air and Space Museum, opened in 1976. Kennedy Space Center Visitor Complex, opened in 1968, drew more than one million visitors per year. In early 2009 China was developing a space theme park on Hainan Island, near a new launch facility then in early development.

More pragmatically, space industry and government workers needed technical education services to learn and keep up with rapidly changing technology. Universities expanded undergraduate and graduate aeronautical and other engineering programs to meet this need. NASA created its own internal training programs on a variety of technical subjects that allowed employees to continue to maintain technical currency. They also entered into arrangements with local universities to provide academic courses as part of degree programs. Space industry did the same. Space advocacy and professional groups, such as the American Astronautical Society and the American Institute of Aeronautics and Astronautics, offered technical training courses and published textbooks. New profit-oriented companies successfully marketed professional training courses in various technical and managerial subjects.

In the early twenty-first century spaceflight's benefits to and need for technical education continued to be one of its main justifications.

Stephen B. Johnson

See also: Effect of Sputnik, National Aeronautics and Space Administration

Bibliography

Mike Gruntman, "The Time for Academic Departments in Astronautical Engineering," *American Institute of Aeronautics and Astronautics Space 2007 Conference and Exposition* (2007).

V. V. Malyshev, *Space Education in Russia* (2003).

Barnes McCormick et al., *Aerospace Education during the First Century of Flight* (2004).

Careers

Careers in spaceflight have often been viewed as limited to engineers, wielding slide rules and sporting pocket protectors, the quintessential rocket scientist, and the astronauts who present the public face of the U.S. space program. Regardless of the caricature, there would be no space program without the individual and group efforts of engineers and scientists from across the spectrum of disciplines—engineers from the traditional brick-and-mortar disciplines (for example, chemical, civil, electrical,

Mobility engineers Christopher Voorhees (left) and Brian Harrington test the Mars Exploration Rover B (Opportunity) suspension and wheel capability on staggered ramps in the spacecraft assembly facility at NASA's Jet Propulsion Laboratory. (Courtesy NASA/Jet Propulsion Laboratory)

mechanical) and more recent disciplines (for example, biomedical, computer, human factors) in addition to scientists ranging from specialists in fluid and combustion phenomena in low-gravity conditions to astronomers peering into the far reaches of space to life scientists who examine the effects of long-term exposure of living organisms to the space environment. Less prominent but just as important are a variety of other workers. For example, political scientists and lawmakers establish and maintain a supportive political environment; medical professionals ensure the health and safety of space travelers and the general workforce; broadcast and print-media journalists chronicle the successes and challenges of spaceflight; artists and writers provide the left-brain perspective associated with exploring the cosmos; technicians and trade specialists ensure the day-to-day operations of infrastructure, including test and launch facilities; teachers (educators) use the space program as a catalyst for inspiring the space industry workers of tomorrow; and historians document the successes (and failures) of the past to provide guidance for the future.

The general public interest in and awareness of the need for renewed emphasis on scientific and technical training in the United States is commonly linked with the launch of *Sputnik* in 1957 and the perceived challenge to U.S. technical preeminence. The U.S. space program was catalyst for and beneficiary of the new emphasis on technical education and careers. The U.S. National Defense Education Act of 1958 was enacted to stimulate educational interests of U.S. students, with a particular interest in math and physical sciences, by providing federal funding for low-interest student loans.

After dramatic growth up to the mid-1960s, employment in NASA programs was in general decline from a peak of 36,000 civil servants (and more than 273,000 contractor employees) in 1967 to just more than 19,000 civil servants (and 158,000 contractor employees) in 2006. Civil servants have generally comprised less than 20 percent of the total employment in NASA programs, and that percentage dropped to about 10 percent and held steady at that level from the 1990s through at least 2006. Between 1960 and 1988, the number of scientists and engineers rose from one-third to more than one-half of the NASA workforce. Aside from the initial burst of the space race in the early 1960s, which was unique to the United States and the Soviet Union, similar trends have been evident in the European space industry, which saw a decline in direct employment from more than 35,000 workers in 1997 to fewer than 29,000 workers in 2006.

The efforts of the U.S. civil rights movement succeeded in expanding the pool of available talent to include women and minorities, which had previously been largely absent from the professional ranks of a predominantly white male workforce. The percentage of women and minorities in the professional ranks within NASA steadily rose from about 2–3 percent each at the height of Apollo in 1969 to 10–15 percent each by 1988.

Since the 1980s the private space sector has offered growing employment and career alternatives for professionals outside the traditional government and military-industrial complex. In addition the commercial space industry has provided space-related employment opportunities associated in areas such as telecommunications (for example, cellular telephone systems and satellite television service) and global positioning systems.

The explosive growth of the World Wide Web through the 1990s, which could hardly have been envisioned during the early years of spaceflight, further expanded professional career opportunities in areas such as special-interest websites and news outlets, Earth imaging (such as Google Earth), weather forecasting, and distance learning.

Michael L. Ciancone

See also: Effect of Sputnik, World Wide Web, Economics of Space

Bibliography

Carsbie Adams and Wernher von Braun, *Careers in Astronautics and Rocketry* (1962).
Aerospace Industries Association. www.aia-aerospace.org.
Eurospace. www.eurospace.org.
NASA Historical Data Books.

International Space University

International Space University (ISU) is an interdisciplinary space education institution founded in 1987 by Peter H. Diamandis, then an undergraduate at the Massachusetts Institute of Technology (MIT); Todd B. Hawley, an undergraduate at George Washington University in Washington, DC; and Robert D. Richards at the University of Toronto. Diamandis had founded the international Students for the Exploration and Development of Space (SEDS) at MIT. Hawley and Richards were active leaders in the group.

In 1985 the three men founded the Space Generation Foundation (SpaceGen), a group focused on conducting space-related projects to bring together the young space generation and help further humankind's movement into space. One of the projects conceived by SpaceGen was the creation of a space university.

In 1987 Diamandis, then an MIT graduate student, organized Spacefair '87, a national conference hosted at MIT. The theme for the conference was the creation of a space university. Hawley and Richards assisted with the conference, which gathered more than 60 space specialists in space-related disciplines from academia, industry, and space agencies around the world. This became the founding conference for ISU, which was established on 12 April 1987 under the auspices of SpaceGen. Visionary author and inventor Arthur C. Clarke accepted the position of chancellor of ISU at its inception.

The founders of ISU wanted to create a unique institution to develop the world's future leaders in space—men and women enthusiastic about the future and educated in all disciplines comprising space endeavors. They wanted these future leaders, representing a variety of national and professional cultures, to have the opportunity to learn together and develop international networks. Their vision was a university dedicated to the principle of the "three I's," the interdisciplinary, international, and intercultural approach to education that has become ISU's hallmark.

In 1988, ISU established its two-month-long annual Summer Session Program, renamed the Space Studies Program (SSP) beginning with the 2008 program, to provide students and space professionals with an international, intercultural academic experience covering many disciplines related to space activities—space and life sciences, engineering, space policy and law, business, economics, and management, and space and society. Through 2007 nations in North America, South America, Europe, and Asia, as well as Australia, had hosted the SSP.

In 1993 the university selected Strasbourg, France, as the location for its central campus. In 1995 the master of space studies (MSS) program was initiated, and in the 2004–5 academic year ISU offered a master in space management (MSM) program. Like the SSP, both masters programs were based on the ISU principle of three I's.

The SSP, MSS, and MSM offered graduate-level programs for students who have completed their undergraduate education. By the end of 2007 ISU had graduated more than 2,500 students from 96 countries with many alumni later assuming leadership positions throughout the global space community.

Peggy Finarelli

See also: Space Generation Advisory Council

Bibliography
ISU. www.isunet.edu.

Space Camps

Space camps started as an initiative of communist youth organizations in the Soviet Union, copying the structure of the Young Pioneers, or Young Leninists, soon after the spaceflight of Yuri Gagarin in 1961. They were to provide science training and political indoctrination programs. These organizations, called Young Cosmonauts Groups and named after a cosmonaut (usually Gagarin), depended on a school board or political group in each Soviet region and offered classroom instruction in science and space-oriented political indoctrination. Special preparatory tests in sciences and math were required to enter these space camps. The Orenburg School, for example, conducted a Young Cosmonauts program for boys and girls to encourage them to become cosmonauts (with the implicit goal of becoming military officers). Members built and used basic simulators and donned helmets during presentations. They gave talks about cosmonautics in their schools and were a recognized part of the Soviet Union indoctrination system. Children of both sexes from elementary to high school were encouraged to join.

During the Interkosmos program, other socialist countries, such as Czechoslovakia, the German Democratic Republic, Hungary, Latvia, Romania, and the Ukraine developed similar programs. These space camps were popular until the demise of the Soviet Union in the 1990s. However some of them, such as the All-Russian Youth Aerospace Society "Soyuz," remained active. The Soyuz organization, also called "VAKO Soyuz" by its initials in Russian, was founded in 1998 as a fusion of several space camps and rocket clubs in the Soviet Union. Its founder and president was cosmonaut

U.S. astronaut Walter (Wally) Shirra speaks at the grand opening of U.S. Space Camp California, 1996. (Courtesy NASA/Ames Research Center)

Alexander Serebrov. More than 200 students participate in the VAKO Soyuz activities every year.

The Ukrainian Youth Aerospace Association "Suzirya" (Constellation) offered camps, as did others across the former Soviet republics. Also active was an important space camp in the Urals region of Siberia, operated by Krasnoyarsk State University that annually enrolled several hundred students. The program included workshop activities (in mathematics, chemistry, physics, astronomy, rocket modeling, and space history), meetings with aerospace specialists, games and competitions, and bus excursions to aerospace museums.

In the United States, space camps began as an initiative of Wernher von Braun in 1968 when he noticed that although there were camps for children interested in sports, there were no camps for children interested in science. Von Braun selected Edward Buckbee, a NASA public affairs specialist, to be the first director of the U.S. Space and Rocket Center (USSRC), in Huntsville, Alabama. The USSRC (originally named the Alabama Space and Rocket Center) was created as a state agency overseen by the Alabama Space Science Exhibit Commission and endorsed by NASA's Marshall Space Flight Center.

Buckbee realized the educational value of space and followed von Braun's ideas further by forming the U.S. Space Camp. The first U.S. Space Camp was held in the summer of 1982 at the USSRC. Initially there were 12 sessions of 20–30 children each. In 2007 Space Camp reported that more than 500,000 guests had attended Space Camp since its inception. Under Buckbee's management U.S. Space Camp programs grew to a $25 million education business with high-quality Space Shuttle simulators, a neutral buoyancy tank for extravehicular activity (EVA) familiarization, multiple-axis simulators, robotic arms, and state-of-the-art training systems. U.S. Space Camp also began offering U.S. Space Academy for older children and adults in two levels.

In1987 the Astronaut Scholarship Foundation (formed by the Mercury astronauts to foster space science education through scholarship awards) and the U.S. Space Camp Foundation formed a partnership to develop Space Camp–Florida. In 1988 Space Camp–Florida opened in Titusville, Florida, and hosted 2,500 children that year. The U.S. Space Camp Foundation administered space camps in California and Florida and franchises in Belgium, Canada, and Turkey. Space Camp–Florida closed in 2002 to settle a multimillion-dollar foreclosure lawsuit against the U.S. Space Camp Foundation and also partly because of the effects to tourism following the 2001 World Trade Center terrorist attacks and the fear of further terrorist attacks. In 2003 Delaware North Parks, the company operating the Kennedy Space Center (KSC) Visitor Complex, signed a lease on the property that includes Space Camp–Florida. Beginning in 2003 space camp programs in Florida, renamed Camp KSC, offered a five-day educational program to youths at the KSC Visitor Complex and a one-day experience called Astronaut Training Experience, which included a tour of KSC.

By the early twenty-first century, space camps existed worldwide. With the beginning of private spaceflight, there was a trend to create more space camps for adults and to provide training for suborbital spaceflight. From 1993 the Gagarin Cosmonauts Training Center in Russia offered realistic space camp programs for high school and college students and adults using actual training hardware. Some of these programs included aircraft flights that provided brief periods of microgravity, underwater EVA training in Orlan space suits, and centrifuge familiarization and training in the Soyuz spacecraft.

The Norwegian Space Camp, a weeklong summer camp for young European scientists (17–20 years old) held at the Andøya Rocket Range in northern Norway, was organized by the Norwegian Association of Young Scientists with the National Centre for Space-Related Education in cooperation with the Norwegian Space Centre and the European Space Agency. It was another example of the international growth of space camps.

Pablo de León

See also: Interkosmos, Tourism in Space

Bibliography
Anne Baird and Robert Koropp, *The Official U.S. Space Camp Book* (1992).
V. V. Malyshev, *Space Education in Russia* (2003).

Space Museums and Science Centers

Space museums and science centers in the United States have three major antecedents. The first is museums of natural history, a movement that began in earnest in the nineteenth century to collect and systematize knowledge about the worlds around humankind. The second is planetariums, the earliest date from the 1930s, which provided the opportunity to present information about the cosmos. The third is aviation museums, began soon after the beginning of heavier-than-air flight in 1903 when practitioners and aficionados began to establish displays of artifacts and memorabilia in many locations and venues throughout the United States.

Natural history museums in the United States, such as the scientific collection owned by the Philosophical Society in Philadelphia and the Yale University museum of "natural and artificial curiosities," date from the late eighteenth century. Other organizations in major cities developed natural history collections throughout the nineteenth century, including the one at the Smithsonian Institution in Washington, DC.

Planetariums first began to be established with the opening of the Adler Planetarium and Astronomy Museum in Chicago (1930), followed by the Fels Planetarium at the Franklin Institute (1933), the Griffith Observatory and Planetarium in Los Angeles (1935), the Hayden Planetarium at the American Museum of Natural History in New York City (1935), and the Buhl Planetarium and Institute of Popular Science in Pittsburgh (1939). These five pre–World War II planetariums set the standard for a host of postwar planetariums established throughout the United States.

Aviation museums emerged first as informal displays, and several then evolved into major museums. The Franklin Institute in Philadelphia began to accession artifacts and documents associated with flight in the latter nineteenth century and made a concerted effort to collect material from the Wright brothers and other aviation pioneers beginning in the first decade of the twentieth century. What became the National Museum of the United States Air Force, located at Wright-Patterson Air Force Base, Dayton, Ohio, is the world's largest and oldest military aviation museum.

Each of these provided impetus for, and in many cases the arranging collocation of, space history exhibits. Some existing organizations extended their missions to include the new realm of spaceflight in their exhibits and public programs. In a few instances, however, specially created space museums or centers arose. While there had been earlier efforts to collect and display artifacts of rocketry, the launch of *Sputnik* in 1957 served as a catalyst for many museums that came to dominate the presentation of space history. For example the collections of the Smithsonian Institution's National Air Museum had been housed in a shed known as the Air and Space Building and outdoors in an adjacent Rocket Row during the early years of spaceflight. But the space race of the 1960s helped establish the National Air and Space Museum (NASM) through congressional action in 1971. This museum, opened 1 July 1976, presented a comprehensive history of spaceflight in numerous rotating and permanent exhibitions. It also led to the development of the largest and most diverse collection of historic spaceflight objects in the world. Through a special agreement with NASA, NASM may acquire on a first-offer basis all artifacts no longer necessary for programmatic purposes. This has led to the acquisition of virtually all the Mercury, Gemini, and Apollo spacecraft, in addition to space suits, equipment, and other key spaceflight objects from the NASA programs.

In the 1960s NASA began to create small visitor centers supporting its field organizations. These focused on telling the story of NASA: mixing history, science, and current activities in a popular manner. For example, the John F. Kennedy Space Center

(KSC) Visitor Complex became an early tourist destination. Opened in 1968, the complex soon enjoyed an average of more than one million visitors per year. KSC's Visitor Complex evolved from a simple and straightforward set of displays on NASA and its launch function at Cape Canaveral to a multifaceted theme park relating to NASA and its varied missions. In 2002 this complex undertook responsibility for the U.S. Astronaut Hall of Fame, located in nearby Titusville, Florida.

The U.S. Space and Rocket Center in Huntsville, Alabama (owned by the state of Alabama), opened in 1970 after Wernher von Braun approached the Alabama legislature in the mid-1960s with the idea of creating a museum between NASA and the U.S. Army Missile Command that would showcase the hardware of the space program. Lawmakers approved the effort in 1968, and the U.S. Army donated land to the state on its Redstone Arsenal for the facility. Most importantly, it became the home of Space Camp begun in the summer of 1982 as an in-depth experience for children to learn about spaceflight and to explore the possibilities of careers in the space program. The U.S. Space and Rocket Center also provides tours of NASA's Marshall Space Flight Center.

In addition to visitor centers at NASA facilities, several major space museums emerged since the 1960s, and other air museums, natural history museums, and planetariums expanded with spaceflight exhibits, including the following:

U.S. Museums with Spaceflight Exhibits

Adler Planetarium and Astronomy Museum	Chicago, Illinois
California Museum of Science and Industry	Los Angeles, California
Frontiers of Flight Museum	Seattle, Washington
National Museum of the United States Air Force	Dayton, Ohio
Neil Armstrong Air and Space Museum	Wapakoneta, Ohio
San Diego Air & Space Museum	San Diego, California
Strategic Air and Space Museum	Omaha, Nebraska

Note: The above is an alphabetical list.

Begun in 1962 as a planetarium, the Kansas Cosmosphere and Space Center in Hutchinson, Kansas, grew to become a major educational facility and museum. The Cosmosphere's exhibits, which contained a significant collection of U.S. and Soviet/ Russian space artifacts, concentrated on the drama of space during the 1960s. It also developed educational programs, such as the Future Astronaut Training Program and Dr. Goddard's Laboratory, which reached more than 40,000 students per year.

Collectively as of 2007 there were more than 80 museums in the United States with significant collections and exhibits associated with the history of spaceflight, out of more than 200 self-identified air and space museums. Several other museums, historic sites, and state parks contained displays relating to the history of spaceflight.

Internationally a number of space museums and science centers also told different aspects of the story of spaceflight, including the following:

International Museums with Spaceflight Exhibits

Brussels Air Museum	Brussels, Belgium
Deutsches Museum	Munich, Germany
Hermann Oberth Raumfahrt Museum	Feucht, Germany
Hong Kong Space Museum	Kowloon, Hong Kong, China
Konstantin E. Tsiolkovsky State Museum of the History of Cosmonautics	Kaluga, Russia
Musée de l'Air et de l'Espace	Paris, France
Museum of Cosmonautics	Baikonur Cosmodrome, Kazahkstan
National Space Centre	Leicester, United Kingdom
Powerhouse Museum	Sydney, Australia
Science Museum	London, United Kingdom
Space Expo	Noordwijk, The Netherlands
Yuri Gagarin Memorial Museum of Cosmonautics	Zvezdny Gorodok (Star City), Russia

Roger D. Launius

See also: Popularization of Space, Space Art, Space Memorabilia

Bibliography

Jon L. Allen, *Aviation and Space Museums of America* (1975).

Eric C. Anderson and Joshua Pivin, *Space Tourist's Handbook: Where to Go, What to See, and How to Prepare for the Ride of Your Life* (2005).

Universities

Universities have played important roles in the development of space technology and space science leading up to and throughout the space age. Both research and teaching functions of universities have been prominent in creating and sustaining space activities.

Several universities were critical to the foundation of national space programs and projects. In the United States, the California Institute of Technology's (Caltech) Jet Propulsion Laboratory (JPL) was one of the founders of liquid- and solid-propellant rocketry and, by the late 1950s, spacecraft design and systems engineering. JPL became the world's leader in robotic deep space exploration. The Massachusetts Institute of Technology (MIT) hosted the Lincoln and Instrumentation (later Charles Stark Draper) Laboratories that were in the center of developments in computing, radar, space surveillance, and communications at Lincoln, and inertial navigation systems for the Navy's Polaris missiles and NASA's Apollo Guidance System. MIT offered some of the world's first systems engineering courses beginning in the late 1940s. Japan's space program began from experiments in rocketry performed by professor Hideo Itokawa at Tokyo University starting in the early 1950s, and led to the creation of the Institute of Space and Aeronautical Science, which remained a central Japanese institution for both rocketry and spacecraft development into the twenty-first century. Luigi Broglio, one of the founders of the Italian space program, was both a military officer and the dean of the School of Aerospace Engineering at the University of Rome, where he founded the Center for Aerospace Research in the 1950s.

As the space age developed, many universities became prominent for the development of spacecraft scientific instruments and space science research. Early and continuing leaders in space science in the United States included the University of Arizona, with its Lunar and Planetary Laboratory, and the University of Colorado, which began development of solar research instruments for sounding rockets in the late 1940s. In Europe, many nations, particularly the smaller ones, started their involvement in space through university research in space science, which was funded through national budgets and the European Space Research Organisation, later the European Space Agency. By the end of the twentieth century, around the world the majority of space science instruments and experiments were competed between scientists at universities and government research centers. By the 1990s computing and mechanical technology had developed to the point where small and relatively cheap spacecraft could be developed by university students guided by their professors and university staff, leading to the development of many small spacecraft for educational purposes, usually with a small scientific or technology experiment onboard, and flown for free or very low cost as part of other launches.

Universities were also crucial in their educational functions to support the high-technology activities of space engineering and research. The Bauman Higher Technical Institute and Moscow State University supplied many of the leaders of the Soviet and Russian space programs. While Caltech and MIT helped to create major space institutions in their own right, other universities expanded their curricula to include space engineering and space science, often at the behest of nearby government space organizations. The University of Houston–Clear Lake, the University of Alabama–Huntsville, and the University of Central Florida expanded their curricula and set up local educational programs in collaboration with nearby NASA centers. Many aeronautical engineering departments expanded to become aerospace engineering departments, and in a few cases universities created astronautical engineering departments and degree programs.

While initially focused on engineering and science, other aspects of space also developed. McGill University in Montreal, Quebec, Canada, changed its Institute of Air Law to the Institute of Air and Space Law in 1957. By the late 1980s interdisciplinary space degrees began to emerge, including master's degrees in space studies at the International Space University established in Strasbourg, France, and at the University of North Dakota in the United States, and a masters in space architecture at the Sasakawa International Center for Space Architecture at the University of Houston. NASA funded universities through a variety of programs, including its Space Grant program that provided funds to every state for space-related education, not only at universities, but through universities reaching out to elementary and high schools. In the United States, the U.S. Air Force also established its own accredited degree programs with space emphases, most prominently at the Air Force Institute of Technology in Dayton, Ohio, and for undergraduates at the U.S. Air Force Academy in Colorado Springs, Colorado.

Stephen B. Johnson

See also: Amateur Communications Satellites, Economics of Space Science, Space Science Organizations

Bibliography
Brian Harvey, *The Japanese and Indian Space Programmes: Two Roads into Space* (2000).
J. Krige and A. Russo, *A History of the European Space Agency*, 2 vols. (2000).
Peter J. Westwick, *Into the Black: JPL and the American Space Program, 1976–2004* (2007).

FUTURE STUDIES

Future studies of human expansion into the cosmos have appeared since the advent of the industrial revolution as a reflection of increasing confidence in human capabilities. Individuals, organizations, and governmental entities have conducted future studies to assess the potentials and possibilities of human spaceflight programs.

The first ingredient for creating new technologies is the ability to imagine them. Jules Verne's novels (*From the Earth to the Moon*, 1865, and *Around the Moon*, 1870) illustrated the technical promise of the future that was embraced by many people and lit their imaginations. These novels fueled serious efforts to make spaceflight possible, starting first with a few individual visionaries: Konstantin Tsiolkovsky in Russia, Robert Goddard in the United States, and Hermann Oberth in Germany. In the twentieth century's first three decades, these three men pioneered key ideas required to make spaceflight possible, including the use of liquid-fuel rockets, developing the rocket equation, and envisioning the effects of zero-gravity environment. Their efforts inspired other individuals, such as French aviation pioneer Robert Esnault-Pelterie, who further developed the mathematics of astrodynamics, and Herman Potočnik, a Slovenian engineer writing under the pseudonym Hermann Noordung, who designed a wheel-shaped space station and analyzed the problems of long-term habitation in space (1929). They also inspired the first organizations of like-minded individuals, such as the Society for Spaceship Travel (Verein für Raumschiffahrt or VfR) in Germany (1927), the American Interplanetary Society in the United States (1930), the Jet Propulsion Research Groups in the Soviet Union (1931), and the British Interplanetary Society in England (1933). These amateur groups focused on the practical matter of developing a rocket that would provide a means of escape from the terrestrial "gravity well" and sponsored and published various spaceflight studies.

Dreams of spaceflight were on hold during World War II in favor of military applications, with significant government involvement, especially in Germany where Wernher von Braun, an early member of the VfR, oversaw the development of the A-4 (later V-2) missile. Von Braun and his colleagues discussed plans for human spaceflight, at personal risk and to the displeasure of the Nazi government. When von Braun was brought to the United States after the war, he proposed a human-occupied space station for communications and surveillance, the exploitation of lunar resources, and an ambitious, yet technically plausible, human mission to Mars. The von Braun paradigm, as it became known, for this series of missions was a strong influence in the United States.

Based on the technical (though not military) success of the V-2, the United States and the Soviet Union took the lead in the studies and development of spaceflight from 1945 through the 1950s. Government-funded and supported military studies of the future of spaceflight easily dominated individual efforts. Bureaucratic rivalry also spurred some early studies. For example, concerned by the U.S. Navy's interest in satellites, in 1946 the U.S. Army Air Forces funded the RAND Corporation's influential initial study of the potential uses and feasibility of an orbital satellite spaceship, which included the potential for reconnaissance, weather prediction, navigation, communications, science, and national prestige. This and later feasibility studies led to the initiation of scientific and reconnaissance satellite programs by the mid-1950s. The U.S. military also studied possibilities for human spaceflight, as evidenced by the U.S. Army Redstone Arsenal's proposal for a military outpost on the Moon, *Project Horizon Report* (1959), and the Lunex lunar expedition proposed by the U.S. Air Force (USAF) in 1961. With the Soviet launch of *Sputnik* in October 1957, the U.S. armed forces redoubled their efforts, moving rapidly to develop satellites for the purposes outlined in the RAND study. Human spaceflight and space science was largely transferred to NASA between 1958 and 1960, along with the relevant organizations and their ongoing feasibility studies and assessments. In 1961 these studies provided the basis for President John F. Kennedy's direction for NASA to land a man on the Moon, and return him safely, before the end of the decade.

The Soviet Union also had ambitious plans for spaceflight, which were reflected in a 1960 decree, "On the Creation of Powerful Carrier-Rockets, Satellites, Space Ships, and the Mastery of Cosmic Space in 1960–67." In addition to envisioning new generations of rockets that included standard and exotic propulsion techniques, the decree also included plans for piloted missions to the Moon and Mars by the mid-1960s. Much of this was based on ideas about interplanetary flight developed by OKB-1 (Experimental Design Bureau) beginning in the late 1950s. As the Apollo program commenced spaceflight activities, the Soviet Union developed plans for a human mission to Mars that, if successful, would divert attention from Apollo and effectively leapfrog the U.S. effort. Aelita, as the Soviet program was named, envisioned sending six cosmonauts to Mars in 1969 on a 630-day mission. The absence of funding and political support doomed Aelita.

The U.S. decision to discontinue production of Saturn V rockets ended plans for a follow-on to the lunar landings; the *Skylab* orbital station of the 1970s was Apollo's only direct offspring. The post-Apollo era was introduced by the Space Task Group (STG), led by scientist Thomas O. Paine, which envisaged new goals in "The Post-Apollo Space Program: Directions for the Future" (1969). This ambitious plan envisaged an orbiting space station, a space shuttle to the station, and a human mission to Mars by the mid-1980s. Preoccupied with the Vietnam War and violence on U.S. streets, the Richard Nixon presidential administration and Congress agreed only to fund the Space Shuttle program as a way to lower the cost of access to space. Space station studies were updated in the early 1980s, supporting President Ronald Reagan's direction for NASA to develop Space Station Freedom.

NASA's space science program supported a variety of studies to assess the future direction of its science programs. Among the most prominent were those by the Space Sciences Board of the National Academy of Sciences to set science priorities. The Science Advisory Board played a similar role for the USAF with ongoing studies of current technologies and occasional larger-scale strategic assessments. NASA and the

U.S. military services also funded corporations and universities for both near-term and longer-term feasibility and planning assessments to gather new ideas from which to select future priorities and programs.

Gerard K. O'Neill, a Princeton University physicist, wrote *The High Frontier: Human Colonies in Space* (1976), which provided inspiration for human space exploration advocates. O'Neill envisioned the construction of large, self-sustaining human settlements in orbit around Earth and other planets, making use of 1970s technologies, such as recycling empty Shuttle external tanks and using solar energy. He was also among the founders of the Planetary Society, the Space Studies Institute, and the L5 Society (which later merged with the National Space Institute to form the National Space Society), which promoted studies of extraterrestrial habitats.

Science writer Carl Sagan argued in *Cosmos* (1985) for humankind's inherent connection with the universe, anticipating further studies of planetary exploration that occurred in the 1980s–90s. Also in 1985, President Reagan tapped Paine to head the National Commission on Space, which published *Pioneering the Space Frontier: The Report of the National Commission on Space* (1986). As with the 1969 STG study, it outlined an ambitious master plan for space stations and planetary exploration. In the aftermath of the Space Shuttle *Challenger* accident, the report was shelved and a new study headed by astronaut Sally Ride, "Leadership and America's Future in Space: A Report to the Administrator" (1987), was published. The less ambitious Ride report outlined four possible space exploration goals—Earth, solar system, outpost on the Moon, and humans to Mars.

NASA's Human Robotic Systems Project, part of the agency's Exploration Technology Development Program, focused on human and robotic mobility systems for the moon, Moses Lake Demonstrations, 2008. (Courtesy NASA/Langley Research Center)

The last phase of the Cold War saw President George H. W. Bush's proposal for the Space Exploration Initiative on the 20th anniversary of the *Apollo 11* Moon landing. Later studies included "The Augustine Commission Report" (1990) and *America at the Threshold: America's Space Exploration Initiative* (1991—known also as the Stafford Report), which provided NASA with roadmaps for the presidential vision of lunar and Mars exploration. In part as a reaction to the Space Exploration Initiative, Robert Zubrin, Lockheed Martin engineer, proposed "Mars Direct," a new human mission to the Red Planet using existing technologies and in situ resources for the production of propellant and supplies, as he described in *The Case for Mars* (1997).

In a January 2004 speech President George W. Bush directed NASA to implement a new vision for space exploration, based on incremental expansion of human and robotic activities in the solar system. The Presidential Commission Report *Journey to Inspire, Innovate, and Discover* (2004) recommended the development of a new Crew Exploration Vehicle to replace the Space Shuttle, and the return of humans to the Moon by 2020, followed by a human expedition to Mars.

As other nations developed their space programs, they supported ongoing tactical and strategic studies of future technologies and goals similar to those developed by the United States and Soviet Union in the early years of the space age.

Fabio Sau, Michael L. Ciancone, and Stephen B. Johnson

See also: Literary Foundation of the Space Age, Rocket Pioneers and Spaceflight Visionaries, Space Science Organizations

Bibliography
Gerard K. O'Neill, *The High Frontier: Human Colonies in Space* (1976).
David S. F. Portree, *Humans to Mars: Fifty Years of Mission Planning, 1950–2000* (2001).
Carl Sagan, *Cosmos* (1985).
Asif A. Siddiqi, *Challenge to Apollo: The Soviet Union and the Space Race, 1945–1974* (2000).

Lunar Colonization

Lunar colonization has been a dream of space visionaries for centuries. The Moon has been a potential target for human settlement because it represents a natural laboratory for planetary science, an ideal platform for optical and radio astronomy, a source of minerals and energy, and a near-Earth location for humans to live and work.

The modern idea of humans living on the Moon originated in nineteenth- and twentieth-century science-fiction literature. While researchers, notably Konstantin Tsiolkovsky, began to contemplate the technology required to escape Earth's gravity for a journey to the Moon as early as 1890, not until 1938—when the British Interplanetary Society completed the world's first scientific study of a lunar space vehicle— did the idea appeal to a wider audience.

Several detailed lunar base studies published after World War II captured the public imagination. *The Exploration of Space* (1951) and *Conquest of the Moon* (1953) presented realistic plans and illustrations showing how humans could travel to the Moon and construct outposts.

In 1959 the U.S. Army Ballistic Missile Agency selected H. H. Koelle and Wernher von Braun to complete the first serious, technically detailed feasibility study to construct a lunar base. Classified top secret and dubbed Project Horizon, the plan would have used heavy-lift Saturn rockets to place a crew of 12 on the Moon in pressurized underground modules. When von Braun's team transferred to NASA in 1960 and the Apollo program defining near-term lunar goals, the studies were not pursued further.

During the 1960s–70s, the United States and the Soviet Union moved ahead cautiously with other lunar-base concepts. After the first human lunar landing in 1969, the public and political interest in returning to the Moon began to wane. In 1989 U.S. President George H. W. Bush's Space Exploration Initiative called for a small lunar base by 2008, but funding was not approved.

In the 1990s several private organizations, such as the Artemis Project, began exploring lunar colony ideas. Scientists and aerospace engineers of the Lunar Exploration Working Group also met periodically to discuss proposed lunar missions. In 2004 President George W. Bush proposed a new lunar initiative, and NASA began a new series of studies.

Louis Varricchio

See also: Von Braun, Wernher

Bibliography

Peter Eckart and Buzz Aldrin, *The Lunar Base Handbook* (1999).

Howard E. McCurdy, *Space and the American Imagination* (1997).

Anthony M. Springer, "Securing the High Ground: The Army's Quest for the Moon," *Quest 7*, no. 2 (1999): 34–39.

Paul D. Spudis, *The Once and Future Moon* (1996).

Mars Mission Studies

Mars mission studies investigate future scenarios in which humans land on the surface of Mars. Mars has a very challenging environment, with very low atmospheric pressure, gravity only one-third that of Earth's, little protection from cosmic radiation, and is at best several months' distance from Earth, which requires that humans must be able to address any emergencies with resources brought from Earth or from the Martian environment. In all, Mars presents a hostile and isolated environment to prospective explorers or settlers, against which a Mars base must provide protection.

Mars base studies have been conducted by a variety of scientists and engineers throughout several decades. Perhaps the first such study was Wernher von Braun's *The Mars Project*, published in 1952. It required ten 4,000-ton ships and 70 crewmembers. Some of the explorers would land at the polar ice caps, from where they would trek 4,000 miles to the equator to build a landing strip for wheeled gliders to land. They would deploy inflatable habitats in which they would live for one year.

NASA and its contractors performed a series of studies in the 1960s, but these were set aside in the budget contraction in the wake of Apollo. Soviet engineers also assessed Mars crewed landing missions in the 1960s, including a study led by Vostok's designer, Konstantin Feoktistov. The Soviet Union continued studying Mars missions, particularly in the late 1960s and early 1970s with the idea of using Mars missions to neutralize U.S. success in the manned Moon race.

A British Interplanetary Society study in 1963 assumed a 15-month stay on the surface of Mars by 10 men and five women. It included women because the resulting "spirit of competition among the men" would be good for morale, an oblique recognition that psychological factors could be as important as material ones. The proposed base used inflatable habitats, with enough oxygen and water for the duration of the stay brought from Earth. Air and water regeneration equipment would also be provided, in addition to hydroponic gardens to provide some of the food supply and nuclear generators for electrical power.

NASA conducted extensive Mars base studies during the 1980s and early 1990s, leading up to and during the abortive Space Exploration Initiative proposed by the George H. W. Bush presidential administration. The Mars Design Reference Mission featured an inflatable habitat delivered by a robotic cargo lander. The crew would subsequently land close to the habitat and, after configuring it for long-term operations, move in. However, the proposed cost in the range of hundreds of billions killed the initiative.

Partially in response to NASA's allegedly conservative and certainly expensive approach, aerospace engineer Robert Zubrin proposed quite different missions that extensively used local (in situ) Martian resources, which reduced the cost of bringing materials from Earth. Zubrin cofounded the Mars Society, which, under his direction, proposed a Mars base with a crew of five and systems to extract carbon dioxide from the Martian atmosphere and combine it with hydrogen (carried from Earth) in the presence of a catalyst, producing methane and water. The former could be used for rocket fuel and the latter for drinking water, in addition to provide a source of oxygen and more recycled hydrogen to maintain the process. The Mars Society sponsored the development of Mars analog sites on Earth, starting with a site on Devon Island in the Canadian Arctic, first operational in 2000, and a desert site in southern Utah, operational in 2002. These sites tested various Mars mission and habitation concepts.

A 2003 study by C. A. Cockrell and A. A. Ellery, published in the *Journal of the British Interplanetary Society*, proposed a Mars base located at one of the poles. Their base would eventually house up to 26 crewmembers and provide closed-loop life support systems powered by nuclear reactors.

After the announcement of the Vision for Space Exploration by U.S. President George W. Bush in 2004, which proposed an eventual human mission to Mars, the resulting Constellation program began to sponsor further Mars mission and habitation studies. However, Constellation soon deferred Mars mission studies in favor of near-term ISS and lunar missions.

Michael Engle and Stephen B. Johnson

See also: Mars

Bibliography
Stephen J. Hoffman and David I. Kaplan, eds., *Human Exploration of Mars: The Reference Mission of the NASA Mars Exploration Study Team* (1997).
David S. F. Portree, *Humans to Mars: Fifty Years of Mission Planning, 1950–2000* (2001).

GOVERNMENT SPACE ORGANIZATIONS

Government space organizations placed the first satellites in orbit and have remained influential in space endeavors through their first half-century.

Early spaceflight was an offshoot of the development of ballistic missiles, and hence government military and intelligence organizations dominated it. In the United States, the Dwight Eisenhower presidential administration wanted to project an image of peaceful intentions, and it convinced Congress to create a civilian space agency, NASA, to ensure civilian control over many branches of space applications. NASA did not gain control over all civilian applications, as weather and remote sensing satellites came partially under the purview of the National Oceanic and Atmospheric Administration and the Department of Interior, respectively. Competition for space roles and responsibilities, usually with one another, but sometimes with NASA, continued among the military and intelligence services for many years. In 1960 the Eisenhower administration used a similar civilian criterion in establishing a new, secret organization for space-based national intelligence, the National Reconnaissance Office (NRO). In Eisenhower's view, intelligence should not be dominated by the military, and thus the NRO was established as a joint operation of the Central Intelligence Agency and the U.S. Air Force. The Soviet Union's early space efforts were also divided among many organizations, but there was no clear demarcation between civilian and military purposes. The military services, including the new Strategic Missile Forces (RVSN) created in 1959, generally operated Soviet launchers and satellites. In 1992, after the demise of the Soviet Union, Russia created a separate space service, the Military Space Forces (VKS). Development of launchers and spacecraft was the purview of the Ministry of Armaments (MV) until 1965, when space system development transferred to the new Ministry of General Machine Building (MOM).

For large nations with space ambitions, both the U.S. and Soviet models were influential. Not surprisingly, China followed the Soviet model, with a mix of military services and industry bureaus developing rockets and spacecraft. Western nations usually followed the U.S. model, with early development run by military services and the establishment of a civilian space agency somewhat later to develop scientific and civilian application spacecraft. Depending on domestic politics, civilian agencies took time to form: France (1962), India (1969), Japan (1969), United Kingdom (1985), Italy (1988), Canada (1989), and Russia (1992). West Germany and many smaller nations left space activities to a variety of government scientific and industrial organizations.

Multinational space organizations also developed. The United Nations created its Committee on the Peaceful Uses of Outer Space in 1959 to encourage peaceful uses, promote cooperation and information sharing of space research, and study legal issues. Intergovernmental negotiations led to the creation of Intelsat (International Telecommunications Satellite Organization, 1964) and Inmarsat (International Maritime Satellite Organisation, 1976) to run international communications satellite systems. Both were mixed public–private entities mainly because the United States created a private corporation, the Communications Satellite Corporation, to manage its national interests in Intelsat. By the early 2000s, both Inmarsat and Intelsat were fully privatized. The International Telecommunication Union assigned frequency spectrum to communications satellite operators. West European nations developed multinational organizations to develop launchers, space science, and applications satellites, starting with the European Space Research Organisation (1964), the European Space Vehicle Launcher Development Organisation (1964), and merging the two into the European Space Agency (ESA, 1975). The European Union and ESA formed a partnership to develop a commercial navigation system called Galileo in 2002. It soon acquired new members, such as China and India.

Governments of capitalist nations had to work with for-profit and nonprofit, private corporations, and by the 1990s with individuals. Initially the relationship was one of government organizations contracting to corporations. However, in the 1970s–80s the U.S. and European nations encouraged the development of commercial communications, space launch, and remote sensing. These fields brought forth other issues, such as insurance liability, launch licensing, remote sensing resolution, and spectrum licensing that required government control and supervision. For example, in the United States, a number of government institutions regulated these functions, such as the Department of Transportation for launch licensing, the Federal Communications Commission for spectrum assignments, and the Departments of Commerce and State to oversee foreign technology sales. Private citizens as tourists and the possibility of private launch vehicles posed new opportunities and issues, with governments such as the Soviet Union actively selling rides on spacecraft to private citizens to acquire new revenues, and state governments in the United States establishing spaceports to develop local space industry.

Stephen B. Johnson

See also: International Space Politics, Nations

Bibliography
William E. Burrows, *This New Ocean: The Story of the First Space Age* (1998).
Brian Harvey, *The Chinese Space Programme: From Conception to Future Capabilities* (1998).
———, *Russia in Space: The Failed Frontier?* (2001).
John Krige and Arturo Russo, *A History of the European Space Agency, 1958–1987*, 2 vols. (2000).

European Space Agency

The European Space Agency (ESA) was established in 1975 with administrative headquarters in Paris, France. It was the main arena in which senior officials responsible for space and industrial policy from the participating countries set priorities for a joint European effort, committed funds to space science and to the research and development of cutting-edge space technologies, and managed the development and operation

German astronaut Hans Schlegel of the European Space Agency works on the Columbus *laboratory during an extravehicular activity (EVA) on the* International Space Station, *2008. (Courtesy NASA)*

of a variety of space systems, typically supervising industrial corporations that built the launchers and satellites.

ESA assumed the expanded role of the European Space Research Organisation (ESRO) and replaced the European Space Vehicle Launcher Development Organisation (ELDO) in response to problems in the European space program in the early 1970s. In package deals in 1971 and 1973, European governments agreed to share costs on application satellites for meteorology, navigation, and telecommunications, in addition to guaranteeing the long-term viability of the space science program. In 1973 France agreed to pay two-thirds of development costs of the Ariane rocket, while Germany agreed to pay more than half the development cost of *Spacelab*. *Spacelab* was a scientific laboratory that would be launched in the Shuttle's cargo bay and was intended to help German industry gain exposure to U.S. technology and project management skills after its disastrous performance in ELDO's Europa rocket program. Other partners, who provided the balance of costs of these programs, negotiated their share depending on their degree of political solidarity, financial means, and industrial interest. These initial program orientations—science, applications, launchers, and a human presence in space—defined the core of ESA's activities. ESA has also played a substantive role in the development and operation of the *International Space Station (ISS)*. The *Columbus* research laboratory, launched on STS-122 in February 2008, was ESA's largest contribution to the ISS program.

The scope of ESA's membership expanded with the enlargement of the European family. As of 2010 the Member States were Austria, Belgium, the Czech Republic, Denmark, Finland, France, Germany, Greece, Ireland, Italy, Luxembourg, the Netherlands, Norway, Portugal, Spain, Sweden, Switzerland, and the United Kingdom, with some participation from Canada, Estonia, Hungary, Poland, Romania, and Slovenia. ESA's member nations pooled resources to promote scientific research and development and to reap the commercial benefits of space through a dynamic European-wide high-tech space industry. France, Germany, and Italy, by virtue of their relative wealth, technological and industrial power, and deep commitment to the postwar reconstruction of a united Europe, were driving forces in ESA. The United Kingdom balanced its close relationship with the United States against engagement with continental European partners and maintained different priorities in space than France, Germany, or Italy.

The budget of the agency was €3,592 million in 2009 (more than $4 billion). All ESA member states contribute to the infrastructure and the science programs on the basis of their gross domestic product, and to other programs on a negotiated basis. The ESA Council decides programs on which the money is spent and determines their cost. This has been a high-level decision-making body on which every Member State is represented, and each Member State has one vote. Space policy in Europe *is* industrial policy, and the principle of fair return is a key factor determining the contribution that each government makes to a program. By this principle, ESA places contracts for space programs in the industry of each Member State for an amount of money roughly equivalent to that nation's contribution to its budget. This policy has raised the overall industrial capability in the European space sector and provided a crucial technological and managerial platform for European unity and competitiveness. The main program areas supported by ESA include space science, telecommunications, Earth observation, space navigation, and launchers.

ESA established a number of dedicated centers—the European Space Research and Technology Centre (ESTEC) in Noordwijk, the Netherlands, to serve as the technology research site and development manager for most ESA spacecraft; the European Space Operations Centre (ESOC) in Darmstadt, Germany, to control satellite on-orbit operations; the European Astronaut Centre (EAC) in Cologne, Germany, to train men and women for human spaceflight missions; and the European Space Research Institute (ESRIN) near Rome, Italy, to maintain a satellite data distribution center. The French Space Agency (CNES) and Arianespace (for Ariane launchers) have operated an equatorial launch base in Kourou, French Guiana, on behalf of ESA, which owns the launch infrastructure for the Ariane 5, Vega, and Soyuz launchers.

John Krige

See also: Ariane, Astronaut Training, France, Germany, International Astronauts and Guest Flights, Italy, Launch Facilities, *Spacelab*, United States, United Kingdom

Bibliography

Roger M. Bonnet and Vittorio Manno, *International Cooperation in Space: The Example of the European Space Agency* (1994).

John Krige and Arturo Russo, *A History of the European Space Agency, 1958–1987*, 2 vols. (2000).

Kevin Madders, *A New Force at a New Frontier: Europe's Development in the Space Field* (1997).

European Space Research Organisation

The European Space Research Organisation (ESRO) was the idea of two European physicists and scientific diplomats, Italian Edoardo Amaldi and Frenchman Pierre Auger. Both played key roles in the establishment of Conseil Européen pour la Recherche Nucléaire (CERN), an intergovernmental high-energy physics laboratory, near Geneva, Switzerland. They persuaded scientists and politicians that Europe should set up a collaborative civilian space program by pointing to developments in the United States and the Soviet Union and to the limited resources that many European governments could devote to space. They also feared that the North Atlantic Treaty Organization would build a European satellite within the framework of its military alliance if a civilian program was not undertaken at once, and they used the success of CERN as a model to argue that their project was feasible. Their original idea was for a European NASA devoted to both satellite and launcher development. This plan was revised, partly due to opposition from space scientists to the costs of developing a launcher and their fears that some smaller, neutral European countries would not want to become entangled with rocket development, which had heavy military overtones. Two separate organizations were thus formed in 1964 instead of just one. ESRO would be essentially responsive to scientists' needs, while the European Space Vehicle Launcher Development Organisation (ELDO) would build a European civilian launcher. ESRO's founding Member States were Belgium, Denmark, France, Italy, the Netherlands, Spain, Sweden, Switzerland, the United Kingdom, and West Germany.

ESRO's activities were confined from the start by its undersized budget of $300 million over eight years. This reflected the lack of lobbying experience of the

Preparing ESRO 1 for thermal-vacuum testing. (Courtesy European Space Agency)

young space science community, and the alternative provided by NASA's 1959 offer to fly European scientific payloads on American satellites. The plans originally created by scientists were too ambitious, and some important satellite projects had to be canceled. Nevertheless ESRO had a vibrant sounding rocket program, placed several scientifically excellent payloads in orbit, and helped create a European space science community and aerospace industrial consortia. Its first satellite, *ESRO 2*, was launched in May 1968 and studied solar X rays and various aspects of the cosmic radiation and charged particle population in Earth's radiation belts. Subsequent missions, such as *ESRO 1*, *TD 1A* (Thor-Delta), *COS B*, and *Highly Eccentric Orbiting Satellite* (*HEOS*), included studies of the polar ionosphere and auroral phenomena, measurements of solar wind and associated interplanetary magnetic field changes, and cataloging of the celestial sphere in the ultraviolet and gamma-ray regions of the electromagnetic spectrum.

The commitment of governments to ESRO began to decline in the late 1960s, as their interest in applications such as weather and communications satellites increased. The European organization's goals were reoriented to include the development of application satellites, and its reduced science program was taken over and developed by the European Space Agency that replaced ESRO in 1975.

John Krige

See also: European Space Research Organisation Scientific Satellite Program, France, Germany, Italy, United Kingdom

Bibliography
John Krige and Arturo Russo, *Europe in Space, 1960–1973* (1994).
————, *A History of the European Space Agency, 1958–1987*, vol. 1 (2000).

European Space Vehicle Launcher Development Organisation

The European Space Vehicle Launcher Development Organisation (ELDO) was formally established by six European states and Australia in 1964. Its aim was to provide Western Europe with its own satellite launcher, called Europa, comprised of three stages that would be built separately in the United Kingdom, France, and West Germany. Italy would provide a test satellite, Belgium would provide downrange guidance, and the Netherlands would provide telemetry. Australia was included because it had a launching base in Woomera, where the United Kingdom tested its missiles, and to maintain a good relationship between these Commonwealth partners.

The origins of ELDO trace to a decision in January 1961 by British Prime Minister Harold Macmillan and French President Charles de Gaulle. Macmillan wanted to find a use for his obsolete ballistic Blue Streak missile and to demonstrate goodwill toward the European Common Market, which he had opposed until then. De Gaulle thought that a strong Europe that included the United Kingdom was needed to face increasingly menacing communism. They formed ELDO to promote European unity and to share costs and technology in launcher development.

The original aims of ELDO were somewhat vague, reflecting its birth as an essentially political project that started without clear technological objectives. The program was given new direction by the decision in 1967, promoted by France, to upgrade the Europa rocket (later labeled Europa 2) that would enable it to place telecommunications satellites in geostationary orbit.

The British soon became disillusioned with the organization, however, because the financing structure burdened them disproportionately. The budget rose steadily, and though Blue Streak worked impeccably, there were many problems with the stages being built by France and Germany. The United Kingdom negotiated a reduction in its financial contribution. With British influence declining, the French, never happy about the presence of a non-European partner in the venture, successfully promoted the development of an equatorial launch base in Kourou, French Guiana, and squeezed Australia out of the project.

The persistent technical failures of the Europa rocket, which never successfully orbited a satellite, coupled with time and cost overruns and major flaws in project management, sapped the enthusiasm of the main partners in ELDO. The explosion of Europa 2 on its maiden test flight in November 1971, and offers from the United States to collaborate in the development of the reusable Space Shuttle, ended the project. The United Kingdom withdrew from the program after Europa 2. France and West Germany followed in April 1973. ELDO was officially dissolved and its tasks were taken over by the European Space Agency in 1975.

John Krige

See also: Australia, Europa, France, Germany, Italy, Launch Facilities, United Kingdom

Bibliography
John Krige and Arturo Russo, *Europe in Space, 1960–1973* (1994).
———, *A History of the European Space Agency, 1958–1987*, vol. 1 (2000).

Interkosmos

Interkosmos was a program of multilateral space cooperation involving 10 nations: Bulgaria, Cuba, Czechoslovakia, the German Democratic Republic, Hungary, Mongolia, Poland, Romania, the Soviet Union, and Vietnam.

Initially space cooperation among these countries (with the exception of Vietnam) was based on an agreement reached through an exchange of letters among the government leaders in 1965 and on three agency-to-agency agreements concluded in 1965–68. The 1967 agreement contained a long-term program that included five basic areas of cooperation: study of the physical properties of outer space, space meteorology, space biology and medicine, space communications, and remote sensing. In 1970 this program received the official title Interkosmos.

On 13 July 1976 the same nations signed an intergovernmental agreement on cooperation in the exploration and use of outer space for peaceful purposes. The aim of that agreement, which entered into force on 25 March 1977, was to continue and further develop joint activities in accordance with the Interkosmos program and to provide this program with a clearer legal basis. In 1979 Vietnam acceded to the agreement.

The coordination of the work performed in accordance with the Interkosmos program was carried out at the annual meetings of the leaders of "national coordinating organs," most of which were interdepartmental bodies established under the auspices of their respective national academies of sciences. Interkosmos did not have a joint budget. Work performed under the program by national scientific and industrial organizations was financed by the respective countries. The Soviet Union provided rockets, satellites, and launching facilities to its partners free of charge. Formally, Interkosmos had no permanent secretariat. The Interkosmos Council under the Soviet Academy of Sciences (the national coordinating organ of the Soviet Union) performed in practice the functions of the international secretariat of the program, however, and, for instance, regularly published Interkosmos information bulletins.

Participation in Interkosmos did not preclude the right of the partners to conclude among themselves or with third-party nations other bilateral or multilateral agreements in the field of space cooperation. Indeed a number of such agreements were concluded by the Soviet Union and other participating nations.

Under the Interkosmos program, from 1968 to 1990, a wide range of fundamental and applied research and experiments was carried out on some 100 space objects and geophysical rockets, including 25 satellites of the Interkosmos series. In 1976 the program was extended to include flights by cosmonauts from participating countries. In the period between 1978 and 1981, nine international expeditions worked onboard the Soviet space station *Salyut 6*.

With the dissolution of the Soviet Union and the radical transformation occurring in most of the participating nations, the Interkosmos program and its founding agreements fell into disuse. In 1992 the Russian Federal Space Agency assumed Russia's responsibilities in international space cooperation.

V. S. Vereshchetin

See also: International Astronauts and Guest Flights, Space Science Policy

Bibliography

Proceedings of the International Conference "Intercosmos-30" (2003).
V. S. Vereshchetin, "Co-operation in the Exploration and Use of Outer Space for Peaceful Purposes," in *Manual on Space Law*, comp. and ed. N. Jasentuliyana and Roy S. K. Lee, vol. 1 (1979).

National Aeronautics and Space Administration

The National Aeronautics and Space Administration (NASA) is the space agency for the United States. Although one among many such agencies around the world, in the twentieth century it was the best funded and was responsible for many of the most notable achievements of the space age, including six Apollo landings on the Moon, more than 100 flights of the Space Shuttle, the major partner in the International Space

Station program, and a spectacular array of robotic spacecraft, including *Voyager*, the *Hubble Space Telescope*, and various Mars missions. In addition to its headquarters in Washington, DC, NASA facilities include 10 centers around the country. Its budget for fiscal year 2007 was almost $17 billion.

On 29 July 1958 President Dwight D. Eisenhower signed the National Aeronautics and Space Act to provide for research into the problems of flight, both within Earth's atmosphere and in space. The act created NASA, which became operational on 1 October 1958. NASA's birth was directly related to the pressures of national defense and in particular the Soviet Union's launch of *Sputnik* on 4 October 1957. After World War II the United States and the Soviet Union were engaged in the Cold War, a broad contest over the ideologies and allegiances of nonaligned nations. During this period space exploration emerged as a major disputed area, becoming known as the space race.

NASA began by absorbing the earlier National Advisory Committee for Aeronautics (NACA), including its 8,000 employees, an annual budget of $100 million, three major research laboratories—Langley Aeronautical Laboratory, Ames Aeronautical Laboratory, and Lewis Flight Propulsion Laboratory—and two smaller test facilities. It quickly incorporated other organizations (or parts of them), notably the space science group of the Naval Research Laboratory that formed the core of the new Goddard Space Flight Center, the Jet Propulsion Laboratory managed by the California Institute of Technology for the Army, and the Army Ballistic Missile

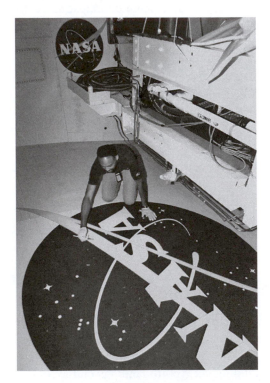

A KSC worker paints the NASA "meatball" logo on the port wing of the orbiter Endeavour, *1998. The logo was designed in the late 1950s and symbolized NASA's role in aeronautics and space in the early years of the agency. (Courtesy NASA/Kennedy Space Center)*

EILENE MARIE GALLOWAY
(1906–2009)

Eilene Galloway had a significant role in drafting and developing the U.S. National Aeronautics and Space Act of 1958 and later U.S. laws and policies involving outer space activities. She was a civil servant of the U.S. Library of Congress, serving as an expert analyst, author, lecturer, editor, advisor, and consultant. Following the launch of *Sputnik*, she was appointed Special Consultant to the Senate Special Committee on Space and Astronautics and as Special Consultant to the Senate Committee on Aeronautical and Space Sciences (1958–77). Galloway wrote many papers, studies, and historical research materials related to national defense, national security, aerospace law and policy, and for international organizations involved in these fields.

Stephen Doyle

Agency in Huntsville, Alabama, where Wernher von Braun's team of engineers was developing large rockets.

Within months of its creation NASA began to conduct space missions and during its first 20 years undertook several major programs, including human spaceflight, robotic space missions, aeronautics research, remote sensing Earth satellites, applications satellites for weather and communications, an early space station, and a reusable spacecraft (the Space Shuttle) for traveling to and from Earth orbit.

NASA did pioneering work in space applications, such as communications satellites in the 1960s. The *Echo*, *Telstar*, *Relay*, and *Syncom* satellites were built by NASA or by the private sector based on significant NASA advances. In the 1970s NASA's Landsat program changed the way humans looked at Earth. Landsat data became used in a variety of practical commercial applications, such as crop management and fault line detection, and to track many kinds of weather, such as droughts, forest fires, and ice floes. NASA engaged in a variety of other Earth science efforts, such as the Earth Observation System of spacecraft and data processing that have yielded important scientific results in such areas as tropical deforestation, global warming, and climate change.

Human spaceflight initiatives began with Project Mercury, a single astronaut program (flights during 1961–63) to ascertain if a human could survive in space. Project Gemini (flights during 1965–66) proceeded with two astronauts to practice space operations, especially rendezvous and docking of spacecraft and extravehicular activity (EVA). These early missions culminated in Project Apollo (flights during 1968–72) to explore the Moon. Apollo became a NASA priority on 25 May 1961, when President John F. Kennedy announced the goal of landing a man on the Moon and returning him safely to Earth by the end of the decade. Despite a deadly fire in 1967, the *Apollo 8* mission orbited the Moon on 24–25 December 1968, and on 20 July 1969 the *Apollo 11* mission fulfilled Kennedy's challenge by successfully landing Neil Armstrong and Edwin E. "Buzz" Aldrin Jr. on the Moon. Five more successful lunar landing missions followed. Twelve astronauts walked on the Moon during six Apollo lunar landing missions. In 1975 the

United States, through NASA, cooperated with the Soviet Union to achieve the first international human spaceflight, the Apollo-Soyuz Test Project (ASTP).

After a gap of six years, NASA returned to human spaceflight in 1981 with the advent of the Space Shuttle program. The Shuttle's first mission, Space Transportation System (STS)-1, launched on 12 April 1981, demonstrated that it could take off vertically and glide to an unpowered airplane-like landing. On 28 January 1986 a leak in the joints of one of two solid rocket boosters attached to the Shuttle orbiter *Challenger* caused the main liquid fuel tank to explode 73 seconds after launch, killing all seven crew members. On 29 September 1988 the Shuttle successfully returned to flight and NASA then flew 87 successful missions before tragedy struck again on 1 February 2003 with the loss of the orbiter *Columbia* and its seven astronauts during reentry. Three Shuttle orbiters remained in NASA's fleet: Atlantis, *Discovery*, and Endeavour. In 1984 in the wake of three Skylab missions in the 1970s, Congress authorized NASA to build a space station as a base for further exploration of space. After many revised plans, the *International Space Station* (*ISS*) program finally emerged. Permanent habitation of the *ISS* began when the Expedition One crew arrived on 2 November 2000.

On 14 January 2004 U.S. President George W. Bush announced a Vision for Space Exploration that entailed sending humans back to the Moon and on to Mars by retiring the Shuttle and developing a new, multipurpose Crew Exploration Vehicle along with new crew and cargo launchers. Robotic scientific exploration and technology development was also folded into this vision. This became known collectively as the Constellation Program, which evolved in scope and detail to focus on missions related to *ISS* and the Moon. In 2010 U.S. President Barack Obama proposed canceling the Constellation program in his budget proposal for fiscal year 2011.

In addition to major human spaceflight programs, NASA managed significant scientific probes that explored the Moon, the planets, and other areas of Earth's solar system. In particular the 1970s heralded the advent of a new generation of scientific spacecraft. Two similar spacecraft, *Pioneer 10* and *Pioneer 11*, launched on 2 March 1972 and 5 April 1973 respectively, traveled to Jupiter and Saturn to study the composition of interplanetary space. Building on earlier Mariner missions, *Voyager 1* and *Voyager 2*, launched on 5 September 1977 and 20 August 1977 respectively, conducted a "Grand Tour" of Earth's solar system. In 1975 NASA launched two Viking spacecraft to look for evidence of life on Mars. The spacecraft, including two orbiters and two landers, arrived at Mars in 1976. Despite tantalizing hints, neither lander found unambiguous evidence for past or present biological activity there. *Magellan*, *Galileo*, and *Cassini* continued NASA's robotic exploration of Venus, Jupiter, and Saturn respectively.

NASA suffered both spectacular successes and failures with its Mars program. The loss of the *Mars Observer* spacecraft in August 1993 resulted in NASA developing a series of "faster, better, cheaper" spacecraft to go to Mars. The first of these spacecraft was *Mars Global Surveyor*, which begin mapping Mars in 1998. Using innovative technologies the *Mars Pathfinder* spacecraft landed on Mars on 4 July 1997 and explored the surface of the planet with its miniature rover, *Sojourner*. After disappointing losses of the *Mars Climate Orbiter* and *Mars Polar Orbiter* in 1999, NASA slowed its pace and recovered by landing the *Spirit* and *Opportunity* rovers in January 2004, proving that liquid water had existed on Mars, to much scientific and popular acclaim. The landers were aided by *Mars Global Surveyor* (launched 1996) and *Mars Odyssey* (launched 2001), which studied Mars from orbit.

In addition to its solar system probes NASA launched a series of Great Observatories. In 1990 Space Shuttle *Discovery* delivered the *Hubble Space Telescope* into Earth orbit during the STS-31 mission. *Hubble* provided a wealth of science, made possible by four Shuttle servicing missions. *Hubble*, which operated in the optical region, was followed by the *Compton Gamma Ray Observatory* (launched 1991), the *Chandra X-ray Observatory* (1999), and the *Spitzer Space Telescope* (2003) in the infrared. Other spacecraft, including the *Far Ultraviolet Spectroscopic Explorer* and the *Wilkinson Microwave Anisotropy Probe* also explored the deep universe.

Building on its roots in NACA, NASA conducted research on aerodynamics, wind shear, and other important topics using wind tunnels, flight testing, and computer simulations. In the 1960s NASA's X-15 program involved flying a rocket-powered airplane above the atmosphere and gliding it back unpowered to Earth. The X-15 pilots helped researchers gain useful information about supersonic aeronautics, and the program provided data for development of the Space Shuttle. NASA also cooperated with the Air Force in the 1960s on the X-20 Dyna-Soar program, which was designed to fly into orbit.

NASA also conducted significant research on high-speed aircraft flight maneuverability that was often applicable to lower speed airplanes. NASA scientist Richard Whitcomb invented the "supercritical wing" that was specially shaped to delay and lessen the impact of shock waves on transonic military aircraft and had a significant impact on civil aircraft design. From 1963 to 1975 NASA conducted a research program on "lifting bodies," aircraft without wings. This paved the way for the Shuttle to glide to a safe unpowered landing, for the later X-33 project, and for a crew return vehicle prototype for the *ISS*. In 2004 the X-43A airplane used innovative scramjet technology to fly at 10 times the speed of sound, setting a world's record for air-breathing aircraft.

Stephen Garber, Steven Dick, and Roger Launius

See also: United States, United States Air Force

Bibliography
Roger D. Launius, *NASA: A History of the U.S. Civil Space Program* (1994).
John M. Logsdon, ed., *Exploring the Unknown: Selected Documents in the History of the U.S. Civil Space Program* (1995–2004).
Howard E. McCurdy, *Inside NASA: High Technology and Organizational Change in the U.S. Space Program* (1994).

Rocket and Space Corporation Energia

Rocket and Space Corporation Energia or RSC Energia was the most important engineering firm in the Soviet Union (now Russia) space program. The earliest Soviet spacecraft manufacturer, it traced its origins back to a May 1946 decree that founded the postwar Soviet missile program. Under Stalin's watchful eye, the Soviet armaments industry established NII-88 (Scientific-Research Institute No. 88) in the Moscow suburb of Kaliningrad (renamed Korolev in 1996) to direct all work on long-range missiles. In August 1946 several departments within the institute were formed, one of which, Department No. 3, was led by Sergei Korolev, widely regarded as the founder of the Soviet space program.

Korolev's department was initially directed to develop copies of the German V-2 missile, but by the early 1950s it had begun developing indigenous ballistic missiles, such as the R-2 (U.S. Department of Defense code name SS-2) and R-5M (SS-3). In 1950 Korolev's department was upgraded to a Special Design Bureau (OKB), and in 1956 it formally separated from its parent NII-88 and became the independent Experimental Design Bureau-1 (OKB-1).

OKB-1's most important work in the 1950s was the creation of the R-7 (SS-6), the world's first intercontinental ballistic missile, which was successfully launched in August 1957. Two months later a modified R-7 successfully launched *Sputnik* into Earth orbit. The design bureau was responsible for many successes in the early years of the space race, including the launch of the first probes to the Moon in 1959, the first human into space in 1961, the first woman in space in 1963, the first multiperson spaceflight in 1964, and the first extravehicular activity by a cosmonaut in 1965. Korolev died in 1966 at the peak of the organization's successes and was succeeded by his deputy, Vasiliy Mishin.

The organization's most expensive space project in the 1960s was the N1-L3 program, designed to compete with NASA's Apollo project to land the first human on the Moon. After four consecutive failures of the giant N1 rocket, the Soviet government canceled the effort in 1974. The same year, after an industry-wide reorganization, the government created the NPO Energia conglomerate with the former OKB-1 at its center. Valentin Glushko, the leading rocket engine designer in the Soviet Union and Korolev's onetime opponent, took control of the giant organization. In the 1970s–80s, NPO Energia worked primarily on the Energia superbooster, the Buran space shuttle, and the Salyut space station programs. In 1986 the organization launched the core of the *Mir* space station. Between 1989 and 1999, NPO Energia kept the station—augmented by six other modules—continuously inhabited for nearly 10 years.

In April 1994 President Boris Yeltsin signed an order that partially privatized the company (the government retained about a 38 percent share in the corporation). The company, which employed more than 20,000 employees at the turn of the century, comprised the main design bureau at Korolev and subordinate enterprises including a factory located in Korolev, the Volga design bureau in Samara, and the Primorsk Scientific-Technological Center in Primorsk. RSC Energia also maintained a branch at the Baikonur Cosmodrome in Kazakhstan. The company's General Designer, Yuri Semenov, steered RSC Energia through its most difficult times in the post-Soviet transitional period but was ousted in 2005 in a bid to salvage RSC Energia's worsening financial situation—it had lost money for three years, largely due to governmental intransigence about paying the company.

Nikolai Sevastyanov, a veteran of Russia's natural gas monopoly Gasprom, took over amid a general reorganization to split the design and management functions of RSC Energia.

RSC Energia retained its role as the prime Russian contractor to the *International Space Station (ISS)*. It created the station's main core module named *Zvezda* and produced the Soyuz and Progress ferry vehicles, which kept the station operational. In 2004 RSC Energia announced the development of a new generation of crewed space (Kliper) and logistics (Parom) vehicles to support *ISS* operations

Besides commitments to ISS, RSC Energia pursued international commercial efforts. As of 2005 its most successful ventures were as partners in Sea Launch and International Launch Services (ILS), two multinational satellite launching services in which Energia provided its Block DM upper stage for the crucial boost to geostationary orbit for the Zenit 3SL (for Sea Launch) and Proton K (for ILS) launch vehicles.

Asif A. Siddiqi

See also: Ballistic Missiles, *International Space Station*, Russian Launch Vehicles, Sea Launch Company, Soviet Manned Lunar Program, Voskhod, Vostok

Bibliography

Rocket and Space Corporation Energia: The Legacy of S. P. Korolev (2001).
Asif A. Siddiqi, *Challenge to Apollo* (2000).

Russian Federal Space Agency

Russian Federal Space Agency (Roskosmos) was established as a branch of the federal government responsible for implementing the civilian space program; coordinating activities of the government, military, and commercial organizations in development, launch, control, and recovery of civilian spacecraft; and managing Russia's contributions to international and commercial space projects.

It was originally established in 1992 as the RKA (Rossiyskoe Kosmicheskoe Agentstvo or Russian Space Agency) in order to split military and civilian space programs. Before then all space activities in the Soviet Union and Russia were controlled by the military and a highly secretive Ministry of General Machine Building (Ministerstvo Obshchego Mashinostroeniya, or MOM), which directed research and development of rocket and space technology. Lack of separation between the military and civilian space efforts hampered Russian scientific space projects because national defense always had a higher priority, which was one of the major reasons for the failure of the Soviet crewed lunar program of the 1960s.

The first attempt to create a civilian space agency was made in the Soviet Union in 1985 with the organization of Glavkosmos to manage international projects and ensure Soviet participation in a growing market of the international launch services. However, Glavkosmos was not entirely successful—as just one of MOM's branches, it was not designed to operate in a commercial environment.

With the collapse of the Soviet political and economic system in December 1991, the RKA inherited most of the research and development infrastructure of the disbanded MOM. In 1998 it assumed control of a larger portion of the Baikonur launch center and began launching space missions separate from the military. In 1999, because of several reshufflings of the government structure, the RKA was additionally assigned to control the aviation industry and was renamed Rosaviakosmos. However, the new responsibilities further distracted the government's attention and diverted funds from the space program, which had already been suffering from the chronic lack of funding.

In 2004 Rosaviakosmos was reorganized into the Russian Federal Space Agency (Roskosmos) and returned to its prime responsibilities in the civilian space program. Simultaneously Yuri N. Koptev, a prominent aerospace engineer who headed the Russian Space Agency since its inception, was replaced by former Space Forces Commander Colonel General Anatoly N. Perminov.

Peter A. Gorin

See also: Russia (Formerly the Soviet Union), Russian Space Forces, Russian Space Launch Centers

Bibliography
Russian Federal Space Agency. http://www.federalspace.ru/main.php?lang=en.

United Nations (UN) Committee on the Peaceful Uses of Outer Space

The United Nations (UN) Committee on the Peaceful Uses of Outer Space (COPUOS) was established as an ad hoc committee by the United Nations General Assembly after launch of the first satellite by the Soviet Union in October 1957. COPUOS comprised 18 Member States tasked to consider the activities and resources of the United Nations, the specialized agencies, and other international bodies relating to peaceful uses of outer space, international cooperation and programs in the field that could be undertaken under UN auspices, arrangements to facilitate international cooperation in the field within the framework of the United Nations, and legal problems that might arise in programs to explore outer space. Because of differences involving the relative number of Member States allied with the United States and with those allied with the Soviet Union, the Soviet Union and other communist nations declined to participate in the committee work. The committee rendered a report in 1959 that contained a number of recommended actions, including practical proposals for international cooperation that included exchange of information on space research, coordination of national space research programs, and assistance in the realization of such programs.

Subsequently General Assembly Resolution 1472 (XIV) established the 24-member UN COPUOS in 1959 to review the scope of international cooperation in peaceful uses of outer space, devise programs to be undertaken under UN auspices, encourage continued research and the dissemination of information on outer space matters, and study legal problems arising from the exploration of outer space. The committee established two standing subcommittees of the whole: the Scientific and Technical Subcommittee and the Legal Subcommittee. The committee and its two subcommittees met annually, starting in 1960, to consider questions put before them by the General Assembly, reports submitted to them, and issues raised by the Member States.

In 1961 the General Assembly, considering that the United Nations should provide a focal point for international cooperation in the peaceful exploration and use of outer space, requested the committee, in cooperation with the secretary general and making full use of the functions and resources of the secretariat, to maintain close contact with governmental and nongovernmental organizations concerned with outer space matters; to provide for the exchange of such information relating to outer space activities as governments may supply on a voluntary basis, supplementing, but not duplicating, existing technical and scientific exchanges; and to assist in the study of measures for the promotion of international cooperation in outer space activities.

In 1962 the General Assembly adopted a resolution declaring a set of principles to guide the activities of nations in the exploration and use of outer space: The Declaration of Legal Principles Governing the Activities of States in the Exploration and Uses of Outer Space. This was the first attempt by the international community to establish a set or rules to govern the activities of nations in space.

The major accomplishment of the COPUOS was the negotiation and completion of five major treaties relating to the use of outer space:

- The Treaty on Principles Governing the Activities of States in the Exploration and Use of Outer Space, including the Moon and Other Celestial Bodies (1967 Outer Space Treaty)

- The Agreement on the Rescue of Astronauts, the Return of Astronauts, and the Return of Objects Launched into Outer Space (1968)

- The Convention on International Liability for Damage Caused by Space Objects (1972)

- The Convention on Registration of Objects Launched into Outer Space (1976 Registration Convention)

- The Agreement Governing the Activities of States on the Moon and Other Celestial Bodies (1984 Moon Treaty)

Committee membership gradually increased to 67 Member States as of 2007. Because of the addition of Member States with no activities involving outer space, the focus of the committee's work shifted after 1980 to consideration and development of declarations of principles, which the General Assembly adopted and promulgated, including:

- The Principles Governing the Use by States of Artificial Earth Satellites for International Direct Television Broadcasting (1982)

- The Principles Relating to Remote Sensing of the Earth from Outer Space (1986)

- The Principles Relevant to the Use of Nuclear Power Sources in Outer Space (1992)

- The Declaration on International Cooperation in the Exploration and Use of Outer Space for the Benefit and in the Interest of All States, Taking into Particular Account the Needs of Developing Countries (1996)

The Committee established annual meetings at its inception to report on the progress of space programs and to survey emerging issues, on which the Committee prepared recommendations for the consideration of the General Assembly.

Stephen Doyle

See also: Space Law

Bibliography
United Nations Office of Outer Space Affairs. www.oosa.unvienna.org.

MEDIA AND POPULAR CULTURE

Media and popular culture have enjoyed a symbiotic relationship with space travel and exploration for more than a century. Space has provided inspiration for short stories, books, and other creative works even before the first airplane took flight. Throughout the decades such cultural expressions have influenced public perception and the ultimate direction of humankind's exploration of space. Although the prevalence of space topics and themes in the media and popular culture has waxed and waned, the public's continued association of space with hope, dreams, fantasy, progress, mystery, and the unknown has ensured that space has remained a popular cultural theme. The space age emerged from the imaginations of artists and writers and the visions of space travel conveyed in their media. Before there were rockets and footprints on the Moon, space exploration was the stuff of science fiction stories, such as Jules Verne's *From the Earth to the Moon* and H. G. Wells's *The War of the Worlds*. Science fiction works such as these inspired German Hermann Oberth and Russian Konstantin Tsiolkovsky to develop the first theories of modern rocketry and American Robert Goddard to design and launch the world's first liquid-propelled rocket. The rocket pioneers were scientists, writers, and cosmologists who shared a common interest in the development of rockets for spaceflight, as a means to an end rather than an end in itself.

Just as literary works influenced the early rocket scientists, a multitude of media creations conveying space themes emerged during the first half of the twentieth century and were critical to fostering public support for and interest in space exploration around the world. Not only did media educate (or at least entertain) the public, it also helped to attract new members (and associated funding from membership dues) to the fledgling space societies. While many books, comics, science fiction stories, and movies of the era gave people the fanciful impression that space was home to intelligent life on Mars and aliens bent on invading Earth, some space advocates of the 1940s–50s harnessed mass media's power to suggest to the public that human spaceflight was credible. The artwork of Chesley Bonestell and others put planets and space stations into visual context for the public for the first time. Wernher von Braun, the

CHESLEY BONESTELL
(1888–1986)

(Courtesy Ted Streshinsky/Corbis)

Widely acknowledged as the father of space art, Chesley Bonestell spent his professional life as an architect, motion picture artist, and illustrator of astronomical and space subjects. He contributed to the design of the Chrysler Building and the Golden Gate Bridge, painted background scenes for more than a dozen films (from *Citizen Kane* to *Destination Moon*), and provided artwork for articles and books on spaceflight in collaboration with such experts as Willy Ley, Wernher von Braun, and Arthur C. Clarke. Bonestell was also head artist on a groundbreaking series of articles about spaceflight published in *Collier's* in the early 1950s. A crater on Mars and an asteroid are named after him.

Melvin Schuetz

German engineer who led the V-2's development during World War II before coming to the United States and pioneering spacefaring rockets, wrote a series of articles in one of the day's most popular magazines, *Collier's*. He produced three television shows with Walt Disney to explain to the public in general terms how humans could and would soon travel into and explore space.

By the late 1950s media and popular culture had convinced many Americans that human spaceflight was in society's future, but the government had not yet stepped up with the necessary funding to realize the space popularizers' visions. Thanks in part to the news media, the political sphere turned its attention to space when the Soviet Union launched the world's first artificial satellite, *Sputnik*, in 1957 and the first human, Yuri Gagarin, four years later. The Soviet Union made Gagarin a national hero and used the media to promote Soviet space triumphs as evidence of the superiority of socialism. With mainstream news media reports of the Cold War rival's space successes, President John F. Kennedy in 1961 overturned the modest, satellite-dominated space program of his predecessor, Dwight Eisenhower, in favor of an all-out program to land humans on the Moon.

While the popular media had long served to promote space exploration, the U.S. national commitment to an ambitious space program unleashed what can only be described as a popular culture phenomenon. The Cold War preyed on American fears, but witnessing the steps toward a U.S. lunar landing encouraged a sense of optimism about technology and the future that became apparent in cultural expressions from architecture evoking the space age via jetting angles and blinking signs to fashions made of silvery, metallic fabrics reminiscent of the astronauts' spacesuits. The television cartoon show *The Jetsons* featured the first space-age family, as *Life* magazine kept the public in touch with the nation's latest space achievements and the lives of their newest heroes, the astronauts. Pan American and Trans World Airlines invigorated the 1960s' space craze by taking reservations for eventual public flights to the Moon, while the Apollo program inspired a generation of space enthusiasts who collected patches, pins, toys, and other memorabilia dating back to the early days of the space age. The first-ever photograph of the entire Earth, taken during *Apollo 8*, yielded an unpredictable, global impact, fueling the rising environment and peace movements of the day while changing public perception of humankind's place in the universe.

After Apollo, space lost some of its prominence in the media and popular culture. In large part, space's popularity declined because the U.S. space program's novelty waned after the Moon landings, without a clear goal around which the public could rally. At the same time the optimism of the space age faded and political realities changed, shifting public attention to seemingly more compelling issues than U.S. space efforts. The media and the public continued to take note of the flights of a few high-profile individuals to the Russian *Mir* space station and the *International Space Station*, but at the turn of the twenty-first century, video games and reality television shows came to dominate American popular culture as premier outlets for living one's desire for adventure and fantasy.

However, space and the American media and popular culture did not part ways. Newspapers, magazines, television broadcasts, and dozens of online news sites routinely reported the successes and failures of U.S. space undertakings, from the successful Mars rovers to the tragedy of Space Shuttle *Columbia*. Three of the highest grossing movies—*Star Wars* (1977), *E.T.: The Extra-Terrestrial* (1982), and *Star Wars: Episode 1, The Phantom Menace* (1999)—featured space themes. Space endeavors have continued to find inspiration in popular works. Paul Allen, financier of *SpaceShipOne*, the first privately sponsored, piloted spacecraft to reach suborbital space, claimed that Robert Heinlein's 1947 science fiction book *Rocketship Galileo* sparked his interest in space projects, while President George W. Bush's 2004 "Vision for Space Exploration" has maintained a leading role for humans in space, consistent with the romantic view espoused by von Braun and others. Russians continued to take national pride in their space programs, but outside the United States and Russia, space played a much smaller popular role.

In the early twenty-first century human adventures in space continued to shape and take shape based on movies, news stories, art, and the possibilities that creative minds envisioned.

Amy Paige Kaminski

See also: Apollo; Mars; *SpaceShipOne*; Von Braun, Wernher

Bibliography
Howard E. McCurdy, *Space and the American Imagination* (1997).
Frederick I. Ordway and Randy Lieberman, eds., *Blueprint for Space: From Science Fiction to Science Fact* (1992).

Conspiracy Theories

Conspiracy theories casting doubt on the veracity of space activities, such as the Apollo lunar landings and planetary exploration, have been a growing trend since the 1970s.

Hoaxes exploiting the general community's scientific illiteracy and fascination with space, for either financial gain or sheer devilment, are nothing new (perhaps the earliest was the celebrated "Great Moon Hoax" of 1835, designed to boost the circulation of the *New York Sun* newspaper). However, the phenomenon of conspiracy theorists claiming that major space achievements, such as the Moon landings, are themselves hoaxes, or that NASA is part of a vast conspiracy concealing the truth about evidence of extraterrestrial life, is a peculiarity of the late twentieth century that seems to have arisen at least in part as a response to the erosion of public confidence in government openness, beginning in the 1960s. While largely, though not exclusively, American in origin, these space conspiracy theories are part of a growing conspiracist subculture, spread globally via the Internet, through books and videos produced by conspiracists, and via pseudo-documentaries (such as the Fox network's *Conspiracy Theory: Did We Land on the Moon?*) produced by a sensation-hungry media.

Space-related conspiracy theories tend to fall into two general categories: those that claim the Moon landings and other space missions (both U.S. and Soviet) did not occur as presented but involved hoaxes and cover-ups in order to achieve status-conferring "space firsts" during the Cold War; and those that assert that the United States (and other governments) have discovered evidence of extraterrestrial life, either ancient or current, on the Moon or Mars and are concealing this evidence from the general populace. The profusion and confusion of these theories is such that some combine elements from both categories, while many are openly contradictory of one another.

Among the best-known conspiracy theories in the first category are the claims that the Soviet Union lost one or more cosmonauts in space before Yuri Gagarin and covered up their deaths and that the United States faked the Apollo lunar landings (or at least the *Apollo 11* landing) to win the space race. In the early 1960s the Judica-Cordiglia brothers, Italian ham radio enthusiasts, claimed to have intercepted signals from Soviet cosmonauts in 1960–61 before Gagarin's flight. Other rumors of cosmonauts lost on unannounced space missions were also reported in the Western press in the 1960s. While the secretive nature of the Soviet space program and Cold War suspicions gave these claims some spurious credibility at the time, they have been thoroughly and repeatedly debunked since the 1980s by James Oberg and by Russian writers.

Claims that the Moon landings were a hoax, staged by the United States to fulfill its Cold War propaganda aim of winning the space race by landing the first person on the Moon, first appeared in *We Never Went to the Moon* (1974), which outlined the basic "proofs" of the hoax that have been the standard fare of all subsequent hoax claims, whatever their additional twists and embellishments. Like most of his later followers, author William Kaysing based his claims on the "analysis" of multi-generation copies of NASA publicity images that lacked the fine detail of the original photographs— detail that immediately dispels some of the supposed anomalies. Other claims, such as the absence of stars in the photographs supposedly taken on the lunar surface and apparently non-parallel shadows on the lunar surface, can readily be debunked by knowledge of photography, physics, and an awareness of the differences between the lunar and Earth environments.

The Moon landing hoax conspiracy theory comes in many versions, from claims that the missions were entirely faked and shot on a secret sound stage somewhere in the United States (a scenario apparently inspired by the 1978 film *Capricorn One*, itself based on early versions of the Moon hoax), to assertions that only the *Apollo 11* Moon landing was faked, in order to beat the Soviet Union to the Moon, and that the later Apollo missions were genuine. Many of these stories place the secret film set in "Area 51," a highly classified U.S. Air Force test facility at Groom Lake, Nevada, which features in a variety of other conspiracies.

Other variants of the hoax claim that Moon landings did occur, but secretly, and that NASA faked the Apollo landings to cover the "real" activities on the Moon (which involve encounters with extraterrestrials in some way). These versions usually cross into the other major category of space conspiracies, that is, that the U.S. government (and others around the world) has evidence of extraterrestrial life that is being kept secret.

Conspiracy theories of a government cover-up of evidence of extraterrestrial life stem originally from the unidentified flying object (UFO) community that has, since the 1950s, asserted the existence of a conspiracy to hide the truth of intelligent life beyond Earth from the general populace. As real space technology reached into the cosmos and gave humans the ability to reach the Moon and send proxy explorers to other planets in the form of space probes, it is perhaps not surprising that UFO conspiracists have attempted to link their theories to real space activities, postulating secret space programs related to contact with aliens behind the cover of the space race.

Similarly, other conspiracy theories allege that images returned by robotic probes to the Moon and Mars show structures that are evidence of ancient, if not current, intelligent life. Foremost among these conspiracy proponents is Richard C. Hoagland, who first publicized the so-called "face" in the Cydonia region on Mars as a sculpture produced by intelligence, rather than the wind-carved mesa that higher resolution images revealed it to be. Supporters of these theories have regularly advanced the notion that any failed planetary probe was deliberately destroyed to prevent its images from revealing some "truth"—when in fact it would have been simpler for a space agency just to not plan a mission that might endanger an existing cover-up.

While an interesting psychological phenomenon, conspiracy theories belittle the significant achievements of the human spirit and ingenuity, in reaching beyond Earth to explore the cosmos, and contribute nothing to the exploration of space.

Kerrie Dougherty

See also: Apollo, Mars, Vostok

Bibliography
Bad Astronomy. http://blogs.discovermagazine.com/badastronomy/.
Robert T. Carroll, *The Skeptic's Dictionary* (2003).
Moon Base Clavius. http://www.xmission.com/~jwindley/.

James E. Oberg, *Uncovering Soviet Disasters* (1988).
Philip C. Plait, *Bad Astronomy* (2002).

Effect of Sputnik

The effect of Sputnik, after its launch on 4 October 1957, on world politics and the direction of space programs was dramatic. The 84 kg *Object PS 1*, as the Soviet Union called it, rode a modified R-7 intercontinental ballistic missile (ICBM) into space and into global headlines.

The sensation was created even though the launch should not have been a complete surprise. Soviet experts and publications openly discussed their International Geophysical Year (IGY) satellite (in general terms), and the U.S. Central Intelligence Agency (CIA) had predicted the possibility a year in advance. Yet it was a surprise. As *Sputnik*'s creator, Chief Designer of the Soviet space and missile program Sergei Korolev, congratulated his comrades for opening the road to the stars, radio operators around the world tuned in the satellite's beep and others scanned the night sky. The satellite was too small to be seen with the naked eye, but the core of the R-7 booster had followed *Sputnik* into orbit and was spotted easily. This visual proof magnified the satellite's impact. Several influential American media outlets, most notably *Life* magazine, published alarmist critiques, which succeeded in raising the public's concern.

Reports that *Sputnik* caused panic in Western nations were exaggerated. However, the satellite did send shock waves through U.S. and allied governments. James R. Killian, a scientific adviser to U.S. President Dwight Eisenhower, wrote that the event violently contradicted the fundamental belief that the United States's technical capacity had no serious rival. Western armed forces had a specific and worrisome concern. Missile experts correctly deduced the launcher was a powerful ICBM. The Soviet Union had announced the first flight of Korolev's ICBM a few months earlier, but U.S. intelligence had been unsure of the validity. Now there was no doubt.

If the little sphere caused consternation among governments, it also excited scientists who knew that the Earth satellite concept, long a theoretical possibility, had at last been proven feasible. British author and space visionary Arthur C. Clarke recalled that it was a complete shock, but he realized it would change the world.

The international impact of *Sputnik* was unexpected even by the Soviet leaders. At first, the official newspaper *Pravda* gave the launch only a brief mention. Only after it became clear *Sputnik* had caused a global sensation did the satellite earn banner headlines. A CIA assessment stated that *Sputnik* had immediately increased Soviet scientific and military prestige among many peoples some governments. Soviet diplomats and politicians made the most of the resulting admiration.

The effect of the *Sputnik* launch on the Western public was raised by the subsequent media coverage and magnified by the 3 November 1957 launch of *Sputnik 2*. *Sputnik 2* weighed 508 kg, was highly visible (thanks to the failure of the

R-7 core stage to detach as planned), and carried the first living creature in space, the dog Laika. Coming at a time when the United States was still scrambling to launch even a 1.5 kg Vanguard test satellite, warnings of Soviet superiority seemed, if anything, too moderate.

President Eisenhower had also been surprised by *Sputnik*. While he reassured the public that the U.S. satellite program had not been conducted as a race against other nations and that *Sputnik* raised no new security concerns, he privately called his advisers on the carpet for an explanation. At the same time, he considered what actions were necessary in response. The president saw reason for concern but not panic. He refused demands for an all-out crash program, but did ask Congress for a $1 billion emergency appropriation to boost American missile programs.

The U.S. government responded to calls from the media and academic leaders to improve education in engineering and the sciences. In 1958 Eisenhower signed the National Defense Education Act to provide funding for science and math programs in colleges and high schools. This federal intervention in education, traditionally a state and local matter, began the transformation of America's system of government. This had consequences in social programs, civil rights, and other areas far removed from space. Another consequence the Soviet leaders did not foresee was the effect of *Sputnik* on international law. Before *Sputnik*, the right of transit through space above a nation's territory was an unsettled question. Donald Quarles, Eisenhower's Deputy Secretary of Defense, pointed out that the Soviets had possibly done the United States an unintentional favor by establishing the concept of freedom of international space. Not one government protested the overflight of *Sputnik*. In July 1959 this acceptance was cited by a United Nations report endorsing "freedom of space"—an idea enshrined by the 1967 Outer Space Treaty.

In the Soviet Union, *Sputnik* made Korolev a powerful man with vast resources to devote to his dreams of spaceflight. The price imposed was the need to keep the successes coming to maintain leadership in this new field. Korolev responded with new satellites, lunar probes, and in 1961 the launch of the first human into orbit.

Sputnik also galvanized the lagging U.S. space program. With the official U.S. IGY satellite program, Project Vanguard, still struggling, the Army missile team headed by Wernher von Braun was given approval to launch a satellite. After a frantic effort, *Explorer 1* was orbited in January 1958. The government was already discussing the options for a long-term space program. On the military side this led to the creation of the Advanced Research Projects Agency (ARPA) and the post of Director, Defense Research and Engineering (DDR&E), beginning a shift of control over research funding and military budgets in general from individual services to the Office of the Secretary of Defense. Civilian space programs, Eisenhower decided, should belong to a new agency. On 1 October 1958 the National Aeronautics and Space Administration (NASA) came into existence. It began pursuing numerous space endeavors, including science and applications satellites and its own human-in-space program. *Sputnik*'s launch was the beginning of the journey to the Moon.

Matt Bille and Erika Lishock

See also: Rocket and Space Corporation Energia, International Geophysical Year, National Aeronautics and Space Administration

Bibliography
Matt Bille and Erika Lishock, *The First Space Race* (2004).
Roger Launius et al., eds., *Reconsidering Sputnik: Forty Years since the Soviet Satellite* (2000).
Walter A. McDougall, *. . . the Heavens and the Earth: A Political History of the Space Age* (1985).
Asif A. Siddiqi, *Challenge to Apollo: The Soviet Union and the Space Race, 1945–1974* (2000).

Literary Foundation of the Space Age

The literary foundation of the space age reflects a defining theme of the 1950s and 1960s. Images of the space age evoke vivid memories for an entire generation. The genre of literature associated with the development of rocketry and human spaceflight concepts is referred to as speculative nonfiction—visionary works by the dreamers and the doers, the practitioners and the promoters of human spaceflight.

As the twentieth century dawned, spaceflight visionaries around the world independently developed the theoretical principles of spaceflight. Inspired by earlier works of fiction, such as the tales of Jules Verne, they dreamed of turning fiction into reality. Foremost among these spaceflight visionaries were Konstantin Tsiolkovsky, Hermann Oberth, and Robert H. Goddard.

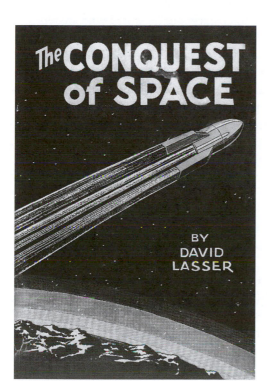

The Conquest of Space *(1931) by David Lasser. (Courtesy Michael Ciancone)*

Tsiolkovsky was a Soviet schoolteacher who developed many of the theoretical principles of spaceflight around the turn of the twentieth century. His article on the theory of spaceflight (1903) went largely unnoticed outside the Soviet Union but received enough domestic attention that he revised, expanded, and published it as a monograph in 1924. Tsiolkovsky belatedly received international recognition in the 1920s–30s due to the efforts of the Soviet Union to garner respect and recognition from the international technical community. Many of Tsiolkovsky's works were translated into English for NASA in the 1970s.

Oberth was a Rumanian scientist who developed many early spaceflight concepts, including ideas for an Earth-orbiting habitable space station and the use of multistage rockets to achieve Earth orbit. Oberth published his ideas in *Die Rakete zu den Planetenräumen* (*The Rocket into Planetary Space*, 1923). In 1929 he expanded his work in *Wege zur Raumschiffahrt* (*Ways to Spaceflight*).

Goddard was a U.S. experimenter who launched the world's first liquid-propellant rocket in March 1926. Much of his early work was funded by the Smithsonian Institution, which published his seminal monograph, *A Method of Reaching Extreme Altitudes* (1920). The only other substantive work published by Goddard during his lifetime was *Liquid-Propellant Rocket Development* (1936) in which Goddard first reported his successful launch of 1926. The American Rocket Society reprinted both monographs together in *Rockets* (1946). Goddard's widow, Esther C. Goddard, and a former colleague, G. Edward Pendray, edited and published his research notes in *Rocket Development—Liquid Fueled Rocket Research, 1929–1941* (1948).

In addition to this trio, there were individuals around the world who shared this vision of the future. The first book of technical nonfiction on rockets for spaceflight in any language was *La Conquête de l'Espace* (*The Conquest of Space*, 1916) by Victor Coissac. However, it was not widely known because Coissac was not actively engaged in the spaceflight movement. In the Soviet Union, Y. Perelman wrote a series of popular books on interplanetary spaceflight (1915–35). Austrian Max Valier conducted a series of spectacular, if not completely successful, rocket car experiments in collaboration with Fritz von Opel and popularized ideas of spaceflight in *Der Vorstoß in den Weltenraum* (*The Thrust into Space*, 1924) and *Raketenfahrt* (*Rocketflight*, 1928). In France Robert Esnault-Pelterie wrote *L'Astronautique* (*Astronautics*, 1930), a comprehensive theoretical analysis of space travel that first introduced the term "astronautics," and a later addendum, *L'Astronautique (Complément)* (1935). In Italy Luigi Gussalli experimented with rocket motors and published *Si puó giá tentare un viaggio dalla terra alla luna?* (*Is It Possible Yet to Attempt a Voyage from the Earth to the Moon?*, 1923) in which he suggested using a multistage rocket to transport two men to the Moon and back. The first British book on space travel was *Stratosphere and Rocket Flight (Astronautics)* (1935) by Charles G. Philip, followed shortly by P. E. Cleator's *Rockets through Space* (1936). In the Soviet Union, Nikolai Rynin prepared a 10-volume series on interplanetary flight (1928–32) that was translated into English for NASA in the 1970s.

Societies devoted to promoting human spaceflight and providing a forum for interested citizens to discuss developments in rocketry began to appear around

WILLY LEY
(1906–1969)

(Courtesy Bettmann/Corbis)

As a founding member of the German VfR (Verein für Raumschiffahrt or Society for Spaceship Travel), Ley, who studied to be a paleontologist, shared his enthusiasm for the prospects of human spaceflight through his avocation as a technical writer. In 1928 Ley edited a compilation of technical essays by members of the VfR that was influential in promoting spaceflight efforts. Ley continued to write about spaceflight after his immigration to the United States in 1936, including books such as *The Conquest of Space* (1949) and *Rockets, Missiles, and Space Travel* (1951), and became an active member of the American Rocket Society. He effectively popularized spaceflight, which led to collaboration with Wernher von Braun and Walt Disney on the *Man in Space* television series. The books and journals of Willy Ley are held in the M. Louis Salmon Library at the University of Alabama in Huntsville, and the personal papers of Willy Ley are held in the National Air and Space Museum Archives in Washington, DC.

Anne Coleman

the world. In Germany a group of spaceflight enthusiasts formed the VfR (Verein für Raumschiffahrt or Society for Spaceship Travel) in 1927 (the first organization devoted to the development of rockets for space applications). The VfR generated public interest in the prospects of space travel through speeches, articles, the society journal, *Die Rakete*, and books, such as *Die Möglichkeit der Weltraumfahrt* (*The Possibilities of Space Travel*, 1928), edited by Willy Ley, which contained chapters written by key participants in the international spaceflight movement. Another early VfR member, Hermann Noordung, published the first comprehensive study of space stations, *Das Problem der Befahrung des Weltraums* (*The Problem of Space Navigation*, 1929). In the United States, a group of space enthusiasts, predominantly composed of writers for early science-fiction pulp magazines, founded the American Interplanetary Society (AIS) in 1930. David Lasser, the first president and founder of AIS, wrote *The Conquest of Space* (1931), which was the first book of nonfiction in English to address the prospect of space travel using rockets (as opposed to Goddard's monograph, which was intended for a limited, technical audience). In the United Kingdom, a group of young rocket enthusiasts formed the British Interplanetary Society (BIS) in 1933. Although enthusiastic, BIS members were prevented from conducting rocket experiments by the Explosives Act of 1875, which forbade such activities. As a result the BIS concentrated its efforts on formulating plans and detailed studies for spaceflight activities, such as the design of an Earth-orbiting spacecraft, which were documented in the society publication, *Journal of the British Interplanetary Society* (*JBIS*). L. J. Carter edited *Realities of Space Travel* (1957), which included selected articles that had previously appeared in the *JBIS*.

In addition to providing a forum for devotees of spaceflight, the early rocket societies also provided fertile grounds for people who would later play key roles in the realization of human spaceflight. G. Edward Pendray, a founder of AIS, wrote *The Coming Age of Rocket Power* (1945) that provided a knowledgeable popular accounting of rocketry state of the art during the immediate postwar years. In the United States, Wernher von Braun (who had been a young member of the VfR) was a vocal and effective advocate of the prospects of space travel and contributed to the growing mass of popular literature. In *The Mars Project* (1953), he outlined the requirements for a human mission to Mars. In addition he contributed to a series of articles in *Collier's* magazine that was later published as *Across the Space Frontier* (1952) and *Conquest of the Moon* (1953). In addition to von Braun, perhaps the most widely recognized name in this genre is Willy Ley, a zoologist by training, a founder of the VfR, and a popular science writer in prewar Germany and later in the United States. His first book on rocketry and space travel to appear in English was *Rockets—The Future of Flight beyond the Stratosphere* (1944). This was later updated and republished as *Rockets and Space Travel* (1947) and *Rockets, Missiles, and Space Travel* (1951), the latter of which went through numerous editions and printings, attesting to the growing public interest in space activities. Ley also worked with noted space artist Chesley Bonestell on *Conquest of Space* (1949). Ley's efforts were instrumental in generating public interest and building public expectations for the human exploration of space. Moreover, any article on space literature would be remiss without mention of the ubiquitous *The Exploration of Space* (1951) by Arthur C. Clarke.

The launches of *Sputnik* and *Explorer 1* ushered in the era of the human exploration of space. This shift from concepts to realities was accompanied by a shift in published material from the practitioners and promoters, to journalists, observers, and others on the periphery.

Michael L. Ciancone

See also: American Rocket Society, Verein für Raumschiffahrt, Rocketry Pioneers and Spaceflight Visionaries

Bibliography

Michael L. Ciancone, *The Literary Legacy of the Space Age* (1998).
Howard E. McCurdy, *Space and the American Imagination* (1997).
Frank H. Winter, *Prelude to the Space Age: The Rocket Societies, 1924–1940* (1983).

News Media Coverage

News media coverage of space exploration has played a critical role in informing the world's people and shaping their perceptions about space exploration and, at times, has even influenced space policy and program decisions. Since the beginning of the space age, newspapers, popular magazines, and televised news broadcasts—and in the late twentieth century Internet news sites—have provided a window and magnifying lens on space exploration activities and the organizations that conduct them. Through the use of selected words, photographs, and video footage, they have publicized achievements along with failures and controversies. Or, in the case of some communist regimes, nonreporting by government-run media served as a means of denying Western space achievements a prominent place in their citizens' consciousness. In doing so, the media have constantly reminded the public that space exploration remains a pioneering and progressive, yet highly complex and challenging undertaking.

The Soviet Union's 1957 launch of *Sputnik* seized the attention of news media organizations around the world. The global media largely recognized that the achievement heralded a major turning point in history, with the *London Daily Express* being the first to name the new era in a headline proclaiming, "The Space Age Is Here." The Western media reported the event with a mix of awe and trepidation, while communist newspapers extolled the action as a sign of socialism's superiority. *Pravda*, and by extension the Soviet government, only belatedly seized on the event as a public relations success in response to the reaction by the Western media. From that point,

U.S. broadcast journalist and TV news anchor Walter Cronkite uses a spacecraft model during his coverage of the Apollo 7 *mission from NASA's Manned Spacecraft Center in Houston, Texas, 21 October 1968. (Courtesy Hulton Archives/Getty Images)*

the Soviet and American space programs were set in motion. The Sputnik experience, combined with the success of space popularizers in the 1950s to exploit popular American periodicals and television to inform people of the imminent prospects of space travel, suggested that media outlets were primed to relate the ensuing American and Soviet space developments to the global public.

Both the American and Soviet governments recognized the value of manipulating the media to reflect specific images of their space programs to their publics during the early years of the space race; however the two nations' news media represented space achievements in markedly different ways. In an effort to shroud plans in secrecy, the Soviet government limited the details the media provided its people about the Soviet space program's development, only revealing goals after they had been achieved and deliberately withholding the names of personnel responsible for the victories. Communist newspapers were saturated with propaganda boasting Soviet pre-eminence in space; as the Soviet Union continued to outperform the United States in achieving one space milestone after the next, *Pravda*, for example, reported in 1961 that "there is evidence again and again of the great conviction in the triumph of Soviet rocket technology," while the Soviet news service's reports of American space accomplishments, when given at all, were highly disparaging.

In contrast NASA worked fervently as the newly appointed steward of American space exploration to create a positive image and sell its programs to the public and the Congress. The agency strained to accommodate reporters, providing them with designated viewing areas at launch events, tours of NASA facilities, and press kits on NASA activities. At the same time the agency carefully orchestrated information flow to the media, even coaching astronauts and NASA officials on what to say during interviews and press conferences. In 1959 NASA flagrantly allied itself with the news media, awarding *Life* magazine exclusive access to the Mercury astronauts' personal stories—but retaining rights to censor articles to propagate a favorable image of the agency.

NASA's public relations strategy worked as the agency had hoped. Throughout the Apollo era, *Life* and other American media sources in large part served as an extension of NASA's public affairs office, covering the agency's activities in glowing, highly optimistic terms. *Life*'s laudatory text and two-page photo spreads glorified America's newest celebrities and their achievements. On occasion periodicals like *Newsweek* and the *Saturday Evening Post* questioned the sudden, expansive growth of NASA's budget, the value of space exploration, and the politics behind NASA contract awards; in 1963, for example, *Time* bemoaned that NASA had "sprouted like Jack's beanstalk, sucking up men and money at a prodigious rate, sending its tendrils into every state." The critics, nonetheless, were overpowered by positive media coverage, which, in addition to rallying public awe of the astronauts' heroism, centered on promoting an American space program that would propel the nation ahead of the Soviet Union for technological supremacy while also opening a new and exciting future on the space frontier. The public and Congress shared the media's passion, putting support behind the efforts that led to *Apollo 11*'s 1969 arrival on the Moon, which the *Chicago Tribune* called "A Magnificent Achievement" and the *Houston Post* termed "A Great Moment of Our Time." American television networks and broadcasters around the world dazzled viewers with ethereal images of the astronauts bouncing around the Moon. While most of the world's media celebrated the feat, the Soviet media report of the American achievement was dry and devoid of televised footage.

The American news media's romance with space exploration faded throughout the 1970s–80s as the American and Soviet human spaceflight programs lowered their sights to low Earth orbit. Controversy about *Skylab*'s uncontrolled reentry over Australia and constant delays of the Space Shuttle, which NASA had claimed would make human spaceflight routine, only fueled the media's increasing disinterest with America's space exploration program. NASA had hoped to reignite the media's passion for its activities by establishing a program to send a journalist on a Space Shuttle flight, but the explosion of *Challenger* dashed the agency's plans to fly a member of the press in addition to its hopes to reinvigorate media coverage of its programs.

Whereas much of the world's news media had portrayed American space exploration activities with a sense of wonder and infallibility, the *Challenger* disaster forever changed NASA's image. Before the disaster, journalists had not questioned Shuttle safety and NASA's management practices; from that moment forward, critical analysis would become a hallmark of worldwide news media treatment of not only American but also global space exploration endeavors. The press relentlessly blasted NASA during the 1990s for the *Hubble Space Telescope*'s flawed optics and a metric-English unit conversion blunder—deemed an "embarrassingly simple mistake" by the *San Francisco Chronicle*—that led to the inflight loss of a spacecraft headed for Mars. At the same time, news reports lamented out-of-control spending on the *International Space Station* and questioned the safety of cosmonauts aboard Russia's *Mir* station, which British Broadcasting Corporation news online referred to as "floating from one crisis to the next" in 1998 after the space station had fallen into disrepair during Russia's economic crisis after the fall of the Soviet Union.

While skepticism and investigative reporting have been common features of many space-related stories, journalists around the world have remained intrigued by space exploration, regarding it as the perfect fodder for news stories: it generates drama and excitement, exposes technological prowess, divulges the unknown, and conveys a spirit of adventure. As a result the news media has continued to follow developments in human spaceflight and robotic space exploration (the latter has received considerably more coverage in the twenty-first century than it did in the first few decades of the space age), seeking to uncover the next big story in space, whether it is about a private citizen's flight into space, the discovery of a planet beyond Earth's solar system, or the rollout of a new launch vehicle. Available to the global public in mainstream newspapers, magazines, and news broadcasts and within numerous space-specific periodicals and websites, space media coverage in the early twenty-first century has been omnipresent and in many cases highly analytical, equipping people with in-depth information and perspective they can use to form opinions and participate in policy debates about space exploration.

Amy Paige Kaminski

See also: Apollo, Mercury, Russia (Formerly the Soviet Union), Space Shuttle

Bibliography
William Boot, "NASA and the Spellbound Press," *Columbia Journalism Review* (1986).
James L. Kauffman, *Selling Outer Space: Kennedy, the Media, and Funding for Project Apollo, 1961–1963* (1994).
Bruce V. Lewenstein, "NASA and the Public Understanding of Space Science," *Journal of the British Interplanetary Society* 46 (1993): 251–54.

Popularization of Space

Popularization of space is, in large measure, a cultural phenomenon. People grant human and robotic space travel more attention than it would receive based strictly on its scientific or commercial merits. They do so because exploration addresses deep-seated cultural needs—the desire for optimism about the future and the need for faith in the principles by which one's society is organized. Especially in the United States, space travel became popular because it dealt with these needs in appealing ways.

Space exploration during the first half of the twentieth century belonged to the realm of fantasy. In 1912 Edgar Rice Burroughs—otherwise famous as the creator of Tarzan novels—began to produce a fantastic series of stories describing the visitation of one Earthling named John Carter to a dying civilization on Mars. In 1926 Hugo Gernsback launched the first inexpensive magazine devoted strictly to science fiction, *Amazing Stories*, successfully imitating similar publications satisfying the public desire for adventure stories and tales of the American West. Part of a genre known as pulp fiction, for the cheap paper that publishers used, Gernsback's magazine inspired a series of competitors and what many have called the golden age of science fiction. Buck Rogers, a famous comic strip character transported 500 years into the future, made his appearance shortly thereafter, followed by the similarly inspired Flash Gordon. Early characters such as these perfected the literary formulas success-fully repeated by creators of spacefaring heroes like Captain James T. Kirk of the starship *Enterprise* and Luke Skywalker of *Star Wars* fame.

Early science fiction stories resembled the fanciful legends found in folk tales and mythical sagas. Set in imaginative societies and exotic locales, the stories were not meant to be real, a feature easily demonstrated by the cavalier manner in which their creators treated the physics of space travel. The spacecraft in E. E. Smith's *Skylark of Space*, one of the first galactic adventure stories to be characterized as a space opera, traversed 237 light-years in 24 hours. John Carter transported himself to Mars (known locally as Barsoom) by falling into a trance outside the mouth of an Arizona cave. The otherwise realistic Jules Verne dispatched his travelers on a trip *From the Earth to the Moon* through the device of a large cannon, the escape velocity of which in practical terms would have reduced the occupants to thin smears on the capsule floor. Given the prevalence of fanciful tales, skeptics often dismissed serious attempts to discuss space travel as "that Buck Rogers stuff." A 1949 Gallup poll found Americans more optimistic about the prospect of atom-powered trains than flights to the Moon.

Though fantastic, science fiction fulfilled an important need. It suggested that sci-ence and technology, which contained both a dark side and a promising side, would produce a future that was better than the past. The dark side of technology was represented by mass industrialization, the dehumanizing effects of the factory, war weaponry, and economic dislocation. Technological pessimism found expression in science fiction stories such as H. G. Wells's *War of the Worlds*, in which an unfriendly collection of Martians used superior technology to make war on Wells's home planet, and Mary Shelley's *Frankenstein*, often cited as the first real work of scientific fiction. Despite such scary visions, most science-fiction writers let good triumph over evil and technology reduce hardship and want. Especially to a depression-weary public, this was a welcome tale. Space travel helped satisfy the need for a vision in which humans could speed away from poverty to a futuristic utopian dream. Like the design of many consumer goods during the late 1930s, spaceships were streamlined, emphasizing the rapidity with which science could deliver material comfort. In America, space exploration represented a new frontier, a metaphor employed with increasing frequency as interest in the subject grew. As presented by the purveyors of popular culture, frontiers were places where settlers started over and began better lives.

The presentation of space travel in popular culture helped to raise expectations in ways that did not require extensive explanation. Space as the "next frontier" contained an unsubtle allusion to the settlement of the American West. Space voyages would rekindle values already thought to make exploring societies prosper, such as innovation and opportunity. It would result in the discovery of extraterrestrial life, in the same manner that leaders of terrestrial expeditions had discovered exotic species in earthly locales. By employing concepts well established in popular culture, promoters of space travel encouraged the venture without having to confuse the public with elaborate technical details.

At the mid-twentieth century, no part of this fanciful vision was real. Space travel belonged to the realm of fiction. In the United States, beginning in the 1950s, a small group of people with access to the popular media began a campaign to convince the public that such activities could actually occur. George Pal released the Hollywood movie *Destination Moon* in 1950, which won an Academy Award for the realism with which it depicted the possibility of a lunar voyage. Chesley Bonestell, the Hollywood special-effects artist who painted the scenery for *Destination Moon*, created a growing body of paintings depicting the startling views that humans might expect to witness in space. The editors of *Collier's* magazine began an eight-part series on extraterrestrial exploration in 1952 with a Bonestell painting of a winged space shuttle and the promise that humans would "conquer space soon." Walt Disney opened his Disneyland theme park three years later, where Tomorrowland featured a simulated space station ride across the United States and a realistic rocket trip to the Moon. Concurrently Disney released a three-part documentary on space travel, broadcast on his popular weekly television program. Disney explained that "great new discoveries have brought us to the threshold of a new frontier," while Wernher von Braun, increasingly viewed as the principal spokesperson for the advocacy group, explained the rocket technology involved.

In Europe, advocates of space travel established rocket societies to promote practical flight. The British Interplanetary Society, formed in 1933, issued elaborate plans for a human flight to the Moon (1939). Members of the German VfR (Verein für Raumschiffahrt or Society for Spaceship Travel), organized in 1927, launched small rockets and inspired engineers, such as the young Wernher von Braun, to contemplate space travel. Hermann Oberth, president of the VfR, helped filmmaker Fritz Lang produce the first realistic movie visualizing a flight to the Moon, *Frau in Mond* (1929). In the Soviet Union, as in the United States, science fiction stories generated interest in the practical reality of spaceflight. Followers of Konstantin Tsiolkovsky, one of the first people to calculate the actual forces required to move objects through space, formed a Society for Interplanetary Travel in the Soviet Union in 1924, but it collapsed the following year. Like others, Tsiolkovsky promoted his ideas using science fiction stories and serious treatises.

What had been popularized as fiction attracted increasing attention from governmental leaders, the only group with sufficient funds to initiate the undertaking. After World War II, governmental leaders needed a means to demonstrate the power of technology without resorting to a display of its darkest side, the production of atomic weaponry. The possibility of nuclear war frightened people around the globe, a prospect made more terrifying by frequent civil defense drills. Governmental leaders understood that the control of such weapons—indeed the outcome of the Cold War—would be determined by technology. The nation capable of producing the biggest rockets and the best reconnaissance satellites would gain an unchallenged advantage over its rivals.

Space travel offered a means to demonstrate that advantage in a benign and unthreatening way. World leaders tacitly understood that any nation capable of great deeds in space could potentially subjugate its foes. Senate Majority Leader (and later President) Lyndon Johnson predicted in 1958 that the nation that controlled space would gain "control of the world." President John F. Kennedy, not a devoted fan of space travel, explained that he embraced the venture precisely because it provided an important test of national capability. In a speech at Rice University in September 1962 he said that America chose to pursue its goals in space "not because they are easy, but because they are hard." In the Soviet Union, First Secretary Nikita Khrushchev pressed engineers to orbit the first Earth satellite and the first human being as a demonstration of his nation's technological prowess and superpower status.

By the latter part of the twentieth century, space travel had become a popular symbol of national achievement in industrialized countries around the world. As U.S. President Dwight D. Eisenhower observed, the scientific and commercial benefits of engaging in ventures such as a "stunt race" to the Moon did not justify their undertaking. Government officials nonetheless directed increasing shares of national revenues to space ventures. Space travel became an expression of hope, a means by which humans could embrace technology, overcome adversity, and work to realize a fantastic and achievable vision. Space missions provided a test of national capability whose successful achievement, unlike the nuclear arms race, promised to create good rather than harm. In the United States, space programs helped maintain confidence in the power of an open society, while in the Soviet Union they served as a symbol of national strength. Space accomplishments provided a much-needed popular diversion from the threat of nuclear war and fulfilled the promise, which indeed came true, that the nation that remained first in space technology possessed the means to win the Cold War.

Space travel in the early twenty-first century was not high on the list of governmental priorities; it was an important but not dominating element of the world economy.

Yet it retained enormous popularity. Space accomplishments have received extensive media coverage; space tragedies have been treated as national failures precipitating extensive analysis. As a cultural phenomenon, modern space travel has been popular not so much for what it achieves, which has been considerable, but because it has symbolized so much of what humans want to believe.

Howard E. McCurdy

See also: Europe, Rocketry Pioneers and Space Visionaries, Russia (Formerly the Soviet Union), United States

Bibliography

Thomas M. Disch, *The Dreams Our Stuff Is Made Of: How Science Fiction Conquered the World* (1998).

Howard E. McCurdy, *Space and the American Imagination* (1997).

Frederick I. Ordway and Randy Lieberman, eds., *Blueprint for Space: From Science Fiction to Science Fact* (1992).

Frank H. Winter, *Prelude to the Space Age: The Rocket Societies, 1924–1940* (1983).

Science Fiction

Science fiction became a popular genre of literature in the nineteenth century after centuries of development. Sometimes technology follows where science fiction leads, and other times science fiction springs from new scientific discoveries. Science fiction includes a variety of subgenres that investigate various ways that science or technology can influence the human condition. However, the most important trait of science fiction is the ability to spark the human imagination to consider what is possible through technology or the scientific process.

Although technological innovation is a frequent ingredient of science fiction, also important to the genre is the story of life beyond Earth. Human literature often discusses gods inhabiting the heavens, but Plutarch (ca. 100 CE) was one of the first to suggest that other mortal creatures inhabited the Moon. As science and technology advanced, the Moon, as Earth's closest neighbor, became more frequently visited in literature.

One early example of proto-science fiction is *Somnium*, written by the renowned astronomer Johannes Kepler and published posthumously in 1634. Although influenced by the science of the day, *Somnium* retains supernatural elements that are typically not considered part of modern science fiction. Science fiction evolved throughout the seventeenth and eighteenth centuries with numerous tales of trips to the Moon, sometimes involving extraordinary treatment of the natural world. For example, Francis Godwin, under the pseudonym Domingo Gonsales, wrote *The Man in the Moone: or a Discourse of a Voyage Thither* (1638), which employs birds for propulsion. This transitional era ended in the mid-eighteenth century with Voltaire's *Micromégas* (1752), which tells the story of how an alien from the star Sirius visited Earth after using comets, sunlight, and other natural means of interstellar transportation.

The nineteenth century saw the rise of modern science fiction with the 1818 publication of *Frankenstein: Or the Modern Prometheus* by Mary Shelly. *Frankenstein* is a

DAVID LASSER
(1902–1996)

(Courtesy Mrs. David Lasser)

David Lasser was introduced to the idea of human spaceflight as the managing editor for *Science Wonder Stories*. Although only briefly involved in the fledgling spaceflight movement in the early 1930s, Lasser left an indelible mark as the founder and first president of the American Interplanetary Society (which later became the American Rocket Society and subsequently the American Institute of Aeronautics and Astronautics) and the author of *The Conquest of Space* (1931), which was the first book of nonfiction in English on the use of rockets for human spaceflight. The David Lasser Papers are maintained in the Mandeville Special Collections Library at the University of California, San Diego.

Michael Ciancone

tale of the "other" often explored with alien encounters in later science fiction. During the nineteenth century science fiction became more scientifically plausible, as seen in Edgar Allen Poe's *The Unparalleled Adventure of One Hans Pfaall* (1835). Poe's plausibility had a profound impact on Jules Verne, who brought space travel to the forefront of science fiction with *From the Earth to the Moon* (1865) and its sequel, *Round the Moon* (1870), which, for the time, described with amazing accuracy a trip to the Moon using technological means. Although Verne fired crew members from a cannon (a launch they could not have survived), he generally took great care to make the story realistic through the extension of the technology and science of his time. During the late nineteenth century science fiction began to appear regularly in periodicals, and the extrapolation of technology continued to predict future developments. The novelette *The Brick Moon* by Edward Everett Hale, which appeared in *Atlantic Monthly* in a serial format during 1869–70, described the myriad of possible functions of artificial satellites.

During the late nineteenth century astronomer Giovanni Schiaparelli published maps of the Martian surface that included what he called channels, *canali* in his native Italian, but were misinterpreted when translated into English as "canals." The consequent belief in artificially created structures on Mars led to mass public interest in the possibility of both water and life on the surface of the Red Planet. Writers helped fuel the popularity of Mars through the early twentieth century with the publication of works like H. G. Wells's *The War of the Worlds* (1898) and Edgar Rice Burroughs's *Under the Moons of Mars* (1912) both of which first appeared as serialized stories in popular publications, such as *Pearson's* and *All-Story* magazines. The twentieth century saw science fiction continue to develop into its modern form, and its popularization continued through the publication of pulp magazines like the *Argosy*, the *Popular Magazine*, and *Amazing Stories* (the first magazine devoted exclusively to science fiction, published by Hugo Gernsback). Gernsback originally included science fiction stories in his magazine *Modern Electrics* to help increase circulation, but found that the popularity of the stories warranted their own publication. In 1930 David Lasser, editor for *Science Wonder Stories*, concluded that the use of rockets for human spaceflight was technically feasible and, with a number of science fiction writers, formed the American Interplanetary Society to encourage interest in spaceflight. This organization eventually became part of the American Institute of Aeronautics and Astronautics.

In the early 1940s science-fiction anthologies began to appear in bookstores, and through the technological innovations of war (radar, ballistic missiles, and atomic weapons) science fiction began looking more like science fact to the general population. Pulps would all but die out by the mid-1950s but would be replaced by magazines like *Galaxy Science Fiction* and the *Magazine of Fantasy and Science Fiction*. Through the latter half of the twentieth century, science fiction became a publishing staple, eventually seeing hardcover books reach the bestseller lists with authors such as Arthur C. Clarke and Isaac Asimov. Science-fiction literature branched out into many subgenres including space opera, alien encounters, time travel, alternate history, and "hard" science fiction.

Science fiction was quick to adapt to new media that developed in the twentieth century. Georges Melie's short movie *Le Voyage dans la Lune* (1902) was adapted from the Jules Verne novel *From the Earth to the Moon* and showed the impact that camera tricks and special effects could have on cinematic science fiction. The first movie to feature a rocket countdown and attempt to portray realistically spaceflight was Fritz Lang's *Frau im Mond* (1929). Radio served as the dominant communication medium throughout the 1930s–40s. A number of familiar characters, including Buck Rogers and Flash Gordon, appeared on dramatic radio programs, and one of the most famous radio broadcasts was Orson Welles's 1938 performance of an adapted *War of the Worlds*. Performed like a news bulletin, the broadcast caused panic among listeners who tuned in after the dramatic disclaimer.

Movies continued to play an important role in the development of science fiction throughout the twentieth century. The 1968 collaboration of author Arthur C. Clarke and director Stanley Kubrik on *2001: A Space Odyssey* produced a true audiovisual masterpiece, while George Lucas's *Star Wars* (1977) popularized science fiction on a level never before seen. Science fiction has been a staple on television since 1949, but most of the shows have been relatively short lived. There have been shows that significantly influenced television audiences, including: *The Twilight Zone*, *Dr. Who*, *Quantum Leap*, and *The X-Files*. However, Gene Roddenberry's *Star Trek* franchise stands out in both the almost utopian portrayal of a future humanity and in the fervent fan base that it created. The expansion from literature to radio, cinema, and television allowed a larger audience to experience the way that science fiction investigates the human experience.

With the advent of the space age, science fiction and science fact truly began to merge. Science fiction began to concentrate more on trips to Mars and humankind's role in the exploration of the larger universe. Science fiction has inspired public interest in space exploration and maintains interest in space travel in the midst of real-world space tragedy and disappointment. Many people directly involved with space exploration have often cited science fiction as an inspiration in their choice of vocation. Science fiction may not have had a direct impact on future planning for space exploration, but it sparked public imagination to demand the continuation of the journey to space.

Howard Trace

See also: Rocketry Pioneers and Spaceflight Visionaries

Bibliography
Paul Allen Carter, *The Creation of Tomorrow: Fifty Years of Magazine Science Fiction* (1977).
Thomas M. Disch, *The Dreams Our Stuff Is Made Of: How Science Fiction Conquered the World* (1998).
Frederick I. Ordway III et al., *Blueprint for Space: Science Fiction to Science Fact* (1992).

Space and the Environment

Space and the environment are inextricably connected. In his address to the U.S. Congress on 25 May 1961, President John F. Kennedy stated that space could provide for a better understanding of Earth's environment. Carl Sagan remarked that space exploration is important for nothing less than species survival and suggested that space exploration, and ultimately settlement, was important as an insurance policy against environmental despoliation of Earth.

Space exploration involves the development of satellites that provide global communications networks and accurate weather and crop forecasting through remote sensing that saves lives by facilitating natural disaster mitigation and by allowing farmers throughout the world to better provide food for people. These satellites are also critical for providing a better understanding of global environmental change issues, such as ozone depletion and climate change, which can threaten the biosphere in which humans live.

As scientists have explored and studied the planets, they have learned more about Earth. Comparative planetology (the study of Earth in comparison to other planets) is instrumental in identifying global environmental problems. NASA scientists trying to understand why the surface temperature of Venus is warm enough to melt lead have proven the validity of the "greenhouse-warming" phenomenon and its potentially devastating effects. Likewise, planetary scientists studying why materials on Mars instantly oxidize due to ultraviolet light exposure identified the cause of ozone depletion on Earth.

Concerns have arisen about the relationship between space and the environment, including global climate change, Earth observations data policy, and environmental issues associated with space exploration and planetary protection. Earth observations by satellites play a key role in assessing global climate change. In the United States, scientific awareness about global change led to formulation of the Global Change

View of the Earth as photographed by the Apollo 17 *astronauts during their return from the Moon in December 1972. (Courtesy NASA)*

Research Program in 1990. The goal of this program was to provide for the development and coordination of a comprehensive and integrated research program that would assist the United States and the world to understand, assess, predict, and respond to human-induced and natural processes of global change. This program recognized that scientific knowledge is crucial to informed decision making on environmental issues. The primary component of this program was NASA's Earth Science enterprise, which involved a constellation of Earth-observing satellites to assess global environmental change.

At the international level, the primary type of remote sensing systems contributing to the acquisition of data for global change research is the specialized and technologically advanced research satellites known collectively as Earth observation satellites. Since 1984 the Committee on Earth Observation Satellites has coordinated efforts of countries involved with Earth observation by satellite. The policy goal of this committee has been to advance scientific knowledge of Earth and its environment to understand and predict natural and human-induced global environmental change phenomena. To achieve this goal, the committee has coordinated missions and addressed such issues as conditions and access to data, data pricing, periods of exclusive use, and data preservation and archiving. The principal purpose was to satisfy the data access and data exchange requirements of all countries involved as effectively and economically as possible for global change research uses.

An important space exploration and development goal of spacefaring nations has been to establish a permanent human presence in space. To this end, NASA and other governmental space agencies have sought to spread life in a responsible fashion throughout the solar system. This responsibility is codified in the 1967 Outer Space Treaty to avoid the harmful contamination of the Moon and other celestial bodies. Of particular interest to the space community is the issue of science and contamination. Astrobiology programs in the United States and abroad have studied the origin and evolution of life, and life-related processes and materials throughout the solar system. The discovery of this connection depends on the extent to which humans can prevent the contamination of any evidence. The context for this involves planetary protection policies as codified by the International Council for Science Committee on Space Research.

Eligar Sadeh

See also: Outer Space Treaty

Bibliography
E. C. Hargrove, ed., *Beyond Spaceship Earth* (1986).
Eligar Sadeh, "Harmonization of Earth Observation Data: Global Change and Collective Action Conflict," *Astropolitics* 3, no. 2 (2005).
Carl Sagan, *Pale Blue Dot* (1994).

Space Art

Space art has been a crucial means for space visionaries and engineers to portray their concepts about the nature of humanity's future in space. Science fiction has been one of the primary vehicles for portrayals of the future in space. Jules Verne's classic novel *De la Terre à la Lune* (*From the Earth to the Moon*, 1865) transformed spaceflight from the realm of the fantastic to an exercise in mathematics and technology. The illustrations (by Henri de Montaut) that accompanied the book were the first attempts to depict realistically spaceflight and the conditions that might exist beyond Earth's atmosphere. A later novel by Verne, *Hector Servadac, voyages et aventures à travers le monde solaire* (*Off on a Comet!*, 1877) features what may be the first true astronomical artwork, an illustration by Paul Philippoteaux depicting Saturn's rings as seen from the surface of the planet.

James Hall Nasmyth and James Carpenter's *The Moon* (1874) featured photographs of meticulously constructed plaster models of lunar craters and mountains. Because of their extraordinary realism and cachet of authority, these photographs influenced artists and illustrators for more than half a century and were probably responsible for the misconception that the Moon's landscape was one of Alpine cragginess. French science popularizer Camille Flammarion published the first major landmark in astronomical art, *Les terres du ciel* (*Worlds in the Heavens*, 1884). It contained hundreds of woodcut illustrations created by a team of artists, including Paul Fouché, perhaps the first specialist in creating scenes of extraterrestrial landscapes.

The popularity of interplanetary science fiction among Victorian readers resulted in publication of a number of important illustrated works, including *Letters from the Planets*

The Conquest of Space *(1948) by Chesley Bonestell. (Courtesy Bonestell Space Art)*

(1893) by Wladyslaw S. Lach-Szyrma; *Les Exiles de la Terre* (*The Conquest of the Moon*, 1887) by Andre Laurie; *A Journey in Other Worlds* (1894) by John Jacob Astor; and *A Honeymoon in Space* (1901) by George Griffith, which contained some of the first depictions of space-suited astronauts. H. G. Wells wrote a long article for *Cosmopolitan* magazine (1908), accompanied by paintings by W. R. Leigh, speculating on the nature of life on Mars. These stories were important in that they reflected current understanding of the conditions on other worlds and approached the concept of spaceflight in realistic terms, at least in terms of contemporary technology and science.

Until then, however, no artist had specialized in astronomical art to the exclusion of other subjects. Perhaps the first of these was Scriven Bolton, whose work appeared often in the *Illustrated London News* of the 1920s–30s. He created many of his illustrations by painting over photographs of detailed plaster models—an effective hybrid technique later employed to even greater success by Chesley Bonestell. British artist G. F. Morrell and French astronomer Abbé Theophile Moreau also specialized in space art. The latter's drawings appeared near the end of the nineteenth century and continued into the 1940s. In the United States, Howard Russell Butler produced a series of paintings of heavenly bodies for the American Museum of Natural History in the 1920s. However, the first real specialist in space art whose work had wide circulation and influence was the French astronomer-illustrator Lucien Rudaux. Originally a commercial artist, he eventually became an accomplished and respected astronomer. Combining these two skills, he produced hundreds of paintings and drawings that remain among the best astronomical art created. He wrote and illustrated many popular books and magazine articles. His masterpiece was *Sur les autres mondes* (*On Other Worlds*, 1937), with more than 400 remarkable illustrations. For the first time the landscapes of the planets began to look like real places and not merely figments of imagination.

The first magazines to specialize in science fiction appeared in the 1920s, and for the next 30 years, their covers featured some of the best space art published at the time. For example, Frank R. Paul painted the first illustration of a space station published in the United States (*Science Wonder Stories*, 1929), while Hubert Rogers created the first depiction of the gravitational lens effect (*Astounding*, 1940).

If Rudaux is considered the grandfather of astronomical art, then Chesley Bonestell is the father. Born in 1888 Bonestell received early art training as an architect. A lifelong interest in astronomy led him to create a series of paintings depicting Saturn as it might look from its satellites. When these paintings were published in 1944, they created a sensation because they were as realistic as photographs. Willy Ley collected these and other paintings in *The Conquest of Space* (1949). The believability of these paintings was almost more important than any scientific information they may have conveyed—indeed, most of Rudaux's earlier illustrations were more accurate than scientific information. In Bonestell's extraordinary renderings, however, the worlds of the solar system took on an unprecedented reality.

A bevy of Bonestell-illustrated magazine articles and books followed *The Conquest of Space*. Although many space scientists trace their interest in astronomy and spaceflight to these books, perhaps the work that had the most influence on the early development of the U.S. space program was a series of articles published in *Collier's* magazine between 1952 and 1954. Under the general direction of Wernher von Braun, the series outlined a systematic space program in unprecedented detail and was accompanied by paintings by Chesley Bonestell, Fred Freeman, and Rolf Klep. These articles were later collected in *Across the Space Frontier* (1952), *Conquest of the Moon* (1953), and *The Exploration of Mars* (1956), which had an immediate and far-flung impact on the public perception of space travel—it no longer represented the stuff of fantasy, but a reality that depended only on money and resolve.

Another important space artist during this period was Ralph A. Smith, an engineer-artist associated with the British Interplanetary Society. He was responsible for illustrating the Moon rocket that the British Interplanetary Society proposed in 1939 andthe space station, which he primarily designed, that it championed in the 1940s–50s. These were the first attempts to describe the development of a space program not only in logical, incremental terms but also in light of contemporary technology. Smith had less interest in the worlds of the solar system than in the machines that would reach them. A U.S. artist who influenced almost as many youthful enthusiasts about space travel during the 1950s as Bonestell was Jack Coggins, whose solidly rendered spacecraft illustrations had a ring of truth to them. Space art collectors covet such beautiful children's books as *Rockets, Jets, Guided Missiles and Space Ships* (1951), *By Space Ship to the Moon* (1952), and *Rockets, Satellites, and Space Travel* (1958). Perhaps the best and most influential of all post-Bonestell-era space artists was Ludek Pešek, the author-illustrator of more than a dozen books. His landscapes possess a powerful sense of reality, a naturalism that leads the viewer to believe that these landscapes exist somewhere. He epitomized a successful astronomical artist who was first a fine landscape painter.

Even before Bonestell's death in 1986, there were dozens of artists specializing in space art—even sub-specializing as some concentrated on astronomical landscapes and others on spacecraft. In 1981 they organized as the International Association of Astronomical Artists and represented a wide spectrum of artistic techniques, philosophies, and schools, from the realist followers of Bonestell to pure abstractionists, in addition to sculptors, musicians, and poets. Space artists have been effective social historians, recording and documenting people's activities and achievements.

Ron Miller

See also: Rocket Pioneers and Spaceflight Visionaries

Bibliography

David A. Hardy, *Visions of Space* (1989).

Roger D. Launius and Howard E. McCurdy, eds., *Imagining Space: Achievements, Predictions, Possibilities, 1950–2050* (2001).

Ron Miller and Frederick C. Durant, *The Art of Chesley Bonestell* (2001).

Frederick I. Ordway, *Visions of Spaceflight* (2000).

Space Memorabilia

Space memorabilia are objects that evoke the history of space exploration. They include inexpensive objects created with a brief intended lifetime (such as patches, pins, buttons, and posters); items created to be collectibles (such as trading cards, comics, photographs, autographs, stickers, stamps, collectible sets, and coins); items intended for use by the general public (such as books, phonograph records, recordings, medals, models, model kits, toys, hats, T-shirts, and jewelry); and unique rarities (such as objects associated with astronauts, engineers, or administrators, and actual spaceflight hardware, such as training equipment and discarded programmatic objects). Although space memorabilia may include artifacts (the material evidence of space exploration), not all artifacts are memorabilia (objects that serve primarily as touchstones for memories). Space memorabilia are sought by collectors worldwide and include items associated with various space programs, including those of China, Europe, Japan, the Soviet Union and Russia, and the United States.

Pieces of space memorabilia hold value (as do any commercial collectibles) based on their relative scarcity, demand, and condition. Space memorabilia, however, are also evaluated based on characteristics specific to their space context. Objects flown in space hold special distinction, as they are both rare and sentimental. Objects directly associated with specific events, programs, or people have additional value because they commemorate or illustrate that achievement (for instance, the *Apollo 11* Moon landing). Having an object autographed by a participant (especially a space traveler) also increases its value. Because the space age coincided with an explosion in affordable consumer products after World War II, space memorabilia became widely available. The tendency to judge anything associated with space as inherently

Apollo 11 *mission patch. (Courtesy NASA/Marshall Space Flight Center)*

valuable (both as a historical artifact and as an appreciating collectible) also means that many space-related objects have been saved.

From the beginning of the space age NASA controlled the types, numbers, and distribution of flown memorabilia. Astronauts recognized that people would want flown objects: Gus Grissom brought rolls of dimes on board *Liberty Bell 7*, intending to use them as souvenirs. After John Young brought a corned beef sandwich on board *Gemini 3* as a good-natured joke in 1965, NASA instituted stricter rules about what could be flown. A furor about stamp covers flown on board *Apollo 15* (but retracted before their public sale) led to additional restrictions on objects eligible to be carried on spaceflights. The (relative) extra space and frequent flights afforded by the Space Shuttle allowed more commemorative items to orbit, while the process became more codified. Launches have included memorabilia in the Official Flight Kits and astronauts' Personal Preference Kits.

NASA helped create new kinds of memorabilia. Mission patches, first created for *Gemini 5*, became an affordable souvenir through reproduction. Beginning with

Apollo 7 in 1968, the Robbins Company created precious metal medallions to be flown on space missions for the astronauts. NASA employees and space contractors often received commemorative items (for example, lapel pins, tie tacks, or desk ornaments) to foster morale. The creation of space memorabilia has served NASA's core mission by publicizing space achievements, solidifying alliances, and fostering support. International space cooperation has also been reinforced through memorabilia. For instance, Space Transportation System (STS)-91, the final Shuttle-Mir missions flown by the Shuttle *Discovery* in 1998, carried small U.S. and Russian flags and souvenir patches that were distributed afterward to dignitaries, political supporters, schools, and space workers.

Even as NASA created memorabilia, restrictions continued. NASA has limited the use of its name and symbols through licensing agreements. An agreement between NASA and the Smithsonian Institution's National Air and Space Museum has allowed the Smithsonian a first look at any soon-to-be-discarded space objects. Lunar material returned by the Apollo missions may not be privately owned.

Space memorabilia became a booming business. Stores at the Smithsonian, NASA centers, and museums, such as the Kansas Cosmosphere and Space Center in Hutchinson, Kansas, successfully marketed space-themed objects. In the early 1990s auction houses, such as Christie's, Sotheby's, Aurora Galleries International, and Superior Auction Galleries, began devoting sales to rocketry and spaceflight materials, including major pieces of spaceflight hardware, such as capsules and space suits, from the remnants of the former Soviet Union, the personal effects of astronauts and cosmonauts. Online auction sites, such as eBay, and collector forums made space memorabilia available globally while online resources, such as bulletin boards and discussion centers, fostered a virtual community for collectors.

Collectors of space memorabilia have also played an important role in preserving the material history of the space age by rescuing items that governments, space agencies, and private companies discard after a program ends. Many retired space workers, or their heirs, have enriched museum collections with generous donations from a lifetime of space collecting.

Margaret A. Weitekamp

See also: Astronauts, Johnson Space Center, National Aeronautics and Space Administration, Space Museums and Science Centers

Bibliography

CollectSPACE. www.collectspace.com.
Howard McCurdy, *Space and the American Imagination* (1997).
Russ Still, *Relics of the Space Race* (2001).

Space Music

Space music is music written about space topics, such as astronomy, space travel, and science fiction. The first documented connection between music and outer space was the Greek concept of the music of the spheres. Around 400 BCE Greek astronomer and mathematician Eudoxus of Cnidus created models of the motions of the planets. Around 500–400 BCE the Greeks, most notably Pythagoras, created models of the motions of planets. Pythagoras envisioned seven crystal spheres separated by distances analogous to the harmonic length of strings. Thus the spheres produced musical tones as they moved in reference to the sphere representing Earth. In the twentieth century Danish composer Rued Langgaard and German Paul Hindemith wrote orchestral pieces inspired by the music of the spheres. Hindemith's work paid tribute to astronomer Johannes Kepler. Kepler hoped to validate the Pythagorean system by looking for perfect geometrical forms, but he discovered that the planetary orbits were elliptical, not circular. This led in 1916 to his calculation of the laws of planetary motion. Laurie Spiegel sent her computer-realized composition of Kepler's planetary motions onboard the *Voyager* spacecraft in 1977 as one of the "Sounds of Earth" on the Golden Record.

Space music has often been written when events in space were occurring. E. T. Paull published the parlor piano sheet music, "A Signal from Mars—March and Two Step" in 1901 shortly after Percival Lowell's "discovery" of canals on Mars. The Happy Hayseeds, country artists of the 1920s, recorded "The Tail of Haley's [*sic*] Comet" to commemorate the passage of Halley's Comet in 1910.

The Moon has always been popular in song. Forty-three of the hit songs between 1892 and 1964 had "Moon" in the title. Bart Howard's "Fly Me to the Moon," sung by Frank Sinatra in 1964, is an example of one of the most famous. The rock band Pink Floyd's album *Dark Side of the Moon* (1973) is another. *The Planets* (1916) by Gustav Holst is an orchestral suite imagining the personalities of each planet. The jazz saxophonist John Coltrane does the same on *Interstellar Space* (1967) and synth pioneer Sun Ra on *Space Is the Place* (1972).

Science-fiction themes in music merged into the space age, but there were earlier examples. Jacques Offenbach wrote *Le Voyage dans la Lune* (1875) based on Jules Verne stories. Jeff Wayne's rock opera *War of the Worlds* (1978) was based on the H. G. Wells novel of the same name. After the purported crash of an unidentified flying object near Roswell, New Mexico, in 1947, flying saucer themes hovered over pop music as "Two Little Men in a Flying Saucer" by Ella Fitzgerald (1951) and "Flyin' Saucers Rock 'N' Roll" by Billy Riley and His Little Green Men (1957). A strange eerie instrument called the theremin became associated with early science-fiction film scores for *The Day the Earth Stood Still* (1951), *The Thing* (1951), and *Forbidden Planet* (1956). Richard Strauss's "Also sprach Zarathustra" (1896) became identified with the 1968 film *2001: A Space Odyssey*. The rock band Hawkwind made a career of albums, such as *Space Ritual* (1973), based on the writings of author Michael Moorcock.

Filk, the folk music of science fiction and fandom, began in the 1950s as late-night post-convention song circles adapted original lyrics to well-known melodies. Filk songs were written for major science fiction and fantasy work in addition to laments for the demise of the Apollo program. One of the most beloved filk songs is "Hope Eyrie" written by Leslie Fish in 1975. This ode to the Apollo program and to a hopeful future became the anthem of the pro-space movement.

In the 1950s the space age also ushered in a subgenre known as incidental lounge or exotica. These light orchestral albums included pop tunes (Hoagy Carmichael's "Stardust"), electronic effects, and jazz combos, for example, *Fantastica: Music from*

Outer Space by Russ Garcia (1958). The 1983 soundtrack to the film documentary *For All Mankind*, titled *Apollo Atmospheres and Soundtracks*, written and played by Brian Eno, though not a direct descendent of exotica, is an ambient electronic music masterpiece.

Even before the advent of the space age, pop music reflected the new futuristic themes in speculations about lifestyles in outer space: Hank Snow's "Honeymoon on a Rocket Ship" from the early 1950s, David Bowie's "Space Oddity" (1969), and Elton John's "Rocket Man" (1972). *Sputnik* in 1957 ushered in the space age proper and songs inspired by it such as the rockabilly "Sputnik Girl" by Jerry Engler, and the pop "Beep! Beep!" by Louis Prima and "Satellite Baby" by Skip Stanley. Tributes to space events followed. Duke Ellington wrote "Moon Maiden" to honor the *Apollo 11* mission in 1969. Jerry Rucker wrote "Blast-Off *Columbia*" to celebrate the advent of the Space Shuttle era in 1981, and jazz synth Jean-Michel Jarre penned the album *Rendez-Vous* dedicated to the fallen *Challenger* crew. "Blue Sky" by Big Head Todd and the Monsters became the official Space Transportation System-114 return-to-flight tribute in 2005. Astronauts honored have included John Glenn, "Happy Blues for John Glenn" by Lightnin' Hopkins (1962); Sally Ride, "Ride, Sally, Ride" by Casse Culver (1983); and Eileen Collins, "Beyond the Sky" by Judy Collins (1999).

In 2001 NASA announced the discovery of the "music of creation" by the presence of acoustic notes in the sound waves generated by the Big Bang 12–15 billion years ago. The cosmic microwave background contains remnants of the intense heat that existed in the infant universe and structures present in this background have a harmonic content. Although the Greeks' music of the spheres was proved to be inaccurate by Kepler, the universe does sing after all.

Colin Fries

Bibliography
Colin Fries, "Flying for Us, Space Age Milestones Celebrated in Music," *Quest* 9, no. 3 (2002).
Hobby Space. www.hobbyspace.com.
Roger Launius, "Got Filk? Lament for Apollo in Modern Science Fiction Folk Music," *Quest* 12, no. 4 (2005).

World Wide Web

The World Wide Web (often referred to simply as the Web) has had a major impact on astronautics since the early days of computer networking. A key role of astronomy and space science has involved sharing and publishing data, and they have often been at the forefront of computer science and networking. Astronomers readily embraced the World Wide Web, in addition to earlier protocols, such as file transfer protocol and gopher. The Web soon became crucial to modern astronomical research, becoming the primary channel for dissemination of data, tools, and information to scientists and the public. With the Web amateur astronomers could access current work and contribute in a meaningful way to the scientific dialogue.

Space science has been a popular topic since the beginning of the Web. In 1993 a Conseil Européen pour la Recherche Nucléaire (CERN or European Council for Nuclear Research) space science information site was one of nine websites noted by the National Center for Supercomputing Applications (provider of the first browser) in its "What's New" webpage. By the 11th day of this first website index, NASA's Langley Research Center had initiated a Web server, NASA's Astrophysics Data Facility followed one month later, as did the Space Telescope Science Institute shortly thereafter.

Web interest in space science grew quickly, not just with scientists, and in unexpected ways. In the absence of a public Internet, the linking of early networks for research would arguably still have come about on an ad hoc basis. But reaching the public via a network was something few people predicted. One side project started in 1995—the Astronomy Picture of the Day (APOD)—received nearly half of all Web traffic for NASA Goddard Space Flight Center in 2004 and was mirrored in 16 countries.

The NASA Mars Pathfinder website was the most frequently accessed website in 1997 with more than half a billion hits in one month. In July 2005 NASA's Return to Flight launch of Space Transportation System-114 had more than 190,000 simultaneous live streams 40 minutes before the launch—more than America Online's "Live 8" concert broadcast just days before. For the period of February 2003 to September 2005 NASA websites tallied more than one-quarter of a billion visitors, and NASA was just one of the many space science sites reaching a large audience.

The Web produced a leveling effect, providing individuals users and small colleges with access to the same data, tools, and literature as that of a well-funded university. Data analysis tools were publicly available for standard personal computers, and raw and reduced data products were available in openly accessible databases.

Journal access via the Web became more common than paper for accessing peer-reviewed literature. Individuals could access and speedily search the equivalent of a university library stack using services such as the NASA Astrophysics Data System, increasing the odds that any given article would be cited. Journal article preprints became widely distributed via the Web. Pragmatically it was easier to find a known paper via the Web than to retrieve a paper copy from the library—or one in an office file cabinet. While this was a boon to most astronomers, it caused difficulty for print journal publishers, as that market sought ways to remain relevant and viable.

Popular astronomy websites proliferated, serving the amateur and the curious, and Education and Public Outreach became a mandatory component in most research grants and in NASA funding. With the advent of the Web, astronomy's half dozen mass media print publications changed little, in part because such magazines have historically concentrated more on impressive visual imagery, data presentation, and in-depth articles. Timely delivery of current news, a function usually left to the science column of the local paper, became largely a Web activity.

The Web has also increased the dissemination of space science information. In addition to popular sites such as APOD, outreach and public affair departments were able to generate large amounts of material without incurring printing costs. This meant that they could choose a wider selection of science to feature and let the audience decide what to view, rather than being restricted to only the top 10 they could afford to print.

Any website can disseminate news and opinions on space, and the difference between "legitimate" sources, such as NASA, and independent astronomer sites, watchdog sites like NASA Watch, blogs, and podcasts (such as Slacker Astronomy) has been largely due to their reputation and academic accreditation. Amateurs have often taken an authoritative stance, ranging from Henry Spencer's Usenet posting in the early 1980s to Mark Wade's Encyclopedia Astronautica Website. Amateur spy satellite spotters shifted from paper to e-mail lists and websites, and conspiracy theorists traded images of never-proven military black projects. The Web improved information exchange and occasionally brought in the attention of the wider public to niche interests.

As a science, citation (in particular reference to peer-reviewed work) and presence of a NASA or university stamp of approval have strong value. The lack of citation on amateur efforts has usually been counterbalanced by the active nature of the amateur community, and any incorrect work that achieves note has been zealously pilloried by its audience. Because the community is composed of either academics or amateurs, any factual basis is observationally verifiable and has helped maintain accuracy and credibility.

The Web enabled dissemination of science products—notably data and analysis software— to the academic community. This has allowed reanalysis and data mining by people other than the principal investigator (PI), yielding new science and the reuse of data in ways not originally intended at the time of its collection. The Web provided a cost savings for engaging in space science work, especially with regard to travel by allowing PI involvement without travel and without interrupting the science operations team.

Educational outreach, particularly with respect to primary and secondary education levels, was able to reach more people at lower cost. Teachers with browsers could access and print information on space science, fun projects, and complete lesson plans from sites such as Imagine the Universe or Earth Observatory. Scientists and the public began a dialogue, particularly at the primary and secondary education levels and in the many Ask an Astronomer websites. The Web also connected amateur and professional astronomers. In some niches—for example, *SOHO* (*Solar and Heliospheric Observatory*) was also good for spotting comets—amateurs dominated in the number of discoveries because of the fact that they had access to the raw data and official tools and had more time to devote to side projects than did the mission scientists.

Whether educational or voyeuristic, space-based Earth observing developed a public component. It became commonplace to view current weather conditions or recent satellite images of one's neighborhood (such as through Google Earth). Although the Weather Channel on cable provided interpretation and weather forecasting, there was a strong entertainment value in being able to access astronautical data that was personally relevant.

The Web enabled easy access to astronomy for researchers, amateurs, students at the primary and secondary education levels, and the curious. It has provided equal access to astronomy and space science; better, faster, cheaper dissemination of information; and cost savings to research and education; and it has fostered international and professional–amateur collaboration.

Alex Antunes

See also: National Aeronautics and Space Administration

Bibliography
Astronomy Picture of the Day. http://apod.nasa.gov/apod/.
Encyclopedia Astronautica. www.astronautix.com.
NASA Watch. www.nasawatch.com.

NATIONS

Nations prioritize their domestic space efforts based on domestic and international concerns and opportunities. In general, citizens, nongovernment organizations, and government organizations interact in complex ways, leading both to explicit and implicit choices by each. The sum and interaction of their choices creates a unique mix of space activities in each nation.

Space activities in the 1950s and 1960s were dominated by the Soviet Union and the United States. In political, military, and economic competition with each other, these two superpowers were in many ways mirror images. Both governments placed their highest priorities and funding on military, intelligence, and prestige programs. Launchers, ballistic missiles, reconnaissance spacecraft, military communications and navigation satellites, robotic space probes to the nearest lunar and planetary destinations, and human spaceflight were emphasized in both countries. Government organizations and funding dominated both countries' efforts, though the United States contracted with private industry and nonprofit organizations to perform many activities. The United States differed from the Soviet Union in funding many more Earth-orbiting science spacecraft and in supporting the development of a privately operated satellite communications system through Intelsat (International Telecommunications Satellite Organization). Both superpowers harbored global aspirations, and the breadth and depth of their space efforts matched those aspirations. In terms of government spending, the United States dominated national efforts, particularly after the demise of the Soviet Union in 1991. In 2006 the U.S. proportion of spending compared to all other governments combined was roughly 80 percent.

A primary motivation for later spacefaring nations was technological and economic development. Western European nations, Japan, Canada, India, and China all began to develop their own space programs in the 1960s, and these grew to rival those of the superpowers during the next three decades. Each nation had to choose its level of dependence or independence from the United States and the Soviet Union. France, Japan, China, and India decided early on relatively independent courses, signified by the decision to develop their own space launch vehicles. Space launchers allowed those nations to put any kind of spacecraft in orbit, even those not approved or supported by the superpowers. Western European nations, led by France, decided to collectively fund a European launcher, initially Europa through the European Space Vehicle Launcher Development Organisation, and later Ariane through its successor the European Space Agency (ESA). ESA allowed both large and small European nations to fund large projects beyond the means of any of them individually. Most other nations decided that they did not have the resources or desire to expend major funding on the development of indigenous launchers, instead purchasing launch service from the United States, the Soviet Union, or later, Europe's Ariane. Many medium and small nations developed or contributed to science missions, as these enabled cooperation and learning with major space powers and with one another. The United Kingdom and Indonesia both focused on communications satellites due to their dispersed national or colonial territories. China, India, France, and Israel emphasized remote sensing satellites for civilian and military purposes.

In most spacefaring countries, commercial interests grew over time, so that by the 1990s and 2000s it rivaled government-run efforts. Communications satellites (comsats) were the first major commercial application; private communications satellite operators came into being starting in the 1970s and early 1980s. These private operators, initially established in the United States and Europe but soon developing in other nations, needed to purchase satellites and launch services, thus spurring the creation and development of private comsat and space launch sectors and corporations. When the Soviet Union fell, its successor states, Russian and Ukraine, partially privatized some government organizations and successfully sold launch services. By the late 1980s China also formed semi-capitalist organizations to sell launch services.

In the 1960s only the United States and the Soviet Union emphasized human spaceflight and military activities. By the 1980s China and Israel aimed for military uses in their space programs through reconnaissance satellites. European nations, Japan, and India shied away from military uses until the late 1990s after the end of the Cold War.

Stephen B. Johnson

See also: Government Space Organizations, International Space Politics

Bibliography

William E. Burrows, *This New Ocean: The Story of the First Space Age* (1998).
Brian Harvey, *Emerging Space Powers: The New Space Programs of Asia, the Middle East, and South America* (2010).
———, *Russia in Space: The Failed Frontier?* (2001).
John Krige and Arturo Russo, *A History of the European Space Agency, 1958–1987*, 2 vols. (2000).

Australia

Australia's involvement with space-related activities began with the establishment of the Woomera Rocket Range, a joint British-Australian project, in 1947. By the 1960s Woomera was one of the largest and most active spaceports in the world, with British, Australian, European, and American investigators conducting research there. Australia was a member of the European Space Vehicle Launcher Development

The Prospero *satellite is launched from a test range in Woomera, South Australia, by a British Black Arrow rocket on 28 October 1971. (Courtesy Hulton Archive/Getty Images)*

Organisation (ELDO)—between 1964 and 1970 Woomera hosted 10 ELDO Europa rocket launches before the program was transferred to Kourou, French Guiana. The United Kingdom also tested the Black Arrow satellite launcher at Woomera (1968–71) and successfully launched the *Prospero* satellite in 1971.

Since 1957 Australia has been home to the greatest number of U.S. tracking and ground control stations outside North America, providing command and control for NASA's Earth-orbiting satellites, deep space missions, and NASA's human spaceflight programs, in addition to American military and intelligence satellites. Ground stations in Australia received the first images from the *Apollo 11* lunar landing before retransmittal for broadcast to the rest of the world. Although the military/intelligence ground stations at North West Cape, Pine Gap, and Nurrungar were vital to U.S. interests during the Cold War, they were also controversial and frequent targets of protests. By 2005 only the Joint Defence Space Research Facility at Pine Gap (a Signal Intelligence, or SIGINT, facility for the U.S. Central Intelligence Agency) remained operational.

Australian space research commenced with sounding rocket programs at Woomera in 1957 using the Australian designed and built Long Tom sounding rocket. Australia developed more than 10 other sounding rockets, and Australian scientific instruments made important contributions to X-ray and ultraviolet astronomy and studies of the physics of the upper atmosphere, before the government shut down the program in 1975. During this period Australian experiments also flew on NASA spacecraft and on the satellites of other nations.

Australia's first satellite, *WRESAT* (*Weapons Research Establishment Satellite*), was launched from Woomera on 26 November 1967. Designed and built by the WRE and carrying instruments developed by the University of Adelaide, *WRESAT* used a Redstone launch vehicle donated by the United States. In 1970 Australia became the second country to fly an amateur radio satellite, *Australis-OSCAR 5* (*Orbital Satellite Carrying Amateur Radio*).

Australian space activities declined through the 1970s due to changing government priorities, but in 1984 the Commonwealth Scientific and Industrial Research Organisation (CSIRO) established the CSIRO Office of Space Science and Applications to focus on remote sensing and space science research. In 1987 the Australian government established the Australian Space Office (ASO) to promote the growth of a domestic space industry and to manage the National Space Program. Although Australia developed expertise in such areas as space instrumentation, signal and data processing, ground positioning system applications, ground station equipment and design, and hypersonics, and it developed a few satellite instruments and small Space Shuttle payloads, the ASO was unable to achieve its goals due to inadequate funding and was terminated in 1996.

Aussat, a domestic satellite communications system that commenced operations in 1985 using U.S.-built satellites, was privatized in 1992 to become Optus. After 1986 there were several proposals to develop commercial launch facilities in Australia, most envisioning the use of Soviet rockets. Locations on Cape York, Christmas Island, and other northern parts of Australia were proposed as equatorial launch sites, with plans for Woomera to be revived as a polar-orbit launch site. By the early twenty-first century, none had come to fruition.

After 1997 the Australian government did not support a centrally funded space agency or space program, relying instead on user needs and market forces to drive space-related activities, with research and business development funding provided through generic industry and science support programs. Small offices within the Department of Industry provided space policy advice and oversaw licensing and safety matters relating to launch activities in Australia or by Australians. The Co-operative Research Centre for Satellite Systems was established in 1998 to develop a technology demonstrator microsatellite, *FedSat*, as a project for the Centenary of Federation in 2001. A Japanese H2-A rocket launched *FedSat* on 14 December 2002; it operated until September 2007.

Kerrie Dougherty

See also: Defense Support Program, Electronic Intelligence, European Space Vehicle Launcher Development Organisation

Bibliography

Kerrie Dougherty and Matthew L. James, *Space Australia: The Story of Australia's Involvement in Space* (1993).

Peter Morton, *Fire across the Desert: Woomera and the Anglo-Australian Joint Project, 1946–1980* (1989).

Ivan Southall, *Woomera* (1962).

Brazil

Brazil has devoted some of its limited resources to space activities since the early 1960s, from the creation in 1961 of GOCNAE (Grupo de Organização da Comissão Nacional de Atividades Espacias), an organization devoted to space exploration, to the signing of an agreement with Russia in 2005 to launch the first Brazilian astronaut into space.

Brazil followed a plan, "The Complete Brazilian Space Mission" (1979–80), in which the Brazilian government proposed to design and build its own satellites, create and operate an indigenously manufactured rocket, and develop its own launch center. It was the first legislative document signed by the president of Brazil officially recognizing the Brazilian space mission.

The Brazilian space program focused on low-Earth-orbiting satellites at a time when geostationary satellites were the goal for most nations. The first Brazilian-made satellite, *SCD 1* (*Satélite de Coleta de Dados*, or *Data Gathering Satellite*), was launched from NASA's Kennedy Space Center in February 1993 on a Pegasus rocket carrying the Brazilian flag into space.

Throughout the 1970s–80s Brazil developed the Sonda family of small rockets. The fourth variation of the Sonda, or Sonda 4, became the base or initial stage for its orbital launcher, the VLS (Veículo Lançador de Satélite, or Satellite Launch Vehicle). The first attempt to launch the VLS in 1997 failed 65 seconds after liftoff when one of its four first-stage boosters failed to ignite and the rocket veered off course. In the second attempt in 1999 the VLS flew for 200 seconds before failing when the second stage did not ignite. The third VLS rocket exploded on the launch pad in 2003 when one of its engines accidentally ignited due to an electrical discharge, killing 21 engineers and technicians.

Brazil created two launch facilities for its space programs—the Hell's Barrier Rocket Launch Center, built in 1965, and the Alcântara Launch Base located three degrees from the equator. Hell's Barrier was the site of more than 2,000 successful sounding-rocket launches. With the opening of the Alcântara facility in 1986, Hell's

The Brazilian VLS 1 rocket launches from the Alcântara launch base in northeastern Brazil on 2 November 1997. Shortly after takeoff, one of the rocket's four engines failed, forcing the Brazilian ground control to activate the self destruction safeguard of the rocket. (Courtesy AP/Wide World Photos)

Barrier became a satellite tracking station. The Alcântara base provided Brazil a modern, cost-saving, launch site.

To achieve its space goals, Brazil has entered into cooperative agreements with countries such as China, France, Ukraine, and the United States. One result was the China Brazil Earth Resources Satellite program, which between 1999 and 2007 launched three satellites (*CBERS 1*, *CBERS 2*, and *CBERS 2B*). Brazil created a national civilian space agency (Agência Espacial Brasileira, or Brazilian Space Agency) in 1994 and became a member of the Missile Technology Control Regime in 1995. Brazil also joined the International Space Station program as a participant through the INPE (Instituto Nacional de Pesquisas Espaciais, or Brazilian National Institute for Space Research), which evolved from GOCNAE.

Marcos Pontes became the first Brazilian astronaut when he launched to the *ISS* on a Russian vehicle in March 2006.

Ricardo Huerta

See also: International Space Science Organizations, *International Space Station*

Bibliography
Brazilian National Institute for Space Research. http://www.inpe.br/ingles/index.php.
Brazilian Space Agency. http://www.aeb.gov.br/.
Brian Harvey, *Emerging Space Powers: The New Space Programs of Asia, the Middle East, and South America* (2010).

Canada

Canada entered the ranks of spacefaring countries with the launch of *Alouette 1* atop a U.S. Thor-Agena rocket on 28 September 1962. *Alouette 1*, built under an agreement with NASA, was the first of four Canadian Alouette and International Satellite for Ionospheric Studies (ISIS) satellites that investigated the properties of Earth's ionosphere. Defense scientists built *Alouette 1* under the leadership of physicist John Chapman, who also headed a team that wrote a major report in 1967 on Canadian space activities. The Canadian government wanted to strengthen Canada's communications infrastructure and in 1969 it established Telesat Canada, which began work on building Anik communications satellites for Canada. The world's first domestic geosynchronous communications satellite, *Anik A1*, was launched on 9 November 1972, the first of the long-running Anik satellite series. Telesat's Nimiq satellites provided direct broadcast services. Canada supported communications research by building the *Hermes* communications technology satellite launched in 1976.

An interdepartmental committee ran Canada's space program in the 1970s–80s with a priority of establishing Canadian space industry, overseeing the space efforts of government agencies, such as the National Research Council of Canada and the Department of Communications. During the 1980s Spar Aerospace of Toronto was a prime contractor for communications satellites. When Spar withdrew, Canadian firms such as MacDonald Dettwiler and Com Dev became major space subcontractors. In

1974 Canada accepted NASA's invitation to take part in the Space Shuttle program by building the Shuttle Remote Manipulator System (Canadarm), which has flown on Shuttle flights and moved satellites, payloads, and astronauts. At the invitation of NASA, Canada established an astronaut program in 1983, and on 5 October 1984 Marc Garneau became the first Canadian in space when he lifted off onboard the Space Shuttle *Challenger*. After Garneau, Roberta Bondar, Steven MacLean, Chris Hadfield, Robert Thirsk, Bjarni Tryggvason, Dafydd "Dave" Williams, and Julie Payette became Canadian astronauts. Hadfield was the only Canadian to visit the *Mir* space station (1995) and the first to walk in space (2001). In 1985 Canada agreed to take part in what eventually became the *International Space Station* (*ISS*), for which it built the mobile servicing system that included a robot arm, a base system, and a manipulator system.

On 1 March 1989 the Canadian Space Agency (CSA) began operations with its headquarters in St. Hubert, Quebec, cooperating closely with NASA and the European Space Agency (ESA). Canadian cooperation with Europe began in the 1970s, when Canada took part in the European Symphonie communications satellite program. Canada signed a formal cooperation agreement with ESA in 1978, making Canada the only non-European country to be a cooperating member of ESA. In 1995 *Radarsat 1*, Canada's first remote sensing satellite, was launched. In 2003 Canada launched two scientific satellites: *MOST* (*Microvariability and Oscillations of Stars*, an astronomy research satellite carrying a small telescope) and *SCISAT* (investigating Earth's ozone depletion problem). While as of 2007 no satellites had been launched from Canadian soil, the Black Brant sounding rocket was developed in Canada, and sounding rockets were launched from Churchill, Manitoba, for many years.

Canadian experiments have flown on board spacecraft from many nations, including a Canadian ultraviolet imager for auroral research flown on the Swedish *Viking* satellite in 1986 and a thermal plasma analyzer experiment to examine Mars's atmosphere on the Japanese *Nozomi* spacecraft that was launched to Mars in 1998.

Chris Gainor

See also: European Space Agency, *International Space Station*, MacDonald Dettwiler Corporation, Space Shuttle

Bibliography
Lydia Dotto, *Canada in Space* (1987).
Chris Gainor, *Canada in Space: The People and Stories behind Canada's Role in the Exploration of Space* (2006).
———, "The Chapman Report and the Development of Canada's Space Program," *Quest* 10, no. 4 (2003).
Doris H. Jelly, *Canada: 25 Years in Space* (1988).

China

China began development of its modern space program in the 1950s, led by American-trained Qian Xuesen (Tsien Hsue-Shen). Qian, with approximately 100 other Chinese scientists, was expelled from the United States in 1955 during the

McCarthy era. Initially through the civilian Chinese Academy of Sciences (CAS), Chinese scientists and engineers worked on updated German V-2 rockets, the R-1 and R-2, given by the Soviet Union and on the development of satellites. Between 1960, when the Soviet Union severed relations with China, and the late 1990s, when collaboration with Russia on human spaceflight began, Chinese space efforts were primarily indigenous. International politics and the self-imposed isolation of the 1966–76 Cultural Revolution (the growth of the program slowed as scientists and engineers were persecuted and/or worked as laborers) left China working independently. China later built a comprehensive, albeit austere, space program.

The Chinese military has played a significant role in Chinese space activities since the mid-1960s. While much of the policy-setting process has remained opaque, the State Council and, until it was merged into the newly created State Administration for Science, Technology and Industry for National Defence (SASTIND) in 2008, the Commission for Science, Technology, and Industry for National Defense (COSTND), as well as the Central Military Commission (CMC) have been considered key players, with implementation left to various parts of the military or the military-industrial complex. For example, by 2008 the head of the General Armaments Department of the CMC, responsible for armaments for the People's Liberation Army, was also the commander of the human spaceflight program. Military involvement in the organizational structure allowed for political control in addition to maximizing resources and talents working on space research and application with utility to the civil and military sectors.

Nevertheless, at the start of the twenty-first century, the Chinese government recognized that some bifurcation of efforts was necessary to increase space engagement with other countries. External organizations were understandably reluctant to work with the Chinese military on space activities. As a result, the Chinese government made bureaucratic changes to separate civilian and commercial programs from military programs. That split continued into the twenty-first century, though the Chinese military continued to provide a variety of services, such as security at launch sites. A multitude of Chinese organizations have worked on space activities, with structures and practices often confusing to Westerners. Since its establishment in 1999, China Aerospace Science and Technology Corporation (CASC), which has also been responsible for other national defense and aerospace endeavors, has managed China's space program. CASC was formed from part of the China Aerospace Corporation, a large state-owned enterprise under direct supervision of the State Council, as part of a concerted government effort to make its industries more competitive (the other portion became China Aerospace Machinery and Electronics Corporation). When dealing externally, CASC has continued to manage most programs, but under the name of the China National Space Administration.

CASC has controlled more than 130 organizations, including five large research academies, the Chinese Academy of Launch Vehicle Technology, the Chinese Academy of Space Technology, the Shanghai Academy of Space Flight Technology, the Chinese Academy of Space Electronic Technology, and the Academy of Space Chemical Propulsion Technology; two large research and manufacturing bases, the Sichuan Space Industry Corporation and Xi'an Space Science and Technology

Industry Corporation; a number of factories and research institutes under the direct supervision of the headquarters; and companies in which it has had various levels of ownership. CASC had 100,000 employees as of 2008. All foreign space launch orders have been handled by CASC's international cooperation platform and launch-marketing company, China Great Wall Industry Corporation (CGWIC), established in 1980. Since CGWIC first offered commercial launches in 1985, it has conducted more than two dozen international commercial launches of more than 30 satellites and six piggyback payloads. In 2004, in a sign of diversifying commercial capabilities, CGWIC contracted with international customers to deliver communications in-orbit satellites based on the newly developed Dong Fang Hong (DFH)-4 satellite bus.

China has three areas of publicly acknowledged space activity: launch vehicles, satellites, and spacecraft; defense systems; and satellite applications. China's first satellite was a DFH communications satellite. *DFH 1*, launched in 1970, broadcast the song "The East Is Red." Since then, China has developed and launched dozens of military, civilian, and dual-use satellites of various types and sizes; including Feng Yun meteorological satellites, remote-sensing satellites (e.g., Fanhui Shi Weixing [FSW] recoverable satellites, Feng Yun weather satellites, and Yaogan disaster relief/military satellites), Shijian scientific experimental satellites, and Beidou navigation satellites. The FSW satellites deserve particular note as they clearly illustrated the dual-use nature of space technology developed and employed in China. Although FSW satellites were originally developed for military reconnaissance purposes and first launched in 1974, the technology was adapted for use in Earth resource observation and crystal and protein growth experiments in the late 1980s.

China developed the Chang Zheng (CZ)/Long March (LM) series of launch vehicles. The LM vehicles were derived from earlier CZ military counterparts, not an unusual practice (U.S. civilian workhorse vehicles Delta, Atlas, and Titan were originally military vehicles). The spacecraft used in China's human spaceflight program, for example, was the CZ 2F. China has used three launch sites: Xichang in Sichuan Province (all geosynchronous satellite launches), Jiuquan in the Inner Mongolia Autonomous Region (Chang Zheng 2F), and Taiyuan in Shanxi Province (polar-orbiting spacecraft); a fourth, Wenchang on Hainan Island at only 19 degrees North, is scheduled to be completed in 2013. The Chinese have significantly increased their heavy-lift launch capability. In the mid-1980s China began offering commercial launch services using the LM series. That effort resulted in the beginning of bifurcation efforts to separate its program into civil and military sectors, with the military oversight less visible. A series of launch accidents in the mid-1990s and subsequent demands by Western insurance companies for accident reports caused political issues for China. Allegations of technology theft by China were made in the United States in connection with the launches and the accident reports.

While commercial launch opportunities decreased in the late 1990s, China continued to seek cooperative opportunities with other countries. Under the China-Brazil Earth Resources Satellite (CBERS) program, for example, China launched three jointly developed satellites, with more planned. Again, the dual-use nature of the technology raised considerable speculation about the military potential of the satellite technology for high-resolution military reconnaissance. Additionally, having been limited in its access to the European Union (EU)'s Galileo, its future alternative to the U.S. Global Positioning System (GPS), China began developing its own positioning system based on Beidou satellites. An initial four-satellite Beidou-1 navigation constellation deployed in 2007 offered basic positioning over China and its immediate periphery; a 35-satellite Beidou-2/Compass system will provide improved regional coverage by 2011 and global coverage by 2015–20; three satellites have been launched as of 2010.

China's second human spaceflight effort was Project 921. Because the capsule that carried taikonauts into orbit atop the CZ 2F was known as Shenzhou (divine or heavenly ship), human spaceflight missions have been designated as Shenzhou missions. An earlier human spaceflight program was started in the 1970s but was halted in 1980 due to lack of funds, technological barriers, and a pragmatic decision to prioritize applications satellites. By 2003, 14 taikonauts had been selected, drawn from the elite ranks of military fighter pilots. Two taikonauts (Wu Jie and Li Qinglong) trained in Russia and became instructors for those later trained at a facility north of Beijing. While the Shenzhou capsule was reportedly able to carry three taikonauts, the first launch on 15 October 2003 took only one taikonaut, Yang Liwei, into space for a 21-hour, 16-orbit trip, making China the third country to develop human spaceflight capability. *Shenzhou 6* launched on 12 October 2005 with two taikonauts, Fei Junlong and Nei Haisheng, for a five-day flight. During their stay in orbit, they were able to change into lighter, more comfortable space suits, for ease of movement while conducting experiments and living in space. During this flight, for the first time, the

Chinese soldier stands guard near the China National Space Administration's Shenzhou 6 *space vehicle. The rocket launched in October 2005, carrying two taikonauts, China's second human spaceflight mission. (Courtesy AP/Wide World Photos)*

taikonauts were able to enter the orbital module. *Shenzhou 7*, launched on 25 September 2008, included the first extravehicular activity by taikonauts. *Shenzhou 8* is planned for 2011.

Shenzhou 6 completed phase one of step one of the officially approved three-step human spaceflight program—step one to demonstrate technical capabilities (with phase one testing life support and crew capabilities, and phase two, beginning with *Shenzhou 7*, focusing on docking maneuvers and related activities, including extravehicular activities or space walks); step two to develop a small space laboratory, likely from Shenzhou modules; and step three to develop a larger space station, which would also require development of a new Long March 5 heavy-lift launcher.

China has announced ambitious plans for robotic lunar and Mars exploration. In addition, although China has expressed interest in *International Space Station* (*ISS*) participation, the United States, which has held a veto within the *ISS* partnership, by 2010 had not extended or approved an invitation. Chinese prospects to participate in *ISS* were not helped by the January 2007 test of an antisatellite weapon, which destroyed a Feng Yun weather satellite and created many thousands of pieces of space debris. Chinese space activities have avoided space spectaculars in favor of incremental progress in a broad range of programs designed to meet key national objectives at affordable cost.

Joan Johnson-Freese and Andrew S. Erickson

See also: Long March, Russian Launch Vehicles, Shenzhou

Bibliography
Iris Chang, *Thread of the Silkworm* (1995).
Brian Harvey, *The Chinese Space Programme: From Conception to Future Capabilities* (1998).
Joan Johnson-Freese, *The Chinese Space Program: A Mystery within a Maze* (1998).
Gregory Kulacki and Jeffrey G. Lewis, "A Place for One's Mat: China's Space Program, 1956–2003," American Academy of Arts & Sciences, 2009.
Evan Medeiros et al., *A New Direction for China's Defense Industry* (2005).
Yu Yongbo et al., *China Today: Defense Science and Technology*, vols. 1 and 2 (1993).

Europe

Europe's space program has been primarily driven by economic and technical considerations, as opposed to the military and political factors that drove the space efforts of the superpowers: the United States and the Soviet Union/Russia. After World War II, Europe consisted of medium-sized and small nations, most of which did not have the money, technical resources, or industrial capability to go to space. Competing with the superpowers was not practical. One of the results was a drive among Western European nations for economic integration, particularly through the Common Market, which led to the European Union. Integration was also fruitful in high technology efforts, such as nuclear research and power generation and, in the 1960s, in space. For many of these nations, the only way to secure a presence in space was to pool resources and build transnational collaborative structures. In Western Europe, this occurred with military space programs through the North Atlantic Treaty Organization (NATO), weather satellites through EUMETSAT, telecommunications through Intelsat and Eutelsat, and a variety of scientific and civilian programs through the European Space Agency (ESA). In addition European nations sought bilateral programs with one another, with the superpowers, and later with other nations.

Space activities in Europe provided a civilian, high-tech platform to aid the cause of European political unity, while also providing a crucial tool of industrial and economic policy. This partly explains why European governments placed space science conducted through the European Space Research Organisation (ESRO) at lower priority than launchers. Space science was the affair of universities and research institutes that could fly payloads of scientific interest, but of little industrial importance, on European, American, or Soviet rockets. The social and commercial stakes were high, however, in applications satellites for telecommunications, meteorology, and Earth observation. Stakes were also high in rocketry, where many thought it was essential that Europe build its own launcher to retain control over access to space, especially when commercial interests were involved. Rocketry's ties to the technology of ballistic missiles that were crucial to the military also contributed to its importance.

Participation in the European program did not stop some of the more powerful countries, notably the United Kingdom, France, and Germany, from developing strong national space programs in parallel. This enabled them to benefit from their European effort and to steer them along paths consistent with domestic priorities, which could be different. The United Kingdom, for example, championed space science and was reluctant to support other ventures that did not promise a commercial return. The United Kingdom consistently opposed major programs such as the International Space Station (ISS), while Germany, which for three decades sought to learn by collaborating closely with the United States, committed heavily to the ISS program. Italy and Spain, being less technologically advanced than northern European nations, viewed space as a crucial domain in which to develop technical, managerial, and industrial skills. France saw space as a means of maintaining independence from the United States. Integration of space activities among European nations was always balanced with national interests.

While governments set the basic agenda, most European space vehicles were constructed by European industry. ESA's principle of fair return was crucially important for the formation of consortia, and later consolidation of private companies. The principle of fair return dictated that the percentage contribution that a government made to a specific program was reflected in the percentage, by value, of technologically important contracts that were allocated competitively to industries on its soil. This meant that the industrial corporations had to organize to work together in similar percentages, establishing relationships that paved the way for waves of consolidation, so that by the early twenty-first century, many of Europe's private space corporations had united under the European Aeronautic Defence and Space (EADS) Company.

Developments in the United States have always loomed over the evolution of the European space effort and influenced the priorities that the Europeans have set and the niches they have sought to occupy. Thus the United States may have been generous in providing support for science, but far more hesitant about encouraging rival programs in the commercially important area of telecommunications satellites or in the even more strategic domain of rocketry. European nations, perceiving themselves as less technologically advanced than the United States in these areas, sought to build a competitive capability in such key sectors. By the mid-1960s it was apparent to European nations that satellite communications would become a profitable and critical industrial force. While the United States sought to control this technology, it was limited by the fact that at this time the primary use of satellite telecommunications was for long-distance telephone calls between the United States and Europe or Japan. Knowing this, European nations banded together to negotiate with the United States. They successfully negotiated the Intelsat accords with the United States and ultimately obtained a distribution of economic benefits in proportion to the usage of the system by each nation. However, this integrated success was counterbalanced by national satellite communications programs as France, Germany, Italy, and the United Kingdom sought to benefit individually, while smaller European countries worked for integrated efforts that they could influence. By the early twenty-first century, European nations and corporations had developed a mature satellite communications capability comparable to that of the United States.

Similarly, as their capabilities matured, European nations individually and collectively pursued launcher technologies (Ariane, Vega), and Earth remote sensing (*SPOT— Satellite Pour l'Observation de la Terre*, and *ERS—European Remote Sensing*), navigation (*Galileo*), and reconnaissance (*Helios*) capabilities, which could and did compete with American capabilities. In other arenas, such as human spaceflight, and on various military systems, Europeans did not pursue an independent path so vigorously, and remained subordinate to U.S. and Russian ventures.

Prior to 1991, space efforts in East European states were closely linked to those of the Soviet Union, primarily through building scientific experiments to be placed on Soviet spacecraft, and through the Interkosmos guest cosmonaut program. After 1991 many East European states sought collaboration with or membership in the European Space Agency.

Stephen B. Johnson and John Krige

See also: Ariane, Europa, European Space Agency, European Space Research Organisation, International Space Politics, Economics of Space

Bibliography

Brian Harvey, *Europe's Space Programme: To Ariane and Beyond* (2003).
John Krige and Arturo Russo, *A History of the European Space Agency, 1958–1987*, 2 vols. (2000).
Kevin Madders, *A New Force at a New Frontier: Europe's Development in the Space Field in the Light of Its Main Actors, Policies, Law and Activities from Its Beginnings up to the Present* (1997).

France

France's involvement in spaceflight activities predates World War II, notably through the achievements of Robert Esnault-Pelterie, who tested rocket motors and wrote

L'Astronautique (1930). During World War II Jean-Jacques Barré built the first French liquid-propellant rocket, EA 1941, which was launched on 15 May 1945. In July 1946 the French Air Ministry initiated a prototype program that encompassed all types of missiles for all military forces. In 1948 it opened the Colomb Béchar/Hammaguir test range in Algeria. With the help of a German team, the LRBA (Ballistics and Aerodynamics Research Laboratory) of DEFA (Armaments Studies and Construction Authority) developed a 4-ton thrust liquid engine, which was tested on an unguided Véronique rocket. Véronique was later used as a sounding rocket, in addition to biological suborbital flights with rats and cats. In January 1959 a space research committee, led by Professor Pierre Auger, initiated a series of scientific sounding rocket tests at Colomb Béchar.

In response to the space achievements of the Soviet Union and the United States, the French government created an organization dedicated to space-related scientific and technological research, the Centre National d'Études Spatiales (CNES, literally "National Center for Space Studies," but usually translated as "French Space Agency"), created with a decree signed by President Charles de Gaulle on 19 December 1961. It coordinated with and drew from existing industry and government organizations including SEREB (Société pour l'Étude et la Réalisation d'Engins Balistiques or Ballistic Missiles Research and Development Company) created in 1959. CNES assumed responsibility for sounding rocket activities, developing Dauphin and Eridan with Sud and Vesta for monkey launches with LRBA. On 26 November 1965 France became the third nation to launch its own spacecraft when a Diamant rocket placed *Asterix 1* in orbit. CNES selected a new launch site in Kourou, French Guiana, in 1964 because the Colomb Béchar base was scheduled to close in 1968. The first operational launch from Kourou occurred in 1968 with the launch of a Véronique sounding rocket. CNES also operated three other facilities, including one in Toulouse (opened in 1974) where work ensued on programs ranging from telemetry coordination to Earth observations including the SPOT program (Satellite Pour l'Observation de la Terre), which launched four Matra-built satellites from 1986 to 1998, French Telebroadcasting (Télédiffusion de France or TDF), and telecommunications.

CNES sought to finance existing government and industry laboratories rather than create its own. As a result, proposed space experiments, for example, remained the responsibility of the project originator and were added to the launch manifest just before launch. This emphasis on industrial scientific procedures and cooperation for the sake of efficiency helped explain CNES's decision to sign cooperative agreements with NASA in 1963 and later with the Soviet Union. The immediate result for CNES was the successful launch of the *FR1* television satellite in December 1965 on a NASA Scout 6 rocket from Vandenberg Air Force Base, California. It allowed CNES to acquire technical and scientific knowledge necessary to design its own satellites. While initial French satellite programs were managed and integrated by CNES, aerospace companies later took over prime contractorship, starting with the Franco-German Symphonie telecommunications satellites.

Starting in the early 1960s France contributed the Coralie second stage to the Europa launcher for the European Space Vehicle Development Organisation (ELDO), became a member of ESRO (European Space Research Organisation) to build scientific satellites, and cooperated with other European nations in negotiations with the United States on satellite communications in the European Conference on Satellite Communications. When ELDO appeared headed for failure despite a strong Franco–German commitment, France modified the L-3 rocket project into a new ESA booster, Ariane, beginning in 1973. ELDO's failure led to the creation of ESA (European Space Agency) in 1975 through the merger of ELDO and ESRO. France also collaborated on a number of bilateral space projects including the Symphonie communications satellites with West Germany and the *TOPEX/Poseidon* ocean monitoring satellite with the United States.

At the beginning of the space age a variety of private companies began development of satellites, launchers, and their components, including Sud Aviation, Nord Aviation, Matra, and Alcatel. In 1970 Sud Aviation, Nord Aviation, and SEREB joined to form Aérospatiale. Matra Espace merged with the United Kingdom's General Electric Company Marconi Space Systems in 1989 to form Matra Marconi Space. In 2000 Matra Marconi joined German DaimlerChrysler Aerospace AG to form Astrium. That same year Aérospatiale merged with DaimlerChrysler Aerospace, Dornier, and Construcciones Aeronáuticas to form the European Aeronautic Defence and Space Company (EADS), which after the merger owned 75 percent of Astrium. Separate from these events, in 1980 the French government helped create the first private space launch company, Arianespace, which marketed Ariane launches. Two years later the government helped create SPOT Image to market data from government-funded SPOT satellites. As intended, both Arianespace and SPOT Image competed successfully with U.S. companies in commercial launch and remote sensing markets.

CNES's activities at times placed French space policy in an ambiguous position. For example, Franco-Soviet cooperation in space during the administration of French President Charles de Gaulle caused considerable concern in the mid 1960s, shortly after CNES's agreement to work with NASA. Cooperation with the Soviet Union was useful to CNES (despite extreme mistrust on the part of the Soviet Union) and resulted in French experiments on Soviet sounding rockets in 1967, Soviet experiments on Dragon rockets in 1969, a French laser reflector on *Lunokhod 1* in 1970, French SRET (Satellite de Recherches et d'Études Techniques) satellites on Molniya launchers in 1972 and 1975, and the flight of the first French astronaut, Jean-Loup Chrétien, in 1982. Though a human presence in space was generally welcomed by the French public, critics charged that such a commitment signaled a qualitative shift away from the original space policy mission that was not warranted for a midsized power. Evidence to that effect included the commitment to develop a miniature shuttle, Hermes. The practical limits of the machine, in addition to cost overruns, led to the project's cancellation in 1992 and continued dependence on American and Russian capabilities for human access to space.

Because CNES played an integral part in defining French space policy, it was also able to defend its programs effectively, partly thanks to the success of international cooperative efforts, but also to privatization of some endeavors. CNES continued to work with the United States and in 2000 issued a statement of intent for a Mars sample return mission. The following year the United States launched the Franco-American oceanographic satellite *Jason 1*. The private company SPOT Image marketed unclassified high resolution photographs, which were also useful for military purposes, as in 1991 when *SPOT 1* and *2* provided images to coalition forces in the First Gulf War. Similarly the launch of the military observation satellites *Helios 1A* and *1B* in 1995 and 1999 reflected the need to serve the French ballistic missile program's targeting needs.

At the turn of the millennium CNES published a modified mission statement that included the planning of French space policy to support space applications serving the public interest at the civilian and military levels. This reinterpreted the text of the 1961 law that established CNES into a specific set of missions including meteorology, surveillance, and the development of practical uses for space instruments. It echoed NASA's "Looking at Earth" statement of the 1980s, though the French statement included a military dimension.

Guillaume de Syon

See also: Ariane; Arianespace Corporation; European Space Agency; European Space Research Organisation; European Space Vehicle Launcher Development Organisation; Hermes; SPOT, Satellite Pour l'Observation de la Terre

Bibliography
Claude Carlier and Marcel Gilli, *The First Thirty Years at CNES* (1995).
Philippe Jung, "The True Beginnings of French Astronautics 1938–1959," *History of Rocketry and Astronautics* 28 (2007).
Walter A. McDougall, "Space-Age Europe: Gaullism, Euro-Gaullism, and the American Dilemma," *Technology and Culture* 26, no. 2 (1985).
Peter Redfield, *Space in the Tropics* (2000).

Germany

Germany's space policy and associated programs can be traced to the early 1960s. Following the end of World War II and the dispersal of scientific know-how, some interest remained among amateurs and scientists returning from abroad. The legacy of Peenemünde, however, prompted a clear reorientation of space interests by framing it within exploratory science. While politics and economic issues would dominate German space activities henceforth, the emphasis on research remains paramount. While East Germany generally followed and supported the space program of the Soviet Union (with East German cosmonauts invited to fly on board Soyuz missions), West Germany displayed considerably more autonomy vis-à-vis Europe and the United States. In 1961 the Deutsche Forschungsgemeinschaft (DFG or German Research Foundation) published "White Paper on the State of Space Research." Although the document emphasized science and the civilian use of space, the German government was unable, for political reasons, to designate a single national space agency, so it created a special commission to advise the Ministry for Research and Technology. This commission was also responsible for drafting the first German four-year plan for space (1964–68).

However, in the course of the first program, which saw involvement in the European Space Research Organisation (ESRO), and the European Space Vehicle Launcher Development Organisation (ELDO), Germany felt the returns on its investments

German astronaut Thomas Reiter from the European Space Agency during a space walk on the International Space Station, *3 August 2006. (Courtesy European Space Agency/NASA)*

showed considerable imbalance and impacted the small discretionary budget left for other national endeavors; this became all the more obvious as its share of funding increased after the United Kingdom began pulling out of ELDO. Nonetheless the dependence of Europe on the United States for launch capability encouraged the Germans to maintain commitment to developing a European rocket while using American assistance to launch their first satellite, *Azur*, in 1969. Subsequent four-year plans saw the nation involved in primarily bilateral programs with the United States (the Helios solar research satellites—launched 1974 and 1976) and France (*Symphonie* communications satellite—launched 1974) along with multilateral programs through the European Space Agency (ESA). In the case of the latter, Germany focused primarily on *Spacelab* in the 1970s, contributing the majority of the costs, with Entwicklungsring Nord (ERNO) as prime contractor. Germany also contributed to the development of launchers. Germany developed the third stage for ELDO's Europa, and invested about 20 percent into the Ariane project. By the 1980s Germany was involved with the development of the *Columbus* module for the *International Space Station* (*ISS*) and had hoped to develop a two-stage orbiter, the Sänger, which was abandoned. In addition it cooperated with France on the Helios/Horus reconnaissance satellite project, which resulted in two satellite launches (1999 and 2004).

Germany, as a member of ESA, nominated candidates for ESA's first astronaut list and German Ulf Merbold was one of four chosen for astronaut training in 1978. Merbold flew as a mission specialist on the D1 and Spacelab Space Shuttle missions.

Other German astronauts later went into space with NASA and the Russian Federal Space Agency.

By the late 1980s the direction of Germany's commitment to space science underwent major restructuring. A semiprivate agency, the Deutsche Agentur für Raumfahrt-Angelegenheiten (DARA), or German Agency for Space Issues, was formed in 1990 as a consortium of governmental and private space interests incorporating the former from both East Germany and West Germany. Its goals paralleled some of NASA's "Looking at Earth" program and emphasized space science and applied technologies. Though DARA has also been the office dealing with ESA, its small size meant that many technical tasks remained in the realm of the Deutsche Forschungsanstalt fur Luft und Raumfahrt (DLR—German Air and Space Research Center), whose staff of 5,000 performs research and operational support tasks comparable to some of NASA's research centers.

Germany did not create a "space center" but maintained an operational command post in Oberpfaffenhofen, while crew training was near Cologne, where ESA's European Astronaut Center is located. It also hosts ESA's European Space Operations Center in Darmstadt.

The German aviation industry expanded in the 1960s to include space activities, but also over time consolidated from a host of relatively small firms to a single entity, now part of the European Aeronautic Defence and Space Company (EADS). Starting with historic firms such as Junkers, Dornier, Messerschmitt and Focke-Wulf, consolidation began soon after World War II. By 1968 Junkers, Messerschmitt, and Heinkel had formed Messerschmitt-Bölkow-Blohm (MBB), while Focke-Wulf, the Dutch firm Fokker, and others formed ERNO-VFW. Famous automobile company Daimler-Benz acquired Dornier in 1985, and in 1989 merged MBB (which by then contained ERNO-VFW) into its aerospace division Deutsche Aerospace AG (DASA). This in turn became part of EADS in 2000. Space programs were important activities in these companies, with their evolution significantly affected by European political consolidation, which spurred the development of consortia and eventual mergers.

The primary barrier to expansion of German space policy has been budgetary. Committed to supporting ESA, Germany's domestic space budget experienced considerable cutbacks starting in 1993. By 2003 of the €1.4 billion earmarked for space, some €300 million, primarily earmarked for DLR, became the target of budget battles between the Social Democratic coalition government and the Christian Democratic–led opposition. That year the proposal to establish a new German space program plan, which would have been the fifth in existence and the first since 1986, failed. Talk of reducing the space research budget to €145 million meant that the primary victims would be local German companies subcontracting work for the DLR. However, as several observers noted, this fall in funding did not necessarily reflect a failure of space politics, but rather a reorientation toward international programs, with Germany remaining firmly committed to its ESA share and to the *ISS*. These also included, among others, ESA's Global Monitoring for Environment and Security (GMES), and the Galileo navigation projects.

Guillaume de Syon

See also: European Aeronautic Defence and Space Company (EADS), European Space Agency, European Space Research Organisation, European Space Vehicle Launcher Development Organisation, *International Space Station*, *Spacelab*, Symphonie

Bibliography
Werner Büdeler, *Raumfahrt in Deutschland: Forschung, Entwicklung, Ziele* (1978).
Ulf Merbold, *Flug ins All* (1986).
Helmuth Trischler, *The "Triple Helix" of Space: German Space Activities in a European Perspective* (2002).

India

India's space program began, through the initiative of Vikram Sarabhai, after India gained its independence from the United Kingdom in 1947. Sarabhai established the Physics Research Laboratory in Ahmedabad in 1947 and the Thumba Equatorial Rocket Launching Station (TERLS) in Kerala in 1962. The India Space Research Organisation (ISRO) was created in 1969 under the Department of Atomic Energy and transferred to the Department of Space (DOS) in 1972. ISRO developed launchers and two separate satellite systems, the Indian National Satellite (INSAT) program and the Indian Remote Sensing (IRS) program. ISRO created Antrix Corporation, Limited in 1992 to handle the worldwide commercial market.

India launched the INSAT satellites into geosynchronous orbit and controlled them from the Master Control Facility (MCF) at Hassan in Karnataka. Sarabhai set up the Experimental Satellite Communication Earth Station (ESCES) in Ahmedabad in June 1967, followed by the Arvi Earth receiving station near Pune, designed for overseas satellite communications via the Intelsat system. By 2005 INSAT satellites were used for domestic long distance, telecommunications, meteorology, direct satellite television, and radio broadcasting. The first-generation INSATs (*1A*, *1B*, *1C*, and *1D*) were built by Ford Aerospace in the United States, whereas the second-generation INSATs (*2A*, *2B*, *2C*, and *2D*) were indigenously built with improved frequency capabilities and an improved radiometer for weather observation. The S-band (aimed at mobile communications), C-band (aimed at television broadcasting), and Ku-band satellites were aimed at business communications.

The IRS satellites served a national interest in the utilization of natural resources for practical applications. Experimental satellites *Aryabhatta*, *Bhaskara 1*, and *Bhaskara 2* were launched from the Soviet Union in 1975, 1979, and 1981 respectively. From these evolved the IRS series—*IRS 1A* and *1B* were launched in 1988 and 1991 respectively, and had a spatial resolution of 36 m. *IRS 1C*, similar to *1D*, had an improved payload, the multispectral Linear Imaging Self Scanning (LISS) 3 camera with spatial resolution of 20 m.*IRS 1D* was the first indigenous launch by the Polar Satellite Launch Vehicle (PSLV). The next generation of satellites, including *IRS P4* (*Oceansat 1*), catered to physical and biological oceanographic studies. *IRS P6* (*Resourcesat 1*) was launched in 2003 by *PSLV C5* to enhance service capabilities in areas of agriculture, disaster management, and land and water resources (spatial resolution of 5.8 m). The IRS satellite capabilities improved throughout the generations,

VIKRAM AMBALAL SARABHAI
(1919–1971)

(Courtesy Hulton Archive/ Getty Images)

Sarabhai published a paper on cosmic rays in 1942 that led to the development of the Indian space program. After India gained independence from the United Kingdom in 1947, Sarabhai established the Physics Research Laboratory. In 1962 the Indian Committee for Space Research, led by Sarabhai, established the Thumba Equatorial Rocket Launching Station that launched the first Indian rocket. He founded the Indian Space Research Organisation in 1969 and established an experimental satellite communications Earth station at Ahmedabad (his home state) in pursuit of his vision of bringing science to the common people.

Shubhada Savant

particularly in their spatial resolution, for example, *Cartosat 1* launched into a polar Sun-synchronous orbit in 2005 with a spatial resolution of 2.5 m.

Satellite launch vehicles progressed from solid-propelled to liquid-propelled rockets. TERLS developed infrastructure for all aspects of rocketry, ranging from rocket design, rocket propellant, integration, payload-assembly, testing, and evaluation. After a difficult start, the PSLV turned out to be ISRO's workhorse with successful launches. Although PSLV was initially designed by ISRO to place 1,000 kg class IRS satellites into 900 km polar Sun-synchronous orbits, its capability was enhanced to 1,600 kg. PSLV was used to launch ISRO's exclusive meteorological satellite, *Kalpana 1*, into a geosynchronous transfer orbit in September 2002. The Geosynchronous Launch Vehicle (GSLV) was used to place heavier communications INSAT satellites into geostationary orbit. The solid first- and liquid second-stages carried over from the PSLV and the final stage was cryogenically propelled, a technology that India worked on for 11 years. Its first operational flight was the launch of *EDUSAT* in 2004.

As a result of a bilateral agreement between India and the Soviet Union, Rakesh Sharma became the first Indian in space when he launched to *Salyut 7* aboard *Soyuz T11* in April 1984. With its budget and ambitions expanding, in the early twenty-first century, ISRO embarked on a robotic lunar orbiter mission and, in November 2006, announced it would embark in an indigenous human spaceflight program to place a human in orbit by 2014 and land on the Moon by 2020.

Shubhada Savant

See also: Indian Launch Vehicles, Indian National Satellite Program, Indian Remote Sensing Satellites

Bibliography
Brian Harvey, *The Japanese and Indian Space Programs: Two Roads into Space* (2000).
Indian Space Research Organisation. www.isro.org.
Gopal Raj, *Reach for the Stars: The Evolution of India's Rocket Programme* (2000).

Israel

Israel's space program dates to the Six-Day War in 1967 and the Yom Kippur War in 1973. As a result of the Six-Day War, Israel's strategic posture vis-à-vis the Middle East was dominant. A transfer of nuclear and ballistic missile technology from France during the 1960s reinforced this posture. The loss of France as Israel's strategic ally in 1967 led to Israel's development of nuclear arms and a ballistic missile program called Jericho. These developments and Israel's strategic posture, however, failed to deter Egypt and Syria from attacking Israel in 1973. In the 1973 Yom Kippur War, Israeli intelligence failures in assessing enemy intentions and capabilities resulted, for a brief time at the beginning of the conflict, in an existential threat to Israel. Israeli officials concluded that an indigenous means of acquiring intelligence data by reconnaissance satellite was essential for national security. This conclusion was reinforced by the lack of satellite intelligence available to Israel during the Yom Kippur War. The development of an Israeli space program began in 1982. Israel subsequently developed a number of indigenous capabilities that entailed most significantly a reconnaissance satellite program, the Ofeq, and a launch vehicle, the Shavit, derived from Jericho.

Israel launched its *Ofeq 1* reconnaissance satellite on a Shavit in 1988, becoming one of the few nations able to launch satellites to orbit independently. Beginning in 1990 Israel had a number of successful Shavit launches of the Ofeq—*Ofeq 2* in 1990, *Ofeq 3* in 1995, and *Ofeq 5* in 2002. The *Ofeq 4* launch of 1998 and the *Ofeq 6* launch of 2004 both failed when the final stage of the Shavit failed to properly place the respective satellites into orbit. The continuation of the Ofeq program was a consequence of events before and during the 1991 First Gulf War in Iraq that demonstrated the strategic and tactical vulnerabilities of Israel to missile attacks.

Other areas of development in the Israeli space program have encompassed the *Amos* telecommunications satellite and a commercial high-resolution remote sensing system, the *Earth Resources Observation Satellite*. Israel has also cooperated in diverse ways with the United States, Europe, Russia, Brazil, China, India, and others to enhance its presence in the civil and commercial aspects of space and to project Israel as a space power on the international scene.

The key players in Israel's space program have included the Israeli Space Agency, which has coordinated cooperative space activities with other nations and scientific programs for Israel; the Israeli military industries, with involvement in the hardware development for the civil, commercial, and military space areas; and the Israeli Ministry of Defense and Israeli Defense Forces, which have been in charge of military space capabilities. Israel's spacefaring capabilities have emphasized advanced technology in the economy and the importance of maintaining a qualitative edge in the conflict with nations hostile to its interests. Israel has come to realize, as have other spacefaring nations, that its economic growth, international legitimacy, and national security are inextricably linked with space.

Eligar Sadeh

See also: Reconnaissance

Bibliography

American-Israeli Cooperative Enterprise Jewish Virtual Library. www.jewishvirtuallibrary.org.
John Simpson et al., "The Israeli Satellite Launch: Capabilities, Intentions, and Implications," *Space Policy* 5, no. 2 (May 1989): 117–28.
E. L. Zorn, "Israel's Quest for Satellite Intelligence," *Studies in Intelligence* 10 (Winter–Spring 2001): 33–38.

Italy

Italy first used rockets in 1823 in Sardinia (part of the Kingdom of Italy) when the army formed its "Compagnia Artificieri" (Artificers' Company). The army continued to use rockets until 1873 and continued testing until 1899. One of the first pioneers in this field was an engineer from Brescia, Luigi Gussalli. In 1923 he published *Si può già tentare un viaggio dalla Terra alla Luna?* (*Are We Ready to Attempt a Journey from Earth to the Moon?*) in which he described experiments conducted with a "double-reaction" engine for a space vehicle. Gaetano Arturo Crocco of the

The Leonardo Multi-Purpose Logistics Module, built by Italy, in Discovery's *payload bay during the STS-102 mission to the International Space Station, 2001. (Courtesy NASA)*

University of Rome initiated systematic studies on a combustion chamber for liquid-propellant rocket engines in 1930. The following year in Milan, Ettore Cattaneo tested an aircraft propelled by solid-propellant rockets. In 1952 Aurelio Robotti of the Polytechnic of Turin built and launched the first Italian liquid-propellant rocket (AR-1). This research, however, soon ended, and it failed to further the development of liquid-propellant rocket propulsion in Italy.

In 1962 while the firms Sispre and Contraves were working on the construction of an experimental sounding rocket, an agreement between the governments of Italy and the United States provided support to start Italy's space activities. Luigi Broglio, an Italian Air Force general and teacher at the University of Rome, was the originator and director of the San Marco Project, a new aerospace program that built five scientific satellites to study the density of the atmosphere using an instrument Broglio devised. *San Marco 1* (the first satellite to be built by a European country) was launched in December 1964 on a U.S. Scout rocket. Other satellites were launched using the same type of rocket from the mobile San Marco launch base, which consisted of two marine platforms anchored in the Indian Ocean off the coast of Kenya near the equator. Four U.S. scientific satellites and one British were launched from the San Marco facility.

Italy continued to collaborate directly with the United States even after the formation of the European Space Research Organisation (ESRO) and the European Space Vehicle Launcher Development Organisation (ELDO) in 1964. Italy also took part in major European programs, building the ELDO test satellites and contributing to *Spacelab*, NASA's Space Shuttle laboratory. Italy built the laboratory structure and developed its internal environmental control system. The first *Spacelab* flight took place in 1983.

Meanwhile the first Italian experimental telecommunications satellite, *SIRIO 1* (Satellite Italiano Ricerca Industriale Orientata), was launched in 1977 to test the characteristics of high-frequency transmissions. The first National Space Plan in 1979 led to the creation of the Agenzia Spatiale Italiana (ASI), or Italian Space Agency, in 1988. ASI was involved in the development of several satellites, including *Italsat 1* for telecommunications (1991) and *BeppoSAX (Satellite per Astronomia X)* for X-ray astronomy (1996). It also developed *TSS 1 (Tethered Satellite System)*, launched on the Space Shuttle *Atlantis* in 1992 accompanied by the first Italian astronaut, Franco Malerba. In the 1990s Italy contributed to the construction of the European Space Agency (ESA) Ariane 5 launch vehicle (supplying parts of the solid rocket boosters and components for the Vulcain engine) and the structure and thermal control for the *Columbus* living-quarters module of the *International Space Station (ISS)*. ASI also built *ISS* logistics modules (named *Leonardo*, *Raffaello*, and *Donatello*) and was involved with NASA and ESA in the Cassini-Huygens mission to Saturn and one of its moons. Within ESA, Italy led the development of the three-stage, solid-propellant Vega launcher, with a financial contribution of 65 percent.

Italy was instrumental in the foundation of ESRO, and later ESA, strongly advocating the principle of just return for the resources invested in the agency.

This principle encouraged the development of the Italian space industry. Italy's major space companies have been Thales Alenia Spazio (previously Alenia Space) (living structures, scientific satellites, and telecommunications) and Avio (space propulsion), both of Gruppo Finmeccanica, and Gavazzi Space (small scientific satellites). Between 1964 and 2007, Italy contracted with NASA, Arianespace, Kosmotras, Indian Space Research Organisation (ISRO), and Cosmos International to launch 19 satellites into orbit. Among these were three types of microsatellite: *Itamsat 1* for radio amateurs, two Unisat experimental satellites for the University of Rome, and three communications satellites owned by private companies.

Giovanni Caprara

See also: Alcatel Alenia Space, European Space Agency, European Space Research Organisation, European Space Vehicle Launcher Development Organisation, *International Space Station*, *Spacelab*

Bibliography
Giovanni Caprara, *The Complete Encyclopedia of Space Satellites* (1986).
———, *L'Italia nello spazio* (1992).
———, *Living in Space* (2000).
M. De Maria, L. Orlando, and F. Pigliacelli, *Italy in Space*, ESA HSR-30 (2003).

Japan

Japan has developed an incremental, risk-averse, but relatively consistent space program that has included launch vehicles and a variety of spacecraft. Post–World War II constitutional limitations on military development, early twenty-first-century economic stagnation and consequent funding limitations, a primarily bottom-up policy making process, and the conservative nature of Japanese culture have limited the scale of program initiatives.

In 1955, in preparation for the International Geophysical Year (IGY), University of Tokyo Professor Hideo Itokawa, working with several colleagues, launched the tiny "Pencil Rocket." Within three years, they had developed the K (Kappa) sounding rocket as part of Japan's contribution to IGY. Many of the scientists working on these programs, and eventually their students, became members of the Institute of Space and Astronautical Science (ISAS), created in 1964–65 as part of the University of Tokyo.

After 1969 Japan's space program took shape and considered more ambitious plans beyond the academic realm. The National Space Development Agency (NASDA) was created in 1969 from the Space Development Promotion Section of the Science and Technology Agency (STA) with responsibility for the applications aspects of space development and a commensurately large budget. It received primary funding from the STA. NASDA reported to the Space Activities Commission, housed in Japan's Prime Minister's Office.

Japanese astronaut Soichi Noguchi waves from the Shuttle payload bay during STS-114. (Courtesy NASA)

While NASDA focused on space applications, science remained the responsibility of academics. ISAS deemed it critical to launch one satellite per year, even if it meant sacrificing scale for frequency, to ensure that graduate students, usually in a five-year education cycle, would have the opportunity to be involved with a flight experiment. It was increasingly clear to government officials that the University of Tokyo was too small to manage this growing enterprise. Consequently, in 1981 ISAS was reborn as a joint research organization among Japanese universities, under the purview of the Ministry of Education.

Another Japanese space participant, the National Aerospace Laboratory (NAL), was founded in 1955 as a national research laboratory, under the authority of STA, with its areas of focus expanded in 1969. NAL's mandate included conducting basic, advanced, and applied research. Together NASDA, ISAS, and NAL formed the nucleus of Japan's space program for several decades, occasionally joined by other organizations, such as the Ministry of Posts and Telecommunications and the Ministry of International Trade and Industry, according to a given program's functional nature. In 2003 NASDA, ISAS, and NAL merged to form the Japanese Aerospace Exploration Agency (JAXA). ISAS became one of JAXA's four departments, which also included Strategic Planning and Management Department, Office of Space Flight and Operations, and Office of Space Applications.

At the time of the merger, ISAS had 301 employees and a budget of $162 million. Programs undertaken by ISAS included development of the L (Lambda) and M (Mu) series launch vehicles; *Ohsumi*, Japan's first satellite, launched in 1970; interplanetary missions such as *Lunar A*; programs undertaken with international cooperation (*Suisei* and *Sakigake* as part of the Halley's Comet mission in 1985–86 and *Geotail* with NASA as part of the ISAS solar terrestrial science program); the *Hayabusa* sample return probe; and a variety of other projects, including research on scientific ballooning. ISAS and its predecessor organization successfully launched 25 scientific satellites and probes.

NASDA, with a 2003 budget of approximately $1.7 billion and 1,090 employees, focused on launch vehicles, participation in the International Space Station (ISS) program, satellites and applications, and general research and development. Initial programs undertaken by NASDA to develop launch technology, specifically the N series, were licensed U.S. technology. With the later H series, Japan began indigenous technology development. As of 2010 Japan had launch sites on Kagoshima, on Kyushu Island (originally called Kagoshima Launch Center but renamed Uchinoura Space Center in 2003), and on Tanegashima Island south of Kyushu. Both could only operate at certain times each year because of pressure from Japan's fishing lobby. These restrictions hampered Japan's launch program and contributed to its difficulties with marketing the H-I and entirely indigenous H-II rockets as commercial launch vehicles. Programs undertaken by NASDA included participation in the ISS program with the *Japanese Experimental Module* (*JEM*) *Kibo*, with the first piece delivered to the *ISS* in March 2008. Although Japan had not independently pursued a human spaceflight program, it participated in those of the United States and the former Soviet Union. Journalist Toyohiro Akiyama flew on the Soviet *Soyuz TM* (Transport Modification) *11* mission in December 1990. NASDA astronaut Mamoru Mohri flew on the Space Shuttle Space Transportation System (STS)-47 mission in 1992. By 2010 an additional six astronauts (Chiaki Mukai, Kouichi Wakata, Takao Doi, Soichi Noguchi, Naoko Yamazaki, and Akihiko Hoshide) had completed space missions.

NAL, with a 2003 budget of approximately $28 million and 424 employees, focused on space technology and aeronautical technology, especially spaceplanes and airplanes. NAL often supported NASDA programs. For example it developed the LE-5 engine for the second stage of the H-I and H-II rockets.

In the early 1990s ISAS, NASDA, and NAL experienced incremental budget growth, with NASDA generally receiving about 75 percent of Japan's space budget. Japan had also traditionally benefited from a strong interest, and subsequent investment, in space activities by industry. The major launch vehicle manufacturers for NASDA and ISAS were, respectively, Mitsubishi Heavy Industries and Nissan Motor Company. Mitsubishi Electric Corporation, Nippon Electric Corporation, and Toshiba Corporation were the main satellite prime contractors—Fuji Heavy Industries, Ltd., and IHI Company, Ltd., supported development of reusable space transportation systems and *JEM* for the *ISS*. Sumitomo Heavy Industries, experienced in launch vehicle support facilities, and construction giants Shimizu and Obayashi were active in the design of long-range facilities, including outposts on the Moon and Mars. The late 1990s economic downturn in Japan negatively impacted public and private space budgets.

In addition to the work done under the auspices of ISAS, NASDA, and NAL, Japan embarked on the development of Information Gathering Satellites (IGS) in 1998. The first two IGS satellites were launched in 2003; as of 2010 four—two optical, two radar—were operational. This program differed from other Japanese space programs in several ways. The IGS system was a dual-use system of satellites to provide information for diplomatic and defense policy decision making and to support crisis management and disaster relief operations. Undertaken soon after North Korea's launch of a Taepo-Dong missile over Japan in 1998, IGS was politically driven and funded through the prime minister's office. That program was considered indicative of the increased importance ascribed to space assets in Japan's security sector and perhaps signaled a change from the past. As a matter of national policy, Japan initially rejected direct engagement regarding military activities in outer space. In May 1969 the Japanese Diet, in compliance with Article 9 of the Constitution, adopted a resolution pledging that the country's space projects would be limited to peaceful (defined as nonmilitary) uses. This was slowly reinterpreted through the years in recognition of the realities of information technology. A strict interpretation of the resolution, for example, could have barred Japanese Self Defense Forces from using communications or navigation satellites.

Following a decade of technical difficulties amid national economic stagnation, on 1 October 2003 a major reorganization of Japanese space activities occurred with the merger of ISAS, NASDA, and NAL to form JAXA. The merger combined the resources, development experience, and technology of these organizations to promote cooperation with industries and the more efficient development of Japan's space program. Despite occasional failures, a variety of missions furthered scientific, technology, and national security objectives. Japan's first attempted Mars mission, *Nozomi*, failed to achieve proper orbit around Mars in December 2003. Following a 2003 launch failure, JAXA successfully orbited *Himawari 6* on 26 February 2005. Between September and December 2005, *Hayabusa* rendezvoused with and analyzed the asteroid Itokawa. During July–August 2005 astronaut Souichi Noguchi flew on the STS-114 mission of the Space Shuttle *Discovery*. On 24 August 2005 the *Kirari* was launched from the Baikonur Cosmodrome. In December 2005 *Kirari* successfully established the first successful optical intersatellite communications with the European Space Agency *Advanced Relay and Technology Mission Satellite*. On 4 October 2007 the *Kaguya* satellite was placed into lunar orbit.

Announced in February 2005, the JAXA Vision/JAXA 2025 Project called for using aerospace technology to address natural disasters and global environmental issues; exploring space, asteroids, and the Moon; establishing competitive, reliable space transportation systems; and developing advanced supersonic aircraft while demonstrating technologies for hypersonic aircraft. Future plans included launching the improved H-IIB rocket and developing improved rockets.

Joan Johnson-Freese and Andrew Erickson

See also: Halley's Comet Exploration, International Geophysical Year, *International Space Station*

Bibliography
Steven Berner, *Japan's Space Program: A Fork in the Road?* (2005).
Brian Harvey, *The Japanese and Indian Space Programmes: Two Roads into Space* (2000).
Joan Johnson-Freese, *Over the Pacific: Japanese Space Policy into the Twenty-First Century* (1993).

Russia (Formerly the Soviet Union)

Russia (formerly the Soviet Union) has maintained one of the two largest space programs in the world since the 1950s, during which time it has launched more spacecraft into orbit than any other nation, beginning with the launch of the world's first artificial satellite, *Sputnik*, in 1957 by the Soviet Union.

The Soviet space program grew out of the desire to build long-range ballistic missiles after World War II. In 1953–54 the Soviet government assigned OKB-1 (Special Design Bureau-1 of Scientific Research Institute (NII) 88, which became Experimental Design Bureau-1 (also OKB) in 1956), the leading missile development organization, to develop an intercontinental ballistic missile (ICBM) to strike targets in the continental United States. Simultaneously, Mikhail Tikhonravov, a senior scientist who worked at the NII-4 (Scientific Research Institute) military institute, was directing studies to evaluate the possible uses of an artificial Earth satellite. OKB-1 Chief Designer Sergei Korolev, who had worked with Tikhonravov in the 1930s, assembled a powerful coalition of forces, which included scientists from the Soviet Academy of Sciences, to propose a satellite that could be launched by the R-7 ICBM. In January 1956 the Soviet government formally approved the development of such a scientific satellite, which would be equipped with an array of instruments to study Earth during the International Geophysical Year (IGY). Fearful that the United States would upstage this project, Korolev proposed a "simple satellite" that could be launched on quick notice in 1957. After two relatively successful launches of the R-7 ICBM in August and September 1957, Korolev's team launched the simple satellite (*PS 1*), popularly known as *Sputnik*, into orbit on 4 October 1957. *Sputnik 2*, launched a month later, carried a dog named Laika into orbit. A larger scientific satellite, *Sputnik 3*, was launched in 1958, ending the first stage of the Soviet space program.

Despite the successes of Sputnik, it took a while for the Soviet government to establish long-range goals in spaceflight. In this vacuum, the early years of the program were essentially driven by the whims of leading Chief Designers, such as Korolev, or the short-term needs of the military. Korolev played a key role in proposing the first human spaceflight project, Vostok, which he offered to the government for dual use as a robotic spy satellite platform known as Zenit. Using the 3KA version of the Vostok, the Soviet Union took a qualitative step forward with the launch into orbit of Yuri Gagarin, the first human in space, in 1961. A number of important achievements in the human spaceflight program followed, including the first group flight (*Vostok 3* and *4* in 1962) and the first woman in space (Valentina Tereshkova on *Vostok 6* in 1963). Responding to political imperatives to upstage the U.S. space program, OKB-1 refitted the 3KA Vostok spacecraft with marginal changes to carry out the first multiple cosmonaut mission (on *Voskhod* in 1964) and the first extravehicular activity (Alexei Leonov on *Voskhod 2* in 1965).

The Soviet human spaceflight program was beset by a number of unprecedented setbacks through the late 1960s and early 1970s. Following Korolev's death in 1966, cosmonaut Vladimir Komarov was killed during a reentry accident on the first flight of the new Soyuz spacecraft, stalling the momentum of the program. As the parallel lunar landing program was facing failure, Soviet policy makers instructed OKB-1 management to combine elements of the Soyuz and a military station concept, known as Almaz, to quickly build a modest Earth orbital station, known internally as *DOS 1* (*Long Duration Orbital Station*) and publicly as *Salyut*. Unfortunately, the first crew was killed on reentry in 1971, crushing any hopes of a revitalized space program. In the mid-1970s, after this spate of failures, OKB-1, now called NPO (Scientific-Production Association) Energia, redirected much of its efforts to developing improved versions of the DOS space stations. Meanwhile, Vladimir Chelomei directed the Almaz project, which launched three stations, also under the public Salyut moniker (*Salyut 2*, *3*, and *5*) to disguise their explicitly military character. In 1978 the Soviet military redirected Almaz toward only robotic operations, realizing that humans were a poor substitute for automated reconnaissance.

The high point of the Soviet space program in the mid-1970s was the joint Apollo-Soyuz Test Project (ASTP), known in the Soviet Union as the Apollo Soyuz Experimental Flight (EPAS). In 1975 a Soyuz spacecraft and an Apollo vehicle docked in Earth orbit in a historic meeting of cosmonauts and astronauts. The Soviet Union clearly had more to gain from the project, because it provided equal status with the United States in the public eye at a time when it was lagging behind the United States in most areas of space exploration.

After regaining its footing with the successful *Salyut 6* and *7* stations in the late 1970s and early 1980s, the Soviet Union's human-spaceflight activities culminated in the launch of the *Mir* (or *DOS 7*) space station in 1986. Between 1989 and 1999, the Soviet Union (and later Russia) kept *Mir* continuously occupied for a period of 10 years using 28 crews (main expeditions) on missions ranging from six months to one year. During that time, *Mir* was augmented by specialized modules (*Kvant*, *Kvant 2*, *Kristall*, *Spektr*, and *Priroda*), each about the size of *Mir*, which significantly improved the station's functioning capacity. Cosmonaut Valery Polyakov set the long duration record in space in 1994–95 with a continuous mission lasting 438 days. The *Mir* complex was deorbited in 2001.

During the late 1980s the Soviet Union began planning for a bigger station, *Mir 2*. In 1993 ideas for *Mir 2* were merged with those of *Space Station Freedom* to form the basis for *International Space Station* (*ISS*). In 1998 the Russians launched the first element of *ISS*, the *Zarya* control module (based on the design of TKS—Transport Supply Ship), a module owned by NASA but produced by the new Khrunichev company, which was a conglomerate formed out of remnants of the former Chelomei empire. Since *Zarya*'s launch, the *ISS* has been augmented by other Russian hardware including the *Zvezda* service module (2000) and the *Pirs* docking compartment (2001). *ISS* has been continuously occupied by joint Russian-U.S. crews since 2000.

Probably the most spectacular abandoned project from the early years was the N1-L3 project, designed to compete with the Apollo program to land a human on the Moon. Managed by Korolev's OKB-1, the massive project involved development of the N1 super-booster, a behemoth whose creation was mired in interpersonal rivalries between Korolev and Valentin Glushko, the two giants of the Soviet space program. Personal conflicts, military priorities, poor funding, inefficient management, and technological limitations eventually derailed the project; four launches of the N1 booster in 1969–72 ended in catastrophic failures, one of which, in 1969, destroyed one of two giant launch pads at the launch base at Tyuratam. Despite headway in the development of the L3 payload—comprising the LOK lunar orbiter and the LK lunar lander—the program was suspended in 1974 and canceled two years later. In parallel, a project known as L1 (or Zond) to send cosmonauts around the Moon (without landing) also failed to preempt the United States and was canceled in 1970 after several partially successful robotic missions.

A similar fate befell the bigger Energia-Buran project, initiated in 1976 in the wake of N1. Energia was a super-booster designed to launch payloads of 100 tn into Earth orbit while Buran was a Soviet version of the U.S. Space Shuttle. Both were approved as part of a military plan to develop an array of defensive and offensive systems for operation in Earth orbit. These plans were spurred by a perception that the Space Shuttle (approved in 1972) was some sort of offensive space weapon designed to attack the Soviet Union. These concerns were reinforced when U.S. President Ronald Reagan announced the Strategic Defense Initiative (SDI) in 1983. Energia was successfully launched in 1987 (although its payload failed to reach orbit) while Buran performed a completely automated space mission from takeoff to landing in 1988. These projects did not survive the economic collapse that followed the dissolution of the Soviet Union and were canceled by 1993.

The Soviet Union (and later Russia) has had a wide-ranging robotic space program since 1957, overwhelmingly dominated by military objectives. Although research on the military applications of spaceflight began in the mid-1950s, such programs were formally initiated by government decree in 1961. The decree defined new systems for photoreconnaissance (film, television), electronic intelligence, communications, navigation, meteorology, ground radar calibration, and geophysical experiments. These projects were flown under the catchall Kosmos label to disguise the true intentions of any particular satellite. More than 2,400 satellites have been launched under the Kosmos label since 1962 and the "true" names of most of these missions only came to light in the 1990s.

In terms of numbers of satellites, the largest programs have undoubtedly been for optical reconnaissance programs dedicated to spying on U.S. assets in addition to global hotspots. The Russians divided their reconnaissance programs into three broad generations of satellites, Zenit, Yantar, and Orlets. Under OKB-1's direction, the first Soviet reconnaissance satellite, a Zenit 2, resulted in a launch failure in 1961. The system was declared operational in 1964 and used in conjunction with Zenit 4, declared operational one year later. Both systems used film capsules that were returned to Earth after the end of their short missions. The Soviet Union (and Russia) continued to use upgraded versions of these satellites into the 1990s. Early second-generation satellites, such as Yantar 2K, launched in 1974–83, had two film return capsules and longer lifetimes. Later models, such as Yantar 4K1 (Oktan) and Yantar 4K2 (Kobalt) flown in 1979–83 and 1981–96 respectively, had improved high-resolution capabilities. The Soviet Union adopted digital imaging in 1982 with Yantar 4KS1 (Terilen), a vehicle that might be considered a counterpart to the U.S. KH-11 Kennan reconnaissance satellites introduced in 1976. In addition topographic mapping duties (for targeting) were performed by the Yantar 1KFT (Kometa), whose photographs were marketed commercially in the post-communist era.

Orlets, launched beginning 1989, combined the benefits of wide-area survey and detailed photography via return capsules, of which as many as 22 were used on advanced models such as *Orlets 2* (*Yenisey*). A new fourth-generation reconnaissance satellite known as Araks, capable of digital photography and transmission, had been launched only twice in the decade after 1997. In the early 2000s, Russia continued to operate five types of reconnaissance satellites—Kometa, Kobalt 1, Yenisey, Neman, and Araks—with possible new systems such as Kobalt M, Condor, and Persona in development.

Signals intelligence (SIGINT) constituted a major goal of Soviet intelligence operations in space. The first dedicated electronic intelligence (ELINT) satellites were the Tselina O and Tselina D series of satellites for area and detailed surveillance respectively, declared operational in the early 1970s. These were succeeded by a more capable and unified second-generation system known as Tselina 2. A high-security program was the ocean reconnaissance satellite program known as Legenda, whose goal was to monitor U.S. naval operations and provide targeting data for Soviet tactical and strategic cruise missiles. The system included an "active" portion (using radar) known as US-A (Controlled Satellite-Active) and a "passive" portion (using ELINT equipment) known as US-P (Controlled Satellite-Passive). The former—known in the West as radar ocean reconnaissance satellite (RORSAT)—was notable for being the only operational Soviet spacecraft to use a nuclear reactor as a power source.

The Soviet Union deployed a space-based early warning system following years of internal disagreement about the specifications of a first-generation system. Beginning in 1972 the Soviet Union tested the US-K and US-KS (Oko) systems; despite many serious problems, the systems were activated 10 years later. A newer system known as US-KMO (Oko 1) was designed to have a "look-down" capability from geostationary orbit to detect launches against the background of Earth's surface. The system was put in service in 1996, but its operational capability was rather limited.

Besides intelligence, the Soviet Union experimented with several offensive programs, including an operational coorbital antisatellite (ASAT) project using an interceptor known as IS against small targets (known as DS-P1-M and Lira). The first successful interception of a target was performed in 1968 and by the late 1970s the system was declared fully operational. It was stood down in 1983 after nearly two dozen Earth orbital tests, although ground testing continued into the 1990s. An uprated system (IS-MU), apparently capable of interceptions at high altitudes, was declared operational in 1991, but the system was decommissioned in 1993 on orders from Boris Yeltsin. Another parallel ASAT system known as Naryad was tested in the early 1990s but abandoned before service duty. One of the more unusual systems tested was the Fractional Orbital Bombardment System (FOBS), basically a space-bombing system designed to put nuclear bombs in partial Earth orbit for deorbit at any given point. After two dozen tests of the system in 1965–71, FOBS was declared operational with the full knowledge that deploying nuclear weapons in space violated the Outer Space Treaty. FOBS was fully decommissioned in 1983.

To support military communications, the Soviet Union flew several different systems providing different capabilities, including the military Molniya, Tsiklon, Strela, Geyzer systems and the geostationary Raduga, Ekran, Gorizont, and Luch systems. Several generations of Meteor weather satellites provided meteorological information for both civilian and military users. The construction of a dedicated navigation satellite system—similar to the U.S. Global Positioning System (GPS)—was begun in 1982 using Uragan satellites. The system, known as GLONASS, has operated using modernized satellites, although the system had not reached its optimal 24-satellite coverage as of 2007.

During the Soviet era, international cooperative programs were limited largely to other socialist or "friendly" nations. Under the Interkosmos program, for example, Soviet bloc countries provided experiments for a series of scientific missions beginning in 1969. On occasion the Soviet Union also launched satellites for other countries, such as France. Under a joint program, Soviet and Indian scientists built two satellites, which were launched in the 1970s.

In the early years of the Soviet space program, the most striking successes in robotic spaceflight were in lunar exploration. Using the first and second generation of lunar programs, the Soviet Union preempted the United States in almost every significant milestone: the first probe to reach escape velocity (*Luna 1*), the first probe to impact the Moon (*Luna 2*), the first probe to take photographs of the lunar far side (*Luna 3*), the first survivable landing on the Moon (*Luna 9*), and the first artificial lunar satellite (*Luna 10*). With the third-generation Ye 8 class lunar probes, the Soviet Union fared less well, although it accomplished two outstanding achievements: the first robotic sample return from the Moon (*Luna 16* in 1970) and the first mobile surface vehicle on the Moon (*Lunokhod 1* in 1970). After three successful sample return missions, the Soviet Union discontinued its lunar program in 1976.

The most successful string of successes in deep space came in the exploration of Venus, where the Soviet Union achieved the first survivable landing on another planetary body (*Venera 7* in 1970), the first photographs from the surface of Venus (*Venera 9* and *10* in 1975), the first Venus orbit (the same vehicles), and the first color photographs of the Venusian landscape (*Venera 13* and *14*). The successes in Venus exploration came at the cost of the extraordinarily poor results of the Soviet Mars exploration program. Despite as many as 15 launch attempts, at best only a couple of probes returned significant data from Mars and none from its surface. After a failed attempt in 1996 to send an ambitious probe to Mars, the Russians abandoned their Mars program.

Behind the scenes, the organization of the Soviet (and later Russian) space program comprised a Byzantine structure impenetrable often even to the people running it. On a macro level, the program had four constituencies: policy makers (Communist Party officials), managers (ministers), developers (Chief Designers), and clients (the military). Although top political officials exploited many Soviet achievements post facto, program implementation decisions were largely the result of often dysfunctional negotiations between the four major constituencies, fueled by personal acrimonies among various Chief Designers vying for influence and prestige. One of the most important tools for Chief Designers was information on the U.S. program that designers used to bolster already existing proposals, by both invoking American superiority and by redirecting the trajectories of broadly conceived policies to benefit their positions in the industrial hierarchy. The secret creation and cancellation of many programs for political or technical reasons implied a government throwing enormous amounts of money into spaceflight without sufficiently clear ideas about reasonable objectives.

At the development level, the Soviet Union used a vast system of "scientific-research institutes" and "design bureaus." The former carried out applied research while the latter developed spacecraft, launch vehicles, and ground segment equipment. The most important primary contractors for spacecraft were typically identified with a single patriarchal chief designer although the organizations used a confusing numbered designation system (for example, OKB-1) that was abandoned in the late 1960s.

The client for most Soviet military programs was the military. A division of the Strategic Rocket Forces—formed in 1959—accepted delivery of all space systems and operated all ground operations (launch, ground control). In 1981 this division separated from its parent service and became an independent body and in 1992, after the collapse of the Soviet Union, it became the Military Space Forces (VKS). After a number of structural changes, in 2001 the VKS became the Space Forces (KV), which was responsible for overseeing the three Russian space launch centers (at Plesetsk, Baikonur, and Svobodny), a control center for military satellites (the Titov Main Center for Testing and Controlling Spacecraft), the ground communications segment, and a number of educational institutions. The Russian Air Force supervises the Gagarin Cosmonaut Training Center at Star City outside of Moscow.

After the breakup of the Soviet Union, many of the design bureaus and research institutes that produce spacecraft were brought under a civilian agency, the RKA (Rossiyskoe Kosmicheskoe Agentstvo or Russian Space Agency, RSA) established in 1992. The agency guided the civilian space program through the economic depression of the 1990s. RKA was renamed the Federal Space Agency (or informally Roskosmos) in 2004 and identified as the governmental agency responsible for overseeing all the major organizations that build spacecraft, launch vehicles, and their subsystems.

Much of the economic power of the Russian space program has come from its highly efficient launch vehicles, most of which were derived from military ballistic missiles. For example, by progressively adding upper stages and improved engines, the original R-7 ICBM was transformed during a period of 50 years into a family of variants. Given the same public names as their primary payloads—Luna, Vostok, Voskhod, Soyuz, and Molniya—these launch vehicles had been used for more than 1,700 successful orbital launches by the end of 2007. Similarly the shorter range R-12 and R-14 missiles provided the foundation for small satellite launchers, such as the Kosmos 1, Kosmos 2, and Kosmos 3M while the heavy R-36 ICBM was upgraded into the Tsiklon 2 and Tsiklon 3 boosters used primarily for launching military satellites. Chelomei's older UR-500 "super" missile became the Proton, deployed in three- and four-stage versions. The former was used for launching large space stations in orbit and the latter for geostationary and deep space missions.

The launch site for the Soviet and Russian space program has been the Baikonur Cosmodrome, established in 1955 as an ICBM launch base near Tyura-Tam (Tyuratam) in Kazakhstan. The base officially remained top secret until the late 1980s; from 1961 the Soviet press simply referred to the range as the Baikonur Cosmodrome, named after a town about 300 km northeast of Tyuratam. As of 2005 all Soviet and Russian human spaceflight launches have originated from the Baikonur Cosmodrome, along with a host of other military and civilian spacecraft beginning with Sputnik in 1957. In 2005 Russia and Kazakhstan signed an agreement to extend Russian rental of the launch range through 2050. The Plesetsk launch site in northern Russia was originally formed in 1957 as the first operational ICBM base. From 1966 it has launched the bulk of Soviet (and Russian) spacecraft into orbit and contains launch pads for the Vostok, Molniya, Soyuz, Kosmos 3M, and Tsiklon 3 launch vehicles. Seeking to replace the dependence on Baikonur, Russians have been working to establish a new site at Svobodny at the defunct site of an ICBM base. In 1997 a reconverted Topol ICBM, renamed Start, launched a satellite into orbit from Svobodny.

Since the mid-1990s many Russian research and development organizations have been engaged in commercial activities with foreign partners by aggressively marketing Russian rocket and spacecraft technology. The major cooperative projects, such as SeaLaunch, International Launch Services (ILS), Starsem, and the use of Russian rocket engines on U.S. Atlas launch vehicles, have provided an enormous boon to the Russian space industry, allowing it to rebound after the 1990s downturn.

Asif A. Siddiqi

See also: Human Spaceflight Programs; Korolev, Sergei; Meteor; Military Applications; Outer Space Treaty; Politics of Prestige; Resurs; Russian Launch Vehicles

Bibliography

Nicholas Daniloff, *The Kremlin and the Cosmos* (1971).
Brian Harvey, *Russia in Space: The Failed Frontier?* (2001).
Nicholas L. Johnson, *Soviet Military Strategy in Space* (1987).
Asif A. Siddiqi, *Challenge to Apollo: The Soviet Union and the Space Race, 1945–1974* (2000).

United Kingdom

The United Kingdom's (UK) first military rockets were developed by Sir William Congreve in the early nineteenth century. Rocketry stagnated until after World War II, when the UK developed a variety of missiles, including the Blue Streak intermediate-range ballistic missile (derived from the U.S. Atlas) and the Skylark sounding rocket. Largely independent of these developments, in 1933 rocket and space enthusiasts formed the British Interplanetary Society to promote space exploration and astronautics. Since its research was constrained by a nineteenth-century act of Parliament that prohibited the launching of rockets by unauthorized people, the society focused on theoretical studies.

The UK's first spaceflight carrying scientific experiments was on 13 November 1957 with a Skylark that provided some of the first space science data for the International Geophysical Year. The Skylark program ran for almost 50 years with more than 400 flights for national and international space science researchers, including the formative research program of the European Space Research Organisation (ESRO) in the mid-to-late 1960s.

Also in the 1960s the UK provided the Blue Streak first stage for the European Space Vehicle Launcher Development Organisation's (ELDO) Europa launch vehicle. Originally intended as an intermediate-range ballistic missile (IRBM), the Blue Streak IRBM program was canceled in 1960. To salvage the huge investment and help the nation's chances of joining the European Common Market, British Prime Minister Harold Macmillan promoted Blue Streak as a first stage in a proposed Anglo-French launcher. The resulting negotiations led to the development of ELDO, within which the UK was burdened with a disproportionately large financial contribution. As the organization suffered through a series of launch failures (which were not due to Blue Streak), UK support rapidly diminished. In the resulting renegotiations, the UK's financial contributions decreased and eventually stopped altogether. Initially the UK insisted on launching ELDO's Europa 1 rocket from its Blue Streak facilities in Woomera, Australia, but with the UK's support waning, ELDO transferred operations to the French launch site in Kourou, French Guiana.

Another British rocket—Black Knight—was also launched from Woomera as part of a testbed for Blue Streak and to assess reentry issues for the warhead. The Black Knight design was heavily utilized in development of Black Arrow—the UK's only satellite launch vehicle. Black Arrow orbited the *X3* (*Prospero*) technology satellite on 28 October 1971. However, the subsequent Black Arrow program had already been canceled, so all other UK satellites were orbited using foreign launchers.

From that first Skylark launch, the UK's space activities were informed by a thriving space science sector. The UK's scientific talent helped build on an existing strong relationship between the United States and the UK, manifest in the transfer of U.S. expertise and resources in rocket technology during the mid-1950s, into several bilateral space programs. The first bilateral (and the world's first international) space mission was the *Ariel 1* satellite, launched on 26 April 1962. The experiments were designed and operated by British universities with the United States providing the spacecraft and the Thor-Delta launch vehicle. Six Ariel satellites were launched during the next 17 years with UK industry taking over satellite design and construction from the United States from *Ariel 3* onward.

Provision of the UK's experiments for *Ariel 1* was organized by the British National Committee on Space Research (BNCSR). The BNCSR had been formed in 1958 by the Royal Society to both liaise with the Committee for Space Research and to rationalize the labyrinthine organization of the various UK national space activities. For the U.S./UK bilateral program a Steering Group for Space Research (SGSR), advised by the BNCSR, was set up within government to look after the financing of the UK's space activities.

In 1964 the responsibilities for space science previously undertaken by the SGSR and BNCSR were shifted to the new Science Research Council, which, through a succession of name changes and reorganizations, has continued to play a pivotal role, including through its main research sites and especially the Rutherford Appleton Laboratory (RAL), in helping to direct UK space science activity. These organizational changes also helped streamline the UK's developing role in shaping the newly formed ESRO space science program. The organization's first satellite, *ESRO 2*, was largely a UK affair with Hawker Siddeley Dynamics (HSD) acting as prime contractor and the universities of London, Leicester, and Leeds contributing the bulk of the scientific payload.

During the 1970s, while national and bilateral programs with the United States continued, the UK placed a greater emphasis on collaborations within Europe. The UK advocated the formation of the European Space Agency (ESA), which would bring ESRO and ELDO activities under one organizational umbrella. As the European space programs developed, the UK space industry was able to tender for contracts, independently at first, but increasingly as members of pan-European industrial consortia, such as STAR, MESH, and COSMOS. However the UK withdrew from all major involvement in ESA launch vehicle programs, opting instead to direct much of its space industry toward development of European scientific and communications satellites. Relative to France, West Germany, and Italy, the UK's investment in space declined significantly, as did its influence on European space programs.

The UK's refocus on satellites both reflected and supported the development of the British Skynet series of military communications satellites, which derived from the U.S. Initial Defense Communications Satellite Program design, and led also to the UK's championing of a European maritime communications satellite program, which evolved to become the MARECS (Maritime European Communications Satellite) program, based on a British Aerospace (formerly Hawker Siddeley Dynamics) satellite design. Hawker Siddeley had also led on the ESRO 4 science satellite, *X-4* (*Miranda*)—the final national technology satellite, and then on *Orbital Test Satellite 1* and *2*. The British Aircraft Corporation (BAC) was prime contractor for *GEOS 1* and *2* (Geostationary Satellite) and had also led on *Ariel 3* and *4*. Marconi Space Systems was prime contractor for *Ariel 5* and *6*.

During the 1980s the successor to HSD and BAC, British Aerospace, won the prime contract for ESA's *Polar Platform* (*Envisat*) Earth observation satellite in addition to the ESA's *Giotto* probe, which passed near Halley's Comet in 1986. The company also produced the Shuttle *Spacelab* pallets and satellite SPELDAs (Structure Porteuse Externe pour Lancements Doubles Ariane) for the Ariane 4 launch vehicle. In 1985 the University of Birmingham's *Spacelab-2* X-ray telescope flew aboard the second Space Shuttle *Spacelab* mission and was used to produce maps of cosmic high-energy X-ray sources, including some of the first ever made of the center of Earth's galaxy, the Milky Way.

The mid-1980s witnessed a flurry of British space policy reconsiderations as ESA proposals for a long-term program were assessed, planned, and balanced with new national proposals. There appeared to be a renewed political appetite for space and the British National Space Centre (BNSC) was unveiled in 1985 to develop and coordinate the UK's space policy. Roy Gibson, a former director general of ESA, was made the BNSC's first director general. This high-profile appointment reflected an informal agreement with ESA to increase the UK's space investment in the European venture. However Gibson's national space plan and the concomitant proposed increase in UK space funding were rejected by the UK government in 1987. So too was any possibility of the BNSC receiving a dedicated space budget; instead it became and has remained a coordinating body for UK space activity only.

A major initiative of the late 1980s was the development of the novel British Aerospace–Rolls Royce single-stage satellite launch vehicle, Horizontal Takeoff and Landing (HOTOL). When the associated budgetary increases proposed by the BNSC as part of its national space plan were rejected by government, HOTOL stalled and failed to progress beyond the concept stage. Elsewhere Surrey Satellite Technology Limited (SSTL) was founded in 1985 as a university campus-based company and became a world leader in the design and production of small satellites.

In the 1990s the UK continued to participate in international programs, including the *Hubble Space Telescope*, *ROSAT* (*Roentgen Satellite*), *Ulysses*, *European Remote Sensing 1* and *2*, *ISO* (*Infrared Space Observatory*), and *SOHO* (*Solar and Heliospheric Observatory*). In 1991 Helen Sharman became the first Briton to go into space when she traveled on *Soyuz TM-12* to the *Mir* space station. The flight was commercially funded through a contest to send a British citizen into space.

By the start of the twenty-first century, there was increasing UK involvement in solar system exploration and, despite the failure of the *Beagle 2* Mars lander, there were other important instrument or subsystem contributions made by university departments, industry, and the government's RAL to the *Cluster 2*, *Rosetta*, *Cassini-Huygens*, *Venus Express*, *STEREO* (*Solar Terrestrial Relations Observatory*), and *Solar-B* missions. Similarly UK institutions also contributed to astronomical missions *XMM-Newton* (*X-ray Multi-Mirror Newton*), *INTEGRAL* (*International Gamma-Ray Astrophysics Laboratory*), and *Swift*.

European Aeronautic Defence and Space Company Astrium's plant at Stevenage was by this time leading on structure, subsystem design, and build for all Eurostar communications satellites and, together with its Portsmouth site, helped develop and manufacture the Inmarsat F4 communications satellites and continued to manage and operate the UK government's Skynet 5 satellite system.

SSTL was prime contractor for *Giove A* (Galileo In-Orbit Validation Element), the experimental satellite for the proposed Galileo European navigation system that was launched in 2006.

Douglas Millard

See also: BAE Systems, British Interplanetary Society, *Cassini-Huygens*, Europa, European Aeronautic Defence and Space Company, European Space Agency, European Space Research Organisation, European Space Vehicle Launcher Development Organisation, International Astronautical Federation, *Mars Express*, Matra Marconi Space

Bibliography
John Krige and Arturo Russo, *A History of the European Space Agency, 1958–1987*, 2 vols. (2000).
Harrie Massey and M. O. Robins, *History of British Space Science* (1986).

Douglas Millard, *An Overview of United Kingdom Space Activity, 1957–1987* (2005).

Peter Morton, *Fire across the Desert: Woomera and the Anglo–Australian Joint Project, 1946–1980* (1989).

United States

United States space efforts began with the rocketry theories and experiments of Clark University physicist Robert H. Goddard, who conducted the world's first liquid-propellant rocket flight test in 1926. Goddard attracted funding from the Smithsonian Institution, the Guggenheim Foundation, and later from the U.S. Navy (USN) and performed rocket tests near Roswell, New Mexico. However, his influence was limited. Other amateur groups, including the American Interplanetary Society, the Cleveland Rocket Society, the Pacific Rocket Society, and others started in the 1930s, developed their own rocket designs and performed tests. These groups helped educate the public and cultivated a number of rocket engineers, who in some cases formed private companies, such as Reaction Motors, founded by four members of the American Rocket Society. In the late 1930s the California Institute of Technology (Caltech) began to develop liquid- and solid-propellant rockets under the guidance of Professor Theodore von Kármán and his student Frank Malina, leading to the formation of the U.S. Army–funded Jet Propulsion Laboratory (JPL) in 1944. The military funded a variety of rocket projects during World War II, leading to working rocket designs used for jet-assisted aircraft takeoff and short-range missiles.

After World War II the Army brought Wernher von Braun's V-2 rocket team to Fort Bliss, Texas, where they helped assemble and launch V-2s from rocket parts salvaged in Germany. U.S. engineers and Army officers learned from von Braun's team, and the V-2s launched a variety of scientific experiments such as radiation counters and spectrometers into the upper atmosphere. The Army also continued funding JPL and the USN contracted with the Martin Company to create ballistic missiles. Intercontinental missiles were the purview of the newly formed U.S. Air Force (USAF), which contracted for North American's winged Navaho and Convair's ballistic MX-774. The Soviet Union's progress in nuclear and missile programs spurred the United States to increase the pace and priority of missile programs. By the mid-1950s USAF Brigadier General Bernard Schriever commanded three high-priority, rapid-development ballistic missile programs: the MX-774 (later called Atlas), Douglas Corporation's intermediate-range Thor, and Martin's long-range Titan.

The development of long-range ballistic missiles made possible the creation of artificial Earth satellites, as a RAND Corporation study in 1946 recognized. RAND analysts predicted a number of satellite applications, including science, communications, weather observation, and military reconnaissance. Reconnaissance was a high priority for the Cold War United States, as the Harry Truman and Dwight Eisenhower presidential administrations attempted to discern the status, pace, and direction of Soviet military developments through a variety of clandestine activities, including aircraft overflights to photograph Soviet infrastructure. Fear of a Soviet nuclear "Pearl Harbor" attack led a high-level committee, the Technological Capabilities Panel, to

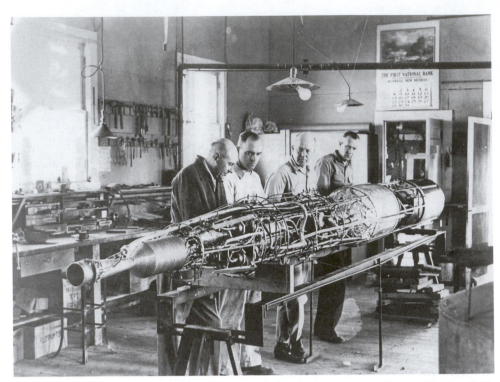

Rocket without its casing at the workshop of Dr. Robert Goddard in Roswell, New Mexico, 1940. With Dr. Goddard (far left) is Nils Ljungquist, machinist; Albert Kisk, brother-in-law and machinist; and Charles Mansur, welder. (Courtesy NASA)

recommend in 1954 development of a reconnaissance satellite: the USAF WS-117L program.

Separately from this effort scientists proposed that the United States launch a scientific satellite as a contribution to international collaboration in the International Geophysical Year (IGY) of 1957–58. Based on scientific criteria the Stewart Committee tasked with selecting the best proposal for IGY, in 1956 chose the Naval Research Laboratory's proposal to develop an upgraded Viking rocket known as Vanguard to launch a small scientific payload into orbit. In response the Soviet Union announced it too would launch a satellite for the IGY, while von Braun's Army team quietly prepared to launch a satellite in case the USN failed. When the Soviet Union successfully launched *Sputnik* in October 1957 and *Sputnik 2* the next month, Vanguard's initial test launch failure in December became an international disgrace. Von Braun's team redeemed U.S. prestige with the successful launch of *Explorer 1* in January 1958.

The strong national and international reaction to the Sputnik launches caught U.S. officials by surprise and galvanized the Democratic leadership in Congress to use this issue against the Republican Eisenhower administration. By 1958 the ensuing political fray led to the creation of the National Aeronautics and Space Administration (NASA) to separate civilian and military space activities, the passage of the National Defense Education Act to improve U.S. science and technical education, and the creation of the Advanced Research Projects Agency (ARPA) to coordinate military space programs. Massive increases in funding poured into space programs as NASA, ARPA, and the military services jockeyed for position. NASA and the USAF were the main winners of this competition. By 1960 NASA had acquired von Braun's team, JPL, personnel from the Army Signal Corps, USN Vanguard and scientific satellite personnel, and a few USAF programs. The USAF claimed all military launchers and most military satellite programs. ARPA lost the battle for control of military space and was diverted instead to handle advanced research that none of the other services wanted. The National Weather Service (eventually part of the National Oceanic and Atmospheric Administration) operated weather satellites.

By the early 1960s U.S. military institutions had established their respective space programs. The Army remained active in communications, while continuing development of ballistic missile defense (Safeguard—operational in mid-1970s) and antisatellite (Program 505 Mudflap—early 1960s) systems. The USAF deployed an operational antisatellite system (Program 437) in the 1960s and restarted active programs in antisatellite systems and ballistic missile defense with the 1980s Strategic Defense Initiative. Naval activities included fleet communications and the Transit program for submarine and ship navigation. Both the USN and USAF contributed to monitoring satellites and debris in space with ground-based radar and optical tracking programs centralized at North American Aerospace Defense Command (NORAD), while the USAF centralized the analysis of this data. The USAF developed programs to detect Soviet missile launches with the ground-based (Ballistic Missile Early Warning System—approved in 1957) and space-based (Defense Support Program—first launched in 1970) systems. It developed and operated the triad of launchers (Thor, Atlas, and Titan) and operated an eastern launch site at Cape Canaveral, Florida, and a western site at Vandenberg Air Force Base, California. The USAF developed a host of communications satellites and funded the winged, piloted Dyna-Soar program and later the Manned Orbiting Laboratory (MOL). The Secretary of Defense ultimately canceled these human military flight projects (Dyna-Soar 1963, MOL 1969) because their reconnaissance and antisatellite missions competed with and also potentially endangered U.S. robotic satellite intelligence activities by threatening diplomatic efforts to keep weapons out of space. The U.S. diplomatic approach succeeded with the signing of the Outer Space Treaty in 1967. All three services contributed to the Global Positioning System (GPS) program started in the 1970s. By the 1990s GPS was fully operational, increasing military effectiveness and became commercially important as private companies sold receivers to help in a variety of applications including emergency response and recreation.

Space-based intelligence and reconnaissance became a shared activity between the military and the intelligence agencies: the Central Intelligence Agency (CIA) and the National Security Agency (NSA). In the late 1950s the CIA pushed for a near-term reconnaissance program using a film-return method, known as Corona, which was supported by the Defense Meteorological Satellite Program to ensure it took pictures of the ground, not cloud tops. Corona's first successful mission in 1960 and the Eisenhower administration's determination to keep civilian control of this activity led to the formation of the CIA-USAF-managed National Reconnaissance Office (NRO). The

NRO, CIA, and NSA remained active in space, developing electronics, signals, and optical intelligence systems. The USN contributed a system (White Cloud—first launched in 1976) to monitor Soviet fleet movements.

NASA has been the public face of the U.S. space program, most conspicuously with its human spaceflight program. Project Mercury successfully aimed to place a man in orbit (accomplished 1962), but was second to Soviet achievements in the early 1960s. Catching and surpassing Soviet achievements became the primary objective of the human flight program. The United States achieved this goal with a number of firsts, including the first rendezvous and docking during Gemini (1966), and in 1969 when the Apollo program placed Neil Armstrong and Buzz Aldrin on the Moon. Skylab leveraged Apollo technologies while NASA designed and built the Space Shuttle, which first flew in 1981. NASA hoped the Shuttle would dramatically reduce the costs of going to space, but development phase cost constraints and design compromises frustrated these plans. In 1984 President Ronald Reagan directed NASA to build a space station, which after several iterations became the International Space Station (ISS) program (first launch 1998). In the Shuttle and ISS programs NASA progressively increased international participation from Canada, Europe, Japan, and Russia to share costs and to improve diplomatic relations. The Constellation program began as a NASA-only activity, although it held discussions with international partners about future participation. With similar diplomatic and cost objectives, NASA also supported international participation in a number of human and robotic missions.

Although human spaceflight consumed most of NASA's budget, NASA aimed to create and maintain a balanced program of science and technology development. Scientific goals included exploring the solar system with robotic probes, understanding the universe through space-based observatories, studying near-Earth space, observing Earth from space, and using microgravity for experiments in life and materials sciences. Robotic missions included the Viking Mars landers (landing on Mars 1976) and the Voyager mission of the 1970s–80s that scouted the outer planets of the solar system. NASA has orbited solar observing satellites such as the *Solar Maximum Mission* (launched 1980), the *Hubble Space Telescope* (launched 1990), and other lesser-known observatories that sense different portions of the electromagnetic spectrum. The Explorer satellite series has provided information about the near-Earth environment, and the Landsat (first launch 1972) and Seasat (launched 1978) programs, among others, helped scientists understand climatic and geologic processes on Earth. NASA has also launched technology demonstrator programs for communications and for advanced space technologies of various kinds.

From the 1950s onward the U.S. government has worked with and through U.S. private industry and to support private activities in space. Both NASA and the military have contracted with private industry for the vast bulk of their programs. Many of these contractors were established aviation firms, such as Vought, North American, Douglas, McDonnell, Boeing, and Martin. Some contractors, such as Thompson-Ramo-Wooldridge Corporation (TRW) and Aerojet, were created by entrepreneurs to support the newly created rocket and satellite industries. Others, such as Rockwell, Ball, Ford, and Chrysler, diversified into space from other industries. Below these major contractors have been a host of contractors supplying a variety of components and subsystems. Throughout time, many firms consolidated, so that by 2007 Boeing, Lockheed Martin, and Northrop Grumman controlled a majority of prime contracts. Many of the large companies created joint ventures with other U.S. or foreign firms to market specific systems and services, such as United Space Alliance (Boeing and Lockheed Martin) to operate the Space Shuttle, and Sea Launch (Boeing and Khrunichev in Russia) to market launch services. However, small new space firms such as Aero-Astro, Orbital Sciences, and Spectrum Astro arose to compete with the established firms in new technologies and for smaller systems.

The U.S. government has actively regulated and controlled private industry. In 1962 it created the Communications Satellite Corporation (COMSAT) to prevent American Telephone and Telegraph (AT&T) from extending its monopoly on telephone services into space. In the 1970s NASA's strategy to lower operational costs of the Space Shuttle meant it had to launch as many payloads on the Shuttle as possible. To do this the U.S. government planned to cease the manufacture of expendable launch vehicles, which would have created a U.S. government monopoly on space transportation. In 1984 the Ronald Reagan administration countered this idea by authorizing the Department of Transportation to license commercial satellite launches. This policy became effective only after the *Challenger* disaster in January 1986 rendered NASA's strategy irrelevant. The U.S. government enacted a 1984 law to commercialize Earth remote sensing data sales, but this law required that Congress approve all business agreements, making it impossible for Earth Observation Satellite Company, which marketed the data, to succeed. In 1992 the Department of Commerce gained authority to license commercial remote sensing satellites, but government policies set regulations to control the resolution of the satellites and to limit U.S. satellite picture-taking in wartime. Arms control regulations hampered U.S. communications satellite sales to foreign countries in the early twenty-first century.

Commercial (nongovernment) space institutions have played an important role in the communications sector. The U.S. government ensured that COMSAT was the manager of the International Telecommunications Satellite Organization (Intelsat) communications consortium, and maneuvered to dominate Intelsat through the 1970s. In the 1970s the United States allowed private companies to create domestic and then regional communications systems. Companies such as Panamsat and Hughes created profitable businesses operating satellite communications systems by the 1980s. In the 1990s Iridium and two other companies went bankrupt building massive satellite constellations for mobile communications. The prospective boom in satellite sales led to the formation of new launch companies, which also went bankrupt as their market disappeared. Also in the 1990s commercial remote sensing companies launched imaging satellites, but by the mid-2000s remained dependent on government sales. In 2004 Scaled Composites placed the first private citizen into space on a privately funded vehicle, *SpaceShipOne*. This raised the prospects for space tourism.

By the early twenty-first century the United States was the leading space power, with strong military, civilian, and commercial programs. Other nations such as France and Russia competed with the United States in certain space domains, but only the U.S. program maintained the size and diversity of its programs throughout the space age.

Stephen B. Johnson

See also: American Private Communications Satellites, Effect of Sputnik, Human Flight and Microgravity Science, Military Applications, National Aeronautics and Space Administration, Outer Space Treaty, Space Shuttle, United States Department of Commerce, United States Department of Defense, United States Department of Transportation, *V-2* Experiments

Bibliography

Joan Lisa Bromberg, *NASA and the Space Industry* (1999).
William E. Burrows, *This New Ocean: The Story of the First Space Age* (1998).
Walter A. McDougall, ... *the Heavens and the Earth: A Political History of the Space Age* (1985).
NASA History. http://history.nasa.gov/.
David N. Spires, *Beyond Horizons: A Half Century of Air Force Space Leadership* (1997).

SPACE ADVOCACY GROUPS

Space advocacy groups have had an influence on humankind's entry into space since the 1920s. With the birth of modern rocketry in the early twentieth century, the idea of space travel has united people of all ages, heritages, and backgrounds. Space advocacy groups have used this bond to promote space science and exploration and to develop new technologies. From their beginning, the work of space advocacy groups has shaped space policy, educated the public about various space issues, and inspired students and adults to learn more about science.

The first groups to promote human spaceflight were amateur societies dedicated to the development of rockets. This included the German VfR (Verein für Raumschiffahrt or Society for Spaceship Travel) founded in 1927; the American Interplanetary Society, founded in 1930 and later renamed the American Rocket Society; and the British Interplanetary Society (BIS), founded in 1933.

Soviet rocket societies also emerged during the 1930s. Sergei Korolev, who later became the Chief Designer of the Soviet Union's human spaceflight program, was a founding member of MosGIRD, a group in Moscow that promoted the study of rockets and jet propulsion.

By the mid-1950s advocacy groups were focused on gaining public support for their initiatives. Wernher von Braun and Willy Ley, who came to the United States at the end of World War II with other German scientists as part of Operation Paperclip, wrote articles in *Collier's* from 1952 to 1954 that inspired millions of readers about the future of spaceflight. As BIS chair, author Arthur C. Clarke bolstered interest in the group and its mission with publication of *The Exploration of Space* (1951) and *The Conquest of the Moon* (1953).

The International Astronautical Federation emerged in 1951 as a federation of societies, such as the BIS and ARS, which advocated the exploration, development, and utilization of space. Their ranks were soon joined by the American Astronautical

Society (AAS), founded in 1954, which promoted itself as the U.S. counterpart to the BIS.

During the 1960s as industries continued to evolve with the burgeoning U.S. space program, so did more advocacy groups that supported the development of space technologies, including the American Institute of Aeronautics and Astronautics—which arose from the merger of the ARS and the Institute of Aeronautical Sciences (IAS) in 1963—and the National Space Club. In 1969 the Aerospace Industries Association began advocating stable funding for space projects after more than 50 years, primarily promoting only aviation interests.

Several advocacy groups appeared after the end of the Apollo program with efforts to strengthen public support for NASA's exploration initiatives. Von Braun agreed in 1974 to lead the National Space Association, which changed its name to the National Space Institute (NSI) in April 1975 in hopes of gaining greater membership.

The L5 Society, formed in 1975, and the Planetary Society, formed in 1980, became outspoken supporters of sending humans into space and exploring the cosmos. By the first Space Shuttle flight in April 1981, several hundred space advocacy groups or clubs supporting robotic and human activities in space were active globally.

A majority of the L5 Society chapters and NSI merged in 1987 to form the National Space Society (NSS), which advocated stronger political action through groups such as Spacecause, a political lobbying organization, and Spacepac, a political action committee to research space policy.

New advocacy groups formed during the 1990s. The most vocal was the Mars Society, formed in 1998, which lobbied Congress to accelerate NASA's exploration plans, especially those for sending human missions to Mars and for testing a variety of technologies in the Arctic to simulate the Martian environment.

Tim Chamberlin

See also: Literary Foundation of the Space Age, Media and Popular Culture, Space Resources

Bibliography
Howard E. McCurdy, *Space and the American Imagination* (1997).
Michael A. G. Michaud, *Reaching for the High Frontier—The American Pro-Space Movement, 1972–84* (1986).

American Astronautical Society

The American Astronautical Society (AAS) was founded on 22 January 1954 in New York City. The AAS originated as the Staten Island Interplanetary Society, which was created in 1952 by Hans Behm, assistant science curator at the Staten Island Museum, to study the possibilities of interplanetary space travel. Although originally founded as an advocacy group, the AAS evolved into a multidisciplinary professional organization whose members shared a common interest in the exploration and development of space. The AAS grew to embrace nontechnical specialists in space history, education, space law, and public policy, for example. This was unusual because spaceflight organizations have tended to be either the exclusive domain of engineers or advocacy groups composed of people not inherently associated with spaceflight.

In early promotional materials, organizers identified the AAS as the American counterpart to the British Interplanetary Society (BIS). This came as somewhat of a surprise to the BIS, which viewed the American Rocket Society (ARS) as its U.S. counterpart, despite the fact that the ARS largely consisted of engineers focused on rocket development and had explicitly avoided association with all things "interplanetary." The AAS joined the International Astronautical Federation (IAF) in 1954 shortly after receiving official recognition.

The *Journal of Astronautics*, later renamed the *Journal of the Astronautical Sciences*, was established as the official organ of the AAS. Although early issues provided broad coverage of astronautical topics, later issues increasingly covered more specialized topics. To meet this need, the AAS began a bimonthly membership newsletter, which evolved into *Space Times*, a bimonthly magazine. Beginning in 1961 the AAS has conducted the Goddard Memorial Symposium in conjunction with the annual Robert H. Goddard Memorial Dinner of the National Space Club in Washington, DC. Together with an annual national meeting, these meetings have provided an opportunity for open discussions of major programmatic and technical issues in the astronautics community.

The AAS has traditionally maintained an active spaceflight history committee. In 1977 the AAS began publication of the AAS History Series, which included selected papers from the proceedings of every space history symposium of the International Academy of Astronautics since their inaugural symposium in 1967. Since 1982 the AAS History Committee has annually presented the Emme Award for Astronautical Literature, named in honor of Eugene Emme, the first NASA historian to recognize the outstanding book serving public understanding about the impact of astronautics on society and its potential for the future.

In 1984 the Chinese Society of Astronautics (CSA) hosted an AAS delegation on a trip to observe the Chinese space program and related activities. This event led to an agreement between AAS and CSA, which was later expanded to include the Japanese Rocket Society. The AAS has since established strong ties with Pacific-rim space organizations under the auspices of the International Space Conference of Pacific-rim Societies.

In 2004 the AAS signed a memorandum of understanding with Students for the Exploration and Development of Space (SEDS), an undergraduate organization that lacked a professional affiliation. This arrangement also provided AAS with an opportunity for infusion of new members.

Michael L. Ciancone

See also: United States

Bibliography
American Astronautical Society. http://astronautical.org/.
Eugene Emme, ed., *Twenty-Five Years of the American Astronautical Society, 1954–1979* (1980).
Frank H. Winter, *Prelude to the Space Age: The Rocket Societies 1924–1940* (1983).

American Institute of Aeronautics and Astronautics

The American Institute of Aeronautics and Astronautics (AIAA) was formed in 1963 with the merger of the American Rocket Society (ARS) and the Institute of the Aeronautical Sciences (IAS). The AIAA has served as a professional, not-for-profit association that supports the aerospace industry. As of 2007 AIAA membership was about 30,000 professionals, mostly aerospace engineers, and 7,000 student members in more than 60 professional sections and 162 student branches throughout the world.

The ARS, originally named the American Interplanetary Society, was formed in 1930 by a group of science fiction enthusiasts eager to expand on the possibilities of space travel. Many early society members were involved in the first public rocket launch on Staten Island, New York, in 1933. Later the ARS became more of an engineering society, and by the end of World War II it almost exclusively supported the engineering community.

Founders of the IAS, later called the Institute of the Aerospace Sciences, modeled their society after the Royal Aeronautical Society. The IAS was "designed to provide a coordination of specialists in these many fields of science and engineering devoting themselves to aeronautical developments." One early project of the IAS, done through the Works Progress Administration, was to collect a copy of every article and/or book that had been written on aviation up until that time. This collection was placed with the Library of Congress and the National Air and Space Museum in Washington, DC. The 50-volume *Bibliography of Aeronautics* was an outgrowth of this project.

The AIAA, with headquarters in Reston, Virginia, and a small office in southern California, has supported the aerospace industry through conferences, publications, awards, public policy initiatives, standards, and educational activities. Its publications program has included seven peer-reviewed technical journals, from the *AIAA Journal* to journals covering propulsion and power, computers, spacecraft and rockets, and aircraft. AIAA has also produced an Educational Book Series that has been used throughout academia for classroom texts, and the Progress in Astronautics and Aeronautics book series bringing current topics to engineers and scientists.

Anthony Springer

See also: Rocket Pioneers and Spaceflight Visionaries, Science Fiction, United States

Bibliography

AIAA. http://www.aiaa.org/.
Tom Crouch, *Rocketeers and Gentlemen Engineers: The History of the American Institute of Aeronautics and Astronautics ... and What Came Before* (2006).

American Interplanetary Society. *See* American Rocket Society.

American Rocket Society

The American Rocket Society (ARS) was the best known of the American amateur rocket groups of the 1930s. The ARS was founded on 4 April 1930 in New York City as the American Interplanetary Society (AIS). Its founders, consisting of 11 men and

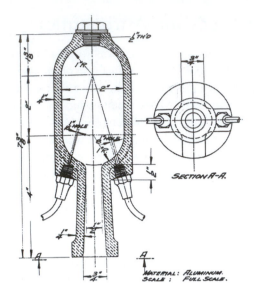

Early rocket motor designed and developed by the American Rocket Society in 1932. (Courtesy NASA/Marshall Space Flight Center)

one woman, were largely idealistic believers in the cause of human spaceflight and were mainly writers or staff of the science fiction magazine *Science Wonder Stories* that regularly featured "interplanetary stories." Like its German counterpart, the VfR (Verein für Raumschiffahrt or Society for Spaceship Travel), it initially sought to promote support for the idea of human spaceflight through lectures, articles, and its mimeograph *Bulletin of the American Interplanetary Society* (later called *Astronautics*). Beyond this, however, it had no specific plans for developing the technology. AIS members naively thought scientists and engineers would flock to them and provide this support. Its first president was David Lasser, editor of *Science Wonder Stories* and later the author of the first English-language book on the subject, *The Conquest of Space* (1931). G. Edward Pendray was the vice president.

Some of AIS's first public meetings were held in the American Museum of Natural History in New York City and attracted attention in the press, especially when Robert Esnault-Pelterie, the French aviation and astronautical pioneer, was the invited lecturer. The AIS asked American rocket pioneer Robert H. Goddard to join, but he declined because he did not wish to reveal details of his experiments, and he remained aloof from the society.

Little progress was made until 1931, when Pendray visited the VfR's Raketenflugplatz (Rocketport). Pendray took notes and, following his return to the United States, spoke before the AIS on the need to conduct experiments. Soon the AIS formed an Experimental Committee. With limited funds at its disposal, the AIS nonetheless built at least five crude liquid-propellant rockets for testflights between 1932 and 1934 and one test stand, but only two rockets flew. On 6 April 1934, in order to sound more "professional," the AIS changed its name to the American Rocket Society. Its rockets thus became known as ARS No. 1 and so on. ARS No. 2 went up 250 ft on 14 May 1933 from Great Kills, Staten Island, New York; while ARS No. 4 rose to 1,338 ft on 9 September 1934 from Marine Park, Staten Island.

However, ARS leaders felt far more could be learned from static tests, and Stand No. 2 was built in 1938. This was proven true, as later that year ARS member James H. Wyld began a series of tests that conclusively showed that his regenerative cooling design effectively cooled rocket motors. In 1941 Wyld, along with fellow ARS members Lovell Lawrence Jr., John Shesta, and H. Franklin Pierce, formed Reaction Motors Inc. (RMI) that used the Wyld regenerative cooling design. They eventually developed the 6000C4 rocket motor (also called XLR11) that powered the Bell X-1, which broke the sound barrier in 1947.

The ARS ceased experimenting with the start of World War II. Following the war, and due to the enormous technological growth of rocketry because of the war, the ARS attracted many engineers and truly became a professional rocket engineering society.

When it merged with the American Institute of the Aerospace Sciences in 1963 to become the American Institute of Aeronautics and Astronautics, the new group had more than 20,000 members and was the largest such group in the United States.

Frank H. Winter

See also: Rocket Pioneers and Spaceflight Visionaries, Science Fiction

Bibliography

G. Edward Pendray, *The Coming Age of Rocket Power* (1945).

———, "Early Rocket Developments of the American Society," in *First Steps toward Space—Smithsonian Annals of Flight*, ed. Frederick C. Durant III and George S. James, 10 (1974).

Frank H. Winter, *Prelude to the Space Age: The Rocket Societies, 1924–1940* (1983).

British Interplanetary Society

The British Interplanetary Society (BIS) is the primary astronautical society in the United Kingdom, founded by a small contingent of visionaries in Liverpool in October 1933. Phillip Ellaby Cleator, initiator of the effort, was elected the first president of the society and edited the *Journal of the British Interplanetary Society* (*JBIS*) from 1934 to 1936. In 1937 the seat of the BIS was moved from Liverpool to London; A. M. Low became president while Cleator continued as a vice president of the society and editor of the *JBIS*. Arthur C. Clarke joined the society in the 1930s and served as its president after World War II.

The BIS did not enjoy the kind of rapid growth in membership experienced by other such societies in the 1930s. Lack of financial resources and a hostile national legal environment prevented the BIS from launching an active rocket experimentation program. The law in the United Kingdom forbade experimentation with explosives or explosive devices by anyone outside the government. The society nonetheless sponsored research in "concepts of astronautics and spaceflight" and organized many meetings and lectures to stimulate public interest and to provide educational information.

During 1937–39, fund limitations reduced the publication frequency of the *JBIS*, but a monthly *Bulletin* kept members informed about society programs. By 1939 preoccupation with the threatening outbreak of war in Europe led to cessation of society

ARTHUR C. CLARKE
(1917–2008)

(Courtesy AP/Wide World Photos)

Sir Arthur Charles Clarke, British author and inventor, is perhaps best known for his science fiction novel *2001: A Space Odyssey* (1968), written concurrently with the Oscar-nominated movie of the same name, an inspiration for many space enthusiasts. During World War II, he served in the Royal Air Force. After the war, he obtained a degree in mathematics and physics at King's College, London. Clarke first proposed the idea of geostationary communications satellites in a 1945 *Wireless World* article. Clarke also wrote a number of nonfiction books on spaceflight, including the ubiquitous *The Exploration of Space* (1951). He moved to Sri Lanka in 1956.

Pablo de León

activities, but a core group of enthusiasts kept the spirit of the society alive during the war years, despite the absence of active programs or publications.

Members of the BIS in the pre-war United Kingdom received copies of *Astronaut*, the publication of the affiliated Manchester Interplanetary Society, which disbanded in 1938. In addition the Astronautical Development Society (ADS), founded in 1938 by Kenneth Gatland, published a magazine titled *Spacecraft*, and the Manchester Astronautical Association (MAA), founded in 1937 by Eric Burgess, published a magazine titled *Spacewards*. The ADS and MAA consolidated in 1944 to form the Combined British Astronautical Societies (CBAS), with Burgess as chair and Gatland as secretary, and continued the publication of *Spacewards* through World War II but, as in many other national societies, program activities were significantly reduced during the war years.

Like other early astronautical societies, the BIS served the critical function of stimulating public awareness and expanding public understanding of the potential of astronautics. After World War II the reinstituted BIS absorbed the CBAS and moderately grew in size. The society restored the publication of its journals: *Spaceflight*, a monthly news journal for society members, was published beginning in 1956 with feature articles on current activities in space; and the *JBIS*, a more scholarly publication, with scientific and research papers describing aspects of the sciences involved in astronautics. The society has hosted monthly lectures by prominent personalities in astronautics at its London headquarters and has maintained an extensive astronautics research library. Occasionally the society has cited and honored personalities making significant contributions to the development of astronautics.

The BIS was a cofounding society, instrumental in forming the International Astronautical Federation (IAF) during 1950–52, and has been a sustaining member of the IAF since its formal establishment in 1952.

Stephen Doyle

See also: United Kingdom

Bibliography
British Interplanetary Society. www.bis-spaceflight.com.
Frank H. Winter, *Prelude to the Space Age: The Rocket Societies, 1924–1940* (1983).

International Academy of Astronautics

The International Academy of Astronautics (IAA) was established by the International Astronautical Federation (IAF) in 1960 under the direction of Theodore von Kármán as a nongovernmental scientific organization devoted to recognizing outstanding contributions to astronautics and to fostering development of astronautics for peaceful purposes. The IAA comprises individual members, rather than an organization of societies (as was the IAF), who have distinguished themselves in a branch of science or technology related to astronautics. The IAA is composed of four sections—Basic Sciences, Engineering Sciences, Life Sciences, and Social Sciences. As of 2007 the IAA had more than 1,200 members from 75 countries.

The IAF formally established an IAA founding committee during the 1959 International Astronautical Congress (IAC) in London, England (the IAC is the annual conference of the IAF). The IAA founding committee proposed a revision to the IAF constitution that would establish the IAA—the IAF approved the revision during the 1960 IAC in Stockholm, Sweden. The IAA officers held their first meeting in March 1961 and conducted their first independent symposium that year. The first regular meeting of the IAA was held in conjunction with the 1961 IAC in Washington, DC. The IAA was officially recognized by the United Nations in 1969.

The IAA has cooperated with the IAF and the Committee on Space Research (COSPAR) of the International Council for Science to organize and conduct sessions, symposia, and meetings on selected topics. The IAA also assumed responsibility for managing and editing the journal *Acta Astronautica* in 1959. The American Astronautical Society began publishing selected papers from the proceedings of IAA history symposia on space history in the inaugural symposium of 1967. The IAA has typically conducted a regular meeting of the full membership biennially, but full membership meetings have been convened annually in conjunction with the IAC to take advantage of the assembly of its members in a common venue. Academicians have typically been invited to meet during IAA regional meetings. When practical, the IAA has cosponsored an academy day, with a host national academy of science, in conjunction with each IAC. A major initiative of the IAA has been the development of a series of studies and position papers dealing with many aspects of international cooperation in space.

Stephen Doyle

See also: Space Science Policy

Bibliography
International Academy of Astronautics. http://iaaweb.org/.

International Astronautical Federation

International Astronautical Federation (IAF) was founded on 4 September 1951 by a small group of astronautics enthusiasts in attendance at the 2nd International Astronautical Congress (IAC) in London (the first IAC was an organizational meeting conducted in Paris in 1950; the 2nd IAC was the first Congress that included presentation of papers). These enthusiasts represented 10 countries (Argentina, Austria, France, Germany, Italy, Spain, Sweden, Switzerland, the United Kingdom and the United States), which are recognized as founder-member nations of the IAF. The IAF constitution, formally adopted in 1952, described the IAF as an international, nongovernmental, nonprofit organization comprising national space agencies, learned societies, and professional organizations. Membership later expanded to include schools and universities, aerospace manufacturers, space service providers, users of space systems, and specialized law and consulting firms. Thus, the IAF is a federation of organizations, rather than of individuals.

Primarily through the efforts of Alexandre Ananoff in France, A. V. Cleaver of the British Interplanetary Society in England, and Hans Gartmann of the Gesellschaft für Weltraumforschung in Germany, the IAF provided a nonpolitical international forum for space program information exchanges. By 2007 the IAF had grown from the original 12 member societies from countries in North America, South America, and Europe to 155 member organizations from 45 countries around the world.

From its headquarters in Paris, France, the IAF has organized annual international astronautical congresses since its formation, symposia, workshops, training and educational programs, and works in collaboration with other international organizations to disseminate public information about programs of astronautics.

Since its creation the IAF has promoted advancement of knowledge about and public awareness of developments involving astronautics and outer space by providing an international forum where space program leaders from the United States, the Soviet Union (until 1992), the Member States of the European Space Agency, and other countries could periodically assemble, report on, and discuss their space programs. Leaders of national and regional programs used this nongovernmental forum to seek initial reactions to ideas for possible cooperative efforts in spaceflight. Despite the Cold War, between 1950 and 1990, U.S. astronauts and Soviet cosmonauts regularly appeared at and presented reports to the IAF membership on the spaceflight accomplishments of their countries. After 1960 senior officials of NASA joined with industrial leaders to describe and discuss their programs with leaders of the Soviet space programs. In addition leaders of U.S., European, Russian (since 1992), and Japanese space programs regularly participate and exchange information on the progress of their respective space programs.

ALEXANDRE ANANOFF
(1910–1992)

Alexandre Ananoff was a Soviet-born, French popularizer of astronautics who significantly advanced understanding of spaceflight activity and international cooperation and sought to raise the dream of human spaceflight in the public consciousness. In 1937 Ananoff and others organized an international rocketry exhibit during the International Exposition in Paris. In 1938 under the sponsorship of Gabrielle Camille Flammarion, widow of Nicolas Camille Flammarion (founder of the Société Astronomique de France), Ananoff established the Section Astronautique of that Society. He wrote *L'Astronautique* (1950) in connection with his role as principal organizer of the First International Astronautical Congress, held at the Sorbonne in Paris, from which arose the International Astronautical Federation, and corresponded regularly with astronautics pioneers around the world.

Stephen Doyle

As an association of organizations focused on engineering, design, and fabrication activities related to space vehicles, the annual IAF meetings complemented the scientific forum of the Committee on Space Research (COSPAR) of the International Council of Scientific Unions (ICSU). Although engineers and scientists write papers and technical articles to describe their work and accomplishments, forums such as IAF and COSPAR have provided the opportunity to discuss, debate the significance of, and learn more details about engineering and scientific advances taking place internationally.

Since 1952 the IAF vice presidents have been selected to provide representation from countries of broad geographical areas and developed to various levels of aerospace achievement. From the 1960s to the 1990s the United States, the Soviet Union (later Russia), Europe, Japan, Canada, and Latin America have provided officers for the IAF. This nongovernmental forum has provided a parallel but separate and informal channel for international discussion of space programs.

Since about 1970 as more developing countries built indigenous space programs for weather monitoring, satellite communications, and remote sensing of national resources, the IAF, in cooperation with the United Nations (UN) Office of Outer Space Affairs (OOSA), has cohosted a training/educational workshop, held in a different country annually, in conjunction with the annual International Astronautical Congress. This has been a successful informal training program, funded by OOSA and program participants, to assist the developing countries in making more productive use of space technology.

The IAF has also collaborated with the International Academy of Astronautics (IAA), which since the early 1960s has organized selected topic symposiums during each annual congress, and with the International Institute of Space Law (IISL), which has organized and convened a colloquium on the law of outer space during each annual congress. Studies developed from these activities have described the rationale and values of programs, such as continued lunar exploration, a mission to Mars, the development of new and better launch systems, and other topics of astronautical interest.

Since 1952 the IAF has created more than 10 technical committees dealing with such matters as astrodynamics, satellite communications, space propulsion, Earth observation, and space exploration. These committees have prepared assessments and recommended developments in a variety of forms, including white papers and technical publications for trade journals.

The IAF has published selected, refereed papers of its congresses in *Acta Astronautica*. The IISL has published a separate, hardbound *Proceedings of the Annual Colloquium on the Law of Outer Space* annually since 1960.

The IAF has worked with COSPAR of the International Council for Science (ICS; formerly the ICSU) to organize two World Space Congresses. The congresses were simultaneous and overlapped meetings of the IAF and the ICS at a single location. The first congress was held in Washington, DC, in 1992, and the second congress was held in Houston, Texas, in 2002. Since about 1990 in cooperation with OOSA, both the IAF and the IISL have provided annual surveys to the UN of *Highlights in Space*, compiled and published by the UN and made available to UN member states. Representatives of the IAF and the IISL have participated as observers during the annual meetings of the UN Committee on the Peaceful Uses of Outer Space (UN COPUOS) and meetings of its Scientific and Technical Subcommittee and its Legal Subcommittee to provide information and updates on organizational programs and activities.

Stephen Doyle

See also: Space Law, United Nations Committee on the Peaceful Uses of Outer Space

Bibliography
International Astronautical Federation. www.iafastro.com.

German Rocket Society. *See* Verein für Raumschiffahrt.

National Space Society

The National Space Society was formed in 1987 from the merger of the National Space Institute and the L-5 Society, two separate private organizations each founded in the early 1970s. These organizations had different beginnings but had similar goals: to increase public awareness and support for human spaceflight.

After public support for human spaceflight reached a crescendo in 1969 with the *Apollo 11* landing of the first humans on the Moon, interest in human spaceflight began to wane. A growing movement of engineers, scientists, and other advocates of human spaceflight started to look away from the government as the place for generating public interest in spaceflight.

In 1957 as a response to the Soviet Union's launch of *Sputnik*—the first artificial satellite—Erik Bergaust founded the National Rocket Club, which became the National Space Club in 1964. Based in Washington, DC, its membership consisted mostly of representatives from aerospace companies and contractors for NASA and the U.S. Department of Defense. By the mid-1980s the group had grown to about 800 members.

When the Apollo program concluded in 1972 the pro-space community in the United States was already experiencing low morale. Gene Bradley, the National Space Club's past president, recommended expanding the club's activities to bolster morale in the space industry. The National Space Association was created in June 1974 by past officers of National Space Club who were employed in the aerospace industry, with Wernher von Braun as its president. Von Braun immediately began traveling throughout the United States to raise funds from the aerospace industry. In response to resistance from prospective donors to contribute to another "association," the organization became the National Space Institute (NSI). The early NSI board included comedian Bob Hope, writers Isaac Asimov and Arthur C. Clarke, explorer Jacques Cousteau, scientist James Van Allen, and U.S. Senator Barry Goldwater.

"The Colonization of Space," written by Princeton University physics professor Gerard O'Neill, inspired the founding of the L-5 Society in August 1975 by Carolyn and Keith Henson. The society's name derived from one of the libration points in the Moon's orbit. Proposed by Arthur C. Clarke in his novel *A Fall of Moondust* (1961) as a place to park a space station, the lunar L-4 and L-5 libration points are stable areas within the Moon's orbit where an object can remain (relative to Earth and Moon) without the use of fuel to maintain its position. In 1980 the L-5 Society urged the United States to not sign the United Nations Moon Treaty for fear of losing the opportunity to develop nonterrestrial resources. In 1982 the L-5 Society organized the first International Space Development Conference (ISDC) in Los Angeles.

The NSI and the L-5 Society merged in 1987 to form the National Space Society (NSS) with Ben Bova (former NSI president) as president. In 1989 the NSS published the first issue of its monthly magazine, *Ad Astra*. The NSS organized grassroots campaigns through the 1990s in an effort to influence public space policy. They also continued to organize the annual ISDC at various locations in North America. On several occasions the society successfully contributed to the political opposition to Congress's proposals to cancel funding for the International Space Station program. In 2004 NSS members urged Congress to support NASA's new Moon/Mars plans as outlined by President George W. Bush in his Vision for Space Exploration. During that decade the NSS also rallied its members to support maintaining the orbiting *Hubble Space Telescope*.

As of 2007 the NSS had approximately 20,000 member-subscribers.

Eric T. Reynolds

See also: Moon Treaty; Space Resources; Von Braun, Wernher

Bibliography
David Brandt-Erichsen, "The L-5 Society," *Ad Astra* (1994).
Michael Michaud, *Reaching for the High Frontier* (1986).
National Space Society. www.nss.org.

Planetary Society

The Planetary Society was founded in 1980 by astronomers Carl Sagan of Cornell University, Bruce Murray of NASA's Jet Propulsion Laboratory (JPL), and Louis Friedman, former project manager of JPL's Advanced Mars Program. Its goals were to unite, educate, and inform space experts, science advocates, and the general public and to convince the U.S. Congress that space exploration had broad taxpayer support. It was most active in the areas of planetary exploration and the search for extraterrestrial intelligence (SETI). The society was outspoken during periods marked by swings in political support, NASA management and technological crises, and the breakup of the Soviet Union.

The Planetary Society carried out its mission by sponsoring conferences, feasibility studies, lecture series, television programs, scholarships, contests, and by lobbying and testifying before Congress for funding and support for specific programs and missions. The Pasadena, California, organization fostered space privatization by subsidizing or cosponsoring projects in asteroid tracking, SETI, the search for planets outside Earth's solar system, comet observation, and the use of balloons to explore Mars. In June 2005 the society launched the world's first solar sail, but *Cosmos 1* was destroyed minutes after liftoff when the Russian Volna launch vehicle failed.

Most of its members resided in the United States and one goal was to maintain and enhance America's technological edge in space exploration. However, the society also collaborated with foreign space agencies and enthusiasts to promote international cooperation for major projects, such as a space station, robotic probes of comets and asteroids, and the human exploration of Mars. It actively recruited from disparate constituencies, thus its eclectic board of directors and advisory council included scientists, policy makers, science-fiction writers, movie producers, poets, social activists, actors, engineers, artists, publishers, and astronauts. With a 2005 membership of approximately 100,000 members in more than 125 countries, it was the largest nongovernmental space interest group in the world.

The Planetary Society raised funds by selling merchandise, hosting paid events, and soliciting large donations from supportive individuals. It syndicated a program called "Planetary Radio" via North American radio stations and its own website. In addition to its bimonthly publication, *The Planetary Report*, the society published the *Mars Underground News* for a subset of members who advocated exploration of that planet in particular. Besides their educational function, these outlets allowed rapid mobilization of members to write Congress in support of programs or missions likely to be canceled or lose funding.

Maura Phillips Mackowski

See also: Mars Mission Studies, Search for Extraterrestrial Intelligence

Bibliography
Bruce Murray, *Journey into Space* (1989).
Planetary Society. www.planetary.org.

Russian Early Rocketry

Russian early rocketry dates back to the use of black gunpowder rockets by the Russian military in the late seventeenth century. In the mid-nineteenth century, Generals Alexander Zasyadko and Konstantin Konstantinov developed explosive barrage missiles, making them a standard issue of the mobile field artillery.

Attempts to perfect barrage missiles continued even after more accurate rifled artillery superseded rockets in the late nineteenth century. In 1921 the Soviet military organized a laboratory to implement ideas of Nikolay Tikhomirov, who proposed to propel rockets by smokeless gunpowder. That facility, named the Gas-Dynamics Laboratory (GDL) in 1928, became a forerunner of rocket research in the Soviet Union. GDL engineers Vladimir Artemiev, Boris Petropavlovsky, and Georgy Langemak developed a successful prototype of a solid-propellant barrage missile, while Valentin Glushko designed a variety of liquid-propellant rocket engines.

In 1931 the widespread popularity of spaceflight ideas in the Soviet Union led to the creation of public rocket clubs, known as Jet Propulsion Research Groups (GIRD), but most of them were short-lived. The GIRD in Moscow (MosGIRD), under the leadership of Sergei Korolev, became a professional research organization by obtaining military funding in 1932. In 1933 MosGIRD launched the first Soviet liquid-propellant rockets, 09 and GIRD-X, designed by Mikhail Tikhonravov, Friedreich Tsander, and Leonid Dushkin.

Members of GIRD shown feeding liquid oxygen into the first Soviet liquid-propellant rocket, 1933. From left to right are Sergei Korolev, Nikolai Yefremov, and Yuri Pobedonostsev. (Courtesy NASA)

In September 1933 GDL and MosGIRD were unified into the Jet Propulsion Research Institute (RNII), renamed Scientific Research Institute 3 (NII-3) in 1936, to work on military rockets. NII-3 created and tested a number of ballistic and winged liquid-propellant rockets with launch masses of up to 97 and 230 kg respectively, Korolev's rocket-propelled glider, RP-318, and dozens of various rocket motors with thrusts up to 300 kg (1934–41). However, its greatest achievement was the development of solid-propellant barrage missiles, M-8 and M-13, which were widely used in World War II and known as "Katyusha." In 1935–39 liquid-propellant rockets were also designed by the military research office, KB-7, headed by the former MosGIRD engineers Leonid Korneev and Alexander Polyarny. KB-7's R-06 was the highest-flown Soviet rocket of the time, achieving an altitude of about 4,000 m (1937).

The Great Terror purges of 1936–39 severely impeded Soviet rocket research. In the atmosphere of mistrust and suspicion induced by authorities, simple disagreements among engineers escalated into mutual accusations of espionage and subversion. As a result NII-3 Director Ivan Kleymenov and Chief Engineer Georgy Langemak were executed as alleged spies, while the leading specialists, Sergei Korolev and Valentin Glushko, were falsely accused of subversion and imprisoned in 1938. A year later, KB-7 was dissolved and its leadership arrested.

World War II limited further rocket research to perfecting the Katyusha missiles and attempts to create rocket-propelled and rocket-assisted aircraft. Alexander Isaev, Alexander Berezniyak from aircraft design bureau, OKB-293 (Experimental Design Bureau), and Leonid Dushkin from NII-3 jointly developed the first Soviet rocket airplane, BI-1 (1941–43). In 1943–44, NII-3 designed an advanced rocket-propelled interceptor, called 302, but the project was behind schedule. That resulted in another reshuffle of NII-3 (1944) and arrest of its director, Andrei Kostikov, whose slanderous allegations against Glushko and Korolev in 1938 had led to their imprisonments. Meanwhile Glushko and Korolev were designing rocket engines for the so-called "hybrid" aircraft (combined piston and rocket propulsion) in a special prison research group in Kazan. In 1941–45, that group created a series of engines with a thrust of up to 1,200 kg using self-igniting propellants. Acknowledging their achievements, authorities released Glushko and Korolev from prison in 1944.

Rocket developments in the Soviet Union paralleled research in various jet engines, such as turbojets (Arkhip Lyulka, 1937), ramjets (Igor Merkulov, 1939) and pulsejets (Vladimir Chelomey, 1942).

Although Soviet rocket research of the 1930s–40s was substantially smaller in scale and less productive than contemporary German rocket activities, it created the necessary scientific potential for the next step: establishment of the Soviet rocket industry, which began in 1946.

Peter A. Gorin

See also: Korolev, Sergei; Russian Space Forces

Bibliography
Peter Alway, *Retrorockets: Experimental Rockets 1926–1941* (1996).
Boris Chertok, *Rockets and People*, vol. 1 (2005).

Asif A. Siddiqi, *Challenge to Apollo: The Soviet Union and the Space Race, 1945–1974* (2000).

Space Generation Advisory Council

The Space Generation Advisory Council (SGAC) was formed in 1999 to support the United Nations (UN) Committee on the Peaceful Uses of Outer Space (COPUOS). That year at the UNISPACE III conference, alumni of the International Space University were invited to organize a forum to express the perspectives of young people. This forum recommended that COPUOS create a council to raise awareness of and exchange ideas about space activities among the world's youth. COPUOS accepted this recommendation and helped young people from around the world to establish SGAC.

SGAC's mission statement is "To enable and promote a continuous and interactive dialogue on space-related issues between the United Nations and young people of the world, and to further enable creativity, enthusiasm and vigor of youth that shall be used to advance humankind through the peaceful uses of outer space."

SGAC has worked to bring the ideas of young people to the UN. The council advises both COPUOS and the UN's Office for Outer Space Affairs regarding ideas and activities of international youth space organizations. SGAC's most significant output has been representing the view of the world's youth on space policy to the UN and other international organizations. SGAC drafted policy documents for and attended the Space Policy Summit in 2002 and provided input to the European Union space policy white paper in 2003. It has also supported international cooperation within a variety of youth space activities and events through annual Space Generation congresses and other innovative activities. These include Yuri's Night, a global space party celebrating Yuri Gagarin's flight; Under African Skies, an educational program to bring science and technology to schoolchildren in rural Africa; and the Space Association of Turkic States, which was created to increase the involvement and cooperation of youth from Turkey, Uzbekistan, Kazakhstan, and Azerbaijan in space activities.

Srimal Wangu Choi

See also: United Nations Committee on the Peaceful Uses of Outer Space

Bibliography
Space Generation Advisory Council. www.spacegeneration.org.

Verein für Raumschiffahrt

The Verein für Raumschiffahrt (VfR), or Society for Spaceship Travel, was one of the most prominent of the early amateur rocket societies of the 1920s–30s. During this period the first rocket societies formed because of interest in space rockets brought about largely by the appearance of Hermann Oberth's book, *Die Rakete zu den Planetenräumen* (*The Rocket into Planetary Space*, 1923).The VfR was founded on 5

July 1927 in Breslau, Germany (later Wroclaw, Poland), by nine men and one woman. Engineer Johannes Winkler was named president.

Typical of such groups, VfR membership consisted of young idealists who vaguely sought to promote the spaceflight idea and perhaps design and construct a spaceship. Initially through membership fees they held lectures and produced the journal *Die Rakete* (*The Rocket*) with Winkler as editor. Through the journal the society grew to about 1,000 members worldwide with members located as far away as Africa and South America. *Die Rakete* featured articles on theoretical aspects of spaceflight, biographical sketches about leading astronautical theorists, and VfR donor lists. Winkler left the society about 1929 to conduct independent rocket experiments and Oberth assumed the VfR presidency. When Oberth resumed teaching, retired German Army Major Hans-Wolf von Dickhuth-Harrach became president.

With the cessation of publication of *Die Rakete* in 1929 and the onset of the Great Depression, VfR membership declined. The remaining members focused on liquid-propellant experiments that began in 1930. The first experiments were conducted on a farm at Bernstadt, Saxony, Germany.

On 27 September 1930 VfR members Rudolf Nebel and Klaus Riedel established the Raketenflugplatz (rocket airfield) at an abandoned German Army garrison in a Berlin suburb. The Raketenflugplatz became the society headquarters. Crude static rocket motor tests were conducted there with Mirak (Minimum Rockets) and Repulsor rockets, usually powered by liquid oxygen and alcohol.

The highest flight went up about one mile (1.6 km) in August 1931. Experimentation continued until 1933 although Nebel, who was more an opportunist than rocket pioneer, created a schism within the VfR and led the last of the experiments away from the Raketenflugplatz. This schism was between the VfR members who chose to follow Nebel, who could offer them continued experimentation (despite his unscrupulous behavior), and those led by Dickhuth-Harrach, who had contempt for Nebel but had no further plans or resources to continue experimentations. Because of this schism the VfR dissolved in early 1934.

Meanwhile the German Army, which had been conducting its own secret rocket program since 1929, hired VfR member Wernher von Braun in October 1932 to help with its experiments. He was soon joined by other former VfR members, such as Klaus Riedel, Helmut Zoike, and Hans Hüter. Von Braun became the program's technical director and eventually his team produced the A-4, or V-2, that appeared in World War II as the world's first large-scale liquid-propellant rocket and forerunner of modern large-scale liquid-propellant rockets.

The VfR influenced the direction of amateur rocketry in the United States. In April 1931 G. Edward Pendray of the American Interplanetary Society (AIS), later called the American Rocket Society, visited the Raketenflugplatz where he witnessed a liquid-propellant rocket launch. On his return, Pendray reported his findings to the AIS and encouraged the group to initiate its own rocket experiments.

Frank H. Winter

See also: V-2; Von Braun, Wernher

Bibliography
Heinz Gartmann, *The Men behind the Space Rockets* (1956).
Willy Ley, *Rockets, Missiles, and Space Travel* (1951).
Frank Winter, *Prelude to the Space Age: The Rocket Societies, 1924–1940* (1983).

SPACE LAW

Space law exists on two levels: national and international. It is the collection of laws, rules, and regulations dealing with the activities of nations, organizations, and individuals related to activities in outer space. Commentary on aspects of space law began early in the twentieth century; however, the body of formal law arose only after spaceflight began in the second half of the century.

The United Nations (UN) developed international space law, working primarily through its Committee on the Peaceful Uses of Outer Space and with nations in cooperation with one another in regional or global organizations. The UN-generated space law is found in treaties and resolutions of the General Assembly.

In 2007 there were five major UN treaties related to space and five UN Resolutions of Principles. These treaties and resolutions relate to the status of outer space and people and objects in space, arms control, freedom of exploration, liability for damages caused by space objects, the safety and rescue of astronauts and spacecraft, the notification and registration of spaceflight activities, prevention of harmful interference, exploration and exploitation of space, and the settlement of disputes arising from spaceflight activity.

Given the dominance of the United States and Soviet Union in space activities in the 1960s, the early UN resolutions and treaties reflected the major policies and interests of the two superpowers. The Outer Space Treaty, for example, reflected the fact that both powers had realized that space-based nuclear weapons were ineffective deterrents compared to ground-based alternatives and that orbital reconnaissance was important to both nations. With the space race to the Moon underway, both

Major United Nations Treaties Related to Space

Date	Title of Treaty
1967	Treaty on Principles Governing the Activities of States in the Exploration and Use of Outer Space, Including the Moon and Other Celestial Bodies (Outer Space Treaty)
1968	Agreement of Rescue of Astronauts, the Return of Astronauts, and the Return of Objects Launched into Outer Space
1972	Convention on International Liability for Damage Caused by Space Objects
1976	Convention on Registration of Objects Launched into Outer Space (Registration Convention)
1984	Agreement Governing the Activities of State on the Moon and Other Celestial Bodies

United Nations Resolutions of Principles Related to Space

Date	Title of Resolution of Principles
1962	Declaration of Legal Principles Governing the Activities of States in the Exploration and Use of Outer Space (GA RES 1962 XVIII)
1982	Principles Governing the Use by States of Artificial Earth Satellites for International Direct Television Broadcasting (GA RES 37/92)
1986	Principles Relating to Remote Sensing of the Earth from Outer Space (GA RES 41/65)
1992	Principles Relevant to the Use of Nuclear Power Sources in Outer Space (GA RES 47/68)
1996	Declaration on International Cooperation in the Exploration and Use of Outer Space for the Benefit and in the Interest of All States (GA RES 51/122)

nations had a strong interest in promoting the peaceful purpose of their human spaceflight programs and in protecting their spacefarers, leading to the agreement to rescue astronauts. Later resolutions and treaties focused increasingly on the economic uses of space for communications, broadcasting, and distribution of potential wealth from space.

Beyond the scope of the UN, treaties and resolutions are international agreements to establish organizations that pursue scientific, technological, and commercial activities in outer space. Organizations of this kind began to appear in the 1960s and continued emerging through the late twentieth century. Some of the major space organizations created by international agreements are listed.

Each of these organizations was established by an international charter or a convention and often a set of supplemental operating provisions (frequently in an operational agreement) and the agreements became part of the body of international space law.

Major Space Organizations Created by International Agreements

Dates	Organization
1962–1975	European Space Research Organisation (ESRO) (replaced by ESA)
1964–1975	European Space Vehicle Launcher Development Organisation (ELDO) (replaced by ESA)
1964–1973	Interim International Telecommunications Satellite Consortium (Intelsat) (Name changed in 1973, see ITSO)
1971	International Organization for Satellite Communications (Intersputnik)
1973	International Telecommunications Satellite Organization (ITSO)
1975	European Space Agency (ESA)
1976	International Maritime Satellite Organization (Inmarsat)
1976	Arab Communications Satellite Organization (Arabsat)
1983	European Telecommunications Satellite Organization (Eutelsat; became a private company in 2001)
1986	EUMETSAT

Major national space laws have existed in more than 15 countries (there have been at least six separate space laws adopted by the United States) and are frequently amended. National laws and regulations have dealt with launch operations, remote sensing of Earth, communications, navigation, satellite direct broadcasting, intellectual properties, exploration, treaty compliance verification, ice and coastal monitoring, and military early warning.

The International Astronautical Federation (IAF) created the International Institute of Space Law (IISL) in 1960 to replace the Permanent Committee on Space Law that the IAF established in 1958. As of 2007 the IISL comprised more than 300 members from more than 40 countries who shared an interest in the development of space law. Although part of the IAF, the IISL has functioned autonomously in accordance with its own statutes. The purposes and objectives of the IISL have included fostering cooperation with international organizations and national institutions in the field of space law, fostering development of space law, and conducting studies of various legal and social aspects of space exploration.

During the Cold War the United States and the Soviet Union established a series of bilateral treaties relating to military uses of space. These included the Strategic Arms Limitation Treaty (SALT 1) and a revised version (SALT 2); an Antiballistic Missile (ABM) Treaty; a Hot Line Agreement providing direct communication between Washington, DC, and Moscow, and other specialized agreements and amendments to these agreements. In December 2001 the administration of U.S. President George W. Bush announced its withdrawal from the ABM Treaty, which was no longer in force six months later. With the exception of the ABM Treaty, these agreements remained in force as of 2007 but with reduced relevance because of changes in the Soviet Union and the bilateral disarmament activities in the United States and Russia.

Stephen Doyle

See also: Government Space Organizations, International Space Politics, United Nations Committee on the Peaceful Uses of Outer Space

Bibliography
Carl Christol, *Space Law: Past, Present, and Future* (1991).
I. H. Diederiks-Verschoor, *An Introduction to Space Law* (1993).
Nandasiri Jasentuliyana, ed., *Space Law Development and Scope* (1992).
U.S. Arms Control and Disarmament Agency, *Arms Control and Disarmament Agreements: Texts and Histories of Negotiation*, 6th ed. (1990).

Agreement on the Rescue of Astronauts

The Agreement on the Rescue of Astronauts, the Return of Astronauts, and the Return of Objects Launched into Outer Space (the Rescue and Return Agreement) of the United Nations went into effect on 22 April 1968. It was negotiated in the context of the 1967 Outer Space Treaty, which frequently required domestic implementing laws to apply the guidelines and principles. The Rescue and Return Agreement also was negotiated amid the fact that the two then-prevailing superpowers, the United States and the Soviet Union, distrusted each other's space activity motives and capabilities, despite humanitarian assertions to the contrary. Each nation wished to ensure that its respective astronauts/cosmonauts, in addition to space objects and component parts possibly in need of rescue by the other nation, were returned quickly without revealing or compromising any technology or national defense–related data.

Nevertheless the Rescue and Return Agreement looked to the Outer Space Treaty and its motivating humanitarian sentiments for interpretation and implementation of its frequently amorphous and unsettled provisions; for assistance to astronauts/cosmonauts in the event of accident, distress, or emergency landing; for prompt and safe return of astronauts/cosmonauts; and for return of objects launched into outer space. Many of the operative words and phrases appearing in the Rescue and Return Agreement are capable of differing interpretations, depending on advancing technology, international politics, economics, national security considerations, and prevailing intelligence-gathering policies.

Indeed, in the instance of the location and recovery in Canada of the Soviet *Kosmos 954*, years of debate in diplomatic, military, legal, and academic fora occurred before settlement of recovery costs was negotiated and accepted by the Soviet Union. In the case of the reentry survival of large components of the *Skylab* habitat, for example, the United States, through its Department of State and Department of Justice, relatively quickly located and identified component parts along the reentry path of the platform and paid for recovery costs in addition to property losses and damages that were incurred. In certain respects, the same approach was followed after the *Columbia* and *Challenger* Space Shuttle accidents.

George S. Robinson

See also: Space Politics

Bibliography
Bin Cheng, *Studies in International Space Law* (1997).
Piotr Manikowski, "The *Columbia* Space Shuttle Tragedy: Third Party Liability Implications for the Insurance Space Losses," *Risk Management and Insurance Review* 8, no. 1 (2005): 141.
George S. Robinson, "Interplanetary Contamination: The Ultimate Challenge for Environmental and Constitutional Lawyers?" *Journal of Space Law* (2005).

Antiballistic Missile Treaty

The Antiballistic Missile Treaty (ABM) of 1972 emerged from the first round of Strategic Arms Limitation Talks (SALT) that took place between 1969 and 1972. Under the provisions of this treaty, the United States and the Soviet Union agreed to limit their missile defense deployments to two sites, each of which could have no more than 100 interceptors. One site was to defend the national command authorities (senior government officials), the other to protect a ballistic missile field. Additionally only fixed, land-based systems were allowed—mobile systems, whether land-, sea-, air-, or space-based, were banned. The treaty also included ambiguous provisions pertaining to future ABM systems that might be developed based on "other physical

Leonid Brezhnev of the Soviet Union (left) and U.S. president Richard Nixon (right) shake hands in Moscow during talks regarding the Anti-Ballistic Missile Treaty, 1972. (Courtesy National Archives)

principles" or OPP. An example of an ABM system based on OPP would be one that employed lasers, as opposed to the nuclear-tipped interceptors that were the standard ABM weapons in 1972. A 1974 treaty protocol reduced the number of ABM sites that each side could have to one.

By 1974 the United States was deploying its one allowed ABM facility north of Grand Forks, North Dakota, where it would protect a Minuteman missile field. Known as Safeguard, this system became operational in October 1975, only to be closed in February 1976. A major reason for Safeguard's closing was that its 100 interceptors could be easily overwhelmed by Soviet rocket forces. The Soviet leaders, on the other hand, chose Moscow as the one site for their A-35 missile defense system, allowing them to protect both a ballistic missile field and their senior leadership. This site remained active into the twenty-first century.

The major logic behind the ABM Treaty was that the deployment of missile defenses would cause the opposing side to build more offensive missiles to overcome the defenses. Severely limiting missile defenses would allow each side to stabilize its offensive forces at the minimum level required to deter an attack. In fact advances, such as improved accuracy and fitting each missile with multiple warheads, created what some analysts referred to as an offensive-versus-offensive arms race. In short the ABM Treaty failed to halt the strategic arms race between the United States and Soviet Union.

By 1983 there were growing concerns in the United States about the vulnerability of American deterrent forces in the face of advances in Soviet strategic rocket forces. These concerns prompted President Ronald Reagan to launch a major new missile defense program known as the Strategic Defense Initiative (SDI). This program triggered a major debate in the United States about the meaning of the ABM Treaty. Those who supported the SDI program took the "broad interpretation" of the treaty, arguing that the treaty's OPP provisions permitted the development of new missile defense systems (including mobile systems). Treaty advocates argued the "narrow interpretation," which held that OPP systems could not be developed without additional arms control negotiations.

While the Reagan administration concluded that the broad treaty interpretation was correct, it agreed that the SDI program would proceed under the narrow interpretation. At the same time the administration used the SDI program as leverage to bring the Soviet leaders back to the arms control negotiations they had abandoned at the end of 1983. To guide American negotiators, Reagan and his senior advisers took as their goal the establishment of a new arms control regime with the Soviet Union. Under this regime the ABM Treaty would remain in effect for 10 years, with the last three years of this period being used to negotiate agreements that would facilitate the transition from a strategic world dominated by offensive nuclear weapons to a world increasingly devoid of nuclear weapons and dominated by nonnuclear missile defenses.

From 1989 to 1992, the administration of President George H. W. Bush continued to pursue the broad strategic aims established during Reagan's presidency. This changed dramatically under President William J. Clinton. From 1993 to 2000, U.S. policy was based on the premise that the ABM Treaty was the cornerstone of strategic stability. As a result the Clinton administration took three major steps to strengthen the ABM Treaty while imposing restrictions on the U.S. missile defense program.

First, within months of Clinton's inauguration, his administration denounced the broad interpretation of the ABM Treaty, abandoning any leverage this interpretation might have provided U.S. arms negotiators. Additionally, negotiations of treaty issues would take place only in the Standing Consultative Commission that had been established under the original treaty.

Next, following the lead of the Bush administration, the Clinton administration moved to multilateralize the treaty. At the end of 1991 the Soviet empire had disintegrated into a number of smaller, successor states. The Clinton administration determined that four of these states—Russia, Belarus, Kazakhstan, and Ukraine—had sufficient strategic interests, including the possession of nuclear weapons formerly controlled by the Soviet Union, to justify their becoming parties to the ABM Treaty.

Finally, the United States and its negotiating partners defined a set of demarcation parameters that could be used to determine what constitutes theater defenses as opposed to strategic national missile defenses that were controlled by the ABM Treaty. As long as missile defense systems fell beneath certain performance thresholds, they would be considered theater systems and not subject to the strictures of the ABM Treaty.

By 1997 when the Clinton administration finally succeeded in working out the details of multilateralization and demarcation, Congress was controlled by pro-missile defense Republicans who opposed these agreements and made it clear that they would use their control of the federal budget to block implementation. As a result the Clinton administration's efforts to strengthen the ABM Treaty came to naught.

In the meantime ABM Treaty provisions had been impeding Department of Defense efforts to deploy the limited national missile defense system that the Clinton administration supported. Clinton's unwillingness to undermine the ABM Treaty was a major reason that he refused in September 2000 to approve measures that would have initiated the deployment of limited national missile defenses.

When President George W. Bush took office in 2001, he made it clear that one of his top national security priorities was to field effective missile defenses as soon as possible. Recognizing the impediments the ABM Treaty posed to developing the best missile defense technologies and then deploying an operational system in December 2001, President Bush notified the Russians, as required by the ABM Treaty, that the United States would withdraw from the treaty in six months. In June 2002 the ABM Treaty ceased to be in effect.

Donald R. Baucom

See also: Antiballistic Missile Systems, Ballistic Missiles

Bibliography
John Newhouse, *Cold Dawn: The Story of SALT* (1973).
Paul H. Nitze et al., *From Hiroshima to Glasnost: At the Center of Decision—A Memoir* (1989).
Thomas W. Wolfe, *The SALT Experience* (1979).

Convention on International Liability

The Convention on International Liability for Damage Caused by Space Objects of the United Nations became effective 1 September 1972 and established liability exposure for countries and certain international organizations conducting space-related activities. It gave relative certainty to risks, rights, duties, and procedural rules relating to damages caused by space activities on Earth, in the air, and in space. Various forums were established in which claimants and defendants could resolve issues of damages incurred, actual wrongdoers, and recoverable losses. This was particularly important to insurance companies beginning to control what the private sector could put into space and how it would be accomplished.

Most important, the convention recognized that the launch and operation of spacecraft involved significant unknowns and was inherently risky and made countries and participants absolutely liable. For example, no proof of negligence was needed for damages caused by spacecraft and parts occurring on Earth and in navigable airspace. Damage caused by space objects to other space objects while in space required proving only simple negligence.

Some important provisions and issues addressed by the convention included compensation apportionment for damages caused by two or more negligent parties; conditions allowing exoneration from absolute liability; conditions wherein no exoneration from absolute liability may occur; the use of diplomatic channels for resolving claims for damage, which is how most claims have been resolved, such as the *Kosmos 954* incident, when a Soviet nuclear-powered US-A (Controlled Satellite-Active)/RORSAT (Radar Ocean Reconnaissance Satellite) disintegrated over Canada in January 1978, leading to the Soviet Union paying Canada $6 million for nuclear cleanup. Compensation is determined in accordance with international law and the confusing and unsettled principles of justice and equity in a global context; and the ability of the parties to establish a claims commission for reasonable and timely resolution of issues.

George S. Robinson

See also: Russian Naval Reconnaissance Satellites, Space Insurance

Bibliography

Alexander F. Cohen, "Cosmos 954 and the International Law of Satellite Accidents," *Yale Journal of International Law* 10 (1984): 78–91.

———, "Protocol on Settlement of Canada's Claim for Damages Caused by 'Cosmos 954,'" *International Legal Materials* 20 (1981): 689–95.

Moon Treaty

The Moon Treaty (formally known as the Agreement Governing the Activities of States on the Moon and Other Celestial Bodies) was one of four treaties drafted by the United Nations (UN) Committee on the Peaceful Use of Outer Space (COPUOS). The treaty, which entered into force for the signatories in 1984, sought to ban weapons from celestial bodies in addition to ensuring that the Moon would be assigned no private ownership. The latter point was to ensure that any wealth obtained from the Moon would be shared internationally; it also reflected the wish for developing countries to have a voice in outer space exploration.

Whereas the Outer Space Treaty of 1967 defined the Moon as "the province of mankind," (which meant no single country could claim ownership, but resources could be exploited), the Moon Treaty argued it to be the "common heritage of all mankind." Most notably, Article 11 called for the establishment of an international regime, whose purposes would be the orderly and safe development of the natural resources of the Moon, the rational management of those resources, the expansion of opportunities in the use of those resources, and an equitable sharing by all nations in the benefits derived from those resources. The rationale for this treaty followed the principles set forth in the UN Convention on the Law of the Sea, the third of which offered provisions similar to the Moon Treaty. This came about in the 1970s when a high degree of mistrust existed between developing and industrialized nations, resulting in challenges to any elements of international law deemed prejudicial to either side. Most notably, the primary concern for nations considering ratification was the need for an international consortium to monitor and hold nations accountable for actions with potential consequence toward any other nation.

By 2007 only 13 Member States had ratified the Moon Treaty: Australia, Austria, Belgium, Chile, Kazakhstan, Lebanon, Mexico, Morocco, the Netherlands, Pakistan, Peru, the Philippines, and Uruguay. Others, including France, signed the treaty, but had failed to ratify it. While some nations did not act since it was unlikely they would enter space as national units, others, especially the United States and Russia, have worried this would unduly restrict their space programs.

Whereas the U.S. delegation to COPUOS issued an interpretation of the wording suggesting private property was still possible under the treaty, intensive efforts by the L5 Society, a group of space enthusiasts supporting space colonization, convinced politicians not to seek Senate ratification of the treaty.

Guillaume de Syon

See also: National Space Society, United Nations Committee on the Peaceful Uses of Outer Space

Bibliography
United Nations Office of Outer Space Affairs. www.oosa.unvienna.org.

Nuclear Test Ban Treaty

(Partial) Nuclear Test Ban (PNTB) Treaty (also known as the Partial Test Ban Treaty (PTBT) or the Limited Test Ban Treaty) was the first successfully negotiated

President John F. Kennedy signs the Nuclear Test Ban Treaty, 7 October 1963. (Courtesy Robert Knudsen, White House/John F. Kennedy Presidential Library)

nuclear weapons treaty among the United States, its allies, and the Soviet Union. Initially derived from the 1946 United Nations resolution creating the International Atomic Energy Commission, the idea of eliminating atmospheric nuclear tests underwent 17 years of upheavals, but accelerated in 1954 after a U.S. hydrogen bomb test produced radioactive fallout that showered a Japanese fishing boat and sickened the crew, ultimately leading to one death. By the early 1960s the United States, the Soviet Union, the United Kingdom, and France had collectively tested nuclear weapons in the atmosphere, underwater, and in space, and they had documented the deleterious effects of atmospheric fallout, electromagnetic pulse, and long-term radiation effects on Earth and on spacecraft. These powers concluded that there was no further benefit and several harmful consequences of such testing, and that further testing should be accomplished underground. The treaty was opened for signature on 5 August 1963. By September the United Kingdom, the Soviet Union, and the United States had ratified it as founding members. Its provisions included a ban on all atmospheric testing in the atmosphere, underwater, and in outer space.

Under Article I, all parties undertook "to prohibit, to prevent, and not to carry out any nuclear weapon test explosion, or any other explosion, at any place under its jurisdiction or control: in the atmosphere; beyond its limits, including outer space: or under water." Such a ban included nuclear fission as part of an antisatellite (ASAT) weapons system or as a component of an antiballistic missile (ABM) system.

The impact of the treaty was notable, though not immediately significant, on projects that called for nuclear-powered rockets, such as Nuclear Engine for Rocket Vehicle Application (NERVA), Orion, Pluto, Rover, and Poodle. In the case of outer space no nation possessed the means to detect an explosion in space at the time the treaty went into effect. Concerned that the Soviet Union and China might explode devices in the upper atmosphere or in space undetected, the United States developed the Vela series of satellites designed to detect gamma rays that a nuclear explosion would emit. While these satellites provided no evidence of such explosions, they did reveal the existence of gamma rays in outer space.

Twelve Vela satellites were built by TRW Corporation and launched in pairs—six of the Vela Hotel design (launched in 1963–65) and six of the Advanced Vela design (launched in 1967, 1968, and 1970). The Vela Hotel series was designed to detect nuclear explosions in space, while the advanced design would also detect such events in the atmosphere.

In 1996 the PNTB was superseded by the Comprehensive Test Ban Treaty (CTBT) but remained open for signature. Of unlimited duration the PNTB had been signed by 108 Member States as of 2007, with the notable exceptions of France and China. Member States that declined to join the CTBT, but which were part of the PNTB, were expected to abide by the latter's provisions. The CTBT had been signed by 177 Member States as of 2007.

Guillaume de Syon

See also: High-Altitude Nuclear Tests, TRW Corporation, *Vela 5*

Bibliography

Wade Boese and Daryl Kimball, "The Limited Test Ban Treaty Turns 40," *Arms Control Today* (October 2003): 37–38.

William Burr and Hector L. Montford, eds., *The Making of the Limited Test Ban Treaty, 1958–1963* (2003).

U.S. Arms Control and Disarmament Agency, *Why a Nuclear Test Ban Treaty?* (1963).

Outer Space Treaty

The Outer Space Treaty is formally known as the Treaty on Principles Governing the Activities of States in the Exploration and Use of Outer Space, Including the Moon and Other Celestial Bodies. As of 2007, 98 Member States had ratified the treaty, including India, China, the Russian Federation, the United States, and the Member States comprising the European Space Agency. The Soviet Union was a party to the treaty until the union dissolved, but only some of the countries that succeeded the Soviet Union later acceded to the treaty.

The principles embodied in the Outer Space Treaty grew out of Cold War disarmament negotiations, and the space race precipitated by the Soviet Union's successful launching of the first satellite, *Sputnik*, in October 1957. International politics were also a factor, as the United States and the Soviet Union vied for the allegiance of many new African and Asian nations. These recently independent former colonies were extremely wary of superpower imperialism. Consequently both the Soviet Union and the United States could expect to gain political influence and prestige if they rejected territorial claims and its overtones of colonialism. National representatives also

President Lyndon B. Johnson, right, prepares to sign the Outer Space Treaty, 27 January 1967. (Courtesy AP/Wide World Photos)

opposed territorial claims because they wanted to ensure freedom of navigation in outer space, analogous to freedom of the high seas.

In 1957 the United States proposed that the United Nations (UN) develop a system of international participation and inspection to assure the use of outer space for peaceful and scientific purposes. Although the Soviet Union publicly favored the concept of peaceful uses of outer space, its leaders recognized that this approach threatened their security. At that time the Soviet Union only had missile bases within its national territory, and it could only reach the United States with intercontinental ballistic missiles (ICBM), which required transit through outer space. The United States, on the other hand, had military bases on the territories of its North Atlantic Treaty Organization (NATO) allies and also in the Near East and Middle East. From these bases the United States could launch short- and medium-range missiles, which would reach the Soviet Union without traveling through outer space.

An international agreement prohibiting military activity in outer space would have precluded all ICBM programs, placing the Soviet Union at a strategic disadvantage. Accordingly Soviet leaders would not agree to ban military activities in outer space unless the United States agreed to eliminate its foreign military bases. This negotiating posture led to a deadlock in disarmament negotiations.

In 1958 U.S. President Dwight Eisenhower and Soviet Premier Nikita Khrushchev each asked the UN to consider the legal problems associated with space activity. On 13 December 1958 the UN General Assembly created the ad hoc Committee on the Peaceful Uses of Outer Space (COPUOS), which the UN made permanent in 1959. COPUOS, in turn, created the Scientific and Technical Subcommittee and the Legal Subcommittee.

The Outer Space Treaty was preceded by three significant statements of legal principles. On 21 December 1961 the UN General Assembly unanimously adopted General Assembly Resolution 1721. Resolution 1721 provided that "International law, including the Charter of the United Nations, applies to outer space and celestial bodies," and that "Outer space and celestial bodies are free for exploration and use by all states in conformity with international law and are not subject to national appropriation."

The COPUOS Legal Subcommittee began its work the following year, seeking to expand on the principles in Resolution 1721. The Soviet Union, the United Arab Republic, the United Kingdom, and the United States submitted to the subcommittee draft statements of principles that provided the basis for negotiations. One issue of contention was private activities in outer space. Paragraph 7 of the Soviet draft said that "all activities pertaining to outer space shall be carried on solely by states." The United States objected to this provision, as did the French and British representatives. Ultimately delegates agreed to permit private enterprise, but only on the condition that nations supervise private activities and assume liability for any damages.

On 13 December 1963 the UN General Assembly unanimously adopted Resolution 1962, titled the "Declaration of Legal Principles Governing the Activities of States in the Exploration and Use of Outer Space." This Declaration set forth nine principles: outer space activities shall be for the benefit of all humankind; outer space shall be free for exploration and use; outer space is not subject to national appropriation or claims of sovereignty; the UN charter applies in outer space; nations are responsible for the private space activities of their citizens; states shall consult with, and not interfere with, one another; the launching state retains jurisdiction over its space objects and personnel at all times; states are liable for damages caused by space activities; and astronauts are due all necessary assistance, and are envoys of humankind.

Also in 1963 the Soviet Union and the United States achieved a disarmament compromise that did not require closure of U.S. foreign military bases. Subsequently, the UN Disarmament Commission proposed, and the General Assembly accepted, Resolution 1884, which called on all states to "refrain from placing in orbit around the Earth any objects carrying nuclear weapons or any other kinds of weapons of mass destruction, installing such weapons on celestial bodies, or stationing such weapons in outer space in any other manner."

From 1964 through early 1966, the United States opposed negotiation of a comprehensive space treaty on the basis that such a treaty was premature. During this period four prominent international organizations submitted draft treaties to the Legal Subcommittee for delegates' consideration: the International Institute of Space Law, the Institute de Droit International (Institute of International Law), the David Davies Memorial Institute of International Studies, and the International Law Association. The Legal Subcommittee also referred to the Antarctic Treaty for prospective treaty language.

As the prospect of humans landing on the Moon appeared increasingly likely, a sense of urgency developed in the U.S. government. It appeared that the Soviet Union might win the space race, and the absence of a binding space treaty became cause for concern. Consequently on 9 May 1966 President Lyndon Johnson reversed the U.S. position and submitted a letter to the UN General Assembly calling for a treaty governing the exploration of the Moon and other celestial bodies. On 16 June 1966 the United States and the Soviet Union submitted draft treaties. By September the Legal Subcommittee had reached agreement on most treaty provisions. The United States and the Soviet Union satisfactorily resolved the few remaining issues in private consultations during the UN General Assembly session.

On 19 December 1966 the UN General Assembly approved by acclamation a resolution commending the treaty. It was opened for signature at Washington, London, and Moscow on 27 January 1967. On 25 April 1967 the U.S. Senate gave unanimous consent to its ratification, and the treaty entered into force on 10 October 1967. The Outer Space Treaty incorporated all the principles set forth in UN General Assembly Resolutions 1721, 1962, and 1884. It has served as the primary source of international law governing civil activities in outer space.

Wayne White

See also: Ballistic Missiles, Russia (Formerly the Soviet Union), United Nations Committee on the Peaceful Uses of Outer Space, United States

Bibliography
Paul Dembling and Daniel Arons, "Space Law and the United Nations: The Work of the Legal Subcommittee of the United Nations Committee on the Peaceful Uses of Outer Space," *Journal of Air Law and Commerce* 32 (1966).

E. Galloway, ed., Senate Committee on Aeronautical and Space Activities, 90th Congress, 1st session, *Treaty on the Principles Governing Activities of States in the Exploration and Use of Outer Space, including the Moon and Other Celestial Bodies: Analysis and Background Data* 1(1967).
James Skorheim and Harold White Jr., "The Law of Outer Space: A Symbol of Social Maturity," *Western State University Law Review* 6 (1979).

Registration Convention

The Registration Convention (Registration of Objects Launched into Outer Space) of the United Nations (UN) opened for signature 14 January 1975 and entered into force in 1976. Prior to that time, beginning in 1962, the UN Secretariat had maintained a registry of launches. The convention recognized that nations bear international responsibility for their activities in space. Any launching authority must provide identifying data, if requested, about a launched object that has returned to Earth and been found beyond the territorial limits of the launching authority. By 2007 48 nations had ratified the Convention, with two international organizations declaring acceptance of the terms.

The Convention mandated the establishment of a national registries and a central UN register containing information about launch vehicles, their payloads, and flight/orbital characteristics that would allow their identification. Space object debris in Earth orbit has been a growing problem for military and civilian users. Nations are required to notify the UN Secretary General when their space objects are no longer in orbit. They are also required to keep the UN register and their own registers current. Military launches and payloads, although not altogether exempt from the requirements of the Convention, were rarely registered in full compliance by the former Soviet Union, the United States, and their respective allies. Convention compliance increased significantly after the end of the Cold War.

The Registration Convention encouraged, if not forced, reasonable communication among member nations regarding space activities; that is, space research, use, and development were expected to be "for the benefit of all mankind." It also was important, given the increasing number of nations involved in space activities, to know which nation or groups of nations owned or were otherwise responsible for which launch vehicles and their payloads for purposes of returning them or component parts if lost and retrieved by a third-party member. Issues relating to rights and duties under the Convention have been resolved primarily through diplomatic comity or according to evidentiary laws of a nation in which a civil action might be brought seeking damages.

George S. Robinson

See also: Expendable Launch Vehicles and Upper Stages, Hypersonic and Reusable Vehicles

Bibliography

"Orbital Debris: A Technical Assessment," Commission on Engineering and Technical Systems, *National Academies Press* (1995).
Robert A. Ramey, "The Law of War in Space," *Air Force Law Review* 50 (2001): 175–213.

Space Ethics

Space ethics designates a philosophical reflection on the goals, conditions, and consequences of the space enterprise and the codes of conduct (especially legal). While the question of the existence of other life-supporting worlds and extraterrestrial life has been a central theme for the precursors of space ethics, questions pertaining to pollution and contamination by spacecraft and the social impacts of funding the space adventure were just as important. In 1982 United Nations delegates to the Second United Nations Conference on the Exploration and Peaceful Uses of Outer Space expressed concern that the principles underpinning the Treaty on Principles Governing the Activities of States in the Exploration and Use of Outer Space (1967) were under threat, calling for greater cooperation among the leading space powers. In 1983 the United Nations Educational, Scientific, and Cultural Organization (UNESCO) assigned V. S. Vereshchetin, vice president of the Interkosmos Council at the Soviet Science Academy, with organizing a roundtable discussion on this issue. In 1984 the Academy of the Kingdom of Morocco organized a session on space ethics.

In 1998 the European Space Agency (ESA) and the French Space Agency (CNES) integrated ethical considerations into their practices. ESA joined forces with UNESCO to set up a working group under the aegis of the World Commission on the Ethics of Scientific Knowledge and Technology (COMEST, for Commission mondiale d'éthique des connaissances scientifiques et des technologies). In 2000 COMEST adopted the theme as a line of investigation within UNESCO, working alongside the Committee on the Peaceful Uses of Outer Space. CNES, meanwhile, published *Icarus' Second Chance: A Road Map for Ethics in Space*. This work examined the scope and issues of space ethics, including production and mitigation of space debris, the uses of remote sensing and telecommunications, commercialization, protection of the planet, and human spaceflight. At the same time, CNES appointed an ethics officer.

Further afield, interest in space ethics was sustained chiefly by individuals working to raise awareness within their organizations. For example, in 2003 and 2004 NASA and the American Association for Advancement of Science organized a seminar on the philosophical and ethical issues surrounding astrobiology research. The International Academy of Astronautics and the International Astronautical Federation invited communication on this theme at their symposia, and the International Space University has offered ethics courses as part of its curriculum. Other aspects of space ethics that have been proposed include ethics for off-Earth businesses, to address anticipated growth of commercial space development and space tourism.

In terms of tangible achievements, the Inter-Agency Space Debris Coordination Committee brought together national space programs to identify future mitigation measures aimed at limiting the proliferation of space debris in low Earth orbit. In 2000 ESA, CNES, the Indian Space Research Organisation, the Canadian Space Agency, and the U.S. National Oceanic and Atmospheric Administration signed the International Charter on Space and Major Disasters, under which they committed to use their efforts to aid disaster-affected regions free of charge. The Argentine Space Agency joined in July 2003, and the Japan Aerospace Exploration Agency became a member in February 2005. This charter was activated more than 60 times by member nations for natural (floods, earthquakes, hurricanes) and technological (train wrecks, oil spills) disasters between 2000 and 2004.

Jacques Arnould

See also: European Space Agency, Planetary Protection, United Nations Committee on the Peaceful Uses of Outer Space

Bibliography

Jacques Arnould, *La Seconde Chance d'Icare. Pour une éthique de l'Espace* (2001).
David Livingston, "A Code of Ethics for Off-Earth Commerce" (2002).
Alain Pompidou, ed., *The Ethics of Outer Space: Policy Document* (2000).

SPACE POLITICS

Space politics influences many facets of space endeavors, including the competition for hearts and minds around the world through science and human spaceflight, in addition to traditional military and economic rivalry.

The Cold War between the United States and the Soviet Union, and their contrasting political and economic systems, was the 1950s setting in which spaceflight came into existence. Before *Sputnik*, U.S. and Soviet leaders thought about space in terms of military power. Both sides were rapidly developing intercontinental ballistic missiles (ICBM) as a way to maintain military power while also reducing the size of conventional forces. ICBMs also had the side benefit of being able to launch spacecraft into orbit. Taking advantage of this expected new capability, the U.S. Dwight Eisenhower presidential administration began to develop the first reconnaissance spacecraft to orbit over and take pictures of denied communist territory, as a logical extension of its provocative and secret aircraft reconnaissance overflights.

In the meantime, by the early 1950s, U.S. and European scientists had conceived the idea of an International Geophysical Year (IGY) and, as part of its contribution to this international scientific collaboration, the United States announced it would orbit a science satellite. Soviet leaders subsequently announced that the Soviet Union would also launch a science satellite for IGY. While the Soviet announcement and program was largely a tit-for-tat response to the United States, the Eisenhower administration had a secret motive; it wanted to ensure that the first U.S. satellite was a scientific craft that would overfly Soviet territory. If Soviet leaders did not protest, this would establish the legal precedent for satellites to traverse national territories, which the United States would exploit by later launching reconnaissance systems to monitor Soviet activities. The Soviet launch of *Sputnik* in October 1957 established this principle, and U.S. leaders secretly recognized that this cleared the way for U.S. reconnaissance satellites. In 1960 the Corona program succeeded in returning the first satellite images of the Soviet Union, which showed that the communist power did not have superiority over the United States in number of ICBMs.

However, *Sputnik* generated a huge public reaction in the United States and around the world. U.S. policymakers, most of whom knew nothing about the Eisenhower administration's strategy, believed that the United States was now suddenly vulnerable to nuclear attack and that the Soviet Union had surpassed the United States in technology development. Domestic critics of the Eisenhower administration used *Sputnik* as a hammer to attack the supposedly indolent administration, while Soviet leaders trumpeted their newfound superiority over the capitalists. These events launched the space race, which dominated space activities through the 1960s. Democrat John F. Kennedy won the 1960 U.S. presidential election in part by attacking the Republicans as being weak on the space issue and behind the Soviet Union in ICBM deployment. The Republican Party candidate, Richard Nixon, did not reveal the secret reconnaissance strategy or the knowledge that the United States was ahead in the ICBM race.

Continuing the ideological assault on the United States, Soviet Premier Nikita Khrushchev pressed for spectacular space firsts to mask the Soviet Union's actual military weakness in nuclear capabilities compared to the United States. After Soviet cosmonaut Yuri Gagarin's orbital spaceflight in 1961, President Kennedy announced that the United States would land a man on the Moon by the end of the decade. For the rest of the 1960s, the United States and the Soviet Union competed in crewed missions. By 1965 the U.S. Gemini program seized the lead, and the Apollo program subsequently fulfilled Kennedy's goal. The Soviet Union never publicly acknowledged it was racing the United States to the Moon, but in fact its crewed lunar programs failed. The Soviet leaders announced that they had only planned to send robotic craft to gather samples more cost effectively (which they achieved in the 1970s) and were focusing instead on long-duration flights in Earth orbiting space stations. The Soviet Union, and later Russia, stuck to its space station strategy in the Salyut and Mir programs through the 1990s.

Both the United States and the Soviet Union also used crewed and robotic science missions to generate support among their allies and across the ideological divide. The United States developed bilateral programs with dozens of nations, from joint scientific missions to establishment of satellite tracking stations. It helped train European, Canadian, and Japanese scientists and engineers in scientific satellite development. The Soviet Union did the same with its allies and some capitalist nations (such as France and India) through scientific missions and Interkosmos flights for "guest cosmonauts" aboard Soyuz spacecraft to the *Salyut* and *Mir* space stations. Scientific and human spaceflight collaboration became major foreign policy tools of both nations, and after the demise of the Soviet Union in 1991, the United States and Russia collaborated to build the *International Space Station* (*ISS*), along with European nations, Canada, Japan, and Brazil.

The 1960s also saw the beginning of economic collaboration and competition in space. The United States proved in the early 1960s that communications satellites (comsats) were economically viable and created the Communications Satellite Corporation in 1963 to manage U.S. interests in the new Intelsat (International Telecommunications Satellite Organization), created in 1964. Responding to U.S. initiatives, European nations banded together to bargain with the United States, ensuring that there would be new negotiations in the early 1970s to rework the terms of the Intelsat agreements. These new terms established the division of shares of Intelsat based on national usage and also allowed for regional satellite communications systems. The Europeans lost no time in creating their own comsat systems, following in the wake of Canada, which in 1972 had launched its first national communications satellite, *Anik A1*. Fearing that the United States would not launch a European satellite that would compete with U.S. interests, European nations united in the 1960s to develop their own indigenous launcher, Europa, which failed. Subsequently ten of these nations (Belgium, West Germany, Denmark, the United Kingdom, Denmark, Italy, the Netherlands, Sweden, Switzerland, and Spain) organized to form the European Space Agency in 1975 with a major goal of developing a European launcher, Ariane, a European comsat, and to collaborate with the U.S. Space Shuttle program by building *Spacelab*.

By the 1980s commercial space activities were becoming more prominent worldwide. Ariane proved to be a successful competitor against the Space Shuttle for commercial comsat launches. More nations and private companies were purchasing comsats and launches, such as Indonesia with its Palapa series and PanAmSat Corporation, which provided comsat service to Latin America. A very profitable new application, direct-to-home satellite television broadcasting, developed in the late 1980s in Europe and the United States, by the 1990s was rapidly expanding. In the 1990s, the U.S. military's Global Positioning System (GPS) became a commercial and military success as hundreds of companies worldwide developed applications to use its signals for a variety of purposes, ranging from automobile and trucking navigation to scientific applications, such as seismology and wildlife tracking, to recreational uses such as fish finders and golf aids. The success of GPS drew competition, as Russia struggled to maintain GLONASS (Global Navigation Satellite System), and ESA began to develop the Galileo global navigation satellite system in the 2000s, as a commercial navigation service. By the 2000s space tourism became a reality with tourist flights on Soyuz spacecraft to *ISS*, and the first successful suborbital flight of Space-ShipOne in 2004.

Military uses of space continued to generate political dialog and debate. By the mid-1960s the Soviet Union and the United States had tacitly accepted the mutual use of reconnaissance satellites, obliquely referred to as "national technical means" in various treaties. However, antiballistic missile (ABM) and antisatellite (ASAT) systems continued to generate heated discussion and debate. The United States and the Soviet Union agreed on the ABM Treaty in 1972 to limit the development of ABM systems, but the two nations never reached such an agreement with ASAT systems. Both nations tested ASATs in the 1960s, and the Soviet Union's tests in the 1970s helped fuel a renewed Cold War intensity in the 1980s. China tested its first ASAT in 2007, fueling renewed concerns about space weaponization and the increasing amount of space debris generated by the on-orbit destruction of targets. In the 1980s the United States began the Strategic Defense Initiative (parodied as "Star Wars") to develop ground- and space-based ABM systems, while the Soviet Union had secretly been developing its own version in the 1970s and 1980s. The end of the Cold War in 1991 reduced tensions and military budgets, but in 2001 the U.S. George W. Bush presidential administration abrogated the ABM Treaty in order to be able to deploy new ABM systems against perceived new threats from North Korea and Iran.

Stephen B. Johnson

See also: Economics of Space, Government Space Organizations, Nations, Space Law

Bibliography
William E. Burrows, *This New Ocean: The Story of the First Space Age* (1998).
John Krige and Arturo Russo, *A History of the European Space Agency 1958–1987*, 2 vols. (2000).
Walter A. McDougall, *. . . the Heavens and the Earth: A Political History of the Space Age* (1985).
Eligar Sadeh, ed., *Space Politics and Policy: An Evolutionary Perspective* (2002).

Economics and Technology in Space Policy

Economics and technology in space policy are important elements in the political strategies of spacefaring nations. The economics of space relate the values and costs associated with space activity. Matching "wants" with "abilities" is what is meant by the economics of space. Because technology is an important element of national economic development, it is closely tied to these economic policies in addition to national security issues.

The "abilities" encompass the technology, people, and natural resources that decision makers can organize to satisfy the wants. How nations have determined their wants and satisfied these with technologies has varied markedly, from the approach taken in free-market economies like the United States to the planned economies in Russia, the former Soviet Union, and China. Political goals largely drove the early history of civilian space programs such as Sputnik and Apollo. Since then, other national objectives have played increasingly important roles in matching wants and abilities.

UNITED STATES

The United States balanced competing priorities, including supporting the commercial aerospace industry and safeguarding national security.

During the Cold War, the driving force behind the space program was largely an extension of national security policy. The Apollo program coincided with a high tide of national interest in engineering and science, which sped the technological development of satellite and launcher technologies. The U.S. government directed that the vast majority of contracts for these systems largely go to industry, not to government "arsenals," modifying the practices of the U.S. Army and consolidating those of the U.S. Air Force. Even in that era of dueling superpowers, the United States promoted significant international economic cooperation to ensure a role for private industry. A consortium of nations, led by the United States, formed the International Telecommunications Satellite Organization (Intelsat), an organization designated to coordinate and manage a constellation of communications satellites, in addition to corresponding orbit and spectrum rights. The United States created the Communications Satellite Corporation explicitly to represent the United States and insisted that it manage the consortium. The United States unsuccessfully tried to perpetuate U.S. domination of telecommunications and launcher technologies and markets through control of Intelsat and selective support or denial of support to foreign efforts in these areas.

Post–Cold War U.S. space policy saw an increase in emphasis on economic and technological issues. Following a surge in commercial satellite enterprise in the mid-1990s—coinciding with the telecommunications boom—Congress clamped down after investigating Space Systems/Loral for allegedly transferring sensitive technology to China. This resulted in transfer of reviews from the Department of Commerce to the Department of State (DoS), which significantly hampered U.S. commercial satellite sales as satellite operators purchased European satellites to avoid U.S. strictures. These strictures centered on the International Traffic in Arms Regulations (ITAR), a set of U.S. government regulations that authorize the president to control the export and import of defense-related material and services. The regulations are described in Title 22 (Foreign Relations), Chapter 1 (DoS), Subchapter M of the Code of Federal Regulations.

Russia's participation in the International Space Station (ISS) program was justified in part due to national security concerns that a failed Russian aerospace industry would be a source of weapons technology proliferation and also to better integrate Russia's economy into the capitalist system.

The world's first global timing and navigation satellite constellation was originally a U.S. military project, but by the early twenty-first century the Global Positioning System (GPS) served both commercial and military users, including international consumers. However, European nations, wary of relying solely on the United States for such an important space application, began building their own navigation constellation, the Galileo Joint Undertaking. Multihundred-million-dollar contracts awarded by the National Geospatial-Intelligence Agency to U.S. commercial remote sensing companies for purchase of space-based imaging have been another means of government support of commercial space.

CHINA

China's space program played an important role in China's economic and technological development. The successful launch of rockets, satellites, and—with *Shenzhou 5*—human missions served to demonstrate China's technological capability and advertised China's satellite launch industry. The space program also tied to other high-tech industries, such as China's burgeoning Internet-based economy, advanced chemical and metals production, and rapidly growing biotechnology industry. Under Deng Xiaoping's plan to build a "healthy well-off society," such economic maturation was seen as a necessary step toward building a prosperous middle class.

China adopted a strategy combining the nation's technological development and international technology transfer. For instance, the success of the Shenzhou crewed spaceflight program was accomplished by basing the design for the Shenzhou capsule on the Russian Soyuz, with Chinese modifications.

In 1998 the U.S. DoS placed restrictions on U.S. satellite manufacturers from using Chinese launch services after concerns about whether American companies gave sensitive technology to the Chinese. That policy helped satellite makers of other countries that do not use American components to win Chinese contracts that might have gone to American firms.

China's economic and technological policies regarding space also tied to China's foreign policy priorities. In 2003 the Asia Pacific Space Cooperation Organization was founded in Beijing, with member states that included most of southeast Asia, Brazil, Russia, and Iran. The organization was established as a way for these countries to pool their space resources and helped give China a leadership position in the region regarding space activities, including selling launch services to or possibly exchanging technology with other member nations. China also committed to investing $200 million in the European Galileo navigation satellite project.

INDIA

The focus of India's space projects has been to deliver practical applications for national development, like weather forecasting, tele-education, and urban planning. This focus on economic development defined India's civilian space program as distinct from India's military space program.

Though India's space activities began by purchasing hardware and launch services from other spacefaring countries, it later developed capability for launching to low Earth orbit, polar orbit, and geosynchronous orbit, in addition to satellite manufacturing. Attempts to purchase Russian cryogenic rocket technology necessary for geosynchronous launch vehicles—but perceived by the United States as important for ballistic missiles—in the early 1990s met with stiff U.S. opposition and sanctions. The Russians agreed to sell the engines without the manufacturing technology. Consequently India pushed ahead with a domestic research and development effort. India also became a financial participant in the European Galileo project.

JAPAN

Through the twentieth century Japan mostly refrained from military and human flight projects, preferring robotic explorers and projects with direct economic applications, such as communications satellites and microgravity materials and life science experiments. That policy direction was called into question between 1998 and 2003 following the North Korean test of its Taepodong launcher and the successful Chinese crewed mission, the 2004 declaration from the United States of returning astronauts to the Moon and going to Mars, and a prominent failure of Japan's workhorse H2-A rocket. By 2005 Japan expanded its programs to include military reconnaissance and crewed missions.

Japan based its launch vehicles on a combination of Japanese technology and licensing from U.S. designs. The heavy-lifter H-series of rockets were designed to compete in the international commercial launch market, but by the early twenty-first century had not done so effectively. Technological cooperation also played a part in Japan's contributions to the ISS program, including a three-part research module named Kibo, or hope in Japanese.

BRAZIL

The Brazilian aerospace program was started as part of the "Great Brazil" campaign of the 1970s, but the dreams of fostering high-tech industries, competing in the international commercial launch market, and building national prestige met with the realities of a constrained national budget. Brazil developed indigenous rocket technology beginning with the SONDOS sounding rocket series up to the Veículo Lançador de Satélite (VLS) series, designed for launching satellites into low Earth orbit. However three launch failures of the VLS design ended with a catastrophic explosion on the launch pad at Alcântara in 2003, killing 21 people. Brazil subsequently withdrew from contributing to the ISS due to budget shortfalls. Brazil's international cooperation has included a partnership with China to launch Earth remote sensing satellites and with Russia and Ukraine to gain rocket technologies. However, the United States has been opposed to unfettered rocket technology exchange with Brazil due to proliferation concerns. The China-Brazil Earth Resources Satellite (CBERS) program produced valuable commercial satellite imagery data on the Amazon, croplands, and fisheries.

EUROPE

Europe's space activities after 1975 were conducted through the European Space Agency (ESA) and national programs. Europeans developed policies to allow technical and economic cooperation through mechanisms such as "just return" that ensured that countries that contributed to the ESA program received sufficient value of contracts in return. ESA is an independent organization although it maintains close ties with the European Union (EU) and Member State national governments.

Europe developed its own technologies, such as the Ariane series of launchers and Galileo to foster independence from the United States, in addition to technology transfer, such as developing a cooperative agreement with Russia to use Soyuz technology. In the European Space White Paper (2003), the ties between technology applications and economic objectives were linked through projects like Galileo and the Global Monitoring for Environment and Security (GMES) program. Europe has sought international participation with Galileo from many countries including Israel, India, and China.

Many European policy makers viewed space as a critical part of Europe's economic development and social integration. For instance, as the EU expanded to include more developing countries, policy makers saw satellite-based Internet service as a way to bridge the "digital divide" by providing access where terrestrial infrastructure was not expected to be built in the short or medium term. Europe's established capabilities in remote sensing provided information for EU member nations to better manage urban planning, agriculture, transportation, and other economic activities.

SOVIET UNION AND RUSSIA

In the 1950s Soviet Premier Nikita Khrushchev reduced expensive conventional forces in favor of expanded missile forces to reduce military expenditures and also add to the Soviet labor pool, spurring economic growth. Soviet leaders vowed to develop indigenous technologies to catch up and surpass the United States in military technologies of all kinds, including space. To do this, they developed a vast spying apparatus to acquire and improve on Western technologies.

After the demise of the Soviet Union in the early 1990s, Russian space programs operated at a fraction of the budget of the Soviet era. However, Russia maintained many of its space capabilities, including its contribution to the ISS program and helped to foster space tourism. To acquire the cash needed to keep its space programs operating, Russia commercialized many civilian activities and sold some of its technologies.

Based on its experiences with international guest flights to *Mir* under Interkosmos, Russia sold flights to *Mir* to the United States, France, and the United Kingdom. In partnership with U.S.-based Space Adventures, Russia launched the first space tourists in 2001–2, each reportedly paying about $20 million. Other commercial space station endeavors included construction of *ISS* modules paid for by the United States, astronaut flights on Soyuz to *ISS*, and Progress cargo shipments to *ISS*. Russia also sold rocket engines to India and the United States.

Russia privatized some of its government institutions, such as RSC Energia, which supplied *ISS* modules and other space products. However it kept others as government entities while selling launch services on the commercial market, such as Khrunichev, which partnered with the U.S. Lockheed Martin Corporation to form International Launch Services to market Proton and Atlas launchers.

David Chen and Molly Macauley

See also: Brazil, China, Europe, India, Interkosmos, Japan, Russia (Formerly the Soviet Union)

Bibliography
John M. Logsdon et al., *Exploring the Unknown: Selected Documents in the History of the U.S. Civil Space Program*, 6 vols. (1999–2004).
Walter A. McDougall, . . . *the Heavens and the Earth: A Political History of the Space Age* (1985).

Critiques of Spaceflight

Critiques of spaceflight began within a few years of the first space launches in 1957. These critiques were more prominent in western nations in which open criticism of government activities was allowed, though they also appeared in the Soviet Union. Critiques by conspiracy theorists are not addressed in this article.

In the United States, the first, most persistent, and most prominent political critiques were from the political left. Already by the mid-1960s, the human spaceflight program, particularly Apollo, attracted criticism from academics, politicians, and civil rights activists for being a gigantic waste of money, drawing precious funding and human resources away from more pressing problems on Earth, such as poverty, disease, civil rights, and education. Sociologist Amitai Etzioni argued that in comparison to these needs on Earth, scientific discoveries and the exploration of space were not worth the cost. Left-wing political opposition to Apollo existed from the inception of the program, and every year the opposition, led by liberal members of Congress (usually from districts and states without significant space contracts), argued that the funding going to NASA should be spent on social programs instead. From the John Kennedy presidential administration onward, spaceflight supporters had to deploy a variety of arguments to advance their cause, including national prestige and exploration, scientific and technological advancement, and economic development. Examples of NASA's poor cost and technical performance, such as the Space Station Freedom program in the late 1980s and early 1990s, aided this argument. The Space Station program survived by a single vote in the U.S. House of Representatives in June 1993 but was soon rescued by a new political argument—as a tool of foreign policy to support the newly democratizing Russia.

From the other end of the political spectrum, conservative and libertarian politicians and scholars criticized the space program for being a tool of big, activist government, to which they were philosophically opposed. The most sophisticated conservative argument was made by historian Walter McDougall, who argued that the U.S. embrace of technocracy was a step in the direction of its Soviet enemy in the Cold War. Conservative critics supported the privatization of space endeavors and, at its extreme, the elimination of NASA. Their efforts were an important component of the Ronald Reagan presidential administration's space policy and programs, which included having private companies assume many of the tasks, both technical and nontechnical, previously performed by federal employees. Critiques from the right continued in the 1990s and 2000s, favoring private space endeavors, such as commercial launch systems, space tourism, and mining the Moon for resources.

In the Soviet Union, the political leadership constructed a picture of Soviet cosmonautics and cosmonauts as a perfect harmonization of technology constructed in a socialist society, with perfect missions flown by flawless cosmonauts. During the Khrushchev era in the early 1960s, Soviet newspapers occasionally allowed a critique of spaceflight in the form of worker's letters that complained that cosmonautics did not help them directly. Allowance of a few dissenting views provoked much larger numbers of supporting letters, thus supporting the regime's goals. Only after the fall of the Soviet Union did differing views appear, starting with memoirs from space leaders and cosmonauts, who offered their views about which organizations were the most important and which others were obstructive or less significant. By the 2000s, support in Russia for human spaceflight had waned in comparison to practical space applications.

Another prominent critique in the United States was environmentalist objection to the use of nuclear power. These objections, accompanied by protests and lawsuits, peaked during launches of nuclear-powered spacecraft, such as *Galileo* and *Cassini*. Environmentalists argued that radioactive nuclear materials posed a major health hazard should an accident occur in Earth's atmosphere, as occurred with the Soviet *Kosmos 954* spacecraft that disintegrated over Canada in 1978. Through the first 50 years of spaceflight, these objections had not prevented any launches, but they slowed space nuclear power and propulsion development programs.

A third type of critique regarded objections to specific programs. Two U.S. attempts to initiate a human mission to Mars, in the late 1960s and again in the late

1980s, both foundered politically, largely due to huge costs and concurrent concerns with large budget deficits. In the decades after Apollo ended, many space analysts believed that Apollo had sent NASA in the wrong direction, away from a more evolutionary program that built on previous projects. In this view, Apollo led nowhere, because NASA did not use Apollo hardware for any follow-on programs. The Space Shuttle and International Space Station programs drew fire for costs that greatly exceeded initial predictions. These two programs also received criticism for diverting NASA's human spaceflight programs away from exploration. This argument emphasized the ability of space exploration programs, such as Apollo, to inspire, whereas repetitive trips to low Earth orbit were uninspiring and literally going nowhere except in circles. Europeans had similar arguments about specific programs, exacerbated by different national interests. In the events leading up to the formation of the European Space Agency, France argued against participation in the U.S. Shuttle program, while the United Kingdom declined to participate in development of a European launcher. In the 1980s, France supported a crewed spaceplane program, Hermes, to provide an independent European capability to put humans in space, but objections from other European nations eventually prevailed and the program was canceled.

Stephen B. Johnson

See also: Media and Popular Culture, Nuclear Power and Propulsion

Bibliography

Matthew Cunningham, "But Why, Some Say, the Moon?" *Quest* 16, no. 1 (2009): 32–45.

Steven J. Dick, ed., *Remembering the Space Age: Proceedings of the 50th Anniversary Conference* (2008).

Amitai Etzioni, *The Moon-Doggle: Domestic and International Implications of the Space Race* (1964).

Walter A. McDougall, . . . *the Heavens and the Earth: A Political History of the Space Age* (1985).

Military Space Policy

Military space policy, developed originally through interactions between the United States and the Soviet Union, encompasses several aspects, each subject to change throughout time: who should have responsibility for military space activities, what activities are conducted, and what activities should not be conducted. Before the launch of *Sputnik* in October 1957, both U.S. and Soviet military leaders expected to administer their respective space programs, because the military had long played a leading role in exploration and because space launch vehicles were essentially converted missiles. Early proposals for space applications included military space stations, weapons in orbit, and fortifications on the Moon and celestial bodies. Driven by the dual factors of technology development and evolving government directives, military space policy matured along with civil and commercial space policies.

Four general schools of thought or policy perspectives on the military use of space emerged. The sanctuary school, the space equivalent of President Dwight Eisenhower's open-skies approach, favored "peaceful" uses of outer space for reconnaissance and intelligence-gathering purposes and opposed so-called "weaponi-zation" of that realm. Alternatively, the survivability school envisioned conducting space operations across the military spectrum, strategic to tactical, to effect force enhancement on the ground, at sea, and in the air without actually placing weapons in orbit. Another school of thought emphasized controlling space as a fourth area of operations, where military supremacy would permit more efficient, effective applica-tion of power in all realms. The high-ground school perceived space as the ultimate battlefield, domination of which could decisively affect the outcome of terrestrial con-flicts. These four perspectives all involved an expansion of military space operations.

In the United States, President Eisenhower quashed an incipient interservice space race by transferring the U.S. Army long-range rocket program to then newly estab-lished NASA, a civil agency, and designating the U.S. Air Force (USAF) as the lead service for defense-related space activities. The USAF, in turn, was forced to transfer primary responsibility for human spaceflight to NASA. Early efforts to place military crews in space ended in the 1960s with cancellation of the Blue Gemini, Manned Orbiting Laboratory, and Dyna-Soar programs. Consequently a bifurcated U.S. space program with distinct civil and military sectors stood in stark contrast to the military-controlled Soviet program.

Military control of the Soviet program, however, did not preclude space exploration and science activities similar to those conducted by NASA. Unlike the United States, the Soviet military built Salyut space stations into the mid-1980s. Its extensive space-related investments and the extent to which it saw strategic value in an orbital presence set these two superpowers apart from the rest of the international community for sev-eral decades. Although that distinctive gap remained at the beginning of the twenty-first century, a handful of other nations, most notably China, had begun expanding its military space efforts.

An expansive policy approach encountered two obstacles: cost, and comparative strategic value. Neither the United States nor the Soviet Union chose to extend the nuclear arms race into space. Both assumed the cost would be prohibitive, with stra-tegically little advantage. Analysts on both sides concluded that basing weapons in space or on the Moon would add little military value compared to Earth-based weap-onry. Orbiting weapons, in fact, became more vulnerable to attack than ground-mobile intercontinental ballistic missiles (ICBM) or those concealed and protected in hard-ened, underground silos. Rather than placing weapons in space, military leaders in both nations focused on how space applications could enhance their ICBM capabil-ities, especially with regard to ensuring survivability and a second-strike capability. With those aims in mind, Soviet political and military officials initially rejected in principle the legality of U.S. reconnaissance satellites passing over Soviet territory but, as they acquired similar capabilities in the mid-1960s, their protests became less vocal.

Beyond cost considerations a desire existed to halt the proliferation globally of weapons of mass destruction (WMD), which in the 1960s primarily meant nuclear weapons. Politically, preventing WMD proliferation rather than removing already deployed weapons became the goal. The military effects were immediate. The Lim-ited Test Ban Treaty (1963) prohibited testing nuclear weapons in outer space; the

Outer Space Treaty (1967) prohibited deployment of WMD in orbit, on celestial bodies, or in any other manner in space. Weapon testing, military bases, installations, or fortifications on celestial bodies also were prohibited. The 1972 Antiballistic Missile (ABM) Treaty further restricted testing or placing in orbit weapons to intercept attacking missiles or warheads. Consequently, in policy terms, outer space became a weapon-free sanctuary, albeit one open to unarmed military spacecraft. While international space agreements did not specifically ban conventional weapons, political expectations and resulting policy decisions allowed for no weapons whatsoever.

Banning weapons did not preclude other military applications, some not entirely defensive. Satellites provided secure communications, signals intelligence, reconnaissance and surveillance, early warning, navigation, meteorological data, and geodetic information that enhanced military effectiveness and the lethality of air, land, and sea power. Ensuring the legality of space reconnaissance was a primary goal of U.S. military space policy from the mid-1950s until 1972, when the Strategic Arms Limitation Talks agreements between the United States and the Soviet Union formally sanctioned the use of "national technical means" (a euphemism for reconnaissance satellites) to verify compliance with ballistic missile arms control agreements. Space systems became the backbone supporting U.S. military forces globally. Their growing importance had earlier prompted both superpowers to explore development of antisatellite (ASAT) weapons, including ground-based, air-launched, and co-orbital versions. Neither country went beyond initial deployment of an ASAT system, however, due to cost and the potentially destabilizing effects on the nuclear balance of terror. Even as the vulnerability of satellites to electronic interference or physical destruction became increasingly apparent, and as the increased reliability of launch and guidance technologies improved the probability of satellite interception, the superpowers' respective policies rejected adoption of a fully operational ASAT capability.

During the 1990s the U.S. conceptualization of military space capabilities in terms of several broad categories influenced how it crafted space policy. The first category, space force support, involved the launch and on-orbit operation of military satellites, with "launch on demand" or "responsive space," as it was later dubbed, being the highest priority. To sustain space launch into the twenty-first century, the USAF undertook the Evolved Expendable Launch Vehicle program, while the Russians employed upgraded versions of earlier Soviet rockets. Space force enhancement, a second category, involved direct or indirect support to strategic and theater forces through the timely provision of weather data, navigational capabilities, early warning, communications, and reconnaissance and surveillance. A third category, space force applications, referred to the ICBM wings and nuclear submarines, which could launch missiles through space in response to an enemy attack. Space control, the last category, involved protecting U.S. space assets and negating an enemy's use of space. Protection meant preserving the operational status of U.S. satellites passively by shielding them from the effects of electromagnetic pulse or maneuvering them to evade interception; more actively, it meant intercepting and destroying enemy ASAT systems. Negation included attacking enemy satellites with kinetic-energy weapons, lasers, or other weapons; electronically jamming onboard sensors and receivers or ground equipment; and preventing the launch of new spacecraft. At the end of the twentieth century, the United States experimented with various ways, including a Space Maneuver Vehicle, to improve performance in all four categories.

Space control generated much discussion, because actually pursuing it represented a significant shift in military space policy away from the traditional sanctuary school and, potentially, an irreversible step toward weaponization of space—actual placement and employment of weapon systems in outer space. The earlier international understanding embodied in the sanctuary approach had expanded to the general expectation that no weapons of any type would be placed in space even though certain classes, kinetic energy specifically, were not explicitly prohibited. In the 1980s advocates for President Ronald Reagan's Strategic Defense Initiative had argued that Earth orbit provided the most effective location for interception of nuclear missiles or their warheads during the launch and midcourse phases. One proposed system, Brilliant Pebbles, would have employed thousands of small, autonomous spacecraft to attack ICBMs with minimal human intervention. When President George W. Bush unilaterally abrogated the ABM Treaty in June 2002, it signaled that the United States again considered weaponization of space on the agenda for missile-defense purposes. While most other nations, including China and Russia, objected to that eventuality, survival of the sanctuary concept clearly depended on creating new international policy agreements.

After the First Gulf War in 1991, the U.S. military increasingly used satellite-based technologies to enhance its effectiveness and lethality, thereby allowing quicker and more efficient deployment of forces than any other national military; the force-multiplier effect of space-based resources became so great that it increased other nations' anxieties about their comparative military inferiority. As integration of space capabilities into overall U.S. military operations increased, concern about U.S. satellites' vulnerability to attack grew. When the Chinese publicly stated their intention to attack U.S. satellites immediately in the event of conflict, and as the various means of attacking—electronic jamming or spoofing (tricking a system into accepting false commands), blinding sensors with lasers, high-powered microwave bursts, kinetic-energy ASATs, high-altitude nuclear blasts, or clouds of pellets strewn along orbital paths—became more sophisticated, U.S. military interest in space control intensified.

By early 2006 U.S. military space policy focused on space control. As defined by lawmakers, such control included three facets: space situational awareness, defensive operations to protect U.S. assets, and offensive operations to deny enemies use of their space assets. The last two facets significantly increased the near-term probability of U.S. weaponization of space. Liberal politicians in Canada sought international support for a total ban on weapons in space, but the policies of Russia, China, India, and other space powers remained uncertain in that regard. Clearly, amid multiple rather than bipolar participants, military space policy had entered a period of turbulence when previous understandings based on a strictly sanctuary model were in flux. Questions of cost tended to slow the process of implementing policy changes, however, because deployment of fully operational, cutting-edge space systems remained expensive and technically challenging.

Roger Handberg

See also: Antisatellite Systems, Ballistic Missiles and Defenses, Doctrine and Warfare

Bibliography
W. Henry Lambright, ed., *Space Policy in the 21st Century* (2003).
David E. Lupton, *On Space Warfare: A Space Power Doctrine* (1988).
Matthew Mowthorpe, *The Militarization and Weaponization of Space* (2003).
Michael E. O'Hanlon, *Neither Star Wars nor Sanctuary: Constraining the Military Uses of Space* (2004).
Eligar Sadeh, ed., *Space Politics and Policy: An Evolutionary Perspective* (2003).

Politics of Prestige

The politics of prestige has many similarities with the term soft power. Both capture the concept of one nation impressing other nations with positive achievements rather than displays of raw power, and, for a brief period in the mid-twentieth century, space exploration became an important means of portraying technological leadership to a new world of multilateral institutions and cooperation. The key lies in a nineteenth-century concept with a double meaning that, thanks to an unusual combination of circumstances, provided the most acceptable motivating force for turning space exploration from a dream to reality.

The term prestige underwent a transformation during the twentieth century. When the century began, the word still retained much of its original Latin meaning of "quick fingered" or "sleight of hand," from which derived the French word for conjuror. But by the 1950s this slightly disreputable connotation had morphed into a new sense, with prestige now used to describe a positive quality that fostered a desire to associate with or emulate its possessor. This might have remained an etymological curiosity in most periods of history, but three factors made it a crucial concept for world politics in the decade after the Korean War.

The first factor was the suspended animation of the Cold War. The growing nuclear arsenals in the United States and the Soviet Union made a direct, or even proxy, confrontation potentially catastrophic. Superiority could no longer be demonstrated by the direct application of military might and, by the mid-1950s, much of the world had settled into two blocs that shared an uneasy acceptance of this situation. In this world, new weapons could intimidate, but had limited influence.

The second factor that would link prestige with spaceflight was another unprecedented aspect of the international situation following World War II. The opposing blocs were not simply geographic or ethnic groups but represented two fundamentally different approaches to the organization of modern society summed up as collectivism versus consumerism. Communism and capitalism both claimed to represent the way to prosperity, better living standards, and a brighter tomorrow. With direct military confrontation between the United States and the Soviet Union ruled out, the importance of presenting a positive image of a society that others might want to emulate took on great importance.

This set the stage for the third factor, the presence of an audience of newly independent states. Decolonization had accelerated after 1945, notably in India, Indochina, and Indonesia, all prominently represented at the first meeting of the nonaligned movement in 1955, and in 1957 Ghana became the first major African country to achieve independence. Not only were there two blocs adamantly claiming to represent the best path to development, but a growing number of new nations, many in strategic locations or rich in mineral resources, was now assessing these claims as they made their choice of economic models. Against this background, prestige became the short-hand term for the competition between the United States and the Soviet Union to inspire adherents in what was then called the Third World.

Meanwhile, people working in the aerospace communities in the United States and the Soviet Union had a parallel evolution in their arguments for the funding of space-flight. In the United States the potential of both the intercontinental ballistic missile (ICBM) and the reconnaissance satellite was appreciated by the mid-1950s, with crash development programs underway. The attempt to tie space to science looked promising when both blocs announced they would launch a small Earth satellite as part of the International Geophysical Year (IGY) in 1957–58. However, in the case of the U.S. Vanguard program, a small budget and a determination that this civil venture should not distract the highest-priority development work on ballistic missile programs showed that science carried little weight in the policy world.

That left the prestige argument, which gained strength as the decade progressed. In the 1940s more than one RAND Corporation study had pointed to the political capital that the United States could gain from being first in space. Several others in the 1950s—including some associated with Wernher von Braun—reversed the argument to suggest the impact of the shock, should the Soviet Union be first into orbit.

Until fall 1957 the prestige argument kept discussion of orbiting a satellite going among those unaware of the top-secret WS-117L reconnaissance satellite system but cognizant of the slow progress being made by Vanguard. Only those at the top of the Dwight Eisenhower presidential administration appreciated that allowing the Soviet Union to establish the principle of freely orbiting over countries would be of great help in establishing the legitimacy of spying from space. Moreover, Eisenhower, who had led the liberation of Western Europe barely a decade before, refused to accept that polling reports of falling U.S. prestige in many of its allies represented anything of significance.

All this changed with the Soviet Union's launch of *Sputnik* on 4 October 1957. What would have been remarkable if achieved by the United States proved truly shocking when accomplished by the Soviet Union. *Sputnik* enabled the Soviet Union to claim that it had gone from a peasant-based economy to outer-space pioneer in just four decades, surely proving that the Soviet model offered the quickest route to economic development for Third World countries. It redefined the currency of progress and gave the United States no choice but to follow. Despite his refusal to admit that the United States was in a race, Eisenhower accepted the need to respond, though he insisted in placing U.S. space exploration in civilian hands, partly to protect the military programs from interference and partly to keep the budget of the newly created space agency in check.

In the Cold War context, prestige became the main rationale of the U.S. space effort, and pressure was suddenly relentless. By October 1959, just two years after *Sputnik*, the Soviet Union's *Luna 3* had returned the first images of the far side of

Commemorative postcard from Armenia of the launch of Sputnik by the Soviet Union. The postcard says, "The Soviet Sputnik—the first in the world" in six languages. (Courtesy Radio Moscow)

the Moon, even as U.S. lunar probes struggled to leave Earth orbit. In the 1960 U.S. presidential election, Senator John Kennedy lambasted Richard Nixon, Eisenhower's vice president, for declining American prestige and for falling behind in space, two issues that he presented as inextricably linked. By the end of the presidential campaign, space exploration had become an accepted political issue, a domestic measure of an administration's ability to get things done and acknowledged as the leading international standard for judging the relative merits of claims from the United States and the Soviet Union to represent the future.

The high point of prestige as the motivating force for space exploration came in the six weeks between 12 April 1961, when Yuri Gagarin became the first human in orbit, and 25 May 1961, when Kennedy publicly committed to a U.S. lunar landing program. Caught between the public failure of an American-backed attempt to overthrow Cuban President Fidel Castro and an upcoming summit meeting with Soviet Premier Nikita Khrushchev, Kennedy moved to trump Gagarin's success by pledging to send men to the Moon by the end of the decade. There had been intensive discussion within the Kennedy administration about whether there were any acceptable alternatives, but in the end the choice was clear. The need to restore American prestige required a bold and public acceptance of an impressive challenge, and there was nothing that compared to this. At that moment, prestige required a show of self-belief—commitment now, rather than achievement later, was the key factor in a situation where the ability of the United States to attract the political support of newly independent nations was being compromised by its failure to match exploits in space by the Soviet Union.

Kennedy was fully aware that his commitment to space exploration was first and foremost a political tool. When relations with the Soviet Union had modestly improved by 1963, he twice tried to persuade Khrushchev to pool resources in a joint lunar landing program that would symbolize cooperation as much as the 1961 pledge had symbolized competition. His lack of success was due as much to opposition in the U.S. Congress as it was to hesitancy in the Soviet Union. By then, what had started as a low-key approach to space exploration with Vanguard had become a major government program in which all the major aerospace companies and most regions of the United States held a significant stake.

The talk of prestige remained, but the reality had moved. Space spending had a constituency of political support that could prevent any dilution of either national honor or, most important, federal appropriations. By the late 1960s, when prestige was mentioned in connection with the U.S. space program, the connotation had once again swung back to a "sleight of hand"—the talk was of space exploration, but the goal was the protection of domestic political interests. By then the idea of prestige had played its part in turning spaceflight into a reality and was already fading into the background.

For no other country had it proved such a motivating force, not least because the concept of a race between competing political systems to reach the same technological goal had never been repeated so starkly. For the Soviet Union a version of regional prestige lived on in the Interkosmos program of the 1980s in which cosmonauts from communist bloc countries (Afghanistan, Cuba, Vietnam, and nations of Eastern Europe) flew as passengers on brief Soyuz missions. In Europe early French efforts to use *la glorie* as a rallying cry for European space efforts brought few results as its partners steered the European Space Agency toward a steady and unspectacular investment in space science. Japan's space program had largely been justified, both at home and abroad, as a way of assisting research and development efforts with commercial applications.

By the early 1990s the main relationship between space and prestige came from U.S. efforts to use space cooperation to bolster the international image, and even more important the self-esteem, of the new Russian state that was struggling to emerge from the collapse of the Soviet Union. Space Shuttle missions to the *Mir* space station and Russia's partnership with the United States on the International Space Station (ISS) program were intended to present Russia as retaining that ultimate mark of a great power, an active space program. Yet by the end of the decade, ISS had already come to symbolize less the dynamic of the future than the high cost of simply being in space for the sake of being in space, costs that took on a new dimension with the loss of the Shuttle *Columbia* in 2003. Instead it had become apparent that the space-related news best able to capture worldwide public attention came from the discoveries made by robotic planetary missions, pictures from which were explained by scientists often working in large, multinational teams rather than by politicians claiming credit on a national basis.

With no one racing to become the third country to launch its own citizens into orbit, China pursued human spaceflight at a slow pace, buying much of its technology from Russian companies that suffered in the early 1990s as a result of the collapse of the

Soviet Union. China's presentation of its space program as a symbol of achievement was tailored to a domestic audience rather than an international one. Brazil and India, huge countries with clear uses for space-based communications and Earth-observation resources, chose to pursue space exploration for work on technology with direct relevance to the needs of their economies.

In the early twenty-first century, national investment in space activity was no longer the forum for displaying the products of a political system that it briefly had been in the 1950s–60s. Instead it increasingly represented a means to stimulate technological development with rewards coming from sales in the global marketplace rather than coverage in the international media.

Giles Alston

See also: Apollo, Effect of Sputnik, Soviet Manned Lunar Program

Bibliography

Brian Harvey, *China's Space Program* (2004).
Roger Launius and Howard McCurdy, eds., *Spaceflight and the Myth of Presidential Leadership* (1997).
Walter A. McDougall, . . . *the Heavens and the Earth: A Political History of the Space Age* (1985).

Space Debris

Space debris comprises the by-products of human exploration and exploitation of near Earth space, plus natural objects like meteoroids and planetary particles that travel through the solar system. Because they possess tremendous kinetic energy, many such objects pose a serious threat to active, Earth-orbiting satellites and spacecraft or to humans performing extravehicular activities. The 1946 RAND report titled "Preliminary Design of an Experimental World-Circling Spaceship" devoted a substantial portion of one chapter to the probability of a meteorite hitting a satellite. Later that year Harvard astronomer Fred Whipple proposed designing satellites in a way that would shield their interiors from the worst effects of meteorite strikes.

After several decades of spaceflight, however, humans became more concerned about threats from artificial objects that remained in Earth orbit. Generally called orbital debris or space junk, those by-products included spent rocket bodies, aluminum oxide (alumina) particles from the exhaust of solid-rocket motors, inoperative satellites, fragments from on-orbit explosions (usually from heating of propellant in spent rockets) or collisions, spring-release mechanisms, spin-up devices, explosive bolts, release cords for solar panels, protective covers for payloads or attitude-control sensors, and paint flakes or bits of insulation from day-to-day deterioration of objects in the space environment. A 1995 U.S. interagency report estimated the quantity of orbital debris in low Earth orbit (LEO): 15,000 objects larger than 10 cm, 150,000 fragments larger than 1 cm, more than one million pieces larger than 1 mm, and more than one billion particles of at least 0.1 mm. A Whipple shield (an outer shell that absorbs the initial impact and dampens the kinetic energy of residual particles that penetrate it, thereby preventing damage to the spacecraft's main body) would afford an active satellite unsatisfactory protection against debris larger than 1 cm. The U.S. military Space Surveillance Network, operated by the North American Aerospace Defense Command (NORAD), and the Russian Space Surveillance System regularly tracked nearly 13,000 objects, 10 cm in diameter or larger, most of which were debris. Consequently, how to protect against debris measuring 1–10 cm became a troublesome issue for spacecraft designers.

Evidence of active spacecraft being damaged by collisions with junk caused worldwide concern from the mid-1990s onward. NASA spent more than $5 million to replace debris-scarred cockpit windows in the Space Shuttle, and Russia's *Mir* space station showed visible dents from encounters with debris. During a February 1997 mission to service the *Hubble Space Telescope*, astronauts found a hole in the high-gain antenna dish and suspected debris as the cause. Several months earlier, a 10-year-old, suitcase-size fragment from an exploded Ariane rocket destroyed the active, British-built *Cerise* microsatellite. Some analysts even predicted a catastrophic scenario dubbed "collisional cascading" in which debris generated from one collision in LEO would cause multiple secondary events that would trigger even more tertiary fragmentations, making LEO unusable.

In the mid-1990s, specialists began to address the orbital-debris problem in several ways. To better understand the scope of the problem, civil and military agencies sought to improve their modeling of the debris environment by various means: on-orbit experiments, refined application of ground-based surveillance systems, and computing or software simulation packages. Building on results from NASA's Long Duration Exposure Facility experiment during 1984–90, orbital projects like the European Retrievable Carrier in 1992, NASA's Orbital Debris Radar Calibration Spheres experiments in 1994–95, and the U.S. Air Force (USAF) *Advanced Research and Global Observation Satellite* (*ARGOS*) in 1999 contributed to that endeavor. Both the Haystack radar, operated by the Massachusetts Institute of Technology Lincoln Laboratory, and the Advanced Electro-Optical System, operated by the USAF Phillips Laboratory on Maui, Hawaii, worked on tracking debris fragments in the 1–10 cm range.

Meanwhile, other experts contemplated ways for mitigating the problem: better protective technologies for future satellites, reduction of junk through on-orbit garbage-collection systems or terrestrial-based laser "sweeping" to decelerate orbital fragments and hasten their atmospheric reentry, and minimizing the creation of new junk by voluntarily modifying both operational procedures and hardware designs. Some of the operational procedures used to prevent creation of new debris included venting propellants from spent rocket bodies, transporting refuse from human spaceflight back to Earth, and maneuvering dying satellites from their regular orbits to altitudes not normally used by operational spacecraft.

By 2005 as space junk continued to accumulate in LEO, many of the world's spacefaring powers either had instituted or were discussing regulatory standards to limit its further creation. The Inter-Agency Space Debris Coordination Committee, formed in April 1993 by the European Space Agency, Russian Space Agency, Japanese Aerospace Exploration Agency, and NASA, was working with the United Nations

Committee on the Peaceful Uses of Outer Space on voluntary, international guidelines for debris mitigation. Still, after a decade of discussion, the United Nations had not even agreed on a legal definition of space debris.

Rick W. Sturdevant

See also: North American Aerospace Defense Command, Space Surveillance, United Nations Committee on the Peaceful Uses of Outer Space

Bibliography

John R. Edwards and William H. Ailor, "Space Environmental Protection: The Air Force Role," *High Frontier: The Journal for Space and Missile Professionals* 2, no. 2 (February 2006): 41–46.

Theresa Hitchens, *Future Security in Space: Charting a Cooperative Course* (2004).

NASA Orbital Debris Program Office. http://orbitaldebris.jsc.nasa.gov/.

Space Industrial Policy

Space industrial policy deals with national and international policies and regulations about utilizing the space environment to perform industrial activities. Industrial processes conducted in outer space use its microgravity, radiation, temperature extremes, and/or vacuum in order to enhance product quality. The products produced must exhibit characteristics that render them significantly more useful than an equivalent terrestrial product in order to justify the higher costs, that is, what additional marginal value is provided that justifies the great expense of reaching and operating in orbit? As of 2008 no economically competitive product had yet been developed. Space commerce in its successful sectors earned income from processing information either collected in space (remote sensing) or sent through space (communications in its various forms). One must distinguish between national policies for industrial processes and the general concept of space activities.

Before *Sputnik*, there existed expectations that industrial processes would be carried out in outer space, but those processes remained vague. These expectations were premised on previous trends from other industries, that accessing and operating in space would follow traditional commercial patterns in becoming less costly over time. When that decrease did not occur, states desiring economic return from their space activities embarked on a series of subsidy programs aimed at fostering such industrial applications. U.S. initiatives were the largest for several reasons. One, the ideological context within which U.S. nonmilitary space activities operated emphasized reducing government involvement and funding, especially the latter, regarding activities that in principle could be operated by the private sector. This debate was at the heart of early American controversies about commercialization, especially of communications satellites. The result was a general government willingness to foster space commercialization endeavors, including the industrial sector. Second, NASA in the early 1980s needed to increase the number of payloads to launch on the Space Shuttle, the goal being to amortize the Shuttle's costs over a greater number of flights. Establishing industrial processes in orbit and possibly beyond would help meet those needs for additional payloads and revenue. For example, the *Apollo 14* lunar mission conducted the first materials processing experiments in a microgravity environment—a process continued on the Skylab and Apollo-Soyuz Test Project in 1975 and expected to continue on Shuttle missions in the 1980s.

The problem became establishing what products could be developed in orbit that would meet the price competition from terrestrial products. One of NASA's goals was to encourage industrial partners to research products in cooperation with the agency. Establishing this community of interest was comparatively easy before the Space Shuttle become operational, because the Shuttle's goal was a significant reduction in operational costs from $10,000 per pound to low Earth orbit to $1,000 per pound or less and a quick turnaround between flights. This drop did not occur as the Shuttle proved expensive to operate with an erratic schedule. The erratic schedule was the most troubling because commercial applications demand regularity and certainty in scheduling manufacturing tasks. A flight delay could mean loss of a market or a major client. Meanwhile NASA's quest for additional payloads led to extensive subsidies to prospective customers, including flights at heavily discounted rates and developmental contracts to companies producing test items for market. The applications seen as most likely were unique high-value and low-volume products for the pharmaceutical, electronics, chemical, and advanced alloy industries. Among the applications were crystal growth and materials solidification in the absence of gravity.

NASA engaged in a series of subsidy programs; the most important were those that utilized the microgravity environment for materials processing. The Space Shuttle was not truly useful for production, although its flights proved valuable for research. The Ronald Reagan presidential administration's 1984 space station program announcement changed the dynamics by proposing a long duration and more stable location for conducting such activities. The originally proposed station concept envisioned automated free flyers orbiting the space station, whose crew would fly to the human-tended fliers to supply materials and remove the products for return to Earth. Free-flyers eliminated the effects of human presence, including vibrations and contamination by out-gassing from life-support systems. Minimizing contamination effects was the goal of the Wake Shield Facility, which was deployed from the Shuttle cargo bay first in 1994, flying in formation with the Shuttle as processing took place in the vacuum of space. This experiment cost $125 million, of which only about $15 million was for the experiment itself. Human rating the experiment—that is, making it safe to travel on the Space Shuttle with humans—accounted for the remainder.

The European-built *Spacelab*, carried in the Shuttle cargo bay, provided the Shuttle's primary experimental capability. On each mission, different experiments could be carried inside *Spacelab* or attached to its outer surface on pallets, exposing the experiments to the effects of vacuum when the cargo bay opened. *Spacelab* flew on repeated Shuttle missions in 1983–97. Among them were the Advanced Gradient Heating Facility; Advanced Protein Crystallization Facility; and the Bubble, Drop, and Particle Unit. Other instruments characterized the space microgravity environment in order to allow comparison of the samples produced to terrestrial products.

The materials processing program was repeatedly criticized for lack of analytic rigor by the National Research Council, but for NASA the real goal was always keeping the Shuttle operational. When NASA subsidies declined after the 1986 *Challenger* accident, corporate interest decreased. However, interest resurfaced as the *International Space Station* (*ISS*) drew closer to construction. The Europeans with their *Columbus* module and Japan with its *Japanese Experimental Module* or *Kibo* envisioned using the *ISS* as the platform for experiments in materials processing and biotechnologies. Likewise the United States through its *Destiny* module envisioned a large microgravity materials processing program employing, for example, the Space Station Furnace Facility. U.S. difficulties with the Space Shuttle, which led to plans to shut it down by 2010, put the two international partner modules at risk of being left on the ground. It also hampered use of the research facilities already on orbit because the crew shrank to two crew members, meaning less time for research. By 2007 the United States was moving in other directions—the Vision for Space Exploration, which deemphasized microgravity research for its own sake.

Roger Handberg

See also: *International Space Station*, Space Shuttle, *Spacelab*

Bibliography

Office of Technology Assessment, *Civilian Space Policy and Applications* (1982).
———, *Civilian Space Station and the U.S. Future in Space* (1984).
Space Studies Board, National Research Council, *Assessment of Directions in Microgravity and Physical Sciences Research at NASA* (2003).
———, *Microgravity Research Opportunities for the 1990s* (1995).

Space Navigation Policy

Space navigation policy became increasingly important to the United States, Russia, and Europe as the commercial aspects of space navigation grew in significance in the 1990s. From the 1970s–90s, the United States through its NAVSTAR Global Positioning System (GPS) developed the capacity to pinpoint locations with great accuracy on Earth's surface or in the air. That information, when converted to civilian use, also possessed enormous economic value when the navigation satellite signals were transformed into data useful for customers. As a result of the economic value of the location and timing signals, serious policy questions arose about maintaining routine civilian access to the GPS signal. The U.S. military feared that unlimited access enhanced the targeting information provided America's adversaries. The U.S. Department of Defense (DoD) demanded its independent ability to control or deny others access to the GPS signal, a position resisted by international users and troubling for U.S. civilian users. The Soviet Global Navigation Satellite System (GLONASS) system was available to civilian users, but after the end of the Cold War it fell into difficulties when failing satellites were not replaced. The Russian Federation desired to rejoin the United States as a global power but lacked this important capability.

The DoD originally provided two GPS signals, GPS L1 and GPS L2. The former contained the Standard Positioning Service, the coarser signal was available to all users at no cost. Access began after the Soviet Union shot down Korean Airline flight 007 in

Illustration depicting the orbits of global positioning system satellites. (Courtesy National Oceanic and Atmospheric Administration)

September 1983, when the aircraft strayed into Soviet airspace due to a navigational error. That month U.S. President Ronald Reagan signed National Security Decision Directive 102, making the GPS signal accessible. The GPS system was incomplete in 1983 but provided sufficient navigation value to prevent entering restricted air space. The Standard Positioning Service provided a predictable positioning accuracy of 100 m horizontally and 156 m vertically with time transfer accuracy within 340 nanoseconds. The Precise Positioning Service, the highly accurate signal on both channels, was available only to authorized users.

The William Clinton U.S. presidential administration pushed to create greater commercial potential and stave off possible international competitors. That included upgrading the signal quality. Although the DoD totally funded and operated the GPS system, 2005 estimates showed that the number of civilian GPS users outnumbered military users by a ratio of 100 to 1 and growing. Commercial demand grew despite restrictions placed by the military on the signal.

Selective Availability (SA) occurred in two forms: through signal degradation, or turning it off. First, the DoD deliberately degraded the GPS signal. This distortion reduced the accuracy to a 95 percent level, one clearly adequate for civilian requirements. However reflecting the growth of inexpensive computing power, commercial GPS vendors quickly developed software eliminating much of that inaccuracy. Accuracies became fine enough that surveyors could employ the GPS signal in their work. DoD's growing awareness of this reality heightened its growing security concerns. Second, the DoD could turn off the GPS signal to all unauthorized users except for the encrypted signal in a particular geographic region. During the 1991 First Gulf War, the SA function was turned off completely so that all receivers in the region accessed an accurate signal. That decision reflected the fact that the U.S. military lacked sufficient quantities of military grade GPS equipment for its troops, and it scoured the west coast of the United States for civilian receivers that were shipped directly to units.

The user community's impact was minimal until Cold War restrictions gradually loosened. Until the system went fully operational in 1995, DoD held the commercial sector readily in check. A more aggressive civilian posture came in a 1995 report, *The Global Positioning System: A Shared National Asset*, from the National Research Council. Its title signaled the new era in which military constraints were not sacrosanct.

On 28 March 1996 President Clinton signed Presidential Decision Directive/National Science and Technology Council 6 (PDD/NSTC-6), which explicitly acknowledged a civil sector stake in the GPS system; GPS was then a dual-use governmental function. This overrode the military's exclusive focus on military missions—that remained its primary purpose—but GPS was recognized as becoming a public utility for the emerging global economy and communications infrastructure. On 2 May 2000 DoD turned off SA. American leaders expected that SA removal would further enhance the overall U.S. competitive position by reducing economic incentives to develop independent systems, such as the Europeans' proposed Galileo system. The American military could still switch the system off or not during crises and actual armed conflicts. The free civil signal became more robust on the L2 channel, improving signal acquisition and accuracy, while the new military signal (M-Code) operated on both channels.

U.S. policy did not succeed because critics remained skeptical of DoD's absolute control of the GPS signal, which given its global responsibilities was not subject to compromise. European nations responded in the early twenty-first century by accelerating the development of their Galileo navigation system. Galileo is analogous to GPS without the security cutoff, plus the program incorporates non-European states such as China and India, which desire independence from U.S. control. The Chinese have orbited a three-satellite regional system, the Beidou, supplementing GPS. Galileo, which was planned to be operational in 2010, represents a commercial and military challenge to GPS dominance. The U.S. response, signed by President George W. Bush in December 2004, was the U.S. Space-Based Positioning, Navigation, and Timing Policy, which reaffirmed that the United States would not compromise military necessities. However the necessity to incorporate commercial ramifications also remained a national priority. U.S. navigation policy worked within the limits of its security.

Roger Handberg

See also: First Gulf War, Global Positioning System, Navigation

Bibliography

Scott W. Beidleman, "GPS vs Galileo: Balancing for Position in Space," *Astropolitics* 3 (Summer 2005): 117–61.

Galileo: European Satellite Navigation System, European Union, Directorate of Energy and Transportation. http://ec.europa.eu/dgs/energy_transport/galileo/faq/index_en.htm.

Michael Russell Rip and James M. Hasik, *The Precision Revolution: GPS and the Future of Aerial Warfare* (2002).

U.S. Naval Observatory, NAVSTAR Global Positioning System. www.usno.navy.mil.

Space Science Policy

Space science policy among nations has followed a structured path carefully developed by each nation and as part of a larger framework in which overall scientific priorities were set and programs defined. These pathways varied from nation to nation, but in nearly every case where space activities have been a priority, scientific research objectives were contained within a governmental body or agency tasked by the nation's leadership. Historically these objectives have often become secondary space research goals when compared to more ambitious and more expensive human space exploration programs.

Whatever its stated public purpose, space exploration has always been subservient to broader national geopolitical policies, depending on the political issues of the moment as each country addressed its space technologies. In the United States and the Soviet Union, satellites and launch vehicles evolved primarily from technology funded and directed for missile and warhead research. In China defense applications also drove the priorities behind its space policies. But other nations had more civilian orientations for their space activities. India and Japan gave space development funding priority because of national scientific gains to be derived from space technology. In India remote sensing became the primary initial objective, drawn to the needs for water and land allocation resources needs. Japan sought Earth science applications in addition to communications satellite technology development. In Europe both science and commercial applications were early priorities, partly hampered by the lack of an indigenous launch vehicle capability. Eventually the restrictions imposed by the United States on access to its launch vehicles gave rise to independent development programs and political organizations in Europe, first the European Space Vehicle Launcher Development Organisation (ELDO) and later the European Space Agency (ESA). Japan also sought its own capability to launch satellites and spacecraft, giving rise to the N-1 and N-2 launcher programs, themselves evolved from U.S. Delta rocket technology. The largest and most ambitious space program of the early space age, the competition between the United States and the Soviet Union to develop a human lunar capability had little to do with the perceived value of lunar science. Soviet leaders sought to evolve their nascent human space capability to further strengthen their global reputation as a superpower, helping underdeveloped nations choose communism. The United States chose to "race" the Soviets to the Moon to provide a nonconfrontational way to reassert U.S. geopolitical hegemony in a peaceful but competitive setting. Lunar science and commercial space technologies would also be derived from this effort, but were never its central purpose.

Space research, however, was the original stated objective behind the early space activities of the two initial space powers, the United States and the Soviet Union. As such it was often characterized so in public, but the true priorities for space activities were usually national security or military in nature. The development of reentry vehicles for ballistic missiles often drove materials and flight-control research for space applications, and the development of deployable ballistic missiles was the first, initial objective of the U.S. and Soviet space programs. The first satellite launches conducted by both nations in 1957–58 had as their central purposes the gathering of data on the space environment and the Moon, utilizing instruments carried on board small satellite bodies launched by modified intercontinental ballistic missiles (ICBM) with some form of second stage that inserted the satellites into Earth orbit or to escape velocity toward the Moon and planets. Scientific research objectives or communications research also were the stated purposes of the first satellites launched by China, France, Japan, and the United Kingdom. In many cases science became a primary focus of collaborative endeavors among nations to enhance diplomatic relationships, because it was not directly tied to politically sensitive economic and military concerns. Thus the United States in the 1960s offered to launch two scientific satellites for free as a goodwill gesture to its allies. Many scientific satellites were developed through bilateral or multilateral agreements among nations, including such famous spacecraft as *Galileo* and the *Hubble Space Telescope* in addition to lesser-known spacecraft such as *TOPEX* (*Topography Experiment*), *Ulysses*, and *Phobos*.

Political organizations were formed to select specific science goals and administer the missions that would be flown. In the Soviet Union a series of political agencies reporting directly to the state central authority oversaw the design and construction of satellites, launch vehicles, and their instruments, which were assigned to different design and production bureaus. These assignments for the development of scientific research satellites varied, often determined by the political relationship among the national leadership and the heads of the design bureaus. These organizations were associated with their initial leaders, among them OKB-1 (Experimental Design Bureau), led by Sergei Korolev, and competing organizations, headed by Mikhail Yangel and Valentin Glushko, each of whom had varying vehicle designs and propellants under development. With the fall of the Soviet Union in 1991 space science missions were conducted by the Russian Space Agency (RKA), later renamed the Russian Federal Space Agency, which contracted with many of these same bureaus to construct the spacecraft and launchers. The Russian Federation also engaged in multilateral space research missions with France, Germany, and the United States. Human spaceflight programs, which contained microgravity research activities and experiments, have been conducted by Russia on board the *Mir* space station and the *International Space Station.*

For European nations the European Space Research Organisation, followed by its successor the European Space Agency (ESA), became the multinational umbrella organization that planned and conducted space research missions on behalf of its Member States. Assembled by 17 European nations and Canada, ESA conducted space science missions as approved and assigned by a board at the ministerial level. National ministers meet every three to four years to approve ESA's budget, in addition to new and continued space projects. Each member is assigned a specific minimum, mandatory funding level based on the national Gross Domestic Product of each Member State. This mandatory funding supported ESA's space science program, technology development, and infrastructure, while commercial applications and launchers systems were funded on a case-by-case, à la carte basis.

In Japan the government established a space activities commission to plan scientific research programs and the development of new spacecraft. The commission agreed to a space plan that assigned space research and engineering development to three agencies: the Institute of Space and Aeronautical Science, the National Aerospace Laboratory, and the National Space Development Agency. In 2003 Japan merged these agencies into the Japanese Aerospace Exploration Agency, which became responsible for all its space activities.

In China the central government established the China National Space Agency to conduct all space activities in a national program that began in 1956. By 1970 China launched the first of 88 Earth satellites in a series of scientific, communications, Earth observation, and remote sensing programs. Among these, scientific satellites carried a low priority. The central government established a broad series of intergovernmental agreements with other entities. Separate agreements for space research were signed with Brazil, ESA, France, and Russia. Under the terms of the agreements the prime minister was responsible for the development and execution of the individual space projects with the signatory nations. China established the Convention on Asia-Pacific Space Cooperation in 2005 to manage joint space activities, most of which were scientific in nature.

India began development of space research planning with establishment of a space commission and the Indian Department of Space in 1972. The department created the Indian Space Research Organisation, a joint agency of the space department and four other agencies of the Indian government responsible for overseeing advanced space development programs in remote sensing, communications, navigation, launch vehicle technology, and several launch sites in the country.

In the United States, Congress established the Office of Science and Technology Policy (OSTP) in 1976 as the principal federal agency to manage and coordinate the federal government's domestic and international space and science policies. OSTP was tasked to work with all federal agencies with science programs and also with Congress and the private sector to assure that presidential directives in science and technology were implemented. The director of the office usually served as the president's science advisor and has been considered the U.S. chief scientific adjunct to the president. A Council of Advisors on Science and Technology established in September 2001 coordinated science program planning. Three main federal agencies conducted space science research in the United States: the Department of Defense (DoD), the National Oceanographic and Atmospheric Administration (NOAA), and NASA. The DoD's scientific programs supported military purposes; NOAA promoted civilian applications, such as weather and ocean monitoring; and NASA conducted more pure research, stimulated by scientific agendas and publicly popular programs useful to motivate students in science and engineering. NASA and the DoD have been designated as the launch agents for other U.S. government agencies that develop scientific satellite programs.

As of 2007 only China, Europe, India, Japan, Russia, and the United States had indigenous space capabilities, defined as the ability to develop, construct, and launch their own space satellites and other spacecraft. Other nations that have sought to utilize space for scientific or telecommunications purposes have had to negotiate with one of these six to either build or launch their satellites. While space science has often become a tool of a larger geopolitical interest among nations, few nations have conducted space activities without the direct involvement of other countries. The flight of instruments developed by other nations aboard European, Russian, and U.S. space-craft and space probes has become common. The same has been true of research conducted by other nations, usually allies of the primary state, in crewed spaceflights in the Space Shuttle, Soyuz, Salyut, *Mir*, and the *International Space Station*.

Frank Sietzen Jr.

See also: European Space Agency, International Geophysical Year, Japan, United States

Bibliography

William Burrows, *This New Ocean* (1999).
Walter A. McDougall, . . . *the Heavens and the Earth: A Political History of the Space Age* (1985).

Space Telecommunications Policy

Space telecommunications policy was initially formed in the 1960s, along with the technical feasibility of communications satellites, and evolved through the twentieth century. The U.S. government did not have a satellite communications policy when the Soviet Union launched *Sputnik* in October 1957. Over the next seven years, Congress, the White House, the State Department, the Federal Communications Commission (FCC), NASA, and other departments and agencies competed to develop a policy. The military had its programs, but commercial satellite communications was different. As early as March 1959 Congress held hearings on "satellites for world communications." The need for action became apparent when American Telephone and Telegraph (AT&T) Company outlined plans for a global communications system just before the successful launch of the NASA/AT&T/Jet Propulsion Laboratory (JPL) *Echo* experiment in August 1960. In October AT&T filed an application with the FCC for permission to launch and operate a communications satellite system. When AT&T asked NASA to launch its satellite, NASA initially refused.

NASA claimed that satellite communications was part of the space program and therefore part of its charter. By contrast the FCC was ready to treat satellite communications in a manner similar to the transoceanic telephone cables. AT&T needed to consider NASA as the "owner" of the civil side of the U.S. government's launch vehicle fleet. After much negotiating, NASA completely funded the Radio Corporation of America (RCA) *Relay* medium Earth orbit spacecraft, agreed to launch—subject to reimbursement—AT&T's *Telstar* satellite, and eventually agreed to fund the Hughes small geosynchronous satellite *Syncom*.

Geopolitics also entered the equation. In April 1961 the John F. Kennedy presidential administration saw the launch of Soviet cosmonaut Yuri Gagarin, the first human in space, and the defeat of the U.S.-backed Cuban fighters at the Bay of Pigs. Kennedy asked Vice President Lyndon Johnson to identify something the United States could do in space that might restore its prestige—especially in underdeveloped countries. Johnson tasked NASA Administrator James Webb with defining far-reaching goals for the U.S. space program. On 25 May 1961 President Kennedy announced a series of goals for the U.S. space program. The most remembered of these goals was landing a man on the Moon; but there were others, including establishment of a global satellite communications system.

Over the next year Congress held hearings, White House staff wrote reports, NASA testified, and industry—telecommunications and aerospace—argued its points of view. The two basic choices were a government-owned system or a privately-owned commercial enterprise. There were arguments, with some arguing for an arrangement of international carriers (AT&T, RCA, Western Union International, and International Telephone and Telegraph), others arguing for ownership that included the aerospace companies, and still others for a compromise. In the end a compromise was reached. The Communications Satellite Act of 1962, signed into law on 31 August 1962, granted a monopoly to a new commercial entity to be owned 50 percent by telecommunications companies (international and domestic) and 50 percent by the public.

The Communications Satellite Corporation (Comsat) was formed after the act was passed, with Joe Charyk—then under secretary of the Air Force and secretly, director of the National Reconnaissance Office—appointed as president of the new organization and Phil Graham—publisher of the *Washington Post*—as chair. Several more "incorporators" were appointed; the company was incorporated in Washington in February 1963, and in June 1964 held its initial public offering of stock. The stock was oversubscribed and rose rapidly in price.

The Europeans—especially the members of the European Conference of Postal and Telecommunications Administrations (Conférence européenne des administrations des postes et des télécommunications or CEPT), which were organizations of postal, telegraph, and telephone (PTT) services—were unsure about the direction of events. CEPT, with occasional help from Japan and other countries, insisted on forming Intelsat (International Telecommunications Satellite Organization) to own and operate the international communications satellites. The interim Intelsat Agreement was more an agreement to disagree, but in the meantime the parties agreed to launch a satellite and to reach a definitive agreement at a later date. While the Communications Satellite Act

U.S. astronauts capture and redeploy Intelsat 6, *a communication satellite for the International Telecommunication Satellite Organization, which had been stranded in an unusable orbit since its launch in March 1990. (Courtesy NASA/Marshall Space Flight Center)*

settled some aspects of policy, Comsat, the U.S. domestic carriers, and the European PTTs were interested in signing a series of bilateral agreements among PTTs, not agreements among governments. On the other hand the U.S. State Department and the European foreign offices saw Intelsat as an international treaty organization.

It became U.S. policy to support a single global commercial communications satellite system. As part of that policy, in 1965 the Lyndon Johnson administration restricted transfer of communications satellite or launch vehicle technology for any purpose other than support of Intelsat; though the next year, the United States decided to encourage and provide financial support to selected less-developed countries. During the next few years, leading up to the 1969 conference to determine the Definitive Intelsat Agreements, it became obvious that many countries saw these policies as an attempt to maintain a technological monopoly on communications satellite technology and to maintain dominance within Intelsat. Less-developed countries did not expect to build communications satellites or launch vehicles in the foreseeable future. As a result they were interested in the most economical, highest performance, lowest risk solution: normally meaning a U.S. satellite. As with their internal negotiations (soon to form the European Space Agency), Europeans sought "just return:" a share of the procurement contracts proportional to their investment in the Intelsat system.

The Definitive Intelsat Agreements (opened for signature August 1971) provided for governmental participation (parties) and PTT participation (signatories). Comsat was the U.S. signatory, retaining its position as manager of the Intelsat system until 1979. The transition from a joint-venture consortium with a secretary general to an intergovernmental organization with a director general was to be complete by 1979.

One point of the agreements was that any entity proposing to build a satellite communications system would need to consult with Intelsat. What was unclear was whether Intelsat had any authority to reject such systems. While France was generally seen as the country most likely to launch a competing system, Canada was the first to actually launch a domestic satellite communications system (Telesat Canada's *Anik A1*) in 1972. The Richard Nixon U.S. presidential administration had taken office just before the beginning of the conference on Definitive Intelsat Agreements and domestic staff member Clay T. Whitehead began to look at the future of satellite communications. Whitehead soon headed the Office of Telecommunications Policy within the White House. He proposed that an "Open Skies" policy for domestic satellite communications (domsats) be implemented. In March 1970 the FCC, which had been talking about domsats since 1965, responded by requesting that all interested parties file applications for such domsat systems. Eight applications were filed. In June 1972 the FCC allowed for multiple entries (many systems) except for Comsat and AT&T, which were seen as dominating the existing market. After imposing certain restrictions, the FCC allowed the Comsat/AT&T project to proceed and later allowed Comsat, with International Business Machines and Aetna, to take over the MCI (Microwave Communications, Inc.)/Lockheed filing.

The U.S. domsat market expanded rapidly. Western Union launched its first satellite in 1974; RCA launched its first satellite in 1975; and Comsat/AT&T launched its first satellite in 1976. Other countries soon joined the United States and Canada with domsats being launched for Indonesia in 1976 and many launches in the 1980s when Australia, Brazil, India, Japan, and Mexico launched domsats—mostly U.S. satellites on U.S. launch vehicles. The Ronald Reagan presidential administration was uncomfortable with intergovernmental organizations such as Intelsat performing commercial functions. Early in Reagan's first term the FCC received several applications to establish international communications satellite systems. While the Intelsat Agreement required consultation before new systems were launched, it had come to be assumed that domestic satellite systems—and later regional satellite systems—were allowed, but that direct competition with Intelsat was forbidden. In November 1984 the Reagan administration decided to allow competition with Intelsat. Thereafter, a Special Interagency Group white paper concluded that new international satellite communications entities should be permitted and that customers should be able to bypass the Comsat monopoly to have "direct access" to Intelsat.

While the United States had been successfully setting domsat and international systems policy, mobile satellite systems arose. The first mobile system to be seriously considered was AeroSat, a high-profile U.S.–European program in the mid-1960s to mid-1970s. It was to provide communications and navigation services to aircraft over the Atlantic and Pacific oceans. A contract was awarded to General Electric to build a satellite. However, shortly after the contract award, the FAA withdrew financial support, infuriating the Europeans. At this same time maritime satellite communications were growing. While the United States and Comsat were first with the 1976 launch of Marisat for the U.S. Navy, the Europeans insisted on significant shares of the procurement contracts and a European headquarters. Inmarsat was finally established in 1982 with headquarters in London and European prime contractors for the spacecraft.

It was U.S. policy from 1962 to 1984 to support Intelsat as the single global satellite communications system. The treatment of Comsat was more complicated. Sometimes Comsat was a monopoly and sometimes it was a "chosen instrument" of the U.S. government. The 1972 "open skies" decision had opened the U.S. market to domestic systems, but Intelsat had a monopoly over international fixed (as opposed to mobile) satellite communications. Because of the original arguments in favor of an international satellite system carrying telephone calls, Intelsat had been designed to carry traffic across the major oceans: Atlantic, Pacific, and Indian. The original justification for U.S. domsats may have been telephony, but by 1976 it was clear that television distribution was most profitable proper market for communications satellites— as Arthur C. Clarke had foreseen in 1945. When the first fiber optic cables were laid across the Atlantic in 1988, Intelsat's telephony market began to disappear.

Comsat and Intelsat could see the problems facing them by the 1990s, but government regulators still treated both as monopolies. Complete loss of the Comsat and Intelsat monopolies would have happened in any case, but the actual course of events was accelerated by Lockheed Martin's decision to get into the satellite communications business by buying Comsat.

As early as 1996 Congressman Thomas Bliley had asked the General Accounting Office to look at restructuring Comsat/Intelsat. By 1998 Bliley's House committee passed a bill allowing both "direct access" (anyone can buy directly from Intelsat) and requiring "fresh look" (all existing contracts must be renegotiated). In 1999 the

Senate passed a bill allowing direct access, but not requiring fresh look. Shortly there-after Lockheed bought 49 percent of Comsat shares but needed legislation to allow it to buy the remaining shares. A revised bill passed both the House and Senate in March 2000, allowing Lockheed to buy Comsat and requiring Intelsat to privatize by 1 January 2003. Within two years Lockheed had shut down Comsat. Within five years Intelsat was privately owned and losing money.

U.S. policy has dominated the satellite communications arena since the 1960s—even when the aim was to subvert it, as with CEPT's efforts. As satellite communications became more commercial and national satellite systems more common, two disturbing trends arose. First, some countries restricted telecommunications traffic to their national systems. Second, many countries objected to the introduction of foreign culture via television programs. In some cases—notably Iran—this led to jamming.

Telecommunications had always been regulated and in the early twenty-first century was still regulated, but relaxation of regulations was becoming the order of the day. Congress created Comsat, but it also destroyed Comsat through regulation. Intelsat merged with PanAmSat—its major competitor—to create a mammoth satellite company.

David J. Whalen

See also: Comsat Corporation, Intelsat

Bibliography
Andrew J. Butrica, ed., *Beyond the Ionosphere: Fifty Years of Satellite Communication* (1997).
Government Accountability Office. www.gao.gov.
David J. Whalen, *The Origins of Satellite Communications 1945–1965* (2002).

Space Transportation Policy

Space transportation policy in the twentieth century primarily involved the launching of payloads, including humans, to Earth orbit and beyond. How payloads reach orbit is what constitutes space transportation policy, because decisions must be made regarding the launch vehicles employed and more critically how or by whom future launch vehicles will be developed and operated. Public military and civilian agencies initially made that choice, joined later by commercial institutions. Over time, space transportation policy became more systematic, signaling recognition of its economic, political, and military importance.

Initially no explicit policy existed because expendable launch vehicles (ELVs) were simply converted military missiles, both intermediate range and intercontinental. The U.S. military conducted all early space launches in pursuit of government objectives, using Redstone, Thor, and Atlas missiles. The earliest U.S. policy involved the partitioning of space launches from ongoing missile programs. This reflected the Dwight Eisenhower presidential administration's requirement that space launch efforts not interfere with the critical development of an intercontinental ballistic missile (ICBM) force. By the late 1950s U.S. ICBMs became solid fueled while space launch vehicles remained liquid fueled. The creation of NASA in 1958 further split the field because that agency assumed control of human spaceflight.

At first commercial space transportation did not exist—governments controlled all launch vehicles, even if they were manufactured by commercial companies. Efficiency was not a priority; rather, the reliability of operations was the goal, also called assured access. In pursuit of assured access, a proliferation of launch vehicles resulted. Within the U.S. space program, NASA pursued human spaceflight while the military focused on ELVs. However, NASA deemed the robust and reliable Saturn V too expensive to continue as its primary launcher. Approval and development of the Space Shuttle was the focus of NASA's efforts. The Shuttle, formally designated the Space Transportation System (STS), was to replace all other U.S. launch vehicles.

NASA's intent was to create a completely reusable vehicle in order to dramatically reduce costs. Cost reduction of expendable rockets is fundamentally limited as stages are discarded during ascent; the Shuttle was to break that paradigm. Unfortunately Shuttle development and operations proved costly and difficult. Reaffirming an understanding when the Shuttle was approved in 1972, NASA convinced President Jimmy Carter to designate the STS as the U.S. launch vehicle, replacing all ELVs and thus increasing the number of Shuttle payloads.

Skeptics doubted that the Shuttle would be able to lower launch costs. France, for one, was convinced that Europe needed its own launch capability independent of the United

Ariane 5 GS launcher lifts off from Europe's Spaceport in French Guiana, on 18 December 2009. (Courtesy European Space Agency/CNES/Arianespace)

States, so as to ensure Europe could launch communications satellites that would break the U.S. monopoly. Despite major problems with the earlier Europa program, in the early 1970s France and several other European countries decided to build the Ariane launcher to compete with the Shuttle. Ariane would be built and marketed by Arianespace, which became the world's first commercial space transportation company, albeit with significant subsidies from European nations through the European Space Agency. In the United States there were proposals to privatize the Shuttle once operational and to purchase a commercial Shuttle. Uneasy about the Shuttle's potential monopoly of space launch, and hence the potential for the loss of ability to launch payloads should it fail, the United States Air Force (USAF) eventually fought the shutdown of the ELVs.

Motivated by free-market ideology, the Ronald Reagan presidential administration aggressively pursued a commercial launch alternative, although its early policy statements placed the Shuttle as the core element in U.S. launch policy. In July 1982 the administration formally announced the end of the U.S. government ELV fleet, directing NASA to stop ordering Delta and Atlas boosters, while the USAF was to terminate Titan production. In October 1984 the Commercial Space Launch Act passed, establishing the Office of Space Transportation, which issued launch permits and licensed nonfederal spaceports. This reduced the paperwork and hindrances impacting private launch vendors. However, the heavy subsidies NASA employed to lure payloads to use the Shuttle undercut economic competition. In 1985 the Reagan administration reversed course and allowed the USAF to build the expendable Titan IV as an alternative to the Shuttle.

On 28 January 1986 the Shuttle *Challenger* exploded during liftoff. Two decisions flowed from that event: the ELV fleet received further support, and the Shuttle was removed from commercial operations. The first decision confirmed the U.S. mixed-fleet policy, which became the bedrock for subsequent U.S. space transportation policy. In 1988 title to the government ELVs was transferred to private ownership—with Shuttle competition removed, ELVs could hope to compete effectively. However, the U.S. ELV fleet had fallen technologically behind Arianespace with its Ariane 4s. With the Shuttle fleet grounded and U.S. ELVs unavailable after 1986, Arianespace grabbed the majority of the commercial launch market.

Adding to U.S. difficulties, by 1986 China and the Soviet Union had announced the commercial availability of their launchers. To protect U.S. launch manufacturers, the United States forced quotas on the number of permitted launches and placed a lower limit on the launch prices of these new "commercial" players. The George H. W. Bush presidential administration forced compliance through U.S. control of the communications satellite market (U.S. manufacturers built most comsats) and therefore an overwhelming share of the launch market. Quotas ended in the late 1990s when the launch market shrank below projections. These agreements triggered a split between U.S. satellite builders, who wanted cheap and readily available launches, and space transportation providers, who wanted to discourage Soviet and Chinese competition.

Within the United States the contending interests of the Department of Defense (DoD) and NASA impacted space transportation policy. A cooperative hypersonic flight program, the National Aero-Space Plane, fell victim to those differences.

U.S. Space Transportation Policy Evolution

Date	Administration	Topic	Synopsis
November 1981	R. Reagan	Space Transportation Policy	
July 4, 1982	R. Reagan	National Space Policy	NASA: end of ELV production; USAF: no Titans
October 30, 1984	R. Reagan	Commercial Space Launch Act	Established Office of Space Transportation: launch permits and spaceport licensing
August 15, 1986	R. Reagan	Presidential Directive	Removed shuttle from commercial operations
December 27, 1986	R. Reagan	U.S. Space Launch Strategy (NSPD 254)	
February 11, 1988	R. Reagan	National Space Policy	Government procure commercial launches
	R. Reagan	Commercial Space Launch Amendments of 1988	Government guarantee of liability limits for commercial launches
September 5, 1990	George H. W. Bush	National Space Policy Directive (NSPD) 2—Commercial Launch Policy	Trade agreements for free and fair markets—quotas
July 1991	George H. W. Bush	NSPD 4 - National Space Launch Policy	Policy on use of excess ICBMs for space launch
August 5, 1994	William Clinton	National Space Transportation Policy	NASA-RLV; DoD-EELV; commercial launch
September 18, 1996	William Clinton	National Space Policy	
1998	William Clinton	Commercial Space Act	
December 2004	George W. Bush	Commercial Space Launch Amendments	Regulation of commercial RLV flights
January 6, 2005	George W. Bush	U.S. Space Transportation Policy	

The DoD engaged in a lengthy series of space transportation studies in pursuit of cheaper and more reliable space access, because the Cold War's end triggered budget cuts, putting pressure on expensive systems such as the Titan IV. On 4 August 1994 President William Clinton's National Space Transportation Policy split the field into an expendable launch sector for the military and a reusable launch sector for NASA. Private companies were encouraged, although truly private operators still confronted the usual government subsidies to competitors. The Americans, Russians, Europeans, Chinese with Long March systems, and the Japanese with the H-2 built or operated government-developed and sustained launch vehicles.

Truly private-sector options had only two real opportunities, neither of which was economically viable in the twentieth century. One was for small payloads, but here the revenues were generally too small to pay for the development costs and compete against subsidized competitors. The second was if the number of satellites to be launched exceeded the capacity of government-developed and subsidized systems to put into orbit. This situation appeared likely in the mid-1990s as several companies were making major investments to build medium Earth orbit communications satellite constellations. The collapse of this demand in the wake of Iridium's bankruptcy led to the collapse of several private launch companies.

Russian competitors often evolved through a process of economic alliances with U.S. or European corporations. The most prominent include International Launch Services (Proton and Atlas rockets), Sea Launch (Zenit rockets off a Norwegian platform in the Pacific Ocean), and the European Starsem (Soyuz). These ventures helped sustain the Russian economy and further saturated an already oversupplied launch market. Because of government subsidies, no major rocket suppliers exited the market despite their inability to find customers. In addition upgrades of existing systems occurred even before they were fully operational, including the Ariane 5 Enhanced Capability A (ECA) and the H-2a.

Within the United States in the late 1990s NASA pursued reusable launch vehicles (RLV) in the form of the X-33 and the X-34, the former an RLV prototype to replace the Space Shuttle. The DoD promoted its Evolved Expendable Launch Vehicle (EELV) program. Both programs promoted consolidation of the U.S. launch industry into two major competitors: Boeing and Lockheed Martin. Orbital Sciences Corporation provided small launch vehicles Pegasus and Taurus, derived from earlier Strategic Defense Initiative programs. Attempts to recycle U.S. military missiles as launchers, such as the Minotaur (formerly Minuteman 2) missiles, were strongly resisted by small rocket developers despite an earlier 1991 National Launch Policy permitting such launches. The Russians found recycling missiles as launch vehicles a useful option enhancing their competitiveness in the marketplace. The X-33 RLV program and its sister, the smaller X-34, were both canceled in March 2001 due to cost overruns, schedule delays, and technical concerns. U.S. space transportation policy regarding RLV entered a period of disarray. The Space Launch Initiative program was a follow-on effort that was folded into the later Orbital Space Plane program.

The DoD EELV program envisioned reducing the costs of launch by more than 25 percent and improving reliability. Part of the cost reduction was to be realized through commercial launches that amortized development costs over more launches.

Boeing with its Delta 4 competed with the Lockheed Martin Atlas 5 for DoD contracts, and both vied with international competitors for commercial launches. However, the collapse of the projected commercial demand for launchers by 2000 effectively compromised these plans. By 2004 the U.S. military subsidized both systems in order to sustain them as viable launch options.

Expanded scope and sensitivities to technology transfer was another result of the Cold War's end. The Atlas drew on 1970s Soviet rocket technology, the RD-170, with a modernized version, the RD-180. Even in this U.S. security-driven program, the field's internationalization must be noted—a more vivid example of change is difficult to imagine. While the United States was happy to tap into Soviet technology, it tried to prevent others from doing so, as shown by U.S. objections to India's purchase of Soviet engines in the 1990s.

The breakup of Shuttle *Columbia* during reentry in February 2003 forced U.S. policymakers to confront a Shuttle replacement. NASA decided in 2005 to use the Shuttle to complete construction of the *International Space Station* (*ISS*), but then terminate Shuttle operations by 2010. While the Shuttle fleet was grounded after *Columbia*, access to the *ISS* depended on Russian Soyuz spacecraft. Another outcome of *Columbia* was the Vision for Space Exploration, announced by President George W. Bush in January 2004, which projected a return of humans to the Moon and later missions to Mars. The new plan envisioned development of a new Orion Crew Exploration Vehicle, an Apollo-like capsule, to transport astronauts into space. It would be hoisted into space by a new Ares 1 Crew Launch Vehicle (CLV), which would use the Shuttle solid rocket boosters and Space Shuttle main engine to reduce development costs. A new heavy-lift Cargo Launch Vehicle, using common design elements from the CLV, would handle cargo. In November 2005 NASA Administrator Michael Griffin announced that NASA would fund private launch vendors to take over cargo launches to the *ISS* on retirement of the Space Shuttle. This became much more of a reality when U.S. President Barack Obama proposed cancellation of the Constellation program, which was to provide the crew transportation replacement vehicle for the Shuttle, in his budget proposal for fiscal year 2011.

Other issues impacted space transportation policy, including the question of launch range modernization and the building of new spaceports. Earlier Reagan-era reforms raised the potential for non-federal U.S. spaceports operated by U.S. state governments. Alaska, California, Florida, and Virginia were the first, although their economic prospects proved difficult due to the collapse of the launch market and the failure of small rocket companies to survive the decline. The state spaceports were focused on small payloads, but that market was partly absorbed by larger rockets modified to carry multiple payloads. New Mexico later joined the field—its prospects tied to RLV efforts, such as the October 2004 flight of *SpaceShipOne*, the first purely commercially produced human spaceflight to suborbital space. To further that effort legislation was passed setting the parameters for future government regulation.

Internationally a number of states by 2005 were pushing various locations as launch sites in addition to existing spaceports, including Woomera, Australia, and Alcântara, Brazil, being the most prominent. China considered building a coastal launch site, while Russia assessed what to do with the Baikonur Cosmodrome, in the nation of Kazakhstan, which charged for use of the site. Space transportation in all nations remained a government, especially military-driven field, with few purely commercial players. Several efforts to build private launch vehicles in the early twenty-first century began, but even these envisioned obtaining government contracts as part of their business plan. Entrepreneurial spaceflight ventures received new interest and attention with the proposed cancellation of the Constellation program in 2010 and the concomitant re-vectoring of U.S. human spaceflight policy with greater focus on the commercial sector.

Roger Handberg

See also: Arianespace, France, National Aeronautics and Space Administration, Russia (Formerly the Soviet Union), Russian Launch Vehicles, Space Shuttle, United States Department of Transportation

Bibliography

Brian Harvey, *China's Space Program: From Conception to Manned Spaceflight* (2004).

John Krige and Arturo Russo, *A History of the European Space Agency, 1958–1987*, 2 vols. (2000).

Roger D. Launius and Dennis R. Jenkins, eds., *To Reach the High Frontier: A History of U.S. Launch Vehicles* (2002).

John M. Logsdon, ed., *Exploring the Unknown: Selected Documents in the History of the U.S. Civil Space Program, Volume IV: Accessing Space* (1999).

Asif A. Siddiqi, *Challenge to Apollo* (2000).

Technology and Engineering

Technology and engineering advances, particularly in rocketry, control systems, and systems engineering, made spaceflight possible.

Rockets originated in China along with the development of gunpowder, with the first written references dated roughly 1045 CE. The Chinese used them as fireworks for entertainment, and with "fire arrows" as weapons. Eventually knowledge of rockets diffused westward, and they were in use before 1400 as weapons in Europe. Gunpowder rockets continued in use as weapons thereafter, but they were relegated to minor roles over time, as artillery became progressively more powerful and accurate.

The three primary pioneers of rocketry, Russia's Konstantin Tsiolkovsky, the United States's Robert Goddard, and Romania's Hermann Oberth, were all motivated by visions of space travel and recognized that rockets were critical both for getting to space and for moving in space. The basis of rocket theory was Isaac Newton's third law of motion, described in his 1687 *Mathematical Principles of Natural Philosophy*. It states that for every action, there is an equal and opposite reaction. As hot gases exit through a nozzle, the launch vehicle or spacecraft moves reactively in the opposite direction. By the mid-1920s all three rocket pioneers had independently determined that the most efficient rocket engines propelled gases out the nozzle at the highest possible speeds and recognized that the combination of liquid hydrogen (LH2) and liquid oxygen (LOX) was an effective chemical propellant. All three also used Newton's gravitational law to calculate the velocity needed to propel a spacecraft into Earth orbit, and recognized that humans in space would experience zero gravity. They also determined that multistage rockets provided significant advantages for placing payloads into space.

Others soon elaborated on these ideas. In 1925 German architect Walter Hohmann discovered the minimum energy trajectory between planets, soon called the "Hohmann transfer." Slovenian engineer Herman Potočnik published a study of large-wheeled space stations in 1929 under the pseudonym Hermann Noordung. In 1930 French aviator Robert Esnault-Pelterie published a comprehensive overview text, called *Astronautics*, that summarized the burgeoning theories of spaceflight. Also in the late 1920s, Austrian engineer Eugen Sänger conceived of a rocket plane, an idea that he would pursue vigorously during the next two decades. In 1945 Arthur C. Clarke, a British engineer and later science fiction writer, published an article describing how three satellites orbiting Earth at just the right altitude would appear to be

ROBERT GODDARD
(1882–1945)

(Courtesy NASA)

Robert Goddard, U.S. physicist and inventor, developed much of the technology of liquid-fueled rocketry working with a small team and funding from the Guggenheim Foundation, arranged with the assistance of aviator Charles Lindbergh. Goddard's seminal paper on using rockets to reach space, "A Method of Reaching Extreme Altitudes," appeared in 1919, causing a worldwide sensation and convincing key technology leaders that rocketry was a viable means to achieve spaceflight. Goddard launched the world's first liquid-fueled rocket in Auburn, Massachusetts, in 1926; by 1940 his experiments near Roswell, New Mexico, had evolved his rockets into sophisticated designs including regeneratively cooled combustion chambers, gimbaled engines, and gyroscopic stabilization. NASA's Goddard Space Flight Center, in Greenbelt, Maryland, was named for him.

John D. Ruley

stationary above Earth's surface and act as communications relays that could provide television and microwave coverage to most of the globe.

However, spaceflight could not become reality until engineers built rockets with sufficient thrust to put an object into space. Reaching space required development of rocket engines to propel payloads to ever-higher altitudes. Goddard began practical experiments, in 1909 making measurements of rocket propellant exhaust velocities, and in 1921 switching from solid to liquid propellants. In 1926 he performed the first successful liquid-propellant rocket launch with funding support from the Smithsonian Institution and later garnered funding from the Guggenheim Foundation to continue his experiments. The Soviet Gas Dynamics Laboratory began experiments with liquid propellant rocket engines in 1930, the same year in which the German VfR (Verein für Raumschiffahrt or Society for Spaceship Travel) began tests. The American Interplanetary Society (later the American Rocket Society) began experiments in 1932, while the next year experimenters from the Moscow region of the Soviet Union, with some government support, launched their first liquid-propellant rocket.

Rocket development could not proceed far with only individual or private funding; only governments could provide the resources needed to make rockets useful. The German Army was the first to provide significant funding to create a ballistic missile with greater range than artillery. Recruiting the young aristocrat Wernher von Braun, in 1932 the German Army began serious missile development. Another important application of the 1930s–40s was rocket-assisted takeoff, in which a rocket helped an aircraft get into the air in a short distance. The U.S. Army and Navy both began funding of rocketry for this purpose. Army Ordnance and Army Air Forces funding went to the California Institute of Technology (Caltech) and the Navy funded Goddard and rocketeers from the American Rocket Society. The Soviet Union's early government funding supported short-range missiles, including Katyusha rockets and later rocket-assisted takeoff applications, but Josef Stalin's purges of the late 1930s led to the arrest of many key rocketeers, which slowed the development of Soviet rocketry through World War II.

Boosted by the Nazi government's rapid rearmament in the 1930s, the German Army developed ballistic missiles quickly. From 1932 to 1942, when the first successful A-4 (dubbed the V-2, as it was more widely known, when it was put into operation) test flight occurred, the German team under Walter Dornberger and Wernher von Braun solved many significant technical issues. Controlling the rocket's pointing direction required development of sophisticated gyroscopes, which controlled vanes in the engine exhaust. The Germans developed supersonic wind tunnels to determine the proper aerodynamic shape needed for the A-4 supersonic flight. The scaled-up A-4 rocket engines required the development of both regenerative cooling (circulating alcohol in tubes built into the combustion chamber wall) and film cooling (injecting alcohol along the internal wall of the combustion chamber). The A-4 required large-scale yet precise manufacturing of lightweight magnesium-aluminum tanks, for the alcohol and liquid oxygen, and turbopumps to drive the fuel and oxidizer into the combustion chamber through 1,224 fuel and 2,160 oxidizer injectors.

The U.S. Army and Navy both began serious ballistic missile and sounding rocket development only after receiving intelligence reports of the German V-2. The Army funded Caltech's Jet Propulsion Laboratory (JPL), founded in 1944, to develop the Corporal missile, which used a storable (noncryogenic), hypergolic (self-igniting), oxidizer-fuel combination of red fuming nitric acid and aniline that had been developed by Navy teams in the early 1940s. The liquid-propelled WAC Corporal sounding rocket, first flown in 1945, also used a solid-propellant booster, as the Caltech team was the first to develop composite grain solid propellants incorporating asphalt. These provided longer duration and more stable burning than gunpowder. Soon replacing asphalt with tar, these new solid propellants could be cast directly into rocket casings and stored for long periods without deterioration. The Army, with assistance from German rocketeers brought to the United States after German defeat, flew rebuilt V-2 rockets from 1946 to 1951. U.S. engineers studied and improved on the V-2 design. North American Aviation developed more powerful rocket engines with improved injector designs, which were used in the Navaho cruise missile, the Redstone, and Thor intermediate-range ballistic missiles (IRBM). Consolidated Vultee (Convair, later General Dynamics) built a prototype ballistic missile known as MX-774 that featured integral tanks in which the walls of the rocket body acted as fuel and propellant tanks. The integral tanks used thinner walls than earlier vehicles by using internal pressure to maintain structural integrity. Both innovations significantly lightened the MX-774, while swiveled engines instead of vanes in the rocket exhaust, and a separable warhead further improved performance. These all were used for the later Atlas intercontinental ballistic missile (ICBM).

Soviet ballistic missile efforts derived directly from the V-2, with the establishment of several rocket research centers in the Soviet zone of occupation in eastern Germany. Stalin commanded that Soviet rocketeers, soon led by Sergei Korolev, copy the V-2. With the assistance of captured German engineers and technicians, the first reassembled V-2 was launched from Kapustin Yar in 1947. The next year the Soviets launched the first R-1, similar to a V-2 but using Soviet components. Soviet engineers developed progressively more powerful rockets with similar improved technologies as their U.S. counterparts, leading to the R-7 ICBM that launched *Sputnik* in October 1957.

Most early U.S. and Soviet ballistic missiles used liquid fuels (typically alcohols or kerosene) with liquid oxygen, but by the late 1950s both nations had begun moving toward storable propellants, which reduced launch preparation times from hours to minutes. Soviet designs, such as the R-12 IRBM, emphasized liquid storable propellants with oxidizers, such as nitric acid and nitrogen tetroxide, and fuels, including kerosene and unsymmetrical dimethyl hydrazine. The U.S. Titan II missile used a similar storable combination. However, U.S. ballistic missiles generally shifted to solid propellants using composite grains but with improved performance from the addition of aluminum to the grain. The U.S. Air Force (USAF) Minuteman ICBM and the U.S. Navy Polaris submarine-launched ballistic missiles used solid propellants and became the cornerstones of the U.S. ballistic missile forces. The early cryogenic missile designs, such as the U.S. Thor, Atlas, Titan, and the Soviet R-7 were converted to use as space launchers, since they had higher performance than solid-propellant designs.

The development of ballistic missiles proved difficult. This was due partly to exotic technologies, such as rocket engines and guidance systems. Another factor was the uniqueness of the space environment in which ballistic missiles and later spacecraft needed to operate. The missiles and spacecraft also had to operate in zero gravity, in which floating particles could create unexpected failures, such as floating metal debris shorting out closely spaced wires, and in which fluids had to be forced to flow using induced pressure instead of gravity. This imposed the need for extreme attention to cleanliness and removal of manufacturing debris. New design mechanisms, such as helium pressurization, were needed to move fluids. The vacuum of space created problems because many engineering designs required convective air currents to transfer heat. Instead engineers had to design systems to use radiation and conductive heat transfer, thermal blankets, and other methods, and had to test the systems in new thermal vacuum chambers to simulate the space environment. Finally the sheer number and diversity of components became an issue, so that tracking manufacturing and design changes was a significant problem. These many concerns led to the development of systems engineering as a discipline to better track and control changes and to integrate and test complex systems.

Ballistic missiles, human spaceflight, and reconnaissance satellites all needed to return their payloads (warheads, humans, and film) to Earth. In the late 1940s and early 1950s, aircraft designs used ever more streamlined, swept-wing shapes with pointed tips. This trend made it seem that high-speed reentry from space would also use pointed shapes. Instead aerodynamicists from the National Advisory Committee for Aeronautics (NASA's predecessor) discovered in the early 1950s that blunt shapes were far more effective at slowing the reentry vehicle and dissipating heat. This discovery and the development of lightweight heat-absorbing and heat-dissipating materials were the bases for the design of nuclear warhead reentry vehicles, reconnaissance satellite film return capsules, and human spaceflight capsules.

While the Soviet Union was the first nation to place an artificial satellite into orbit, the United States was the first to study potential uses of satellites, and in the late 1950s and early 1960s developed spacecraft applications more rapidly than its communist rival. In 1946 the U.S. Army Air Forces funded the first Project RAND study, *Preliminary Design of an Experimental World-Circling Spaceship*. It assessed the potential applications and technical design capabilities of a satellite and the launch vehicle necessary to put it into orbit. It noted that a 500 lb satellite could be used for reconnaissance, weather observation, communications, astronomy, and microgravity experiments. Another RAND report in 1950 recognized that the Soviet Union would consider reconnaissance satellites as a threat and proposed that the United States launch a scientific satellite first to set a precedent for satellite overflight of the communist nation. Separate from this recommendation, that same year scientists from the United States and the United Kingdom proposed an international collaboration for 1957–58 to study the upper atmosphere during the next period of high solar activity. This eventually became the International Geophysical Year (IGY), for which the United States and the Soviet Union proposed to launch scientific spacecraft.

The Soviet Union held a significant advantage in the early years of the space age because of the earlier development and much greater lifting capacity of the R-7 in comparison to U.S. launchers. This was due to the relatively large size of Soviet nuclear warheads in comparison with their smaller but more efficient U.S. counterparts in 1953–54, when both nations decided the configuration of their first ICBM designs. Soviet ICBMs had to be larger to lift much heavier warheads than their U.S. competitors. The difference in capabilities was obvious from the mass of the first satellites launched for IGY: *Sputnik* weighed 83.6 kg and *Sputnik 2* weighed 508.3 kg, compared to *Explorer 1* at 4.8 kg and *Vanguard 1* at a mere 1.35 kg. The initial U.S. satellites were orbited by the Jupiter C and Vanguard launchers, neither of which was based on ICBMs. With the first successful orbital flight of Atlas (based on the Atlas ICBM) in 1958, the United States narrowed the Soviet lead in heavy-lift launchers. Not until the deployment of Titan and Saturn launch vehicles in the mid-1960s did the United States surpass the Soviet Union in heavy-lifting capability.

Launching larger payloads and placing them farther into space required the further development of upper stages. Upper stages had to ignite in flight and, in some cases, in zero gravity. Zero-gravity ignition required new technologies, such as settling motors to provide a temporary acceleration that would push liquid fuel and oxidizer toward the main engine(s). Deep space missions that required significant thrust and sometimes the ability to stop and restart an engine in deep space, such as for orbit insertion maneuvers, required the use of storable propellant stages, usually solids. Very heavy lift missions also required efficient engines that provided maximum thrust for minimum mass, with LOX-LH2 being the most efficient. The first successful LOX-LH2 upper stage was the U.S. Centaur, which first flew in 1963. It was soon followed by the U.S. S-IV and S-II stages used for the Saturn launch vehicles. The technology and operation of liquid hydrogen was a particularly difficult fuel to master. This was due to several factors, including the extreme cold temperatures that made metals brittle and shrink in size; the large boil-off rate that required much-improved insulation methods; and the small size of hydrogen molecules, which leaked through the tiniest of pores. These required extremely precise welding techniques using new metal alloys.

The international reaction to *Sputnik* surprised the leaders of both superpowers and stoked a race to achieve "firsts" in space in a battle for prestige and for military and economic influence. One aspect of the space race that followed emphasized robotic missions aimed at the Moon and planets. Performing missions far from Earth required heavy-lift launchers and the capability to communicate with spacecraft at ever-greater distances. High-performance radio frequency subsystems on board spacecraft, combined with large antennae and high performance amplification equipment on Earth, made long-distance communications possible. The U.S. Deep Space Network became the premier institution and set of technologies, pioneering a variety of new techniques, including high-precision timing equipment, error-correcting codes, and low-noise amplification using cryogenically cooled instrumentation. Both the United States (Surveyor) and the Soviet Union (Luna) successfully developed robotic landers that reached the lunar surface in the mid-1960s. In the 1970s the Soviet Union placed Venera landers on Venus, and the United States placed Viking landers on Mars. The Soviet Union developed the first remotely operated rovers to operate on the lunar surface, the Lunokhods of the 1970s.

The most visible aspect of the 1960s space race was human spaceflight. Human spaceflight required the development of spacecraft life-support systems and full-body space suits that enabled humans to operate in the vacuum of space and in zero gravity. These technologies generally derived from military programs, such as carbon dioxide scrubbers developed for submarines and pressure suits created for military aircraft pilots. Spaceflight required more efficient, lightweight, and extensive versions of these systems, and the solution to unique problems, such as the ballooning of space suits in a vacuum that hampered movement. Food safety was crucial to spaceflight, as illnesses in flight could lead to the premature end of the mission or to critical medical situations with limited capability to return to Earth. Improved methods to kill germs, such as irradiation, were first used for NASA's space programs, and then deployed on Earth by the 1970s to improve national food safety systems. To land a human on the Moon, the capability for two spacecraft to rendezvous was essential. The decision to use "lunar orbit rendezvous," in which a lander deployed to the lunar surface and returned to dock with the Command and Service Module so as to return to Earth, was a crucial decision that enabled the success of the Apollo program. Rendezvous capability, which

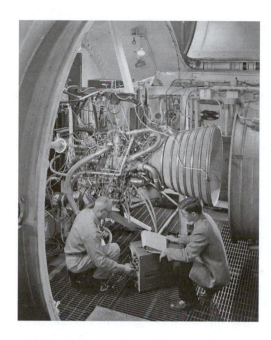

In this image, engineers test the RL-10 engine in NASA Lewis Research Center's (now Glenn's) Propulsion Systems Laboratory. Developed by Pratt & Whitney, the engine was designed to power the Centaur second-stage rocket. (Courtesy NASA/Glenn Research Center)

was a combination of the application of orbital mechanics and the development of orbital maneuvering systems, was developed on NASA's Gemini program and the Soviet Soyuz program in the 1960s. Important additions to the Apollo program were the manned lunar rovers that enabled the crews of the later Apollo missions to explore larger areas of the lunar surface.

Far less visible to the general public was a host of robotic application satellites. These included communications, navigation, reconnaissance, meteorology, missile early-warning prototypes, and Earth-orbiting science spacecraft, all of which had been launched into space by 1960. Most spacecraft consisted of the satellite payload and the "bus" that enabled the payload to perform its tasks and to return its data to Earth. The payloads comprised specialized equipment necessary to relay communications, broadcast navigational signals, observe Earth, or perform scientific observations and experiments. Subsystems provide power, point the spacecraft and payload, receive and store commands from, and store and send data to Earth, control the payload and other spacecraft functions, maintain proper temperatures, and so on. Depending on the spacecraft's mission, these subsystems used different technologies to provide their functions.

For example, spacecraft electrical power subsystems were designed to provide and distribute electrical power in several different ways. Spacecraft electrical power could be provided by solar arrays, radioisotope thermal generators (nuclear power), batteries, or fuel cells (combining hydrogen and oxygen to form water and in the process generate heat). Batteries were the least expensive, but were heavy and, if used as the sole source of power, had limited life. Solar energy worked well as long as the spacecraft was in sunlight, at Mars distance or closer to the Sun, and did not require too much power.

Nuclear sources were necessary for spacecraft on long missions to the outer planets far from the Sun. Fuel cells were sometimes used when hydrogen and oxygen was used onboard for several purposes, such as human spaceflight missions. Power distribution subsystems utilized direct or alternating current and were generally standardized at a specific voltage level. Designers typically chose among these various alternatives.

Other subsystems presented similar kinds of choices. Pointing a spacecraft could be achieved through momentum (reaction) wheels, thrusters, or by spinning the entire spacecraft, and for Earth-orbiting spacecraft, magnetic torque or gravity gradient stabilization using Earth's magnetic or gravitational field. Determining a spacecraft's orientation required measurement of the position of specific stars, the Sun, or Earth's horizon, magnetic, or gravitational field. Communications subsystems required assessments of the amount of information that needed to be sent and received, the frequencies that needed to be allocated for that data (including international and national political agreements and assignments regarding the use of radio frequencies), the appropriate size and type of antennae (helix, deployable, fixed structures, phased arrays), and the kinds of error correction and security encoding that were needed. Thermal control approaches were selected from thermal blankets, paints to absorb or reflect heat, thermal radiation mechanisms, such as the back of solar panels, and heat-absorbing or conducting materials.

Some technologies were particularly suited for spaceflight and developed along with space systems. Inertial measurement units to determine the position, velocity, and acceleration of a system were crucial for ballistic missiles. Photovoltaic (solar) cells were critical for spacecraft from the late 1950s and developed to supply space-craft electrical power. Hydrogen-oxygen fuel cells were developed for Gemini and were later used on the Space Shuttle. Space systems required strong, extremely light-weight structures that could endure extreme temperature and radiation environments, leading to the development of new structural analysis methods and new materials to address these extremes. The long distances required to communicate with deep space probes led to the creation of error-correcting codes and data-compression methods to allow large amounts of information to be transmitted with relatively few data bits. Reconstructing and enhancing that information was important for reconnaissance and scientific missions. These techniques were later adapted in other industries and applications, including communications, medicine, and entertainment.

Space systems also benefited from technologies developed for other applications. The most important of these was computing. Launch vehicles and spacecraft deployed computer systems of ever-increasing capability and complexity. However because the lack of atmosphere made spacecraft computers far more susceptible to cosmic radiation that could change the state of ones and zeros in computer memory and data transfers, spacecraft computers had to be radiation-hardened. This meant that spacecraft computing capabilities, particularly from the 1970s onward, followed and adapted computing advances driven from other applications and industries. These improved computer processing capabilities enabled space missions to perform much more complex functions than would otherwise have been possible. Space telescopes, such as the *Hubble Space Telescope* and the *Chandra X-ray Observatory*, depended on sophisticated flight software that enabled these observatories to observe many different targets with great precision and with smaller mission operations teams than would otherwise be necessary.

Spacecraft payloads were necessarily specialized to support their applications, usually with state-of-the-art technologies. Reconnaissance spacecraft were crucial to the development of Charge-Coupled Devices (CCDs), which were used for sensitive detection of imagery in a digital form. By the twenty-first century, CCDs were utilized in many practical Earth applications, including digital cameras, while continuing to be a core technology used for space remote sensing. Scientific missions deployed specialized instruments that were typically the state of the art for that particular observation or experiment. The Viking landers, for example, used the most advanced techniques to attempt to detect life on Mars in the mid-1970s. The Space Shuttle periodically serviced the *Hubble Space Telescope*, installing more advanced detectors and optical systems with each mission in the 1990s and 2000s. Communications satellites (comsats) deployed the best available technologies to maximize the number and quality of signals to generate profits for commercial systems and to ensure secure and/or high-speed data for military missions. Navigational satellites, such as the Global Positioning System, deployed state-of-the-art time generation technologies to provide precision time and position signals.

Understanding and developing the technologies of very high-speed flight through the atmosphere posed a critical challenge. While blunt bodies using ablative coatings solved the reentry problem for nuclear warheads and capsules containing humans or film, these were not sufficient for spaceplane designs that allowed for a returning spacecraft to land like an airplane. The U.S. X-15 program flight tested a variety of hypersonic technologies, including a new metal alloy Inconel X for surface structure, which could withstand temperatures up to 1200°F. It also used wedge-shaped aerodynamic shapes for the tail to ensure control at speeds near Mach 7. It flew 199 flights from 1959 to 1968. The USAF X-20 Dyna-Soar program, intended to put a military astronaut into orbit and gliding return, proposed using a delta wing with new heat-absorbing metallic alloys and cooling that used liquid hydrogen, also supporting an auxiliary power system. After cancellation of Dyna-Soar in 1963, the United States fielded several experimental vehicles that tested lifting body dynamics and new materials that led ultimately to the design of the Space Shuttle orbiter's aeroform and its thermal protection tiles that insulated the structure. The Soviet Union developed variations on these developments with the Raketoplan spaceplane (a design that never flew), lifting-body test vehicles, and the Buran shuttle.

The design of the Space Shuttle also drew from the development of large segmented solid rocket boosters. To lift heavy payloads, the USAF funded the development of large solid boosters to improve the capability of the Titan III series of launch vehicles. These successfully flew by the mid-1960s, and were the basis for new designs used on the Space Shuttle. While it used two O-rings instead of the single O-ring used between segments on Titan III and seemed intuitively more reliable, the Shuttle's Solid Rocket Boosters (SRB) proved flawed. The *Challenger* disaster in 1986 led to a redesign of the joint connecting the segments. The redesigned SRBs proved far more dependable. Because of their relatively low cost and high reliability, solid boosters were used on many launchers, including the Atlas, Delta, and Ariane launch vehicles, to improve performance.

To further boost performance, several nations developed launch vehicles that used liquid hydrogen in several stages, such as the Japanese H-2, the European Ariane V, and the U.S. Delta IV. Nuclear engines to enable large deep space missions were under development in the 1960s, but in the United States were canceled in the 1970s due to environmental concerns and the lack of interest in large space programs after Apollo. Soviet nuclear engine research continued at a low level for many years.

The development of the first space stations for manned reconnaissance began in the 1960s, with the USAF Manned Orbiting Laboratory (MOL) program and the Soviet Almaz. MOL was canceled in 1969 in favor of robotic reconnaissance systems, such as Corona and the later KH-9. The Soviet Union orbited Almaz as part of the supposedly civilian Salyut program and tested human operation of reconnaissance systems in the 1970s. Human reconnaissance turned out to be inefficient and costly compared to robotic systems, and by the 1980s the Salyut and Mir programs focused on civilian experimentation and long-duration missions to assess the effect of zero gravity on the human body. Two major innovations enabled more sophisticated and longer-duration space station missions. The first was the development of stations with two docking ports, first flown on *Salyut 6*, launched in 1977. This enabled robotic vehicles, such as Progress, to provide supplies and remove waste, while allowing the crew to keep its lifeboat Soyuz docked. The other major innovation was the building block design first implemented on *Mir* in the late 1980s. Much larger and more capable stations could be built from smaller components launched and docked together. The *International Space Station* also used this approach, with modules developed by several nations for different specialized functions. Telerobotic arms, pioneered by Canada and operable from orbit or from Earth, assisted with construction.

From the 1970s through the early 2000s, the engineering of deep space probes took advantage of new technologies and of the characteristics of the solar system to perform more complex and scientifically valuable missions. *Mariner 10* was the first spacecraft to use gravity assist in its successful mission to Venus and Mercury. Gravity assist enabled mission designers to change the trajectory and speed of deep space probes. This allowed spacecraft to fly by multiple planets, such as *Voyager 2*, which flew by Jupiter, Saturn, Uranus, and Neptune in the late 1970s and 1980s. Deep space missions required high levels of autonomy enabled by more powerful computers. By the 1970s JPL became a leader in the development of autonomous fault protection software that could detect and respond to internal spacecraft faults, including pointing the spacecraft antennae to Earth to await further instructions. In 1997 JPL demonstrated the combination of aerobraking and airbags to place *Mars Pathfinder* on Mars. This reduced the weight allocated to performing landings and correspondingly enabled less expensive missions with more capable payloads. *Pathfinder*'s small rover *Sojourner* demonstrated enhanced exploration capabilities, enabling more sophisticated rover missions, such as the Mars

Exploration Rovers in 2004. NASA initially tested high-performance ion engines with the Space Electric Rocket Tests in the 1970s but not in an operational environment until the Deep Space 1 mission launched in 1998. After that time, several more NASA, ESA, and Japanese missions successfully used ion engines.

The Mars program of the late 1990s was also a testbed for a streamlined management system, often coined "faster, better, cheaper" (FBC), in which the rigorous bureaucratic methods of systems management and systems engineering were partially replaced by small engineering teams. *Mars Pathfinder*'s success seemed to prove the validity of the new methodology, but only two years later the failures of *Mars Climate Orbiter* and *Mars Polar Lander* signaled the end of FBC and a return to classical systems management. Similar discussions and events also occurred in the larger NASA and aerospace communities, including the military, regarding the relative merits of bureaucratic processes and documentation versus small engineering teams, with the pendulum swinging back and forth between these two poles.

In the 1990s and early 2000s, new efforts to develop reusable launch vehicles were prominent. An initial attempt was the U.S. National Aero-Space Plane (NASP), a joint NASA-USAF reusable spaceplane program using several new technologies including a scramjet. These technologies proved too technically demanding and NASP was canceled in the early 1990s. Also in the early 1990s the USAF DC-X program tested a prototype launch vehicle that could launch and land vertically, using a small operations team. It too was canceled. Later in the decade, NASA's X-33 spaceplane demonstrator failed, largely due to problems with its aerospike engines and aluminum-lithium tanks for cryogenic hydrogen. More successful was the U.S. X-43 program, which drew from earlier U.S. and Russian developments and successfully tested a scramjet engine in 2004.

After a few attempts in the 1980s, a host of private companies began to develop expendable launch vehicles in the late 1990s, spurred by the lure of profits to launch constellations of communications satellites and the potential for space tourism. While most drew from existing technologies, some featured innovative designs. One was Rotary Rocket's Roton, which was to use helicopter rotors to slow its descent for landing. Others included proposals for reusable systems that returned to Earth using parachutes and airbags, and midair LOX refueling. These efforts collapsed by the early 2000s when the comsat market disappeared, but hopes for tourism were kindled when *SpaceShipOne* used a novel feathering reentry system to perform the first successful suborbital flights of a private reusable vehicle in 2004. It also used solid-liquid hybrid engines originally developed by the American Rocket Company in the early 1990s.

Spaceflight required new technologies, but to operate in space these technologies had to be made reliable. Launchers and spacecraft usually cannot be repaired once in operation, with the exception of crewed vehicles and some spacecraft designed to be repaired. Given the huge costs of putting humans in orbit, the costs of maintaining those systems that can be repaired are very large. Because of these issues, space vehicles have been a primary driver for the development of safety and reliability methods. Spacecraft and launch vehicle failures result in major investigations, the responses to which usually result in renewed emphasis on reliable designs for future vehicles. The investigation of the *Columbia* disaster of 2003 led to the recognition that organizational and social factors were important causes of technical failure.

Technologies and their related engineering disciplines developed rapidly from the 1930s through the 1960s with the initial development of ballistic missiles, launch vehicles, and spacecraft. After that time technology development slowed for most disciplines, with the notable exception of computing hardware and software. Over time, later space systems almost invariably used higher-powered computers and more sophisticated software. Technology development in other fields proceeded, but at a slower pace than in spaceflight's formative years.

Stephen B. Johnson

MILESTONES IN THE DEVELOPMENT OF SPACE TECHNOLOGY AND ENGINEERING

1045 Approximate date of first uses of rockets and gunpowder in China.

1400 First known drawings of rockets appear in Europe, in Conrad Keyser's *Bellifortis*.

1687 Publication of Isaac Newton's *Mathematical Principles of Natural Philosophy*.

1806 Adrien-Marie Legendre publishes method of weighted least squares, which enable improved celestial orbit estimation.

1903 Russian schoolteacher Konstantin Tsiolkovsky publishes "Exploitation of Cosmic Space by Means of Reactive Devices," the first article describing the use of rockets for spaceflight.

1909 U.S. physicist Robert Goddard begins first experiments to measure rocket propellant exhaust velocity.

1911 Tsiolkovsky publishes "Investigation of Universal Space by Means of Reactive Devices" which calculates escape velocity, orbital velocity, orbit insertion, and zero gravity.

1914 July: Goddard granted first patents for liquid and solid propellant propulsion systems and for multistage rockets,

1919 Goddard publishes "A Method of Reaching Extreme Altitudes."

1921 Goddard switches experimentation from solid- to liquid-propellant rockets.

1923 Romanian physicist Hermann Oberth publishes *The Rocket into Planetary Space*.
Goddard performs first tests using bipropellant thrust chambers with live propellants.

1925 German architect Walter Hohmann describes the minimum energy transfer orbit in *The Attainability of Celestial Bodies*.

1926 16 March: Goddard launches world's first liquid-fueled rocket.

1928 French aviator Robert Esnault-Pelterie coins the term "astronautics."

1929 Slovenian Captain Herman Potočnik (Noordung) discusses wheel-like space station in *The Problem of Spaceflight*.

1930 German Society for Spaceship Travel begins first liquid-propellant rocket engine tests.

The Soviet Gas Dynamics Laboratory begins work on liquid propellant rocket engines.

17 December: German Army provides funding to begin a liquid propellant rocket program.

1931   German Society for Spaceship Travel launches first liquid-propellant rocket.

1932   April, Goddard successfully flies the first gyroscopically controlled rocket.

1 December: Wernher von Braun joins German Army and begins army-funded research on liquid propellant rockets.

1933   14 May: American Interplanetary Society launches its first rocket.

17 August: GIRD (Group for Investigation of Reactive Motion) successfully launches first Soviet liquid propellant rocket.

1934   19 December: First flight of German Army A-2 rocket.

1936   Frank Malina and William Bollay start the Rocket Research Project at the Guggenheim Aeronautical Laboratory at the California Institute of Technology (GALCIT).

1937   December: Test flights of the German A-3 rockets fail due to control problems.

Goddard's first rocket tests with gimbaled steering.

1938   Sergei Korolev, Valentin Glushko, and other Soviet rocket designers arrested.

1942   3 October: First successful A-4 (V-2) test flight.

GALCIT group develops composite solid propellants.

1944   1 August: German engineer Eugen Sänger releases final report on rocket-boosted winged spaceplane.

1945   18 July: Institute "Rabe" founded, in which German and Soviet rocket engineers reconstructed and improved V-2 designs.

September: German scientists, including von Braun, arrive in United States under Operation Paperclip.

11 October: First flight of U.S. WAC Corporal sounding rocket.

15 October: In his article, "Extra-Terrestrial Relays" in *Wireless World*, Arthur C. Clarke proposes that three satellites in geosynchronous orbit around Earth could provide communications for the entire globe.

1946   16 April: First U.S. test of a V-2 ballistic missile at White Sands Proving Ground.

2 May: Project RAND publication of "Experimental World Circling Spaceship," which describes the feasibility of spacecraft and their potential applications.

13 May: Stalin signs decree establishing the Soviet ballistic missile program as a top priority and Scientific Research Institute (NII-88) Special Design Bureau No. 1 (OKB-1) as the main rocket research center.

October: German rocket engineers deported from East Germany to Gorodomlya Island, Russia.

1947   August: Tests of the first internally pressurized, integral tank rocket structure, with the MX-774 vehicle.

30 September: First Soviet R-1 (V-2 copy) flown from Kapustin Yar.

Cape Canaveral selected as U.S. missile test site.

1950 26 April: Korolev becomes Chief Designer of NII-88 OKB-1.

August, First tests of the 75,000 lb thrust XLR43 engine for the Navaho cruise missile.

1951 15 August: Soviet Union begins suborbital biological flights with dogs as passengers.

National Advisory Committee for Aeronautics (NACA) Ames Research Center engineer H. Julian Allen discovers that blunt bodies can survive reentry.

1952 11 February: *Collier's* magazine publishes concepts for piloted spaceflight discussed at First Symposium on Spaceflight.

17 April: Bell proposes to build piloted bomber missile spacecraft for the U.S. Air Force (USAF) after hiring German scientist Walter Dornberger.

20 May: First launch of the French sounding rocket Véronique.

19 July: First launch of a rockoon (rocket launched from a balloon), by the U.S. Office of Naval Research.

1954 9 July: X-15 Project begins as joint venture among NACA, USAF, and U.S. Navy.

27 November: The USAF issues requirements for a reconnaissance satellite.

Atlantic Research Corporation in the United States begins studies that lead to greatly improved solid propellant performance by adding aluminum.

1955 12 April: Japan launches its first Pencil sounding rockets.

September: Tsien Hsue-Shen deported from the United States to China.

December: USAF approves high-altitude human-occupied balloon flights under Project Manhigh.

Tyuratam (Baikonur) selected as Soviet intercontinental ballistic missile (ICBM) test site.

1956 14 August: OKB-1 becomes independent organization.

8 October: China creates the Fifth Research Academy of the Ministry of National Defense to develop rockets.

October: China receives two R-1 (V-2 copy) missiles from the Soviet Union.

1957 8 March: OKB-1 establishes Department No. 9 to develop piloted spacecraft.

2 June: First flight of Project Manhigh.

3 August: First successful test flight of the Soviet R-7 ICBM.

4 October: Soviet *Sputnik* becomes first human-made object to reach Earth orbit.

3 November: *Sputnik 2* carries first animal, a dog, into Earth orbit.

17 December: First successful test launch of Atlas ICBM.

21 December: USAF issues directive to implement Dyna-Soar (X-20) program.

Publication of first textbook of systems engineering.

1958 January: China receives several R-2 ballistic missiles from the Soviet Union.

January: Jet Propulsion Laboratory deploys portable tracking stations around the world to monitor *Explorer 1*.

31 January: U.S. launches its first satellite, *Explorer 1*, into Earth orbit.

17 March: *Vanguard 1* is the first satellite to use solar power.

26 March: *Explorer 3* carries the first tape recorder into space.

2 July: Soviet government authorizes development of Raketoplan.

August: U.S. Army Ballistic Missile Agency given approval to build Saturn launch vehicle.

1 October: NASA begins operation; Project Mercury starts.

18 December: SCORE (*Signal Communication by Orbiting Relay Equipment*) carries the first communications payload into orbit.

1959 2 January: Launch of *Luna 1*, the first spacecraft to escape Earth's gravity.

6 February: First successful launch of Titan ICBM.

28 February: First launch of Thor-Agena puts *Discoverer 1* (Corona 1) into the first polar orbit.

28 May: Army Ballistic Missile Agency launches monkeys Able and Baker on suborbital spaceflight.

8 June: First flight of X-15.

7 August: *Explorer 6* uses the first deployable solar arrays and sends first photograph of Earth from space.

4 October: *Luna 3* launched and sends first images of lunar farside.

Publication of first systems management standards in the USAF.

1960 1 April: *Tiros 1* launched, the first dedicated weather satellite.

13 April: *Transit 1B* orbited, the first successful navigation satellite.

22 June: *SolRad 1*, and *GRAB*, the first electronic intelligence satellite, put in orbit.

1 July: First launch of NASA Scout rocket from Wallops Island.

10 August: Launch of *Discoverer 13*, which returns the first capsule recovered from orbit.

12 August: *Echo 1*, the first passive communications satellite, placed into orbit.

20 August: First successful film returned from orbit with *Discoverer 14*.

4 October: *Courier 1B*, the first active repeater communications satellite, placed into orbit.

5 November: China fires its first ballistic missile, Dong Feng 1.

30 November: Launch of *Transit 3A*, which carries the first radioisotope thermal generator into orbit.

1961 12 April: Cosmonaut Yuri Gagarin, first human to fly in space.

5 May: Astronaut Alan Shepard Jr., first American to fly in space.

1962 7 March: NASA orbits the first space-based astronomical observatory, *Orbiting Solar Observatory 1*.

10 July: The first transatlantic television signal sent from the United States to France via satellite transmitted through *Telstar 1*.

11 July: NASA selects Lunar Orbit Rendezvous as mode for Apollo lunar landings.

August: First successful use of magnetotorque control on a Defense Meteorological Satellite.

1963 21 March: The Soviet Union begins testing the M-12 lifting body reentry vehicle prototype.

26 July: *Syncom 2*, first successful geosynchronous communications satellite, placed into orbit.

27 November: First successful test flight of Centaur, the first liquid hydrogen-based upper stage.

10 December: U.S. Department of Defense (DoD) cancels Dyna-Soar (X-20) program; approves Manned Orbital Laboratory.

1964 12 October: First multicrewed spaceflight (*Voskhod 1*).

28 November: Launch of *Mariner 4*, the first spacecraft to use a star tracker.

1965 18 June: First flight of Titan IIIC, which uses the first large segmented solid rocket motors.

16 July: First launch of Soviet Proton launch vehicle.

21 August: Launch of *Gemini 5*, the first spacecraft to use fuel cells.

26 November: First successful orbital flight of France's Diamant rocket with the *Asterix* satellite.

Systems management becomes the standard for the U.S. DoD.

1966 14 January: Korolev, founder of the Soviet space program, dies.

3 February: *Luna 9* performs first successful hard lunar landing.

16 March: First docking in space during *Gemini 8* between Agena and Gemini.

3 April: *Luna 10*, first spacecraft to enter lunar orbit.

2 June: *Surveyor 1* performs first soft lunar landing.

1967 27 January: *Apollo 1* crew dies in launch pad fire.

23 April: Cosmonaut Vladimir Komarov dies during *Soyuz 1* reentry.

9 November: First all-up test of U.S. Saturn V rocket is successful.

1968 24 December: *Apollo 8* the first crewed spacecraft to orbit Moon.

1969 16 January: First successful docking of two piloted Soyuz spacecraft.

11 February: First successful Japanese orbital launch, using the Lambda 4S launch vehicle.

21 February: First all-up test of the Soviet N1 rocket ends in failure.

1970 24 April: China's Long March 1 orbits the first Chinese satellite.

24 September, First automated lunar sample returned from the Moon from *Luna 16*.

17 November: *Luna 17*, with the first automated rover, *Lunokhod 1*, lands on Moon.

15 December: *Venera 7* successfully soft lands on Venus.

1971 16 February: Japan orbits its first satellite using an indigenous launch vehicle.

19 April: Launch of first space station (*Salyut 1*).

31 July: First sortie of the Lunar Roving Vehicle as part of the *Apollo 15* mission.

28 October: First and only flight of the United Kingdom Black Arrow launcher.

2 December: The Soviet *Mars 3* successfully lands on Mars, but fails shortly thereafter.

NERVA (Nuclear Engine for Rocket Vehicle Applications) canceled.

1972 5 January: President Nixon approves Space Shuttle development.

23 November: Last test of Soviet N1 rocket ends in failure.

1973 14 May: Launch of *Skylab*.

1974 24 June: Valentin Glushko cancels Soviet crewed lunar program.

13 August: Glushko cancels N1 program.

1975 September: First orbital launch of Japan's N-1 booster, a licensed Thor-Delta.

1976 7 February: Soviet leaders approve Buran-Energiya Space Shuttle.

20 July: *Viking 1* successfully lands on Mars.

1978 20 January: Use of Soviet automated Progress cargo freighters begins.

22 February: First launch of experimental Global Positioning System (GPS) satellite.

1979 24 December: First launch of Ariane 1.

1980 18 July: First successful Indian launch, using Satellite Launch Vehicle 3.

1981 12 April: First piloted Space Shuttle mission.

1982 30 October: First launch of Defense Satellite Communications System III introduces operational multiple beam antennae.

1983 23 March: U.S. President Ronald Reagan announces the Strategic Defense Initiative.

1984 4 February: First untethered extravehicular activity using Manned Maneuvering Unit.

11 April: First satellite retrieval and repair on orbit during Space Transportation System (STS) 41C.

1985 13 April: First launch of Soviet Zenit launcher.

1986 28 January: Shuttle *Challenger* explodes 73 seconds into flight.

20 February: *Mir* reaches orbit.

1988 19 September: First successful flight of Israel's Shavit launcher.

15 November: Only flight of Soviet space shuttle *Buran*.

1990 25 April: STS-31 crew deploys *Hubble Space Telescope* (*HST*).

1993 18 August: First test flight of DC-X (Delta Clipper), a single-stage to orbit demonstrator vehicle.

2 December, STS-61 crew repairs *HST*.

1995 April: GPS declared fully operational.

1997 4 July: Mars Pathfinder, first lander to successfully use airbags to land on another planet, later deploys the first Mars rover, *Sojourner*.

1998 20 November: First node of *International Space Station* (*ISS*), Zarya Control Module, reaches orbit.

4 December: First *ISS* assembly flight links *Unity* node and *Zarya*.

1999   Failures of *Mars Climate Orbiter* and *Mars Polar Lander* spacecraft begin the demise of NASA's Faster, Better, Cheaper initiative.

2001   1 March: NASA cancels X-33 and X-34 reusable launch vehicle programs.

2003   1 February: Space Shuttle *Columbia* breaks apart during reentry.

   15 October: China's first piloted spacecraft (*Shenzhou 5*) reaches orbit.

2004   14 January: U.S. President George W. Bush announces Vision for Space Exploration.

   21 June: *SpaceShipOne* is first privately funded spaceflight.

   16 November: X-43A sets new airbreathing vehicle record of Mach 9.6 with a scramjet engine.

2009   February: Iran launches its first satellite into orbit.

Stephen B. Johnson

See also: Astrophysics and Planetary Science, Civilian and Commercial Space Applications, Human Spaceflight and Microgravity Science, Military Applications, Space and Society

Bibliography

Mike Gruntman, *Blazing the Trail: The Early History of Spacecraft and Rocketry* (2004).
T. A. Heppenheimer, *Countdown: A History of Space Flight* (1997).
Mark Williamson, *Spacecraft Technology: The Early Years* (2006).

EXPENDABLE LAUNCH VEHICLES AND UPPER STAGES

Expendable launch vehicles and upper stages have been the primary means to propel payloads into and through space up to the early twenty-first century. Since ballistic missiles necessarily involved a one-way trip to deliver their warheads on target, expendable space launch vehicles (ELVs) from which they ultimately derived shared that characteristic.

ELVs derived from the development of ballistic missiles and sounding rockets from the 1930s through the 1950s. While Konstantin Tsiolkovsky, Robert Goddard, Hermann Oberth, and others speculated about, performed calculations on, and experimented with rockets, military organizations provided the resources necessary to turn these early ideas and small-scale experiments into practical devices. The development of the ballistic missile into an operational weapon by Nazi Germany under the direction of Walter Dornberger of the German Army (Wehrmacht) and Wernher von Braun demonstrated the viability and near-invulnerability of ballistic missiles. The United States, the Soviet Union, France, and the United Kingdom quickly began development of their own ballistic missiles by the mid-1940s, using both indigenous expertise and captured and hired German technology and personnel. Sounding rockets were a natural complement to ballistic missiles, as the missiles provided the means to perform scientific experiments in the upper atmosphere and near-Earth space, and knowledge of these environments was important to make ballistic missiles more effective.

Orbital launch capability was a straightforward, though difficult, process of extending ballistic missile performance and range. The motivation to achieve this level of performance was the potential marriage of nuclear weapons with ballistic missiles, able to deliver nuclear warheads at intercontinental distances from the Soviet Union to the United States and vice versa. The Soviet Union's R-7 intercontinental ballistic missile (ICBM) program, led by Sergei Korolev's OKB-1 (Experimental Design Bureau), achieved success in the summer of 1957, and shortly thereafter an R-7 launched *Sputnik* into orbit. The R-7 formed the core of the Soyuz and Molniya space launchers. Soviet ballistic missile programs expanded in the late 1950s and early 1960s, with OKB-456 developing the R-12 and R-14 intermediate-range ballistic missiles (IRBM) and the R-36 ICBM, and OKB-52 developing the "super-heavy" UR-500 ICBM. The R-12 and R-14 IRBMs became the basis of the Kosmos ELVs, while the R-36 led to the Tsiklon. The UR-500 became the Proton ELV. U.S. efforts were divided among the U.S. Army, Navy, and Air Force (USAF), with the USAF developing the Atlas and Titan ICBMs and the Thor IRBM. The U.S. Army developed the Jupiter IRBM, while the U.S. Navy developed its Polaris submarine-launched ballistic missiles (SLBM) and the Vanguard rocket to orbit a scientific payload. By 1964 the United States had launched satellites with ELVs derived from Thor, Atlas, and Titan. The Scout solid-propellant ELV used a first stage derived from the Polaris SLBM and a second stage developed from the U.S. Army Sergeant IRBM. France's Diamant (first orbital launch 1965) and China's Long March (first orbital launch 1970) were based on Sapphire and Dongfeng ballistic missiles, respectively. Japan's N-1 and H-1 launchers were licensed Delta designs from the United States, which in turn derived from the Thor IRBM. The European Space Vehicle Launcher Development Organisation's Europa launcher used the British Blue Streak IRBM as its first stage and the Coralie second stage, which drew from French ballistic missile technologies.

In later decades, the Soviet Union and the United States converted other ballistic missiles into launchers. Upgraded Titan IIs served as launchers for the U.S. Gemini missions in 1965–66, while the USAF used 13 retired Titan II ICBMs as small ELVs from 1988 to 2003. The Soviet Union converted several ballistic missiles into ELVs. The Dnepr ELV was converted from the R-36M2 ICBM and used from 1999. The UR-100 ICBM was used from 2000 as the basis for the Rokot and Strela ELVs, while the Start-1 ELV used from 1993 evolved from the RT-2PM solid-propellant ICBM. From 1995 the R-29R SLBM was used as the Volna ELV.

Not all launchers derived directly from ballistic missiles. A number of launchers were designed as space launchers from the start, though their engines and other technologies usually derived from ballistic missiles. In the United States, the Saturn ELVs drew significantly from U.S. Army Jupiter IRBMs and USAF ballistic missile engine programs. The Vanguard drew from the Viking sounding rocket, which in turn used ballistic missile technologies. The Soviet (later Ukrainian) Zenit rocket stemmed from the R-16 and R-36 technologies at OKB-456, later Scientific-Production Association (NPO) Yuzhnoe. The Soviet N-1 and Energia launchers were designed at OKB-1 (later NPO Energia, then Rocket and Space Corporation Energia) with technologies derived in part from the earlier R-7 and R-9 ICBMs. The European Space Agency (ESA) Ariane continued to evolve designs from the earlier Diamant and Europa launchers. India's satellite launch vehicles drew from Soviet and U.S. sounding rocket designs. Japan's first "Pencil" rockets were built for scientific purposes and evolved into the Kappa sounding rocket and then the Lambda (first successful launch 1970), Mu (first launch 1971), and M-series ELVs.

The first launchers of the 1950s–60s primarily used liquid oxygen (LOX) with either kerosene or alcohol fuels or toxic combinations of chemicals, such as nitric acid or nitrogen tetroxide as oxidizers and unsymmetrical dimethyl hydrazine (UDMH) or aerozine as fuel. The former were ultimately less useful for ballistic missiles than the latter, because the latter combinations were storable at room temperature for long periods, while LOX boiled off. For example, Atlas and Delta in the United States, and the R-7 in the Soviet Union, used LOX-based systems, while the U.S. Titan and Soviet Kosmos and Tsiklon used the storable combinations. The United States was the first nation to successfully develop large, solid-propellant ballistic missiles and launchers, starting with the Sergeant, Minuteman, and Polaris missiles, from which the Scout launcher was derived. Japan's solid rockets, built by the Institute for Space and Aeronautics Sciences of the University of Tokyo, were also successful. The Soviet Union's first solid propellant missiles, the RT-2 and Temp-2S, were deployed in the 1970s and the later Molodets and Topol missiles in the 1980s.

Space launchers often needed much higher performance than could be supplied by the standard liquid and solid propellants, particularly in their upper stages. This led to the development of liquid hydrogen (LH2) as a fuel to be used with LOX. The U.S. Centaur upper stage was the first to successfully develop and use LOX-LH2. This upper stage found use in a variety of missions, particularly deep space missions that required boosts out of Earth orbit altogether. Similar issues led to the development of the LOX-LH2 Saturn S-IVB used as the second stage for the Saturn IB and the third stage of Saturn V. Later upper stages and eventually entire ELVs moved to LOX-LH2 to provide the greatest possible performance, including the Japanese H-2, the Ariane 5, the Delta III upper stage, and Delta IV. Other upper stages, which had to perform many days or weeks after launch, and hence could not afford the boiloff of LOX-LH2 over time, used solid propellants, such as the U.S. Inertial Upper Stage.

During the first 50 years of their existence, ELVs generally evolved to higher performance and reliability. Early launchers had reliabilities in the 60–70 percent range, but by the 2000s, most ELVs had success rates of more than 90 percent. Higher performance was achieved through several means, including more powerful and efficient engines, lighter structures (such as composite or aluminum-lithium structures), and lighter weight (but higher performing) electronics, largely driven by dramatic advances in computing and communications technology.

Through 2007 governments had funded all successful ELVs. In the United States, Europe, and Japan, governments usually funded private corporations to build ballistic missiles and ELVs. The Soviet Union, China, and India developed their ELVs through government institutions. Europe began a partial privatization by creating Arianespace, which would market, build, and operate Ariane ELVs developed by the ESA. U.S. companies Martin Marietta (Titan), McDonnell-Douglas (Delta), and General Dynamics (Atlas) began offering launch services in the mid-1980s.

Privately funded efforts to develop ELVs began in the 1980s with American Rocket Company and Space Systems, Inc., which both attempted to develop purely commercial ELVs. A second wave of private attempts to build both ELVs and reusable launchers began in the 1990s when the demand to launch commercial communications satellites (comsat) appeared much greater than the supply. None of these corporate attempts, which included Beal Aerospace and Kistler Aerospace, succeeded, because the comsat market collapsed after the bankruptcy of Iridium, the first of the companies that had launched dozens of comsats into orbit. The first successful privately funded ELV to make orbit (in September 2008) was the Falcon 1 built by Space Exploration Technologies Corporation. All of these new ELVs drew heavily from technologies developed with government and private funding in prior decades.

Stephen B. Johnson

See also: Ballistic Missiles, Commercial Space Launch, Liquid Propulsion, Solid Propulsion

Bibliography

Brian Harvey, *Russia in Space: The Failed Frontier?* (2001).
Steven J. Isakowitz et al., *International Reference Guide to Space Launch Systems* (2004).
Frank H. Winter, *Rockets into Space* (1990).

Agena

Agena, an upper-stage launch vehicle developed in the 1950s by Lockheed Corporation for the U.S. Air Force (USAF) Ballistic Missile Division. Originally called "Hustler," the Agena's Bell XLR81 rocket engine was designed to provide thrust for an air-to-surface mission, but the USAF dropped this requirement and transferred the engine to the Agena program.

The USAF combined Agena with the Thor, Atlas, and Titan boosters to launch military reconnaissance satellites, such as Corona. NASA began plans to use Agena with the Thor and Atlas boosters in January 1959. The USAF completed the first successful launch of a Thor-Agena on 28 February 1959, placing the *Discoverer 1* satellite into the first polar orbit. In April 1959, the USAF authorized development of an advanced Agena B. NASA coordinated the Agena B development after the cancellation of its Vega launch vehicle program and authorized Marshall Space Flight Center to negotiate with Lockheed for the equipment. With another mutual agreement between the USAF and NASA, development and testing of the Agena D started in 1962 (an Agena C variant was proposed but never built). From 1964–68, NASA employed the Atlas-Agena D and Thor-Agena D in the Gemini program; the Agena launched alone as a spacecraft became the Agena Target Vehicle. The different Agena models used varied chemical propellants, including inhibited red fuming nitric acid (IRFNA), unsymmetrical dimethylhydrazine (UDMH), and nitrogen tetroxide (N_2O_4).

The Agena B and D stages achieved an 89 percent success rate, with four failures in 38 launch attempts.

Giny Cheong

See also: Corona, Gemini, Lockheed Corporation

Bibliography
Linda Neuman Ezell, *NASA Historical Data Book, Volume II: Programs and Projects 1958–1968* (1988).
———, *NASA Historical Data Book, Volume III: Programs and Projects 1969–1978* (1988).

Ariane

Ariane, a successful satellite launcher family developed within the framework of the European Space Agency (ESA), manufactured, launched, and marketed by a commercial firm, Arianespace. The name was derived from the Greek myth of Ariadne, Theseus, and the Minotaur's labyrinth. The labyrinth was a metaphor for the ambiguities that engulfed the European space effort in the early 1970s after the failure of the Europa rocket, the demise of the European Space Vehicle Launcher Development Organisation, and reorientation of the European Space Research Organisation to include application and scientific satellites.

Ariane was based on a French rocket designed by engineers at CNES (French Space Agency). The government was reluctant to develop Ariane because of the enormous cost, and because NASA was promoting the reusable Space Shuttle as the most cost-effective access to space. French authorities risked backing an obsolete launch technology with no hope of competing commercially. They were persuaded to proceed by the fear that they could not trust the United States to launch European telecommunications satellites that threatened U.S. business interests. Europe needed an independent launch capability. To make the project financially palatable at home, France phased out its national launcher program, Diamant, and convinced its European partners to contribute one-third of the Ariane development cost. Project management remained in the hands of CNES.

Ariane was designed to make use of trusted technologies as far as possible. The three-stage liquid-fueled Ariane 1 was first launched on 24 December 1979. This was followed by the more powerful Ariane 2 and Ariane 3, which were quickly replaced by Ariane 4. Ariane 4 was the primary European launcher for many years. From its first flight in June 1988 through its last in February 2003, it flew 116 missions, more than 70 of them consecutively without failure.

Ariane 4 comprised a central structure, propelled by 228 tons of liquid propellants, that could be combined with two or four liquid- or solid-propellant strap-on boosters. The simplest model could place 1.9 tons in geostationary transfer orbit. The most powerful version, using four liquid-propelled boosters, could place 4.2 tons in that orbit. This adaptability to the client's needs, along with its reliability, explains Ariane 4's commercial success.

ROBERT AUBINIERE
(1912–2001)

Robert Aubiniere, the first managing director general of CNES (French Space Agency) between 1961 and 1971, made significant contributions to French and European space programs. He played a significant role in the establishment of Toulouse Space Center and in the choice of the Kourou launch site in French Guiana. With his strong managerial skills and vision, he structured CNES and established close relations with the industry and academia, leading to a highly competent national space agency and industry. Aubiniere also served the European Space Vehicle Launcher Development Organisation as president from 1968 to 1970 and as general secretary from 1972 to 1973.

Incigul Polat-Erdogan

In 1987 ESA members committed themselves to develop Ariane 5, which was designed to launch increasingly heavy telecommunications satellites, beginning with a 6.5-ton payload capacity to geostationary orbit and evolving to 12 tons. A special system inside the fairing permitted the launch of multiple payloads. Ariane 5 was also intended to fly low-orbit missions to service the *International Space Station*. Its first launch, in June 1996, was a failure: the rocket veered off course and exploded less than one minute after ignition due to a software problem with its guidance computer. By July 2008 Ariane 5 had 36 fully successful launches, with two partial successful launches and two failures. With Ariane

Ariane Launch Vehicles

Variant	Dates	Payload Capacity (kg)	Successes/ Launches	Comments
Ariane 1	1979–1986	1,850 to GEO (geostationary Earth orbit)	9/11	Three stage, first four flights qualification launches
Ariane 2	1986–1989	2,175 to GEO	5/6	Increased thrust in first and second stage engines, taller third stage, larger payload fairing
Ariane 3	1984–1989	2,580 to GEO	10/11	Same as Ariane 2, but with two strap-on solid rocket boosters
Ariane 4	1988–2003	2,100 to 4,946 to GTO (geotransfer orbit)	113/116	Several variants, with two to four liquid or solid propellant boosters, 44LP version had four stages
Ariane 5	1996–	6,200 to 10,500 to GTO	36/40 through July 2007	Two stage, LOX-LH2 main stage, early second stage used N204-MMH; later second stage LOX-LH2, two solid propellant boosters

4's final launch in February 2003, Ariane 5 became the centerpiece of the European rocket family.

John Krige

See also: Arianespace, European Space Vehicle Launcher Development Organisation, European Space Agency, France

Bibliography

Emmanuel Chadeau, ed., *L'ambition Technologique: Naissance d'Ariane* (1995).
John Krige and Arturo Russo, *A History of the European Space Agency 1958–1987, Volume II* (2000).

Asterix. *See* Diamant.

Atlas

Atlas was the first U.S. intercontinental ballistic missile (ICBM), designed during the height of the Cold War, to deploy nuclear warheads and deter attack from the Soviet Union. In the early 1960s, it became one of the U.S. main launch vehicles. The original Atlas design by prime contractor Convair-Astronautics specified a missile 110 ft long and 12 ft in diameter with five rocket engines, but a dramatic reduction in the size of the thermonuclear warhead in the mid-1950s enabled the U.S. Air Force (USAF) to reduce Atlas to 75 ft long with a diameter of 10 ft and only three engines. Atlas was unique because it incorporated a pressurized stainless steel "balloon-tank" structure that did not need internal bracing.

The unmanned Mercury-Atlas 3 test in April 1961. This Atlas's inertial guidance system failed in flight. (Courtesy NASA)

Atlas Space Launch Vehicles

Atlas Booster Designation	Booster Length(ft)	Liftoff Thrust (kilo lb)	Upper Stage	No. Flown/Failed	Weight to GTO (lb)	Notes
Atlas-Able	67.3	357	Able	3/3	N/A	In use 1959–60
LV-3A	67.8	389	Agena B	47/5	1,800	Introduced 1960
LV-3B	67.5	367	None	10/3	N/A	Mercury-Atlas
LV-3C	66	389.4	Centaur	12/2	4,000	Introduced 1962
SLV-3	68.9	389	Agena	54/2	1,950	
SLV-3A	78.7	395	Agena D	12/0	2,265	
SLV-3C	69.5	395	Centaur D	17/2	4,500	Introduced 1967
SLV-3D	69.5	438	Centaur D	32/2	4,500	
Atlas G	72.7	438	Centaur D-1A	7/2	5,200	Introduced 1984
Atlas H	69.5	438	Solid motor	5/0	N/A	Introduced 1983
Atlas E/F	67.3	387	Solid motors	50/3	N/A	Refurbished ICBM
Atlas I	72.7	439.3–485	Centaur D-1A	11/3	5,100	Introduced 1989
Atlas II	82	490	Centaur 2	10/0	6,100	
Atlas IIA	82	490	Centaur 2A	23/0	6,400	
Atlas IIAS	82	714	Centaur 2A	30/0	8,000	Four solid boosters
Atlas IIIA	95.1	to 850	Centaur one-engine	2/0	8,800	RD-180 engine
Atlas IIIB	95.1	850	Centaur two-engine	4/0	9,920	Retired 2005
Atlas V 400 series	106.5	860	Centaur one-engine	15/0	10,913	Russian RD-180 engine, introduced 2001, regular aluminum construction
Atlas V 551 series	106.5	2,000	Centaur one-engine	1/0	19,114	Five solid rocket motors flew 1/06

After signing the Atlas contract with Convair in December 1955, the U.S. government invested more than $6 billion in Atlas development, exceeding the Manhattan Project, which developed the atomic bomb. To save time, the Atlas missile and bases were developed and built concurrently. Flight testing of the prototype Atlas A began in June 1957 at Cape Canaveral, Florida, and continued through the B, C, and D models until Atlas achieved Initial Operational Capability in September 1959, when the first armed Atlas Ds were deployed above ground at Vandenberg Air Force Base, California. Deployment of Atlas D at the first operational bases in Wyoming began on schedule in early 1960, and by 1962 the USAF had also deployed Atlas E and F versions.

After the solid-propellant Minuteman ICBM became operational in 1963, the USAF decided to decommission liquid-propelled Atlas D, E, and F in 1964–65. A total of 129 Atlases were removed from alert status, and more than 100 missiles were eventually refurbished for space launch purposes from 1968 to 1995. Atlas's space career began with Project SCORE (Signal Communication by Orbiting Relay Equipment) in December 1958, when an Atlas was shot into orbit to conduct satellite communications experiments. Atlas also served as the orbital launch vehicle for Project Mercury. Astronaut John Glenn was the first American to ride an Atlas into Earth orbit on 20 February 1962, followed by three more Mercury launches.

From 1960 to 1978, Atlas was paired with the Agena upper stage to launch a variety of military and civilian payloads, including the early Ranger and Mariner space probes. By far the most productive space application of Atlas was with the hydrogen-powered Centaur upper stage. After a rocky start (four failures in the first seven test launches), Atlas-Centaur became one of the workhorses of the U.S. space program, starting in 1966. Centaur carried numerous communications satellites and Mariner probes to Mars, Venus, and Mercury, and the Pioneer probes to Jupiter and Saturn.

In the early 1990s General Dynamics developed the enhanced-capability Atlas 2 and Atlas 3 vehicles. When the last balloon-tank Atlas 3 flew in February 2005, a total of 148 Atlas-Centaurs had been launched with only 13 failures. Many of the Atlas 2 and 3s were used for commercial launches and government launches. Atlas 3 used the Russian-built RD-180 engine. Martin Marietta purchased the Atlas launcher when it acquired General Dynamics in 1994 and moved the manufacturing facilities from San Diego to Denver as part of the corporate merger that created Lockheed Martin in 1995. As its contribution to the USAF Evolved Expendable Launch Vehicle program, Lockheed Martin designed the new Atlas 5 family of boosters, which first flew in 2002. Unlike previous versions of Atlas, which were evolutionary upgrades to the original Atlas ICBM, the Atlas 5 was a new design, with little in common with its like-named predecessors.

Joel Powell

See also: General Dynamics Corporation, Lockheed Martin Corporation, Mercury

Bibliography
J. D. Hunley, *U.S. Space-Launch Vehicle Technology: Viking to Space Shuttle* (2008).
Dennis Jenkins, "Stage-and-a-Half: The Atlas Launch Vehicle," in *To Reach the High Frontier: A History of U.S. Launch Vehicles*, ed. Roger Launius and Dennis Jenkins (2002).
R. E. Martin, "A Brief History of the Atlas Launch Vehicle," *Quest* 8, nos. 2, 3, and 4 (2000, 2001).
Chuck Walker with Joel Powell, *Atlas: The Ultimate Weapon* (2005).

Centaur

Centaur, a launch vehicle upper stage, was the world's first liquid-hydrogen rocket, serving the United States reliably for more than 40 years. In 1956 Krafft Ehricke of General Dynamics Corporation conceived of Centaur as a radical new second-stage rocket to sit atop Convair's Atlas intercontinental ballistic missile. Marshall Space Flight Center managed the Centaur program initially, but in May 1962 an explosion of the first Centaur 54 seconds after launch caused Congress to investigate. NASA debated canceling the program but instead decided to restructure it. Convair/Astronautics Division of General Dynamics maintained the industry contract to develop and build the Centaur, but in December 1962 NASA headquarters transferred management of the program from Marshall to Lewis Research Center (later Glenn Research Center) in Cleveland, Ohio.

Lewis Director Abe Silverstein worked diligently with a team of engineers to enable Centaur to become a vital workhorse for NASA, launching space science missions and commercial satellites. Centaur's first successful test launch came on 27 November 1963. Its first operational flight, on 30 May 1966, was the Atlas-Centaur launch of *Surveyor 1*, which performed the first controlled landing on the Moon and returned the first pictures from the Moon. In the 1970s Centaur launched spacecraft that visited every large body in the solar system: With Atlas, Centaur launched *Mariner 6–10*, *Pioneer 10* and *11*, and *Pioneer Venus*. Centaur was able to launch its heaviest and most scientifically complex missions with Titan as its booster: *Helios 1* and *2*, *Viking 1* and *2*, and *Voyager 1* and *2*. NASA interrupted this record of success in the 1980s with a failed $1 billion attempt to integrate the Centaur with the Space Shuttle. Though Shuttle/Centaur was ready to make its first launch in May 1986, NASA canceled the program for being too risky after the *Challenger* explosion in January 1986. Centaur experienced a rebirth in the 1990s as NASA commercialized upper-stage launch vehicle technology.

Centaur's unusual design contributed to its unique capabilities and to its longevity. Pressure from its propellants kept its stainless steel shell from collapsing. This pressure-stabilized structure meant Centaur required no internal supports. When engineers coupled this design with liquid-hydrogen propellants and its Pratt & Whitney RL10 engines, the Centaur became lighter, more powerful, and able to lift heavier and more advanced scientific payloads than rockets using other fuels. Liquid hydrogen also gave Centaur the ability to start and restart its engines in space, providing greater flexibility in designing planetary and lunar missions. In the early twenty-first century liquid hydrogen remained the signature fuel of the U.S. space program and was used as the main engine propellant for the Space Shuttle.

As of 2005 the Centaur was flying as an upper stage for the Atlas rocket, though management shifted from Lewis to Kennedy Space Center in 1998. The taming of liquid hydrogen, made possible through the operational success of Centaur, was one of the most important technical achievements of twentieth-century American rocketry.

Mark D. Bowles

See also: General Dynamics Corporation, Lockheed Martin Corporation, Space Shuttle

Bibliography
Mark D. Bowles, "Eclipsed by Tragedy: The Fated Mating of the Shuttle and Centaur," in *To Reach the High Frontier: A History of U.S. Launch Vehicles*, ed. Roger D. Launius and Dennis R. Jenkins (2002).
Virginia P. Dawson, "Taming Liquid Hydrogen: The Centaur Saga," in *To Reach the High Frontier: A History of U.S. Launch Vehicles*, ed. Roger D. Launius and Dennis R. Jenkins (2002).
Virginia P. Dawson and Mark D. Bowles, *Taming Liquid Hydrogen: The Centaur Upper Stage Rocket, 1958–2002* (2004).
John L. Sloop, *Liquid Hydrogen as a Propulsion Fuel, 1945–1959* (1978).

Delta

Delta, a family of expendable launch vehicles descended from the Thor intermediate-range ballistic missile (IRBM) that first flew during NASA's earliest years and was still in regular use entering the twenty-first century. Thor and later Delta were originally designed and built by the Douglas Aircraft Company, which merged into the McDonnell-Douglas Corporation in 1967. The Boeing Company acquired McDonnell-Douglas in 1997 and continued to build the Delta family.

In November 1955 the U.S. Secretary of Defense directed that the U.S. Air Force (USAF) develop the Thor IRBM. The first (unsuccessful) flight test occurred in January 1957, and after 17 further test flights, in September 1958 the USAF decided the IRBM was ready for operational deployment. Ultimately Douglas built 160 production Thors, which were removed from service in 1963. While solid-propellant missiles quickly made Thor obsolescent as a weapon, the design proved very useful as a launch vehicle first stage.

The first launch vehicle using Thor was the Thor-Able, which the USAF used to launch reentry test vehicles and scientific and meteorological satellites. For the reentry tests Able used a modified Vanguard second stage, to which Able added the Vanguard third stage for its satellite launches. The next Thor version was Thor-Able-Star, where Able-Star was an upgraded Able design by Aerojet Corporation. It launched navigation, communications, and geodetic spacecraft from 1959–65. Thor was also paired with Agena for 188 launches through 1972, most prominently of reconnaissance satellites, but also many other USAF and NASA payloads. The Thor-Agena vehicles underwent a variety of upgrades to engines, electronics, and also increased in length to carry larger payloads.

A Delta II rocket in the Strategic Defense Initiative program lifts off beside its launch tower, 29 September 1989. (Courtesy U.S. Department of Defense)

In 1959 NASA decided to create a launch vehicle from the Thor-Able rocket. The fourth modification of the Thor missile for spaceflight (after Thor-Able, Thor-Agena, and Thor-Able-Star), the rocket was given the name Thor-Delta ("delta" being the radio code word for the fourth letter of the alphabet), later shortened to Delta. The new rocket improved Thor-Able by adding an attitude control system for the coasting phases of flight and the solid-propellant third-stage Altair X-248 from Vanguard. Delta's first launch attempt in May 1960 failed, but the next launch in August 1960 placed the *Echo 1* satellite into orbit. Other early Delta successes orbited Tiros weather satellites, a solar observatory, the first international satellite (*Ariel 1*, a joint U.S./United Kingdom project), and American Telephone & Telegraph Company's *Telstar.*

NASA intended Delta as an interim solution while more powerful rockets were still in development, and initially ordered only 12 rockets. However, the rocket's flight record—11 successes out of 12 flights—was exemplary for the era, and NASA decided to continue the Delta program. Thus began a long series of incremental modifications: upgraded engines, lengthened tanks, and strap-on booster motors all increased payload capacity while maintaining a high successful launch rate. Delta flew 186 times between 1960 and 1990, lofting commercial communications satellites, NASA probes, and National Oceanic and Atmospheric Administration weather satellites. Delta was also licensed to Japan, and flew as the N-1, N-2, and H-1 vehicles.

Thor and Delta Launch Vehicles

Variant	Dates	Payload Capacity	Successes/ Launches	Comments
Thor-Able	1958–1960	350 lb to 300-mile orbit	10/16	Thor intermediate-range ballistic missile first stage, Able = modified Vanguard second stage, and some launches included modified Vanguard third stage.
Thor-Able-Star	1959–1965	110 lb to 385-mile orbit	15/20	Able-Star = improved performance Able upper stage, with in-space restart capability.
Thor-Agena (TA)	1959–1972	TA-A—1,700 lb to 100 × 1,000–mile orbit; Thorad-Agena D—2,000 kg to 150 × 300–km orbit	11/16 TA-A 39/48 TA-B 115/124 TA-D	Thor-Agena-A Thor-Agena-B Thor-Agena-D thrust-augmented Thor-Agena-D Thorad-Agena-D long-tank thrust-augmented Thor-Agena-D
Thor-Burner (TB)	1965–1979	TB 1—550 lb to 300 × 510-mile orbit	24/25	TB 1, 2, and 2A versions TB-1 used Altair third stage. Burner 2 upper stage, developed by Boeing, used Thiokol Star 37B solid. Burner 2A used Thiokol 26B solid motor for third stage.
Delta	1960–1990	200–450 to geostationary transfer orbit	174/186	Many variants (A through N, M-6, N-6). Early versions used Vanguard second and third stages. Later versions improved engine performance, lengthened and expanded stages, added solid Castor strap-on boosters.
Delta II	1989–	2,700–6,100 to low Earth orbit (LEO)	139/141 through 2008	Lengthened first stage, larger fairing. Later versions improve first-stage engine (RS-27A) and solid strap-on (Castor) performance.
Delta III	1998–2000	8,290 to LEO	1/3	Liquid oxygen/liquid hydrogen (LOX/LOH) upper stage using RL-10 engine. Nine Alliant graphite-epoxy motor-46 solid boosters.
Delta IV	2002–	8,600–25,800 to LEO	7/8 through 2008	Five versions. LOX/LOH first stage, R68 engines. Enlarged Delta III upper stage.

When the Space Shuttle entered operational service in the mid-1980s, NASA expected it to replace all expendable launch vehicles. Lacking NASA support, the Delta production line was shut down. Then the loss of Space Shuttle *Challenger* on 28 January 1986 created a sudden backlog of payloads. Unwilling to wait until the Shuttle returned to service, the USAF ran a competition to develop a medium-capacity vehicle to launch its Global Positioning System (GPS) satellites.

The result was the McDonnell-Douglas Delta II rocket, an upgraded direct descendant of previous Delta rockets. Compared to its 1960 predecessor, Delta II was considerably taller (132 ft) and carried some 40 times more payload, yet the core stage remained 8 ft in diameter. Since its first flight in 1989, Delta II rockets have launched dozens of GPS satellites and many significant NASA spacecraft, among them *Mars Pathfinder*, the Mars Exploration Rovers, *Near-Earth Asteroid Rendezvous*, *Stardust*, *Deep Impact*, and several Earth-observing spacecraft.

Delta III was an upgrade of Delta II that used the first liquid-hydrogen second stage on board a Delta. It had more than twice the payload capacity of Delta II. Only three Delta IIIs were launched in 1998–2000, two of which suffered unrelated failures. Delta III's second stage was incorporated into its successor, Delta IV.

Delta IV was Boeing's entrant in the USAF Evolved Expendable Launch Vehicle program and was much larger than, and significantly different from, other Delta rockets. With its powerful heavy-lift capability, it was expected to launch many large satellites for the military, in addition to scientific and commercial applications. The first Delta IV flight, on 20 November 2002, placed a communications satellite into geostationary transfer orbit (GTO) for Eutelsat. Two years later, on 21 December 2004, the first Delta IV Heavy demonstrated its massive capacity. This vehicle, with its three core stages strapped together in parallel, could deliver more than 14 tons of payload to GTO.

Kevin S. Forsyth and Stephen B. Johnson

See also: Ballistic Missiles, Boeing Corporation, Douglas Aircraft Company, McDonnell-Douglas Corporation, Japanese Launch Vehicles, National Aeronautics and Space Administration, United States Air Force

Bibliography

The Boeing Company, *Delta IV Payload Planners Guide* (2002).
Kevin S. Forsyth, "Delta: The Ultimate Thor," in *To Reach the High Frontier: A History of U.S. Launch Vehicles*, ed. Roger D. Launius and Dennis R. Jenkins (2002).
J. D. Hunley, *U.S. Space-Launch Vehicle Technology: Viking to Space Shuttle* (2008).
John M. Logsdon, ed., *Exploring the Unknown: Selected Documents in the History of the U.S. Civil Space Program, Volume IV: Accessing Space* (1999).

Diamant

Diamant, the first orbital launch vehicle built by France, roared into space from Hammaguir in Algeria on 26 November 1965, placing the French satellite *Asterix* (named for a comic strip character) in orbit. This first test launch secured France's place as the third nation in the world with independent access to space, following the United States and the Soviet Union.

Diamant was part of the "precious stones" series of launchers France built in the early 1960s, developed as part of a military program begun in 1958. The government was persuaded that it could not rely on the United Kingdom or the United States to defend it in the event of an attack or to support its strategic objectives. France's security lay, it believed, in having an independent nuclear deterrent and its own means of launching nuclear warheads into enemy territory.

The first precious stones were ground-to-ground missiles intended to carry warheads. Diamant differed: it was conceived as a civilian satellite launcher, but it piggybacked on the missile program. Its first two stages were an adapted Sapphire missile, which was comprised of liquid-propelled Emerald and solid-propelled Agate stages. Diamant's third stage, known as "P-6," derived from the second stage of the Ruby test vehicle, was used to place the satellite in orbit. Diamant was relatively inexpensive, costing only about 15 percent more than Sapphire. The rocket was developed by the rocket consortium SEREB (Société pour l'Étude et la Réalisiation d'Engins Balistiques—Ballistic Missiles Research and Development Company).

Three versions of Diamant were built, each more powerful than its predecessor, with Diamant A able to place 80 kg into low Earth orbit (LEO), Diamant B 150 kg into LEO, and Diamant BP4 180 kg into LEO. Diamant was phased out in 1975 after 10 successes in 12 launches. France, having secured the primary role in the development of Ariane, decided to stop building a national launcher and devoted its resources to the European program.

John Krige

See also: France

Bibliography
Claude Carlier and Marcel Gilli, *The First Thirty Years at CNES* (1994).
Herve Moulin, "A-1: The First French Satellite," in *History of Rocketry and Astronautics*, ed. Donald C. Elder and Christophe Rothmund, AAS History Series, Vol. 23 (2001), 51–72.
Berry Sanders, "The French Diamant Rockets," *Quest* 7, no. 1 (1999): 18–22.

Dnepr. *See* Russian Launch Vehicles.

Energia. *See* Russian Launch Vehicles.

Europa

Europa, the multistage rocket developed in the framework of the European Space Vehicle Launcher Development Organisation (ELDO). The first stage, built by the United Kingdom, was an intermediate range ballistic missile called Blue Streak, stripped of its military characteristics. The second stage, Coralie, was built in France. The third stage, Astris, was built in Germany. In 1967 ELDO members decided to create an upgraded rocket called Europa II to place satellites in a geostationary orbit and enable Europe to benefit from the potentially lucrative telecommunications market.

The initial development program for Europa was divided into three main phases, in which successive stages of the rocket were brought into operation. Blue Streak, first launched in June 1964, worked impeccably throughout. Coralie failed on its first two firings, as did Astris. The first time all the stages worked together, in a launch from Woomera, Australia, in June 1970, the heatshield around the satellite failed to eject and the payload fell into the sea. This disappointment was followed by the spectacular explosion of the first Europa II rocket, two minutes after liftoff in November 1971 from the new equatorial base in Kourou, French Guiana. The Europa program was canceled in April 1973.

Europa was a victim of national rivalry and poor project management. Each country jealously guarded control of its rocket stage, and ELDO did not have the authority to impose a management structure on the participants that would ensure the proper integration of the three stages. These problems led to elementary mistakes in design and manufacturing that resulted in launch failures.

Europeans learned a lesson from Europa's failure. The United Kingdom withdrew altogether from rocketry, preferring instead to rely on the United States. Germany decided to participate in the U.S. Space Shuttle program to learn program management skills. France took responsibility for developing, in collaboration with its European partners, the Ariane rocket, locating the management of the project in CNES (French Space Agency).

John Krige

See also: European Space Vehicle Launcher Development Organisation

Bibliography

Stephen B. Johnson, *The Secret of Apollo: Systems Management in American and European Space Programs* (2002).

John Krige and Arturo Russo, *A History of the European Space Agency 1958–1987, Volume I* (2000).

Peter Morton, *Fire across the Desert: Woomera and the Anglo-Australian Joint Project, 1946–1980* (1989).

Berry Sanders, "The Technical Evolution of the Europa Rocket," *Quest* 8, no. 4 (2001): 28–35.

GSLV (Geosynchronous Launch Vehicle). *See* Indian Launch Vehicles.

H-1. *See* Japanese Launch Vehicles.

H-2. *See* Japanese Launch Vehicles.

Indian Launch Vehicles

Indian launch vehicles trace their history to the 1960s, though Indian rocketry is centuries older. In 1792, Indian troops fired thousands of bamboo-and-iron war rockets on the British army at the battle of Seringapatam. India's involvement in space exploration began with the construction of the Thumba Equatorial Rocket Launching Station (TERLS), a cooperative venture with the United States, the Soviet Union, and France. TERLS hosted its first space launch on 21 November 1963 with a U.S.-built Nike-Apache sounding rocket and began launching indigenous payloads in 1965, first on French Centaure sounding rockets and then on license-built Centaures.

India's first launcher was the Rohini RH-75, the initial project of the new Space Science and Technology Center. The solid-propellant RH-75 debuted successfully on 20 November 1967. The modest RH-75 spawned a family of increasingly larger Rohini sounding rockets, several of which were still flying regularly at the beginning of the twenty-first century, and laid the foundation of India's space launch capability.

In 1972 India entered into an agreement to launch its first satellite on a Soviet booster. The following year, the Indian Space Research Organisation (ISRO) embarked on a program to build an indigenous space launcher. The resulting rocket—the four-stage, solid-propellant Space Launch Vehicle 3 (SLV 3)—was often called a copy of the U.S. Scout, because of the physical and functional similarities between the two vehicles. While India attempted to acquire Scout technology and the United States provided some technical reports, SLV 3 was an Indian design, building on the Rohini heritage. Most of the SLV 3's components, including its solid propellant, were produced in India.

The first SLV 3 launch, on 10 August 1979, failed, but on 18 July 1980 the second SLV 3 placed a 35 kg test satellite into orbit, making India the seventh nation with orbital space launch capability. The third SLV 3 suffered a fourth-stage malfunction and failed to place its payload into the intended orbit, but the launcher's fourth and final flight, on 17 April 1983, was an unqualified success.

ISRO then began developing more powerful launchers. The Augmented Space Launch Vehicle (ASLV) used an improved SLV 3 as its core and incorporated two strap-on boosters based on the SLV 3 first-stage motor. While the SLV 3 payloads had been in the 35 to 40 kg range, ISRO designed the ASLV to place a 150 kg satellite into a 400 km circular orbit. On 24 March 1987 the first ASLV launch failed when its second-stage motor did not ignite. Two more failures followed before ASLV succeeded, in its final flight on 5 May 1994.

By then India had shifted its focus to larger rockets that could service more economically viable payloads and orbits. The fruition of studies begun in the 1980s, the Polar Satellite Launch Vehicle (PSLV) debuted on 20 September 1993. Where its predecessors had been solid propelled, PSLV incorporated hydrazine/nitrogen tetroxide liquid propulsion in the second and fourth stages, the former powered by the Indian Vikas adaptation of the French Viking engine. The first PSLV launch failed, but on 15 October 1994 the second flight successfully put an Indian remote sensing satellite into a Sun-synchronous polar orbit. In 1999 the vehicle's fifth launch made India an international launch provider by lofting satellites from Germany and the Republic of Korea. Entering the twenty-first century, the PSLV was India's primary launcher. It launched the nation's first lunar probe, *Chandrayaan 1*, in October 2008.

India developed a more powerful Geosynchronous Satellite Launch Vehicle (GSLV), which replaced the PSLV's six ASLV-derived solid strap-on boosters with four large liquid-propellant boosters using the same engine as the second stage, and a Russian-made cryogenic third stage. Development efforts to replace the Russian stage with an indigenous cryogenic stage began right away. After two successful developmental flights, in 2001 and 2003, GSLV made its first operational flight on 20 September 2004, successfully launching the *EDUSAT* satellite. With the addition of GSLV, India was well positioned to launch a wide variety of space missions.

Bill Dauphin

See also: India

Bibliography

Peter Alway, *Rockets of the World*, 3rd ed. (1999).
Gopal Raj, *Reach for the Stars: The Evolution of India's Rocket Programme* (2000).
ISRO. www.isro.org.

Inertial Upper Stage

The Inertial Upper Stage (IUS) was a two-stage, internally guided rocket used to place satellites and spacecraft into trajectories above low Earth orbit (LEO). The Boeing Company designed and built the IUS under contract to the U.S. Air Force (USAF)

The Magellan *spacecraft with attached Inertial Upper Stage booster in the orbiter* Atlantis *payload bay, April 1989. (Courtesy NASA/Kennedy Space Center)*

Systems Command, as part of the USAF contribution to the Space Shuttle program. The USAF and NASA used IUS to loft payloads from 1982 to 2004.

IUS was used as an upper stage on U.S. Titan 34D/Titan IV boosters and the Space Shuttle. The upper stage was cylindrical in shape, 9.5 ft in diameter, and 17 ft long. It had a mass of 32,500 lb and used solid propellant. The first stage fired for up to 150 seconds, providing 45,600 lb thrust. The second stage fired for about 125 seconds, providing 18,500 lb thrust.

The Titan/IUS combination was used nine times for military payloads, including Defense Support Communication System satellites and Defense Support Program (DSP) missile warning satellites. The Space Shuttle used IUS for 15 payloads, including several Tracking and Data Relay Satellites, the Galileo mission to Jupiter, the *Ulysses* voyage to study the Sun, the *Magellan* flight to Venus, and at least one DSP. IUS flew 24 times throughout the history of the program.

The last Space Shuttle IUS mission was on Space Transportation System-93 in 1999, which boosted the NASA *Chandra X-ray Observatory* into its orbit. The last IUS on a Titan IV rocket placed a DSP satellite into geosynchronous orbit in early 2004.

Kenneth E. Peek

See also: Boeing Company, *Chandra X-ray Observatory*, Defense Support Program, *Galileo*, *Magellan*, Space Shuttle, *Ulysses*, United States Air Force

Bibliography

"History: Inertial Upper Stage," Boeing. http://www.boeing.com/history/boeing/ius.html.
"NASA Fact Sheet FS-1999-02-08-MSFC, The Inertial Upper Stage: Space Workhorse Boosts Chandra X-ray Observatory," Marshall Space Flight Center. http://www.nasa.gov/centers/marshall/news/background/facts/ius.html.

Japanese Launch Vehicles

Japanese launch vehicles derive from two distinct streams of development. Japan's Institute for Space and Aeronautical Science (ISAS) developed rockets for research applications, starting with the miniscule Pencil rocket and continuing through the Lambda and Mu vehicles that launched Japan's first satellites and interplanetary probes. The National Space Development Agency (NASDA) developed large boosters to launch communications, weather, Earth resources, and other operational satellites. While these organizations combined with Japan's National Aerospace Laboratory in 2003 to form the Japan Aerospace Exploration Agency (JAXA), the two threads of launch technology they generated remained distinct.

The roots of ISAS rocketry date to the 1950s. In 1955 Japan agreed to participate in the International Geophysical Year of 1957–58 and asked Professor Hideo Itokawa of the University of Tokyo Institute for Industrial Science (IIS) to develop sounding rockets for the purpose. Beginning in April 1955 Itokawa's team flew solid propellant Pencil rockets, 18 mm in diameter and from 23–30 cm in length. More than 150 Pencils flew, first in horizontal tests at IIS and then in vertical launches at the new Akita Rocket Testing Center at Michikawa Beach. In August 1955 the team began flying the 75 mm diameter, 134 cm long Baby, a solid-propellant two-stage rocket capable of reaching altitudes up to 6 km. Three models of Baby served to develop competency: Baby-S (simple) trained crews and developed launch support equipment; Baby-T (telemetry) transmitted data to ground stations during flights; and Baby-R (recovery) demonstrated recovery capabilities and carried a 16 mm camera.

Itokawa's group next began work on Japan's first true sounding rockets, the Kappa series, based on solid rocket motors from the Nissan Motor Company. Kappa-1 first flew in September 1956, followed by many Kappa models concluding with the Kappa-10, which first flew in November 1965. Starting in 1962 the group began designing a new series of rockets based on Nissan's 735 mm diameter Lambda motor. The first three Lambda models were suborbital sounding rockets, but in 1964 a merger of the IIS rocket group and the Institute of Aeronautics created the ISAS, whose goal was to create an orbital launch capability. The Lambda-4S, consisting of four solid-propellant stages and two strap-on boosters, was designed both to launch Japan's first satellite and demonstrate technologies planned for the forthcoming Mu orbital launchers. After four launch failures, the final Lambda-4S booster placed Japan's first test satellite, *Oshumi*, into orbit on 11 February 1970.

The first-generation Mu launcher (Mu-4S) was also a four-stage solid rocket vehicle, augmented by eight small strap-on boosters. The first Mu-4S launch attempt, in September 1970, failed, but on 16 February 1971, Mu-4S-2 successfully orbited the *Tansei* technology test satellite. After a total of three successful launches, Mu-4S was replaced by the second generation of Mu launchers, the successively more powerful Mu-3C, Mu-3H, and Mu-3S. By 1984 Mu rockets had successfully launched 13 satellites in 15 attempts. In 1981 ISAS began work on a launch vehicle for missions to Comet Halley. ISAS based this new vehicle, designated Mu-3SII, around the Mu-3S first stage, with newly developed second and third stages (and an optional fourth stage for interplanetary missions), and with two large strap-on boosters replacing the eight small boosters of earlier Mu vehicles. In January 1985 the first Mu-3SII successfully placed Japan's first interplanetary probe, *Sakigake*, into solar orbit.

Mu-3SII remained in service until 1995, launching seven spacecraft successfully out of eight attempts. Its success encouraged development of the much larger M-5 launcher. This three-stage solid rocket vehicle entered service in 1997 with the launch of the *HALCA* (*Highly Advanced Laboratory for Communications and Astronomy*) radio astronomy satellite. M-5 rockets launched Japanese missions to Mars (*Nozomi*) and the asteroids (*Hayabusa*), in addition to radio, infrared, and X-ray astronomy satellites. M-5 was still operational in 2008.

The Japanese government formed NASDA in 1969 to pursue large liquid-propelled boosters. NASDA's predecessor, the National Space Development Center, had already begun development of liquid-fuel rockets; a large indigenous satellite launcher was years away. Rather than wait, NASDA fielded the N-1, a license-built copy of the

McDonnell-Douglas Thor-Delta, which first flew in September 1975. Through 1987 the N-2 and its successor the N-2, also a license-built Delta model, successfully orbited 15 communications, weather, and Earth resources satellites, out of 16 launch attempts.

The next phase in the program was the H-1, which reused the N-2's licensed first stage and boosters but added Japanese-designed second and third stages, the former powered by an indigenous liquid-oxygen/liquid-hydrogen (LOX/LH2) engine. The H-1 successfully flew nine times between 1986 and 1991, but its use of U.S.-licensed components prevented NASDA from selling launches on the world market. In 1984, NASDA embarked on the development of the H-2, a wholly Japanese satellite launcher built by Mitsubishi Heavy Industries consisting of a large LOX/LH2-powered first stage and an upper stage with a restartable LOX/LH2 engine, along with two large solid rocket boosters. NASDA planned to launch its H-2 Orbital Plane Experimental (HOPE) space plane with the H-2, and the first H-2 launch, in February 1994, included a HOPE heatshield materials test. The H-2 proved both expensive and complex, and after consecutive failures on its sixth and seventh flights, NASDA stopped H-2 work in late 1999. A successor, H-2A, based on upgraded versions of the H-2 core stages with new solid rocket boosters, entered service in August 2001. H-2A, which can be configured with two or four additional smaller solid rocket boosters, logged its 12th mission in February 2007, with only one launch failure.

Bill Dauphin

See also: Japan

Bibliography
Peter Alway, *Rockets of the World*, 3rd ed. (1999) and supplements (2000, 2003).
Brian Harvey, *The Japanese and Indian Space Programmes: Two Roads into Space* (2000).
Japan Aerospace Exploration Agency. www.jaxa.jp/index_e.html.

Kosmos (Launch Vehicle). *See* Russian Launch Vehicles.

Long March

Long March family of rockets (Chang Zheng in Chinese, abbreviated CZ), an expendable launch system designed and operated by the People's Republic of China. Named for the Long March of the Chinese Communist Army, these rockets were derived from early versions of the Dongfeng (DF) ballistic missiles, designed by Tsien Hsue-shen, China's father of rocketry. As in the United States and the Soviet Union, the different missions of launch vehicles and strategic missiles caused the Long March development to diverge from that of ballistic missiles. The objective of a launch vehicle is to maximize payload mass, while strategic missiles are designed to launch quickly and to survive a first strike.

Long March 1 was developed for China's first satellite, a telecommunications platform called *Dong Fang Hong 1*. It was placed into Earth orbit on 24 April 1970 from

An experimental science satellite Shijian-7 (SJ-7), atop a Long March 2D carrier rocket, is launched on 6 July 2005. (Courtesy AP/Wide World Photos)

Jiuquan in Gansu province, making China the fifth nation to achieve independent launch capability. Long March 1, based on the DF-3 intermediate range ballistic missile, was developed by the China Academy of Launcher Technology (CALT). It featured two liquid propellant lower stages, using unsymmetrical dimethylhydrazine as propellant and nitric acid as oxidizer, and a solid propellant third stage.

To launch the heavier Fanhui Shi Weixing recoverable satellites, CALT designed the two-stage Long March 2, based on the DF-5 intercontinental ballistic missile. It switched to nitrogen tetroxide as the oxidizer, used computer guidance, gimbaled engines, and lighter–aluminum alloy structures. After a first failed launch in November 1974, the first successful Long March 2 lifted off one year later. Variants later included three-stage versions. Starting in 1999 the Long March 2F was used to launch the Shenzhou human-tended spacecraft into orbit, including the crewed *Shenzhou 5* and *Shenzhou 6*.

The Long March 3 was a major technological leap, as it used a liquid hydrogen-oxygen third stage to launch communications satellites into geosynchronous orbit. Reaching geosynchronous orbit required a new launch site closer to the equator, the Xichang site in Sichuan province. The first launch attempt in January 1983 was only partially successful, but the next attempt in April 1984 successfully placed the Shiyan Tongbu Tongxin Weixing (experimental geostationary communications satellite) into geosynchronous orbit. In October 1985 China announced that it would market

TSIEN HSUE-SHEN
(1911–)

Tsien Hsue-shen was born in 1911 in Hangzhou, China. In 1935 Tsien left China on a graduate scholarship to the Massachusetts Institute of Technology. In 1936 he received a doctorate from the California Institute of Technology. During World War II Tsien helped design missiles for the U.S. Army's Jet Propulsion Laboratory and was later sent to Germany to examine captured rocket technology. Tsien was deported to China in 1955 under suspicion of being a communist sympathizer, where he led development of China's first ballistic missiles and satellites. Considered the founder of the Chinese space program, he retired in 1991.

Phil Smith

(Courtesy AP/Wide World Photos)

Long March 3 rockets to launch commercial satellites. The first commercial mission, *Asiasat 1*, was successful in April 1990. Later failures caused difficulties, as commercial insurers required the Chinese to open their secretive practices, and U.S. security concerns about technology transfer intervened.

Long March 4 was designed to place the Feng Yun weather satellites into polar orbit from a new launch site at Taiyuan near Manchuria. The Long March 4 extended the two Long March 2 stages by 7 m and used a new third stage and engines.

By the twenty-first century the Chinese planned to develop a new series (sometimes called the Long March 5) entirely powered by liquid oxygen and liquid hydrogen). Long March 5 would be used to launch space station components and interplanetary spacecraft. China was also developing a more logistically accessible launch center at Wenchang on Hainan Island for the new CZ-5 series launch vehicles being built at a facility in Tianjin. As of 30 November 2006 China had launched Long March rockets 112 times, nine of which ended in failure.

Phil Smith

See also: China, Commercial Space Launch, Feng Yun, Shenzhou

Bibliography
Philip Chien, *When Dragons Fly: An Overview of China's Manned Spaceflight Program* (2002).
Xin Dingding, "New Carrier Rocket Series to Be Built," *China Daily*, 31 October 2007.
Brian Harvey, *China's Space Program: From Conception to Manned Spaceflight* (2004).

Long March Launch Vehicles (Part 1 of 2)

Model	Manufacturer	Years in Use	Capability	Launch Site	Number of Flights*	Comments
CZ-1	China Academy of Launch Vehicle Technology (CALT)	1970–1971	LEO: 300 kg (660 lb) GTO: 250 kg (550 lb)	Jiuquan	2 (0)	Three-stage vehicle used for low Earth orbital missions. First of the Long March series.
CZ-2A	CALT	1974	LEO: 2,000 kg (4,400 lb)	Jiuquan	1 (1)	Two-stage vehicle used for low Earth orbital missions.
CZ-2C	CALT	1975–current	LEO: 2,500 kg (5,500 lb)	Jiuquan, Taiyuan, Xichang	31 (1)	Two-stage vehicle used for low Earth orbital missions, though some variants carry third stage depending on mission (CZ-2C/SD).
CZ-2D	Shanghai Academy of Space Technology (SAST)	1992–current	LEO: 3,500 kg (7,700 lb)	Jiuquan	8 (0)	Three-stage vehicle used for low-Earth orbital missions.
CZ-2E	SAST	1990–current	LEO: 9,200 kg (20,200 lb) GTO: 3,370 kg (7,420 lb)	Jiuquan	7 (2)	Three-stage vehicle used for low-Earth orbital missions.
CZ-2F	SAST	1999–current	LEO: 8,400 kg (18,500 lb) GTO: 3,500 kg (7,700 lb)	Jiuquan	6 (0)	Three-stage vehicle used for crewed low-Earth orbital missions.
CZ-3	SAST (1st and 2nd stages) CALT (3rd stage)	1984–2000	LEO: 4,800 kg (10,500 lb)	Xichang	13 (3)	Three-stage vehicle used for either low-Earth or geostationary orbital missions.

(Continued)

Long March Launch Vehicles (Part 2 of 2)

Model	Manufacturer	Years in Use	Capability	Launch Site	Number of Flights[*]	Comments
			GTO: 1,400 kg (3,000 lb)			
CZ-3A	SAST (1st and 2nd stages) CALT (3rd stage)	1994–current	LEO: 7,200 kg (15,800 lb) GTO: 2,600 kg (5,700 lb)	Xichang	17 (0)	Three-stage vehicle used for either low-Earth or geostationary orbital missions.
CZ-3B	SAST (1st and 2nd stages) CALT (3rd stage)	1996–current	LEO: 11,200 kg (24,600 lb) GTO: 5,100 kg (11,200 lb)	Xichang	10 (1)	Three-stage vehicle used for either low-Earth or geostationary orbital missions.
CZ-4A	SAST	1988–1990	LEO: 4,680 kg (10,310 lb) GTO: 1,100 kg (2,400 lb)	Taiyuan	2 (0)	Three-stage vehicle generally used for Sun-synchronous and geostationary orbital missions.
CZ-4B	SAST	1999–current	LEO: 2,800 kg (6,100 lb)	Taiyuan	12 (0)	Three-stage vehicle generally used for Sun-synchronous and geostationary orbital missions.
CZ-4C	SAST	2007–current	?	Taiyuan	1 (0)	Similar to CZ-4B, but with a more powerful upper stage, possibly restartable. Details on LEO and GTO capability unknown.
CZ-5	China Academy of Launch Vehicle Technology	2013	LEO: 25,000 kg (55,000 lb) GTO: 14,000 kg (30,800 lb)	Wenchang	0	Several variants of LOX-kerosene-and LOX-hydrogen-fueled modular vehicles. To be launched from Wenchang on Hainan Island.

[*]Total launches followed by number of failures in parentheses.

Molniya (Launch Vehicle). *See* Russian Launch Vehicles.

N-1. *See* Japanese Launch Vehicles.

N-1. *See* Russian Launch Vehicles.

N-2. *See* Japanese Launch Vehicles.

Pegasus

Pegasus was a three-stage, air-launched, delta-winged rocket used to place small payloads into low Earth orbit. Beginning in June 1988 Orbital Sciences Corporation developed Pegasus in conjunction with Hercules Aerospace (later Alliant Techsystems), which supplied the solid-rocket propulsion system. A cost-effective, light-lift launcher for small U.S. government payloads and Orbital's Orbcomm satellites, Pegasus filled the gap left by the discontinued Scout rocket.

Orbital cut development to a total of $55 million and minimized production costs by making extensive use of off-the-shelf components, such as an inertial guidance system borrowed from U.S. Navy torpedoes and commercially available Global Positioning System receivers. The Pegasus rocket, which could place 375 kg into orbit, first flew on 4 April 1990 from a B-52 bomber operated by NASA's Dryden Flight Research Center at Edwards Air Force Base, California. In 1994 Orbital began launching Pegasus from a Lockheed L-1011 Tristar. The original Pegasus had nine launches through 1998, all successful.

A contract with the Defense Advanced Research Projects Agency led to the development of the Pegasus XL, which lengthened the first two stages, boosting the system's payload capacity to 443 kg into low Earth orbit. The addition of a liquid hydrazine–fueled fourth stage enabled the Pegasus XL to achieve higher altitudes and perform more complex maneuvers. The failure of the first two XL launches in 1994–95, followed by a third failure the following year, caused some initial concern, but by December 2007 the XL's record was 26 successful launches out of 29 attempts. A typical XL launch, including associated services provided by Orbital, in 2007 cost around $30 million.

Stages from Pegasus rockets, including Alliant Techsystems's Orion motors, were also used in Orbital's Taurus and Minotaur ground-based launchers.

David Eidsmoe

See also: Advanced Research Projects Agency (ARPA), Orbital Sciences Corporation

Bibliography
David Baker, *Jane's Space Directory 2004–2005* (2004).
Orbital Sciences. www.orbital.com.
Pegasus Fact Sheet, Orbital Sciences Corporation. http://www.orbital.com/NewsInfo/Publications/Pegasus_fact.pdf.

Proton. *See* Russian Launch Vehicles.

PSLV (Polar Satellite Launch Vehicle). *See* Indian Launch Vehicles.

R-7. *See* Russian Launch Vehicles.

Redstone

Redstone, the first operational U.S. intermediate-range ballistic missile, contributed to several satellite launch vehicles, and carried the first U.S. astronauts into space. In October 1948 the U.S. Army established what became the Ordnance Guided Missile Center at Redstone Arsenal in Alabama. In 1950 Army engineers began development of the Redstone missile. Designated PGM-11A, the 300-mile-range Redstone built on technology from the German V-2 and an Army–General Electric development program called Hermes.

The Redstone was the first rocket designed by Wernher von Braun to use the vehicle's outer skin as the walls of the propellant tanks. This integral tank design reduced weight and increased strength. The missile stood 69 ft tall and was 6 ft in diameter. Its Rocketdyne A-6 engine burned liquid oxygen and ethyl alcohol. The Redstone's liftoff mass was 62,750 lb; thrust at liftoff was 78,000 lb. An inertial system controlling graphite exhaust vanes and small aerodynamic rudders provided guidance. One technical first was a flight computer into which a target program could be loaded on tape. The first Redstone was launched on 20 August 1953, and the missile went on nuclear alert in Europe in 1958. Chrysler built about 120 Redstones for the Army.

The Redstone contributed to several military and civilian space programs. It was the core of the Jupiter C (aka Juno I) launch vehicle for *Explorer 1*, the first U.S. satellite. It was also the foundation for the larger Jupiter missile, which in turn became the Juno II satellite launcher. Six Redstone launches from November 1960 to July 1961 carried Mercury capsules on suborbital flights. The last two of these carried the first two Americans into space.

Matt Bille

See also: Explorer; Mercury; Von Braun, Wernher

Bibliography
Paul H. Satterfield and David S. Akens, *Army Ordnance Satellite Program* (1958).
Wernher von Braun, "The Redstone, Jupiter, and Juno," in *The History of Rocket Technology*, ed. Eugene M. Emme (1964).

Russian Launch Vehicles

Russian launch vehicles originated primarily from the Soviet Union's military missiles and were developed by the same military-industrial enterprises. Because the real designations of the Soviet launch vehicles (8K72, 8K78, 11A511) were not publicly available, they became known in the West by pseudonyms, such as A-1, B-1, C-1 (assigned by the U.S. Library of Congress) and SL-1, SL-2, SL-3 (assigned by the

The Soyuz FG launch vehicle carrying the Soyuz TMA-6 spacecraft is transferred and erected on the launch pad at the Baikonur Cosmodrome, Kazakhstan, on 13 April 2005, in preparation for the Eneide mission to the International Space Station. *(Courtesy European Space Agency)*

U.S. Department of Defense). For the same reasons, the Soviet mass media attributed to them unofficial names based on their most famous payloads: Vostok, Kosmos, Proton. When the greater openness of the Soviet space program revealed designations for some of the launchers (N-1, Energia, Zenit) in the late 1980s, the Soviet military adopted the payload names as official for the rest of them. The launch vehicles based on ballistic missiles include several major types.

R-7–based launchers represent the oldest and most widely used family of the Soviet space vehicles, with kerosene/liquid-oxygen propellant on all stages. Four major configurations were developed, each having a number of different models. The first type, called Sputnik, was a modified two-stage R-7 (SS-6) intercontinental ballistic missile (ICBM), which consisted of five rocket blocks. That rocket became the world's first space launcher, orbiting *Sputnik* on 4 October 1957. Addition of a third stage in the Vostok type allowed launch of early lunar probes and crewed spacecraft, including the first human spaceflight, *Vostok*, on 12 April 1961. The third variant, known as Soyuz (which became the most used Soviet launcher) with a more powerful third stage, launched the majority of the crewed spacecraft and military and scientific satellites. The four-stage Molniya launcher was used for the high-apogee satellites and lunar and planetary probes. The rockets of all four configurations were launched from Baikonur and Plesetsk, but as of 2007 only Soyuz and Molniya were still in service. Originally designed by OKB-1 (Experimental Design Bureau), the R-7 launchers were transferred in 1963 to the Central Specialized Design Bureau Progress (TsSKB)

Soviet and Russian Launch Vehicles (Part 1 of 2)

Designation	Length and Diameter	Launch Mass	Number of Stages	Payload on 200 km orbit	Comments
Launchers based on R-7 ballistic missile					
Sputnik	29.17 × 10.3 m	267 tons	2	1.5 tons	Two models used in 1957–58.
Vostok	38.4 × 10.3 m	287 tons	3	5 tons	Four models used in 1958–91. Includes Luna variants.
Molniya-M	43.4 × 10.3 m	305 tons	4	2 tons on high apogee orbit	Three models used since 1960. In service as of 2007.
Soyuz-U	51.1 × 10.3 m	313 tons	3	7.3 tons	Eight models used since 1963. In service as of 2007.
Aurora (Onega)	50 × 10.3 m	376 tons	3–4	12–15 tons	Proposed project of 2000.
Launchers based on other ballistic missiles					
Dnepr	34.3 × 3 m	211 tons	3	4 tons	Based on R-36M2 ICBM. In service since 1999.
Kosmos	32 × 1.65 m	49.4 tons	2	0.45 tons	Based on R-12 IRBM. Two models used in 1961–77.
Kosmos-3M	32.4 × 2.4 m	109 tons	2	1.5 tons	Based on R-14 IRBM. Three models used since 1964. In service as of 2005.
Proton-K	61.2 × 7.4 m	690 tons	3–4	21 tons 2 tons on geo-synchronous orbit	Based on UR-500 ICBM. Five models used since 1965. In service as of 2007.

Soviet and Russian Launch Vehicles (Part 2 of 2)

Designation	Length and Diameter	Launch Mass	Number of Stages	Payload on 200 km orbit	Comments
Rokot	28.5 × 2.5 m	107 tons	3	1.9 tons	Based on UR-100NU ICBM. In service since 2000.
Strela	28.5 × 2.5 m	106 tons	3	1.8 tons	Based on UR-100NU ICBM. In service since 2004.
Volna	14.2 × 1.8 m	35 tons	2	0.2 tons	Based on R-29R SLBM. In service since 1995.
Start-1	22.7 × 1.8 m	47 tons	4	0.2 tons	Based on RT-2PM ICBM. In service since 1993.
Tsiklon-2	40.5 × 3 m	188 tons	2	2.9 tons	Based on R-36 ICBM. Used by Space Forces 1969–2004.
Tsiklon-3	39.3 × 3 m	191 tons	3	3.6 tons	Based on R-36 ICBM. Used by Space Forces 1977–2004.
Custom-built launchers					
N-1	105.3 × 17 m	2, 800 tons	3	95 tons	Developed in 1960–76. Canceled after four failed launches.
Energia	59 × 16 m	2, 400 tons	2	105 tons	Developed in 1976–93. Canceled after two successful launches.
Zenit-2	57 × 3.9 m	459 tons	2	15 tons	In service since 1985.
Zenit-3SL	59.6 × 3.9 m	470.8 tons	3	2 tons on geo-synchronous orbit	Used by international Sea Launch consortium. In service since 1999.
Angara 1.1	35 × 2.9 m	145 tons	2	2 tons	Under development as of 2007.
Angara 5A	64 × 8.7 m	790 tons	3	30 tons	Under development as of 2007.

© Peter Gorin, 2007

in the city of Samara for further development. In 1992 the Russian government adopted an upgrade program to develop the Soyuz-2 to replace older Soyuz and Molniya rockets. As a result several new models (Soyuz-FG, Soyuz-2.1) entered service by 2005. Rocket and Space Corporation (RSC) Energia (former OKB-1) and TsSKB Progress proposed more extensive modernization of the R-7 rocket in 2000. More powerful second and third stages would double the vehicle's payload capacity. One version of that rocket, Aurora, was planned for international commercial use, while the other, called Onega (after the Russian lake), was designed to orbit the new reusable crewed spacecraft, Clipper.

Dnepr was the Russian/Ukrainian joint conversion of the R-36M2 heavy ICBM (SS-18) for international commercial payloads. The rocket, developed by NPO (Scientific-Production Association) Yuzhny (former OKB-586) in Ukraine and named after the Ukrainian river, used two ICBM stages with additional orbital insertion third stage (all with storable liquid propellant). Dnepr was launched from the underground silos in Baikonur.

The name Kosmos (cosmos) designated two different types of lightweight launchers developed by OKB-586 in Ukraine. Kosmos and its modification, Kosmos-2, were based on the R-12 (SS-4) intermediate-range ballistic missile (IRBM) with an additional second stage and were used to orbit small research and early Interkosmos satellites in 1961–77. Kosmos was launched from underground silos in Kapustin Yar, while Kosmos-2 used surface launch pads in Kapustin Yar and Plesetsk. A more powerful Kosmos-3M and its previous models, Kosmos 1 and 3, were based on the R-14 (SS-5) IRBM with an addition of the second stage. Since its introduction in 1964, the storable liquid propellant rocket became known for its reliability, and as of 2007 it still served as the primary lightweight launcher of the Russian Space Forces. In 1970 further development of Kosmos-3M was transferred to NPO Poliyot in the Russian city of Omsk. The rocket was launched from Plesetsk and Kapustin Yar.

Proton-K, the most powerful operational Russian launcher of the late twentieth and early twenty-first century, was based on an abandoned OKB-52 superheavy ICBM project, UR-500. Named after its first experimental payload (launched in 1965), the rocket was the Soviet response to the U.S. Saturn IB launcher and shared similar design: the first stage consisted of six cylindrical tanks around a larger central tank. Yet unlike its U.S. rival, Proton used storable liquid propellant for the first three stages and kerosene/liquid oxygen propellant for its escape stage, called Block-D. Proton played a major role in unsuccessful Soviet attempts to beat the U.S. Apollo to a piloted circumlunar mission in 1967–68. Beginning in the 1970s the three-stage model orbited heavy payloads, such as space station modules, while the more common four-stage configuration was widely used for geosynchronous satellites and lunar and planetary probes. The Proton-M, with a new fourth stage, upgraded control system, and larger payload bay, entered service in 2003. Proton was produced by the Khrunichev Space Center in Moscow and launched from four launch pads in Baikonur.

Rokot (rumble) and Strela (arrow) were different vehicles based on the UR-100NU (SS-19) ICBM. Both rockets used the same first two stages (with storable liquid propellant) but different orbit-insertion third stages and payload sections. Rokot was equipped with a specially designed Briz-KM third stage, while Strela used the converted post-boost bus, which originally aimed and released multiple warheads of the prototype ICBM. The Khrunichev Space Center developed Rokot primarily for commercial launches from the converted former Kosmos-3M launch pad in Plesetsk, while NPO Mashinostroeniya (former OKB-52), which had designed the UR-100NU ICBM, developed Strela mostly for military payloads launched from ICBM silos in Baikonur.

Volna (wave) was the first in a series of space launch vehicles proposed by the Makeev Missile Center in the city of Miass, where most Soviet submarine-launched ballistic missiles were developed. Based on a storable liquid propellant missile, R-29R, Volna was a unique launcher, because it was launched from a submerged submarine. However launches with commercial payloads, starting in 1995, demonstrated low reliability.

Start-1 was a conversion of the RT-2PM Topol ICBM (SS-25) by the Moscow Institute of Thermal Technology. The four-stage satellite launch vehicle (SLV) was the only Russian space launcher with solid propellant on all stages. Mostly used for small commercial payloads, Start-1 was launched from Plesetsk and Svobodny, using mobile launchers of the prototype ICBM. Tsiklon (cyclone) rockets were widely used for more than 30 years by the Russian Space Forces for a variety of military payloads. Two models with storable liquid propellant were developed by OKB-586 in Ukraine, based on R-36 ICBM (SS-9). The two-stage Tsiklon-2 became known mostly for launching signal reconnaissance satellites from Baikonur, while the three-stage Tsiklon-3 primarily orbited meteorological and navigation satellites from Plesetsk. Tsiklon launchers utilized launch systems with minimal human participation to ensure safety of the launch personnel. Although both rockets were highly reliable, the Russian military retired them by 2005 because they were produced outside Russia. Russian Space Agency planned to convert those rockets for civilian use as of 2007. Ukrainian NPO Yuzhny also intended to modify Tsiklon-3 for international commercial applications (Tsiklon-4 project).

Since 1960 the Soviet Union produced several custom-built launch vehicles. The N-1 (from the Russian word nositel or launch vehicle) was designed by OKB-1 and originally intended to start a family of launchers for payloads ranging from 5–75 tons. However the government and the military initially were not sure of the necessity for such a rocket. Redesigned in 1962, the project had to change again to achieve payload capability of 95 tons after it was adopted for the crewed lunar program in 1964. Because Soviet engineers at that time had only started developing powerful rocket engines and liquid hydrogen fuel, N-1 was built around an array of smaller engines with kerosene/liquid oxygen propellant on all stages, while liquid oxygen/liquid hydrogen stages were planned for later models. As a result N-1 had only 68 percent of the payload capacity of its U.S. rival, Saturn V, which severely impeded the Soviet crewed lunar landing project. The rocket had three booster stages and a payload section, which included two more rocket blocks with landing and orbital piloted spacecraft (hence some Western sources called it a five-stage rocket). Because N-1 had 30 engines on the first stage and spherical tanks, it had an unusual conical shape.

This design caused a feud among Soviet space engineers. Some criticized OKB-1 for the multi-engine concept and abandoning the parallel staging, which made N-1 too cumbersome; others demanded switching to less efficient but more common storable liquid propellant. Numerous technical and logistical problems and inconsistent government funding slowed N-1 development; its final concept was approved in 1966. Due to its size, N-1 had to be assembled at Baikonur, where two launch pads were constructed. To save money, a first stage firing test stand was never built, while individual engines were tested randomly. That proved to be a fatal error: all four launches of N-1 (twice in 1969, once in 1971 and 1972) failed due to the first stage malfunctions. The second launch on 3 July 1969 had grave consequences: the rocket crashed onto its launch pad, completely destroying it. By the time of the planned fifth launch in 1974, new, highly reliable engines had been developed. Yet after the successful completion of the U.S. Apollo program, the Soviet leadership decided to abandon N-1. Several N-1 rockets were destroyed, while any information about the program was suppressed until 1989. In 1998, after 24 years in storage, the latest N-1 engines were tested in Russia and the United States and proved their high efficiency. RSC Energia proposed them for the Aurora/Onega project.

Energia (energy), the most powerful Soviet launcher, was developed by RSC Energia in cooperation with other Soviet aerospace enterprises. The program started in 1976 as a replacement for the ill-fated N-1 and the Soviet response to the U.S. Space Shuttle. Unlike N-1, Energia utilized parallel staging, which allowed logistical and applications flexibility. Although similar to the Space Shuttle in size and appearance, Energia was built differently. It used four liquid-oxygen–kerosene strap-on boosters as the first stage and the liquid oxygen–liquid hydrogen central block as the second stage. Those rocket blocks were equipped with the most powerful liquid propellant rocket engines with vacuum thrusts of 800 and 200 tons respectively. Unlike the Space Shuttle's integrated design, Energia's modular construction allowed launching of the reusable orbiter, Buran, and a variety of heavy payloads of up to 105 tons. Those potentially included crewed lunar spacecraft, orbital space stations, high-power communications satellites, and military battle stations. A smaller version, Energia-M with a payload capability of 34 tons, was also under development. Energia was launched from Baikonur, where two launch pads were converted from N-1 pads, while the third one was newly built to serve additionally as a firing test stand of the entire rocket. Unlike N-1, fully fabricated Energia rocket blocks were delivered to Baikonur by railroad and special airplanes. Energia was flight-tested twice: on 15 May 1987 the first two stages worked flawlessly, but the experimental spacecraft, *Polius*, failed to reach orbit due to its guidance system malfunction; on 15 November 1988 Energia successfully launched an uncrewed Buran spacecraft, which landed safely at Baikonur after two orbits. After the collapse of the Soviet Union, Russia was unable to continue the Energia/Buran program due to the shortage of government funding, and it was canceled in 1993.

Zenit-2 (zenith) was the most sophisticated and the only custom-built Soviet launcher that was still in Russian service in 2007. The program began in 1976 at NPO Yuzhny in Ukraine with a broad participation of other Soviet aerospace enterprises. The two-stage launcher with liquid oxygen–kerosene propellant on both stages used a completely automated launch preparation system for the highest launch personnel safety. Its first stage was also unified with strap-on boosters of the Energia launcher, making it a part of the same family. Due to its high energy efficiency and compatibility with Energia systems, Zenit was intended to be universal and potentially replace all other rockets in the payload range of 7–20 tons. First launched in 1985, Zenit entered full service in 1989. However, technical setbacks and the collapse of the Soviet Union limited its use. From the outset Zenit had a reliability problem, which resulted in several costly failures, especially in 1990 when a rocket explosion destroyed one of two launch pads at Baikonur. Zenit orbited mostly signal reconnaissance satellites, but the Russian military announced its intention to abandon that rocket in the future because it was built in Ukraine. Some critical Zenit components produced in Russia also prevent Ukraine from using it alone. Zenit survived because of the international consortium, Sea Launch, with the participation of Norway, Russia, Ukraine, and the United States. The modified Zenit-3SL, with Block-DM-SL third stage, launched commercial satellites to geosynchronous orbit from the energy-efficient equatorial plane, using a unique floating launch pad.

Angara was, in the early twenty-first century, an ongoing development program of a family of five multipurpose launchers, destined to replace most of the older space launches in the payload range of 2–30 tons. The system, named after a Siberian river, was designed around the liquid oxygen–kerosene Universal Rocket Module, which could be launched alone or equipped with additional stages. For maximum payload capacity (Angara-5A), a packet of five Universal Rocket Modules, as the first stage, would be supplemented by two upper stages with liquid oxygen/liquid hydrogen engines. There were also plans, as of 2007, to use reusable rocket blocks for the first stage. All Angara variants would be launched from the same launch pads. The program started in 1993 at the Khrunichev Space Center but was completely redesigned in 1997 and was hampered by inconsistent government funding. As of 2007 two Angara launch pads were under construction in Plesetsk and an additional pad was planned for Baikonur, as part of the Russian-Kazakhstan joint program, called Byterek.

Peter A. Gorin

See also: Ballistic Missiles, Khrunichev Center, Launch Facilities, Rocket and Space Corporation Energia, Russia (Formerly the Soviet Union), Russian Space Forces, Sea Launch Company, Vostok

Bibliography
Steven Isakovitz, *International Reference Guide to Space Launch Systems* (1993).
Russia's *Arms Catalog*, vol. 6, *Missiles and Space Technology* (1997).

Saturn Launch Vehicles

The Saturn launch vehicles were the family of heavy-lift rockets that ultimately sent the first humans to the Moon in Project Apollo and launched *Skylab*. By early

1957, the U.S. Department of Defense was considering the need to place heavy communications, weather and science payloads into orbit, beyond the capability of current launchers. Having lost the competition with the U.S. Air Force to develop long-range ballistic missiles, the Army Ballistic Missile Agency (ABMA) near Huntsville, Alabama, decided to study large space launch vehicles. In the wake of the Soviet Union launch of *Sputnik* in which the United States sought to catch up with Soviet heavy-lift launch capabilities, the newly created Advanced Research Projects Agency in August 1958 directed ABMA to develop a heavy-lift vehicle in the shortest possible time using existing hardware.

ABMA engineers, led by Wernher von Braun, designed a first stage consisting of a 110 in core tank (built from the same tools used to make ABMA's Jupiter ballistic missile) surrounded by ten 70 in tanks (built on the tools for the Redstone rocket). A special plumbing system fed liquid oxygen and RP-1 fuel into eight H-1 engines (Rocketdyne engines slightly modified from Jupiter). This clustering concept, which required control of several engines at once, saved time and money because Rocketdyne's larger engines would not be available for several years. The resulting design, initially called Juno V, became the S-I stage of the Saturn I launcher. Saturn I's second stage, called the S-IV, used six RL10 hydrogen/oxygen engines then under development for the Centaur upper stage. NASA, which absorbed ABMA in 1960 to become the core of the Marshall Space Flight Center (MSFC) eliminated a proposed Saturn I third stage in 1961. The Saturn I used a unified Instrument Unit atop the last stage, which carried the complete guidance system, rather than having a separate system in each stage. The launcher could carry 22,500 pounds to low Earth orbit. MSFC awarded Chrysler Corporation the S-I manufacturing contract, while Douglas Aircraft Company received the S-IV production contract. The first four Saturn I rockets flew with dummy payloads in 1961–63, and the last six flew with live S-IV second stages in 1963–65. The last five missions carried Apollo spacecraft mock-ups, and the last three also orbited Pegasus meteoroid detection satellites. All Saturn I flights succeeded.

In July 1962, MSFC announced an upgraded Saturn I that was eventually (in February 1963) called the Saturn IB. It consisted of a modified S-I stage, called S-IB with eight stabilizing fins and upgraded H-1 engines, and an improved S-IV stage, called S-IVB using a single J-2 engine. Nine Saturn IBs from 1966 to 1974 successfully launched various Apollo spacecraft components, first in four unmanned tests (AS-201 through AS-204) and later with crews on *Apollo 7*; *Skylab 2, 3,* and *4*; and *Apollo-Soyuz Test Project* missions. *Skylab 4* was the heaviest payload, at 20,847 kg.

The Saturn V was a completely new design approved in January 1962, comprising a new S-IC first stage with five F-1 engines, a new S-II second stage with five J-2 engines, and the S-IVB third stage (used earlier as the second stage of Saturn IB) with one J-2 engine. While the S-I stage employed 11 tanks built on the tools for existing rockets, each Saturn V stage employed separate, single tanks for fuel and oxidizer. The S-IC and S-II stages were 33 ft in diameter, and the S-IVB stage was 21 ft, 8 in wide. The entire vehicle, which stood 363 ft high, could deliver a massive 107,000 lb into a lunar trajectory. The Boeing Company built the S-IC, and North American Aviation the S-II.

One of the most daring aspects of the Saturn V program was its rapid advancement to flight. The pressures to put an American on the Moon by the close of 1969 would not allow extended multivehicle testing before human flight. NASA Associate Administrator for Manned Space Flight George Mueller convinced his subordinates to try an unprecedented all-up first test. He reasoned that to simulate flight loads properly, the second and third stages should carry real propellants; and if they had live propellants, requiring real stages, why not real engines; and then, why not test the Apollo spacecraft (redesigned following the *Apollo 204* fire)? The *AS-501* flight (*Apollo 4*) on 9 September 1967 was a resounding success that put the Apollo program back on track. A second unmanned test, *AS-502* (*Apollo 6*) had problems with miswiring on the S-II engines, but cleared the Saturn V for human flight starting with the *Apollo 8* mission (*AS-503*) around the Moon. While all 13 Saturn V flights from 1967 to 1973 experienced problems, each was ultimately successful.

The Saturn program had its share of technical development problems and cost over $9 billion in then-year dollars. These costs appeared far too high in the cost-conscious 1970s, and, despite its technical successes, NASA ended the program. Much later, in the early 2000s, the J-2 engine used in the S-IVB stage was resurrected and modified to become the J-2X that in 2009 was under development by Pratt and Whitney Rocketdyne for NASA's Ares launch vehicles.

Dave Dooling and Stephen B. Johnson

See also: Apollo; Boeing Company; Douglas Aircraft Company; Marshall Space Flight Center; North American Rockwell Corporation; Von Braun, Wernher

Bibliography
Roger E. Bilstein, *Stages to Saturn: A Technological History of the Apollo/Saturn Launch Vehicles* (1980).

J. D. Hunley, *U.S. Space-Launch Vehicle Technology: Viking to Space Shuttle* (2008).

Ray A. Williamson, "The Biggest of Them All: Reconsidering the Saturn V," in *To Reach the High Frontier: A History of U.S. Launch Vehicles*, ed. Roger D. Launius and Dennis R. Jenkins (2002).

Scout

Scout (a name retroactively translated as Solid Controlled Orbital Utility Test) was the first NASA-developed satellite launcher. Scout was the world's first operational all-solid-propellant satellite launch vehicle. In 1957 engineers of NASA's predecessor, the National Advisory Committee on Aeronautics, concluded that a four-stage combination of existing solid motors could provide a modest orbital capability at low cost. On 1 March 1959 NASA and the U.S. Air Force (USAF) jointly announced the development of Scout as an inexpensive launcher suitable for NASA, Department of Defense, and foreign payloads. Vought Astronautics (later part of the LTV—Ling-Temco-Vought—Aerospace and Defense Company) manufactured it.

The base Scout was a four-stage vehicle approximately 75 ft long. Ten major Scout models and a myriad of variations were employed, ranging from three stages to five. The payload to Earth orbit ranged from 130–465 lb. The Scout was assembled and the payload was integrated in the horizontal position, with the vehicle raised to vertical or near-vertical orientation before launch. On 1 July 1960 NASA launched the first Scout from Wallops Flight Facility, Virginia.

The Scout launched many NASA, USAF, and U.S. Navy payloads, including NASA's first Small Explorer Program payload, most of the Transit navigation satellites, and the first military "lightsats" of the early 1990s. A USAF branch of the program in the 1960s used three orbital and suborbital configurations with Blue Scout designations. Scout was also used by the European Space Research Organisation and five European governments. There were 116 Scouts launched, including 21 research-and-development launches (11 successes) and 95 operational missions (91 successes). The final Scout was launched on 8 May 1994 from Vandenberg Air Force Base, California, closing out a successful launch vehicle program.

Matt Bille

See also: Early Explorer Spacecraft

Bibliography

Matt Bille et al., "History and Development of U.S. Small Launch Vehicles," in *To Reach the High Frontier*, ed. Roger Launius and Dennis Jenkins (2002).

J. D. Hunley, *U.S. Space-Launch Vehicle Technology: Viking to Space Shuttle* (2008).

Jonathan C. McDowell, "The Scout Launch Vehicle," *Journal of the British Interplanetary Society* 47 (1994): 99–108.

Joel Powell, "Blue Scout—Military Research Rocket," *Journal of the British Interplanetary Society* 35 (1982): 22–30.

Shavit. *See* Israel.

SLV (Satellite Launch Vehicle). *See* Indian Launch Vehicles.

Sounding Rockets

Sounding rockets are defined as suborbital rocket-powered vehicles carrying scientific instrumentation and were a major tool used by geophysicists, astronomers, and other scientists to loft payloads into or above Earth's atmosphere. Compared to satellites, sounding rockets offer the ability to develop and launch experiments at modest cost. In 2005 a sounding rocket was $10,000 to $200,000, as compared to millions of dollars for a satellite launch. This provides the flexibility to examine transient phenomena, such as solar eclipses, with properly timed launches from almost any part of the globe, and the unique ability to probe the region above the maximum altitude of balloons (about 25 miles) and below that of satellites (about 100 miles). Recovery of payloads is also relatively simple on sounding rockets. The chief limitation is that sounding rockets usually offer only a few minutes of instrument time on each flight.

The term sounding rocket comes from the maritime tradition of sounding the water to measure depth. These rockets likewise probe the ocean of air. Most use solid

The four-stage Black Brant sounding rocket, shown here blasting off from a launch pad at the Wallops Island Flight Facility. (Courtesy NASA/Goddard Space Flight Center)

propellants, and many were based on surplus motors from military missiles. Stages from the U.S. Nike antiaircraft missile have been used in more than 10 sounding rocket configurations, and the Terrier missile in nearly as many. Other sounding rockets, like the Black Brant series from Canada's Bristol Aerospace, were designed and built specifically for their tasks. The two types may even be combined, as in some Black Brant configurations that incorporated surplus missile stages.

Credit for the first sounding rocket launch is often given to U.S. physicist Robert Goddard, who described the use of sounding rockets in a 1919 paper and launched a barometer and a thermometer on a rocket in 1929. Other authorities consider the 1933 meteorological rocket of the Soviet Union's Mikhail Tikhonravov to have been the first sounding rocket.

The first organization created specifically for a sounding rocket program was the Rocket-Sonde Research Branch of the U.S. Naval Research Laboratory (NRL), organized in 1945. The branch participated in the first joint military-civilian effort, the U.S. Army–led V-2 Upper Atmosphere Research Panel, formed in 1946. The first of more than 60 V-2 flights, using rockets captured from Germany in World War II, was launched on 16 April 1946. Many of these flights carried experiments ranging from simple weather instruments to live monkeys. The Soviet Union also flew V-2s, beginning in 1947, and used some for scientific probes. The first rocket developed specifically as a sounding rocket, the WAC Corporal, was a Jet Propulsion Laboratory product, first flown in September 1945. Two more liquid-fueled rockets followed: the

Navy-sponsored Aerobee (first flight 1947) and the NRL's larger Viking, built by the Glenn L. Martin Company and first flown in 1949.

These rockets were fired from land bases, or, in a few cases, from ships. More innovative launch methods included "rockoons," rockets fired from high-altitude balloons, a technique first widely used by U.S. atmospheric physicist James A. Van Allen, and "rockaires," rockets fired from aircraft, the first example being a U.S. Navy test in 1955.

The International Geophysical Year (IGY), an 18-month period of international scientific cooperation, began on 1 July 1957. The IGY is usually remembered for sparking the competition to launch the first satellites, but it also saw an unprecedented global sounding rocket program. More than 200 rockets from the United States and hundreds more from other nations probed the upper atmosphere and the lower reaches of space. Experiments were carried out in meteorology, atmospheric physics, solar physics, ionospheric physics, and astronomy. New launch facilities were created specifically to support the IGY, one example being the U.S.-built range at Fort Churchill on Hudson Bay in Manitoba, Canada. Several new rockets debuted, including the U.S. Aerobee 300 (developed by adding a solid-fuel upper stage taken from the Sparrow air-to-air missile to the Aerobee 150). The IGY laid the foundation for the common use of sounding rockets globally, which continued into the twenty-first century.

One of the longest-running sounding rocket efforts has been the NASA Sounding Rocket Program in the United States, which launched 2,800 flights from 1959 to 2004. During that period the success rate was 95 percent for the rockets and 86 percent based on scientific return. As of 2007 the program, consolidated at NASA's Wallops Flight Facility, provided 20 or more flights per year from several launch sites in the United States and Europe. The largest of the 10 types of rocket used, the Black Brant 12, had four stages, stood 65 ft tall, and could loft 250 lb to 930 miles.

The European Space Agency's (ESA) major sounding rocket program began in 1982. Most launches were from ESRANGE, a launch base in Kiruna, Sweden, operated by the Swedish Space Corporation. ESA's TEXUS Sounding Rocket Programme (Technologische Experimente Unter Schwerelosigkeit), for example, used SkyLark 7 rockets built by British Aerospace. These rockets carried a gross payload of almost 800 lb to an apogee of 160 miles, providing six minutes of microgravity. Several other nations, including India, Brazil, and Japan, also operated sounding rocket manufacturing and launching programs.

Sounding rockets often served as predecessors or testbeds for launch vehicle technology. An example was the liquid-propellant Viking, whose innovative design led to the Project Vanguard launcher, whose stages in turn became part of the Thor-Delta and Delta series. Experiments flown on satellites, the Space Shuttle, and the *International Space Station* have used technology perfected using sounding rocket flights.

A typical sounding rocket includes one or more solid-fuel rocket stages, an igniter housing, a separation system, an experiment, a nosecone and recovery system, a telemetry system, a guidance system for the booster, and an attitude control system for the experiment container (nosecone) section. The affordability and availability of sounding rockets made them the vehicle of choice for investigations by universities and government agencies globally. They will likely be a part of the toolkit for atmospheric and space scientists for decades to come.

Matt Bille

See also: International Geophysical Year, *V-2* Experiments, Viking

Bibliography

William R. Corliss, *NASA Sounding Rockets, 1958–1968—A Historical Summary* (1971).
NASA Sounding Rocket Program Handbook (2005).
Milton Rosen, *The Viking Rocket Story* (1955).

Thor. *See* Delta.

Titan

Titan launch vehicle development began in October 1955, when the U.S. Air Force (USAF) awarded the Glenn L. Martin Company a contract to build an intercontinental ballistic missile (ICBM). The missile was ordered as a backup to the Atlas ICBM being developed by Convair, as the USAF had doubts about some of the technology being incorporated into Atlas. The new missile became known as the Titan I, the nation's first two-stage ICBM and the first designed for underground silo basing. The USAF deployed 54 Titan I missiles, followed by 54 improved Titan IIs.

Although quite similar to Titan I, the Titan II made 22 configuration changes from that of the earlier ICBM during the test-and-evaluation period. The first Titan II ICBMs were activated in 1962, and NASA selected modified Titan IIs to launch the Gemini spacecraft into orbit during the mid-1960s. Titan II used a different propellant combination than its predecessor: A-50 hydrazine (Aerozine) and nitrogen tetroxide. These propellants became the standard for all Titan configurations. As an ICBM, the Titan II could boost a 4,500 lb nuclear device to a range of 9,750 miles or an 8,000 lb reentry vehicle 6,350 miles. It was also able to lift approximately 4,200 lb into low Earth orbit (LEO). The Gemini space launcher version of Titan II incorporated an improved guidance system and a Malfunction Detection System to detect potential catastrophic faults and allow the crew to escape. When as a result of arms and nuclear reduction treaties, the Titan II ICBMs were deactivated during the mid-1980s, the USAF awarded Martin Marietta a contract to convert 14 of them to space launchers. These were used from 1988 to 2003 to carry payloads for the USAF, NASA, and the National Oceanic and Atmospheric Administration.

In 1961 the USAF ordered the development of Titan III to launch heavy payloads. The new vehicle, to be built by Martin Marietta, would use two large solid rocket

Titan Launch Vehicles

Launch Vehicle	Service Dates	Successes/ Attempts	Payload Capability (Lb)	Comments
Titan II– Gemini	1965– 1966	12/12	4,200 to low Earth orbit (LEO)	Human-rated Titan II intercontinental ballistic missile (ICBM) for NASA Gemini program. Malfunction Detection System. Improved inertial guidance system.
Titan IIIA	1964– 1965	3/4	7,260 to LEO	Test vehicle for Titan IIIC, but with no solid rocket boosters
Titan IIIB	1966– 1987	64/68	7,920 to LEO	Titan IIIC without solid-rocket booster (SRB), and Agena upper stage.
Titan IIIC	1965– 1982	30/36	28,600 to LEO	Titan II with strap-on SRB. Titan IIID with no stage 2.
Titan IIID	1971– 1982	22/22	24,200 to LEO	Titan IIIC with no stage 2.
Titan IIIE	1974– 1977	6/7	7,480 to geostationary Earth orbit (GEO)	Titan IIIC with Centaur upper stage.
Titan 34D	1982– 1989	12/15	32,824 to LEO 4,081 to GEO with Transtage	Titan III with stretched first stage plus stretched 5.5 segment SRBs. Transtage or Inertial Upper Stage (IUS). IUS to GEO = 4,070 lb. Seven Transtage launches, one IUS launch.
Titan II SLV (Space Launch Vehicle)	1988– 2003	13/13	4,200 to LEO	Cost $34 million. Converted from Titan II ICBMs. Refurbished engines, upgraded inertial guidance system. Fourteen vehicles refurbished, only thirteen used.
Commercial Titan III	1990– 1992	3/4	Not available	Cost $130–$150 million. Only commercial Titan. Titan 34D with stretched second stage. One flight with Transfer Orbit Stage.
Titan IV	1989– 2005	35/39	Titan IVA 39,000 to LEO 10,000 to GEO Titan IVB 47,800 to LEO 12,700 to GEO	Cost $196–$248 million. Titan 34D with seven-segment SRB. Titan IVB uses Solid Rocket Motor Upgrade SRBs, three segment, but more powerful than Titan III SRBs. IUS or Centaur upper stages.

boosters (SRB) at liftoff to achieve the much higher proposed performance of roughly 30,000 pounds to LEO. The SRB contract was awarded to United Technologies Corporation, and the first test flights began in 1964. Titan IIIs evolved into several configurations, and were used through 1987 for military, intelligence, and science payloads, including the Defense Support Program, Defense Satellite Communications System 2, classified imaging payloads, and Helios, Viking, and Voyager science missions.

Later versions of Titan evolved to launch ever-heavier payloads. The Titan 34D added a half an SRB segment to the five-segment SRBs used on Titan III, along with a longer first stage. This increased performance by roughly 4,000 lb to LEO over the Titan IIIC. Martin Marietta marketed the Titan 34D commercially as "Commercial Titan," but only sold four launches from 1990 to 1992 before withdrawing from commercial competition. They were too expensive compared to competing launchers from Europe and Russia. The USAF, concerned that the Space Shuttle would not meet its needs, awarded Martin Marietta the Titan IV contract in 1985. It was based on the Titan 34D, but enlarged the SRBs to seven segments. The Titan IV also stretched the first stage by 7.9 ft and the second stage by 1.4 ft. A brand new three-segment SRB, called the Solid Rocket Motor Upgrade, further improved performance. A total of 39 Titan IVs were launched between 1989 and 2005, including classified intelligence payloads and NASA's *Cassini* probe to Saturn. The last launch of Titan IV was the final launch of the Titan series, with Lockheed Martin Corporation being directed by the USAF to instead develop and operate the Atlas 5.

Stephen B. Johnson

See also: Ballistic Missiles, Lockheed Martin Corporation, Martin Marietta Corporation, Reconnaissance and Surveillance

Bibliography
J. D. Hunley, *U.S. Space-Launch Vehicle Technology: Viking to Space Shuttle* (2008).
Steven J. Isakowitz et al., *International Reference Guide to Space Launch Systems*, 3rd ed. (1999).
Roger D. Launius and Dennis R. Jenkins, eds., *To Reach the High Frontier: A History of U.S. Launch Vehicles* (2002).
David K. Stumpf, *Titan II: A History of a Cold War Missile Program* (2000).

V-2

V-2 (Vengeance Weapon 2) was the designation the Nazi Propaganda Ministry applied to the world's first ballistic missile in late 1944, after the German Army began launching it against Allied cities. Called the A-4 by its designers, the V-2 was the world's first ballistic missile, the world's first large-scale rocket, and the first human-made object to enter outer space. After the German defeat in World War II, the V-2 greatly influenced the development of rocket and missile technology around the world.

The V-2 had its origin in 1929–30, when the German Army began to research rocketry's military potential, both for short-range artillery and for long-range bombardment. At that time, longer-range rockets were only possible using higher-energy liquid propellants. In the German-speaking world, Hermann Oberth first laid out the theoretical basis for liquid-propellant rocketry for spaceflight in his 1923 book *Die Rakete zu den Planetenräumen* (*The Rocket into Planetary Space*). His followers set up several small rocket societies, most notably in Berlin. Out of these groups came the young engineering student Wernher von Braun, whom the German Army hired in late 1932 to develop liquid-propellant rockets in a secret program.

Von Braun's first vehicle was dubbed A-1 for Aggregate 1. The A-1 was modified into the A-2, two examples of which flew to an altitude of about a mile in December 1934. These rockets used liquid oxygen and water alcohol, a propellant combination that von Braun carried through to the A-4/V-2. Meanwhile, Adolf Hitler's seizure of power meant greatly increased resources devoted to rearmament, leading to opening of the Peenemünde rocket center on the Baltic coast in May 1937. That December, von Braun's group launched four A-3s, 6.5 m long test vehicles. All failed due to control problems, necessitating a redesign as the A-5, whose sole purpose would be to test guidance and control systems.

The A-4/V-2 required breakthroughs in key technologies. Between 1936 and 1941, thanks to the Third Reich's massive expenditures and the brilliant leadership of von Braun and his military chief Walter Dornberger, the Peenemünde group succeeded in scaling up the rocket engine to 25 metric tons of thrust (55,000 lb), producing workable inertial and radio guidance systems, and proving that the missile would remain aerodynamically stable up to burnout velocity of Mach 4.5. As specified by Dornberger in 1936, the A-4's objective was to hurl a one-metric-ton warhead 270 km; as designed in 1939–41, the four-finned missile was 14 m long, with a span of 3.55 m across its fins and a launch mass of 12,900 kg. The warhead could accommodate high explosives or poison gas, although chemical warheads were never completed. Dornberger and his superiors hoped that the A-4 would be a surprise secret weapon that could change the course of a war. Although von Braun and a few leading subordinates were space enthusiasts, and Dornberger was sympathetic, spaceflight had nothing to do with the Nazi state's investment in this weapons system.

The A-4/V-2 was first launched in June 1942, but it was not until the third attempt, on 3 October 1942, that the Peenemünde group scored a success, reaching a range of 190 km and a maximum height of about 80 km—the boundary of space. This success induced Hitler to support the program more strongly, further increasing the pressure to turn this experimental vehicle into a mass-produced weapon. The Armaments Ministry and the SS (Nazi security forces) began to intervene into the program more. To find enough production workers, Dornberger and his subordinates decided to use concentration-camp prisoners in spring 1943. After a British air raid on Peenemünde in August, production went underground in central Germany. Ten to twenty thousand prisoners died during factory construction and V-2 production, with von Braun and Dornberger fully aware that the SS was mistreating the prisoners.

After many delays because of technical problems, V-2 launches against Allied targets began on 8 September 1944. Germany built about 6,000 V-2s; half were fired at Allied targets—predominantly London and Antwerp—causing about 5,000 Allied deaths. But the V-2 program was a military failure, as it was an extremely expensive way to drop a 1-ton bomb on an enemy city. Its true importance was the breakthrough it represented in rocket technology.

In May 1945 Dornberger, von Braun, and the Peenemünde group surrendered to the U.S. forces, whereas the rocket center and the factory ended up in Soviet hands. Both sides quickly moved to absorb the technology, which fed into the Cold War arms race. The United States brought von Braun and about 120 other Germans to El Paso, Texas, and they helped train Americans to launch captured V-2s from White Sands in nearby New Mexico. Between 1946 and 1952, more than 75 V-2s and modified V-2s were fired, most of them with scientific payloads to research the upper atmosphere, near space, and the Sun. In 1949 a WAC Corporal sounding rocket fired from the nose of a V-2, reached a record altitude of 400 km. Meanwhile, U.S. firms and government bureaus had quickly taken over V-2 propulsion and guidance technology.

The V-2 was also seminal for the Soviet Union, with its first captured rocket launched in October 1947, aided by German engineers. Under the leadership of Sergei Korolev, Soviet design bureaus went on to copy the V-2, which entered Red Army service as the R-1 ballistic missile, the starting point for a long line of rocket and missile development. The Soviet Union later transferred the R-1 to the People's Republic of China, where V-2 technology again became the beginning point for long-range missiles and space boosters. Meanwhile the United Kingdom and France had also seized German engineers and technology after World War II, and the V-2 influenced their postwar missile designs, leading to European space boosters in the 1960s. The V-2 was built as a weapon, but it became a major milestone on the road to space.

Michael J. Neufeld

See also: Ballistic Missiles; Russian Launch Vehicles; *V-2* Experiments; Von Braun, Wernher

Bibliography
David H. DeVorkin, *Science with a Vengeance: How the Military Created the U.S. Space Sciences after World War II* (1992).
Walter Dornberger, *V-2* (1954).
Michael J. Neufeld, *The Rocket and the Reich* (1994).
Asif A. Siddiqi, *Challenge to Apollo: The Soviet Union and the Space Race, 1945–1974* (2000).

Vanguard

Vanguard was the United States's first official satellite program. The launch vehicle, developed by the Glenn L. Martin Company as part of this effort, required major advances in launch technology and placed three satellites in orbit. From an initial estimate of $20 million, over its lifetime the project cost grew to $110 million.

The Vanguard launcher originated with the Viking sounding rocket program of the Naval Research Laboratory. Beginning in 1955, Project Vanguard's technical director, Milton Rosen, oversaw the effort to scale up the Viking to form the Vanguard first stage.

Naval Research Laboratory personnel mount the Vanguard 1 *satellite into the launcher. (Courtesy Naval Research Laboratory)*

The Vanguard launcher used several innovations from Viking to save weight and improve performance. These included a steering system based on a gimbaled engine, extensive use of aluminum as the main structural material, and integral propellant tanks.

The first stage was powered by a new engine, the General Electric X-405, with a thrust of just more than 30,000 lb. Aerojet General built the liquid-propellant second stage. For the solid-propellant third stage, the Grand Central Rocket Company and the Allegheny Ballistics Laboratory (ABL) produced alternative designs. The ABL motor used a fiberglass casing, another first for a large rocket. While the Grand Central stage performed adequately, ABL's X-248 allowed Vanguard to more than double its payload (to 52 lb).

There were 11 Vanguard orbital attempts, counting test vehicles, beginning in December 1957. While only three of these succeeded, they made important technical and scientific contributions. The first and smallest satellite, *Vanguard 1*, carried the first solar cells into space, and observation of its orbit established the "pear-shaped" nature of Earth. When NASA was formed in 1958, the Vanguard program moved to the new agency, along with most of its personnel.

The Vanguard booster's greatest legacy, though, was not in its launch record but in its contributions to future launchers. While the satellite program ended in 1959, Vanguard-developed engines, entire stages, and other technology went into the Thor-Delta (later the Delta) and NASA's Scout, making critical contributions to these two long-lived series of launch vehicles.

Matt Bille

See also: Martin Marietta Corporation, United States Navy

Bibliography
Matt Bille and Erika Lishock, *The First Space Race* (2004).
Constance McLaughlin Green and Milton Lomask, *Vanguard: A History* (1970).
John P. Hagen, "The Viking and the Vanguard," in *The History of Rocket Technology*, ed. Eugene M. Emme (1964).
J. D. Hunley, *U.S. Space-Launch Vehicle Technology: Viking to Space Shuttle* (2008).

Vega Upper Stage

The Vega Upper Stage was NASA's first attempt to develop a rocket for lunar and planetary payloads. The Jet Propulsion Laboratory (JPL) in Pasadena, California, first proposed Vega in a December 1958 study of future U.S. goals in space, commissioned by NASA headquarters. JPL's timing was impeccable; its concept for an upper stage developing 35,000 lb of thrust coincided with a plan at NASA headquarters to establish a national launch vehicle strategy.

NASA officials, who intended to use Vega with the Atlas rocket, accepted the JPL proposal in December 1958. NASA signed a Vega production contract with General Dynamics Convair/Astronautics, builder of the Atlas, on 21 May 1959. Vega was 21 ft long (with the same 10-ft diameter as Atlas) and was capable of propelling a 900 lb spacecraft to Mars or Venus.

In June 1959 the U.S. Air Force (USAF) caught NASA managers off guard when it revealed plans for its formerly secret Agena B upper stage, roughly comparable to Vega. On learning of the two competing stages, members of the congressional space committees sided with the USAF/Lockheed Agena, implying that Vega was a redundant program wasting taxpayer dollars. Lockheed's development of Agena B for the secret Corona photoreconnaissance program may have been the deciding factor in favor of Agena.

NASA Administrator T. Keith Glennan canceled Vega on 11 December 1959 and later accepted Agena B as an interim replacement. NASA was counting on the powerful new hydrogen-powered Centaur stage to succeed Agena, but development difficulties delayed the debut of Centaur until 1966.

Joel W. Powell

See also: Corona, Jet Propulsion Laboratory

Bibliography
Joel W. Powell, "Atlas-Vega: An Early NASA Launch Vehicle Development," *Quest* 9, no. 4 (2002): 40–48.

Viking Sounding Rocket

The Viking sounding rocket program made major improvements in rocket technology and was the basis for the Vanguard satellite launcher. In 1946 Naval Research Laboratory

(NRL) engineers led by Milton W. Rosen and H. C. Smith began work on a new sounding rocket to eventually replace the captured German V-2s that the U.S. Army and the U.S. Navy were employing for scientific work. They produced a single-stage design weighing about 10,000 lb, and NRL selected the Glenn L. Martin Company to build it.

The Viking had an advanced control-and-orientation system with thrusters that could orient the vehicle even after the main engine stopped firing. Other innovations included a gimbaled motor to provide steering, integral propellant tanks, and the use of aluminum as the main structural material. The Viking engine used turbopumps and burned alcohol and liquid oxygen to deliver 20,000 lb thrust.

Martin built 14 Vikings. There were 12 official Viking launches (seven of the original Viking Type 7 and five of the heavier, larger-diameter Viking Type 12) from 1949 to 1955. They carried a variety of scientific instruments, and the last two flights conducted reentry nosecone tests. All but two launches were successful, and the record altitude reached was 158 miles.

The Viking became the foundation of the NRL-led Project Vanguard. The Viking was scaled up to form the basis for the Vanguard first-stage design. The NRL redesignated the last two Vikings as Vanguard test vehicles and successfully launched them to test hardware and techniques for the satellite program.

Matt Bille

See also: Martin Marietta Corporation, United States Navy

Bibliography
John P. Hagen, "The Viking and the Vanguard," in *The History of Rocket Technology*, ed. Eugene M. Emme (1964).
J. D. Hunley, *U.S. Space-Launch Vehicle Technology: Viking to Space Shuttle* (2008).
Milton Rosen, *The Viking Rocket Story* (1955).

Zenit. *See* Russian Launch Vehicles.

HYPERSONIC AND REUSABLE VEHICLES

Hypersonic and reusable vehicles have been important goals of flight researchers from the earliest years of the twentieth century, as they sought to reach the hypersonic realm (generally accepted to begin at Mach 5) and to develop reusable winged spacecraft that would enable flight beyond Earth. Another important goal was single-stage-to-orbit (SSTO), the ability to take off like an airplane, perform a mission in space, and return to a runway.

Before the mid-1940s, most aerospace engineers viewed these objectives as unattainable. However, so as to learn how to achieve the speeds necessary to reach Earth orbit, a handful of spaceflight experimenters and advocates, including Robert Goddard, Eugen Sänger, and Wernher von Braun, seriously pursued research and development of a propulsion system capable of generating enough thrust to power flying vehicles to Mach 12 and beyond.

The majority of hypersonic studies awaited the postwar era. First, the German V-2 rockets developed during World War II provided a means to test concepts. In 1946 the U.S. Navy Bureau of Aeronautics learned of Sänger's Silver Bird proposal. The bureau argued that a rocket-powered hypersonic aircraft required only minor advances in technology, and as a result the U.S. military tested hypersonic concepts using V-2s with WAC Corporals as second stages. On 24 February 1949 a WAC Corporal upper stage reached 5,150 miles per hour and 244 miles altitude. That same year California Institute of Technology professor Tsien Hsue-shen proposed a hypersonic rocketliner that could fly from Los Angeles to New York in less than one hour.

In 1951 National Advisory Committee for Aeronautics (NACA) engineer H. Julian Allen discovered and proved that a blunt-body dumped much of its reentry heat into the airflow. Allen's work was so significant, and in such contrast with intuitive thinking, that it fundamentally shaped the course of hypersonic flight research and provided the basis for all successful reentry vehicles after, including nuclear warheads, camera return capsules, and human flight vehicles.

Hypersonic and reusable spacecraft research took a major step forward with the X-15. Between 1959 and 1968, the X-15 made 199 flights divided among three aircraft. The X-15 program achieved Mach 6.7, 354,200 ft altitude, skin temperature of 1,350°F, and dynamic pressures more than 2,200 lb/ft^2, and yielded 765 research reports. It pioneered materials, aerodynamics, guidance and control, propulsion, and avionics associated with high-speed aircraft and spacecraft, in addition to the techniques to construct them.

While X-15 flights were just beginning, the U.S. Air Force (USAF) began studies leading to hypersonic and maneuverable lifting body test vehicles. The *ASSET (Aerothermodynamic/elastic Structural Systems Environmental Tests)* vehicle, flown six times from 1963 to 1965, used hypersonic glides from 200,000 ft to test thermal protection, while the *PRIME (Precision Recover Including Maneuevering Entry)* vehicle, flown in the late 1960s, tested precision maneuvering. The USAF and NASA tested piloted lifting bodies in the same period, including the M2-F1 and the X-24A. The USAF Dyna-Soar spaceplane was to fly into and return from a 300-mile orbit, but Secretary of Defense Robert McNamara canceled it in 1963 for lack of a military mission. Data from these vehicles fed into the design of the Space Shuttle, which first orbited in 1981.

The Soviet Union followed a similar path, mastering reentry vehicles for its nuclear weapons, reconnaissance cameras, and human spaceflight vehicles in the 1950s and early 1960s. In the 1950s, the Soviet Union planned to evolve the Lavochkin-301A Burya intercontinental cruise missile into a piloted spaceplane, but in 1960 it was canceled. Data from Burya fed Vladimir Chelomey's Raketoplan spaceplane concept, which included variants for lunar missions, orbital bombing, passenger flight, and antisatellite applications. Its *M-12* lifting reentry vehicle prototype flew successfully in 1963. Premier Nikita Khrushchev's removal from office in 1965 led to Raketoplan's official cancellation, as Chelomey had been favored by Khrushchev. Nonetheless Chelomey persevered, quietly developing it until its final cancellation in 1983. Chelomey's design had been transferred to the Soviet Air Force, which developed the

Spiral spaceplane design. Several prototype lifting bodies, known as BOR (unpiloted orbital rocket), were tested in the 1960s, and also as an atmospheric flight test vehicle that flew eight times in 1976–78. In 1981 the Soviet leadership transferred key Spiral engineers to the Buran program—the Soviet space shuttle. To prepare for Buran, the Soviet Union tested further BOR-series prototypes in the 1980s. *Buran* flew only once in November 1988; cost issues thereafter grounded it permanently as the Soviet Union fell apart.

NASA and the USAF researched other hypersonic flight concepts, often emphasizing SSTO vehicles, in the 1970s–80s. These included a joint NASA–USAF National Hypersonic Flight Research Facility (NHFRF) study, which was under development from 1975 to 1977 before its cancellation due to cost growth and technical difficulties. Optimistic Lockheed Skunk Works engineers projected $200 million for two Mach 6 NHFRF vehicles that over a decade would fly 200 times. In 1982 the USAF began studies of a Space Shuttle successor known as TransAtmospheric Vehicle, and two years later the Defense Advanced Research Projects Agency began a scramjet study called Copper Canyon. These two projects merged in 1986 to form the National Aero-Space Plane (NASP), which was projected to be able to fly from Washington, DC, to Tokyo in two hours. However, scramjet engine technical problems led to its cancellation in 1992.

With NASP's demise, NASA undertook additional SSTO efforts. In 1995 it took over the USAF *Delta Clipper-Experimental* (*DC-X*) demonstrator vehicle, which tested launcher reusability and vertical landing capability with small teams. This vehicle flew from 1993 to 1996, when a landing accident destroyed the vehicle. The X-34 sought to demonstrate technologies useful to smaller reusable vehicles. The X-33 was more adventurous, using composite cryogenic tanks, graphite composite primary structures, metallic thermal protection materials, reusable propulsion systems, and autonomous flight control. Begun in 1995, neither of these programs reached flight stage, and both were canceled in 2001. By contrast, NASA's *X-43A Hyper-X* successfully tested a scramjet engine, setting a new airbreathing aircraft record in 2004 with a speed of Mach 9.6.

The Soviet Union, and later Russia, and Europe also experimented with further reusable spacecraft and scramjet concepts. In the late 1980s in response to NASP, the Soviet Ministry of Defense began develop the Tupolev-2000 spaceplane, but rising costs and the new Russia's severe cost constraints relegated its development to low priority. In 1993 the Russian Space Agency began studies of reusable vehicles under the Oryol (Eagle) program. The program ended in 2001, with SSTO concepts shelved. Two Stage to Orbit Systems (TSTO) studies then began under the Grif (Vulture) program. The Russians also continued scramjet research, with the Kholod (Cold) program that flew five flights from 1991 to 1998, with CNES (French Space Agency) and NASA participation in some of the flights. NASA used data from Kholod for its X-43 program in the next decade. In 1987 the European Space Agency (ESA) approved development of a reusable vehicle called Hermes, but ESA Member States soon balked in the face of rising costs, and by 1992 the program was effectively dead.

In the late 1990s and 2000s, private companies began to develop reusable vehicles for projected launches of commercial communications satellites and tourists. When the commercial communications satellite market collapsed in the late 1990s, many of the companies developing these systems went bankrupt. Several of these companies and many others competed for the X-Prize competition to develop a purely private prototype suborbital tourist vehicle. Scaled Composites Corporation won the competition in 2004 with its *SpaceShipOne* vehicle. In the late 2000s Scaled Composites teamed with Virgin Galactic Corporation to build a suborbital tourist vehicle, and a Space Adventures Corporation was teamed with Experimental Machine Building Factory (EMZ) in Russia to build a competing vehicle.

Four major challenges have vexed efforts to develop a viable hypersonic flight vehicle: aerodynamics, guidance and control, materials, and propulsion. Through a succession of projects—especially X-15 and the Space Shuttle—effective aerodynamic shapes for a hypersonic vehicle emerged. By the end of the twentieth century, many of the aerodynamic questions were satisfactorily understood, as were guidance and control issues. Researchers continued to study heat-resistant materials and composites to reduce weight. Propulsion remained the biggest issue.

Stephen B. Johnson

See also: Raketoplan and Spiral, Space Shuttle

Bibliography

Andrew J. Butrica, *Single Stage to Orbit: Politics, Space Technology, and the Quest for Reusable Rocketry* (2003).

Michael H. Gorn, *Expanding the Envelope: Flight Research at NACA and NASA* (2001).

Richard P. Hallion, ed., *The Hypersonic Revolution: Eight Case Studies in the History of Hypersonic Technology* (1998).

Bart Hendrickx, *Energiya-Buran: The Soviet Space Shuttle* (2007).

ASSET, Aerothermodynamic / elastic Structural Systems Environmental Tests. *See* Hypersonic and Reusable Vehicles.

Buran

Buran (Russian for snowstorm) was the orbiter of the Soviet reusable space transportation system Buran-Energia, sometimes referred to as the Russian space shuttle. Officially sanctioned by government and Communist Party decree in February 1976, Buran was developed and built during the 1980s by NPO (Scientific-Production Association) Energia as the Soviet Union's response to the U.S. Space Shuttle program. Buran's design was frozen in 1979, and its only flight took place on 15 November 1988, in a robotic mode, without a human crew aboard. It was delivered to orbit by the Energia launch vehicle developed and built by NPO Energia. The launch vehicle could also be used for other payloads in addition to the Buran orbiter. The Buran program was shut down in 1993, largely due to lack of funds.

The orbiter configuration was similar to that of the Space Shuttle: an orbital plane with a low mounted triangular wing with two different leading edge sweepback angles. There was a 73 m^3 crew cabin for up to 10 cosmonauts in the nose section of the fuselage and a cargo bay with manipulators for transportation and servicing satellites and other space objects in its midsection. The orbiter propulsion system acted as the third-stage engine during launch and also fired for orbit corrections, maneuvering around space objects, orientation and stabilization of the spacecraft, and deorbit. This propulsion system consisted of two orbital maneuvering engines using tsiklin synthetic fuel and liquid oxygen in the rear of the fuselage, and three blocks of gas thrusters (one at the nose and two at the tail), a total of 46 thrusters. Buran was capable of staying in orbit up to 30 days.

Although generally similar to that of the Space Shuttle, the Buran design had at least two major distinguishing features: fully automatic deorbit and landing and the absence of main rocket engines on the orbital vehicle. The engines were mounted on the core stage of the heavy-duty Energia launch vehicle instead. The length of the orbiter was 36.37 m, height 16.5 m (with the landing gear extended), and wingspan approximately 24 m. The cargo bay was 4.6 m in diameter and 18 m long. The mass of the orbiter at launch was up to 105 tons, including up to 30 tons of cargo to orbit. The mass of the cargo that Buran could return from orbit was up to 15 tons. The maximum propellant mass was approximately 14 tons.

For about a decade after its first flight, Buran was kept inside the orbiter processing building at the Baikonur launch site. In 1998 it was moved to the Energia assembly building and mated with an Energia mock-up. In 2002 the roof of the complex collapsed as a result of an accident, and the only Buran orbiter that flew into space was destroyed. A total of three flight models of the orbiter were built (the third was not finished). Manufacturing of two more began and was later terminated, with the components already built destroyed. Seven more full-scale engineering and test models of the craft were built and used for various tests.

Stephen B. Johnson

See also: Rocket and Space Corporation Energia, Space Shuttle

Bibliography
Stephen J. Garber, "A Cold Snow Falls: The Soviet Buran Space Shuttle," *Quest* 9, no. 5 (2002): 42–51.
Bart Hendrickx, *Energiya-Buran: The Soviet Space Shuttle* (2007).

Delta Clipper

Delta Clipper (DC-X) was a proposed single-stage-to-orbit (SSTO) reusable launch vehicle (RLV) designed to launch satellites for a fraction of the cost of conventional rockets. In 1990, McDonnell-Douglas Corporation proposed the Delta Clipper, a cone-shaped vehicle that would take off and land vertically under rocket power, to the Strategic Defense Initiative Organization (SDIO), which was interested in developing low-cost SSTO launch vehicles that could deploy space-based missile defense systems. SDIO awarded McDonnell-Douglas a $58.9 million contract in 1991 to develop the DC-X, a robotic subscale prototype intended to demonstrate key technologies for the full-scale Delta Clipper.

The DC-X first flew on 18 August 1993 from the White Sands Missile Range in New Mexico. The DC-X performed five low-altitude flights through June 1994,

The McDonnell-Douglas Delta Clipper Experimental Advanced (DC-XA) reusable test vehicle. (Courtesy NASA/Marshall Space Flight Center)

demonstrating its takeoff and landing technologies and its ability to hover and translate in flight. By this time, however, SDIO, now renamed the Ballistic Missile Defense Organization, had lost interest in developing an SSTO. After partially funding three additional flights, NASA took over the DC-X program in 1995. NASA, which was embarking on its own RLV technology program, modified the DC-X into the DC-XA, incorporating lightweight propellant tanks and a new reaction control system. Nicknamed Clipper Graham after RLV advocate Daniel Graham, DC-XA flew four times in the summer of 1996. On 31 July 1996, one of the DC-XA's four landing struts failed to deploy, causing the vehicle to tip over on landing and explode, ending the program.

Jeff Foust

See also: McDonnell-Douglas Corporation

Bibliography

Andrew J. Butrica, *Single Stage to Orbit: Politics, Space Technology, and the Quest for Reusable Rocketry* (2003).

Entrepreneurial Reusable Launch Vehicles

Entrepreneurial reusable launch vehicles were developed in the mid-to-late 1990s when a number of new companies entered the space industry, seeking to develop new commercial launch vehicles. While such ventures were not unprecedented—several new companies tried to develop new expendable launch vehicles in the 1980s and early 1990s—the new efforts focused instead on reusable launch vehicles (RLVs). With apparent surging demand for launches by new satellite communications companies, these businesses hoped to develop a new generation of vehicles that could launch satellites for a fraction of the cost of existing rockets.

The biggest driver for the development of these commercial RLVs was the emergence of new companies planning to launch constellations of low-Earth-orbit (LEO) communications satellites. In the early 1990s several businesses, including Globalstar, Iridium, Orbcomm, and Teledesic, planned to launch such systems to provide phone and data services worldwide. Teledesic had the most ambitious plan, requiring at one point nearly 1,000 LEO satellites. Other companies announced plans to develop similar systems. By 1998 the Federal Aviation Administration forecast that up to 1,540 LEO satellites, nearly all for communications systems, would be launched from 1998 through 2010.

While many of the satellites would be launched initially on expendable launch vehicles (ELVs), this high demand led a few entrepreneurs to propose new RLVs that would be less expensive to operate than ELVs. They argued that these RLVs would initially make money by launching communications satellites, but eventually their low cost would encourage the development of new markets.

While reusability was a common theme for these vehicles, developers otherwise exploited a wide range of technical approaches to vehicle design. Some companies, trying to make operations as aircraft-like as possible, opted for winged vehicles. Kelly Space and Technology's Astroliner would be towed aloft by a 747 aircraft; once at cruising altitude, the RLV would separate and fire its rocket engines to ascend to a suborbital trajectory, releasing its payload at apogee before returning to a jet-powered runway landing. Pioneer Rocketplane proposed the Pathfinder, which would take off under jet power and fly to a cruising altitude, where it would perform a midair refueling to fill its tanks with liquid oxygen. The vehicle would then fire its rocket engines to enter a suborbital trajectory and jet-powered landing similar to the Astroliner.

Other companies opted for vertical takeoff and landing vehicles more closely related to conventional rockets. Kistler Aerospace proposed the two-stage K-1 vehicle, which would take off vertically, each stage returning to Earth using a combination of parachutes and airbags. Rotary Rocket Corporation's Roton was a single-stage vehicle that took off vertically under rocket power, but used a combination of rocket engines and helicopter-like blades to land vertically.

While these and similar vehicles faced their share of technical risks, these efforts were eventually undone by market and financial problems. The huge market for LEO communications satellites vanished when the initial ventures, unable to compete with terrestrial alternatives such as cellular phones, filed for bankruptcy protection or shut down. Without a clear market for their services, entrepreneurial ventures found it impossible to raise the hundreds of millions of dollars needed to develop their RLVs. Most companies either went out of business or entered a long-term hibernation mode, waiting for market conditions to improve. An exception was Pioneer Rocketplane (in 2006, Rocketplane Ltd.), which survived by shifting its focus to developing smaller RLVs intended to serve the suborbital space tourism market. Rocketplane purchased Kistler in 2006, forming Rocketplane Kistler. Later that year NASA awarded the company a Commercial Orbital Transportation Services contract, but the next year NASA discontinued it.

Scaled Composites Corporation created the *SpaceShipOne* vehicle that won the X-Prize in 2004 by completing two suborbital missions. Its success led to the creation of The Spaceship Company, a joint venture of Burt Rutan (founder of Scaled Composites) and Richard Branson's Virgin Group to manufacture the SpaceShipTwo vehicle that would carry paying tourists into orbit. Virgin Galactic placed orders for five SpaceShipTwo vehicles and two White Knight carrier aircraft, while Northrop Grumman purchased the company. However, a major accident during a propulsion system test in July 2007 killed three Scaled Composite employees, putting further propulsion system development on hold.

Jeff Foust

See also: Commercial Space Launch

Bibliography

Michael Belfiore, *Rocketeers: How a Visionary Band of Business Leaders, Engineers, and Pilots Is Boldly Privatizing Space* (2007).

FAA Associate Administrator for Commercial Space Transportation (AST) website, *1998 LEO Commercial Market Projections* (1998).

Elizabeth Weil, *They All Laughed at Christopher Columbus: An Incurable Dreamer Builds the First Civilian Spaceship* (2002).

Hermes

Hermes was a European Space Agency (ESA) crewed vehicle project of the 1980s–90s. In 1975 the CNES (French Space Agency) launch vehicle directorate identified a potential need for a crewed capsule to be launched from an Ariane rocket into low orbit. Soon, however, it appeared more sensible to develop a small shuttle. The first designs called for a hypersonic glider reminiscent of the U.S. Air Force Dyna-Soar project, carrying up to five passengers. When presented at the 1979 Paris Air Show, Hermes was a delta-shaped glider to be launched on Ariane 4. Within two years there was discussion of increasing the vehicle's payload and using an Ariane 5, then still only a development project. The costs were such that CNES could not undertake the development alone.

With U.S. President Ronald Reagan's 1984 announcement of a space station project, on which he invited European participation, ESA members began to consider Hermes favorably. The initial timetable involved studies through 1987, development and first flight by 1996, and first operational flight in 1997. ESA's council welcomed France's decision to develop Hermes, but did not commit any financial resources to the project. Germany preferred direct cooperation with the United States, while the United Kingdom pushed its own shuttle project, Horizontal Take-Off and Landing.

Despite this opposition, the Hermes project was approved at the ESA ministerial meeting of 9–10 November 1987. For ESA, France's Aérospatiale and Dassault led initial development under CNES leadership. A full-scale model was shown at the 1989 Paris Air Show. A revised timetable called for drop tests by 1999, followed by an uncrewed flight in 2002 and a crewed one in 2003. Because of the staggering costs projected to reach completion of a full-scale vehicle, a proof-of-concept prototype, known as Hermes X-2000, was proposed and discussed at the 1992 gathering of ESA Member States in 1992. Germany argued that this would add to the costs rather than decrease them by adding a prototype to the production copies, and it declined to commit additional funds to the project. The Munich gathering considered alternatives involving cooperation with the United States or Russia or further theoretical studies, putting Hermes out of existence, though never officially abandoning it. In 2005 the weathered full-scale model shown at the Paris Air Show was donated to the French National Air and Space Museum at le Bourget for restoration and possible display.

Guillaume de Syon

See also: Ariane, European Space Agency

Bibliography

"Company Formed to Manage Next Phase of Europe's Hermes Spaceplane Program," *Aviation Week and Space Technology* (12 November 1990): 76–77.
A. Krige et al., *A History of the European Space Agency 1958–1987, Volume II* (2000).
M. Senechel, "Hermes: the French Shuttle," *Spaceflight* 28 (January 1986): 37–38.

National Aero-Space Plane

National Aero-Space Plane (NASP), designated X-30, was an effort by the U.S. government and industry to develop an air-breathing vehicle that could reach orbit or fly around the world in a few hours. NASP had its roots in the U.S. Air Force (USAF) TransAtmospheric Vehicle program and the Defense Advanced Research Agency Copper Canyon project that investigated the feasibility of a single-stage spacecraft that could take off and land from conventional airports. President Ronald Reagan formally announced NASP in his 1986 State of the Union address, calling for development of a "new Orient Express" that could reach orbit or fly from Washington, DC, to Tokyo within two hours. NASP was a national program, with funding from NASA and the USAF, and a contracting team that included General Dynamics, McDonnell-Douglas, and Rockwell International.

NASP's core technology was a supersonic ramjet, or scramjet, that generated thrust from atmospheric oxygen and hydrogen fuel. The scramjet would accelerate the spaceplane to about Mach 20, and a small rocket engine would provide the final boost needed to reach orbit. While the scramjet would eliminate the need for NASP to carry its own oxidizer, it was also a complicated technology that had not been flight-tested. Scramjet development problems and difficulties creating a thermal protection system to withstand the high thermal stresses NASP would encounter led to cost overruns and delays. By the early 1990s the USAF had lost interest in the program, and budget constraints prevented further NASA spending. NASP was effectively canceled in 1993, and its technology development work transferred to a smaller effort, the Hypersonic Systems Technology Program.

Jeff Foust

See also: Defense Advanced Research Projects Agency, National Aeronautics and Space Administration, United States Air Force

Bibliography

Andrew J. Butrica, *Single State to Orbit: Politics, Space Technology, and the Quest for Reusable Rocketry* (2003).

"X-30 National Aerospace Plane (NASP)," Federation of American Scientists. http://www.fas.org/irp/mystery/nasp.htm.

PRIME, Precision Recover Including Maneuevering Entry. *See* Hypersonic and Reusable Vehicles.

SpaceShipOne

SpaceShipOne was a privately developed and owned suborbital spaceplane. It was the first such spacecraft to reach more than 100 km and subsequently won the Ansari X Prize, a $10 million prize awarded to the first privately funded crewed spacecraft to reach 100 km. Innovative aircraft designer Burt Rutan, owner of Scaled Composites, developed the White Knight aircraft to carry *SpaceShipOne*, a hybrid rocket-powered spaceplane. Rutan had worked on this concept since 1997, using his Proteus high-altitude airplane with the hope of including an upper stage to launch small satellites and human-rated systems. Microsoft cofounder Paul Allen signed a contract

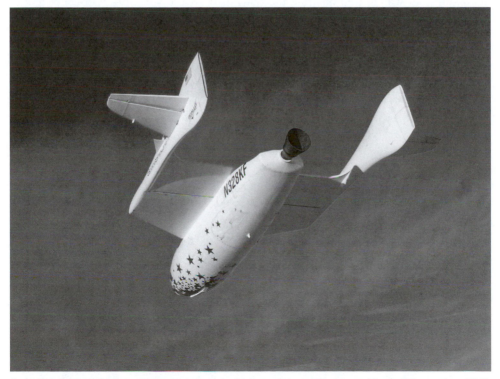

SpaceShipOne *glides down for approach to the Mojave airport. (Courtesy of Scaled Composites, LLC)*

with Scaled Composites in April 2001, providing an undisclosed amount of money (estimated at $27 million) to fund the effort.

Rutan's team primarily used computational fluid dynamics software to evaluate the craft's aerodynamics with some preliminary testing using small-scale models to evaluate the glider characteristics. SpaceDev developed *SpaceShipOne*'s hybrid propulsion system, which used liquid nitrous oxide oxidizer and a rubber compound as fuel. This propulsion system descended from an earlier design by American Rocket Company.

In a nominal flight, White Knight carried the spaceplane under its belly. At about 14 km it dropped *SpaceShipOne*, which ignited its rocket engine for a 65–80 second burn. After this the vehicle climbed ballistically to 100–110 km. *SpaceShipOne* used the revolutionary method of "feathering." This method consisted of tilting the wings forward to create a stable high-drag shape that allowed for atmospheric reentry. This reduced aerothermal loads and at the same time allowed for a return without the need of a pilot's input. Once back in the atmosphere, a pneumatic-actuated system returned the wing boom to the nominal position and the vehicle glided to the Mojave, California, airport.

The first private human-flown spaceflight occurred on 14 June 2004 when pilot Mike Melville flew *SpaceShipOne* to 100 km. On 29 September Melville flew again to 103 km, making the first qualification flight for the Ansari X Prize. On 4 October 2004 pilot Brian Binnie flew the craft a final time to 112 km, wining the prize and breaking the 1963 altitude record of the X-15.

SpaceShipOne was never designed to develop into a commercially viable vehicle; its major goal was to demonstrate the feasibility of the spaceflight concept and to win the Ansari X Prize. Rutan next began working with Virgin Galactic to develop the Virgin SpaceShip *Enterprise* (also called *SpaceShipTwo*), a commercial suborbital space vehicle that would license technology from Paul Allen's Mojave Aerospace Ventures. In 2005 *SpaceShipOne* was installed in the Smithsonian National Air and Space Museum Milestones of Flight Gallery in Washington, DC.

Pablo de León

See also: Tourism

Bibliography
"Scaled Composites Press Releases, 2002–4," X Prize Foundation Commemorative Program, 2004.

X-15

X-15 was a reusable, hypersonic test vehicle that set altitude and speed records in the 1960s. It was the result of advances in hypersonic flight during the 1950s, which stemmed from engineering work during World War II by both sides that contributed to advancements in rocket and jet technology. During the war, many leading aeronautic engineers speculated that building hypersonic aircraft might be possible. These aircraft would reach speeds in excess of five times the speed of sound. Following several paper studies, including several famous hypersonic vehicle designs by German scientists Eugen

Sänger and Irene Bredt, in 1945 the U.S. National Advisory Committee for Aeronautics (NACA) began building a wind tunnel designed for testing hypersonic vehicle concepts.

NACA moved forward with a hypersonic vehicle program in 1954. The military had tested its BoMi concept based on Sänger-Bredt designs and was interested in further exploring the hypersonic realm. On 11 June 1956, NACA awarded North American Aviation a contract, valued at almost $43 million, to build three X-15 aircraft and associated hardware. NACA (and its successor, NASA) and the U.S. Air Force (USAF) jointly conducted experiments with the X-15.

The aircraft's structure was plated in Inconel X, an alloy designed to withstand the high temperatures of hypersonic flight. The forward fuselage (called Bay 1) contained the cockpit, nose wheel well, and avionics compartments. Bay 2, immediately aft of the cockpit, contained auxiliary power units, helium and nitrogen tanks, and equipment. The center fuselage housed two main propellant tanks and an equipment compartment between them (Bay 3). A later version of one X-15 was equipped to carry two external propellant tanks attached to the belly of the vehicle. The razor-thin wings were mounted along the sides of the center fuselage. The aft fuselage contained the engine and provided supporting structure for the wedge-shaped vertical stabilizers and conventional horizontal stabilizers. The aft section also contained a unique deployable skid landing gear system. All X-15s were carried under the right wing pylon of a modified NB-52 bomber.

The first X-15, serial number 56-6670, flew 81 times, reaching a top speed of Mach 6.06 (4,104 mph) and a maximum altitude at 266,500 ft. Serial number 56-6671 flew 31 times before being significantly modified into the X-15A-2, after which it flew an additional 22 times, reaching a record speed for a human-tended aircraft of Mach 6.7 (4,520 mph) and attaining 249,000 ft altitude. Serial number 56-6672 flew 65 times, attaining a maximum speed of Mach 5.73 (3,897 mph) and reaching the highest altitude of the X-15 program at 354,200 ft.

X-15 Vehicles

X-15 Designation	Serial Number	First Flight	Number of Flights	Highest Speed	Highest Altitude
X-15-1	56-6670	6/8/59	81	6,604 kph (4,104 mph, or Mach 6.06)	81,229 meters (266,500 feet)
X-15-2	56-6671	9/17/59	31	6,587 kph (4,093 mph, or Mach 6.04)	66,142 meters (217,000 feet)
X-15A-2	56-6671	6/25/64	22	7,274 kph (4,520 mph, or Mach 6.70)	75,895 meters (249,000 feet)
X-15-3	56-6672	12/20/61	65	6,271 kph (3,897 mph, or Mach 5.73)	107,960 meters (354,200 feet)

Source: Hypersonic: The Story of the North American X-15, Jenkins, Dennis and T. Landis.

Fourteen pilots flew this aircraft, including several who went on to become NASA astronauts. From 1959 to 1968, the program made a total of 199 flights. One pilot, USAF Major Michael Adams, was killed when his aircraft (X-15 56-6672) crashed on 15 November 1967. Eight pilots, including Adams, reached and exceeded the 50-mile altitude judged by the USAF as the boundary of space, and these men were awarded astronaut wings.

The X-15 program contributed greatly to the understanding of high-altitude, high-speed flight environments, helping lay a strong research and development foundation for various government and commercial orbital and suborbital vehicle concepts, including the U.S. Space Transportation System (Space Shuttle). In addition, the program provided the first opportunity to study prolonged human physiological responses to microgravity conditions, observe the behaviors of different ablative coatings and paints, and test a system that blended aerodynamic and reaction control maneuvering capabilities.

On officially retiring in 1968, X-15 56-6671 was delivered to the USAF Museum in Dayton, Ohio. Its sister (56-6670) was installed at the Smithsonian Air and Space Museum in 1976.

Phil Smith

See also: North American Rockwell Corporation

Bibliography

Robert Godwin, ed., *X-15: The NASA Mission Reports* (2000).
Dennis R. Jenkins, *Space Shuttle: The History of the National Space Transportation System* (2002).
Dennis R. Jenkins and Tony R. Landis, *Hypersonic: The Story of the North American X-15* (2003).

X-23A

X-23A (or SV-5D) became the first high-performance maneuvering reentry vehicle. The Martin Marietta Corporation developed the X-23A under contract to the U.S. Air Force (USAF). Made largely from titanium alloy, the vehicle measured 6 ft, 7.5 in, in length and 2 ft, 10 in, in height, with a 4 ft wingspan and 894 lb gross weight. Launched on an Atlas launch vehicle, the uncrewed X-23A used its lifting-body design to glide down before being recovered in midair by parachute. The X-23 flew three times from Vandenberg Air Force Base, California, on 21 December 1966, 5 March 1967, and 19 April 1967. The aerial recovery failed during the first two flights, but succeeded in the third test. Martin Marietta built a total of four vehicles, but the USAF canceled the program after the success of the third test. The USAF retrieved only the third vehicle, which went to the USAF Museum for display.

The fastest flight reached approximately 16,500 mph and the highest flight 500,000 ft. The results from Project Precision Recovery Including Maneuvering Entry (PRIME) under the Spacecraft Technology and Advanced Reentry Test (START) program supported USAF and NASA research: ablative heatshield data from atmospheric reentry and the lifting-body shape contributed to the later design of the Space Shuttle orbiter.

Giny Cheong

See also: Martin Marietta Corporation, United States Air Force

Bibliography
Dennis R. Jenkins et al., *American X-Vehicles: An Inventory—X-1 to X-50* (2003).
Jay Miller, *The X-Planes: X-1 to X-45* (2001).

X-33

X-33 was a joint project by NASA and Lockheed Martin Corporation to develop a demonstrator of a single-stage-to-orbit (SSTO) reusable launch vehicle (RLV). NASA started the X-33 program after the cancellation of the National Aero-Space Plane, the development of the Delta Clipper Experimental (DC-X), and announcement of a new national space policy that gave NASA responsibility for RLV development. After taking over the DC-X program in 1995, NASA issued a request for proposals for a larger suborbital vehicle that would demonstrate SSTO RLV technologies. On 2 July 1996, Vice President Al Gore announced that NASA had selected the Lockheed Martin proposal instead of competing designs by Boeing and McDonnell-Douglas.

Lockheed Martin's X-33 design took the form of a large lifting body that would lift vertically from a new launch pad at Edwards Air Force Base, California. The X-33 would reach speeds of Mach 15 before gliding to runway landings in Utah or Montana. X-33's primary purpose was to test technologies that would be useful for future RLVs, including a linear aerospike engine, liquid-oxygen and hydrogen propellant tanks made of lightweight composite materials, and advanced thermal protection systems. Lockheed Martin, which invested several hundred million dollars of its money in the X-33 under a cooperative agreement with NASA, planned to scale up the X-33 into a large orbital RLV, the VentureStar.

The advanced technologies incorporated in the X-33 proved to be its undoing. Problems with the aerospike engine delayed the first flight, originally scheduled in mid-1999. In late 1999 a liquid-hydrogen tank failed a ground test, further delaying the program. By March 2001 NASA, which had spent more than $900 million on the program without a single flight, decided to focus on a new RLV technology program, the Space Launch Initiative, and announced it would provide no additional funding for the X-33. The U.S. Air Force turned down a Lockheed Martin request for X-33 funding in September 2001, ending the program.

Jeff Foust

See also: Lockheed Martin Corporation, National Aeronautics and Space Administration

Bibliography
Leonard David, "NASA Shuts Down X-33, X-34 Programs" SPACE.com (2001). http://www.space.com/missionlaunches/missions/x33_cancel_010301.html.

"The X-33 History Project Home Page," NASA History Office (2006). http://history.nasa.gov/x-33/home.htm.

X-38 Crew Return Vehicle

The X-38 Crew Return Vehicle was a prototype NASA developed to return up to seven *International Space Station* (*ISS*) crewmembers safely to Earth in case of an emergency situation or an injury. The vehicle was intended to replace the Russian three-seat Soyuz spacecraft, which the *ISS* used as a crew return vehicle.

X-38 was an in-house NASA project, developed and managed by the Johnson Space Center (JSC) starting in 1995, with participation from other NASA centers, including Langley Research Center, Dryden Flight Research Center, and Marshall Space Flight Center. To reduce development cost, JSC engineers used proven commercial off-the-shelf technologies for roughly 80 percent of the vehicle design. Also in 1955 JSC awarded a contract to Scaled Composites, Inc., of Mojave, California, to manufacture three full-scale atmospheric airframes. Full-scale flight tests began in July 1997 and continued until 2002. In these flights, engineers tested parachute deployment and steering systems, along with the automatic flight control system.

The X-38 design was based on an innovative lifting body concept, relying solely on the flat surface of the craft's belly to create the essential lift for a slow descent and on

The X-38 research vehicle descends at a steep angle during a July 1999 test flight at the Dryden Flight Research Center in California. (Courtesy NASA/Dryden Flight Research Center)

small fins to help stabilize the vehicle in flight. Its design heritage was based on the X-24A, which the Dryden Flight Research Center developed in a joint program with the U.S. Air Force from 1963 to 1975, allowing JSC engineers to use X-24A test data to a maximum extent. To accomplish its mission objectives, X-38 was designed to fly automatically from orbit to landing, with little or no input from the crew. Once detached from *ISS*, X-38 would return from orbit unpowered and then use a steerable parafoil parachute for landing. The parachute would enable X-38 to avoid dangerously high landing speeds, a typical issue for lifting bodies because of their lack of wings. The parachute would also enable X-38 to glide great distances, which would give it the opportunity to leave *ISS* anywhere in the orbit and still return to a safe area. X-38 would touch to the ground on skids, rather than wheels.

NASA canceled X-38 in April 2002 because of budget pressures associated with *ISS*. At the time of the cancellation, X-38 was roughly two years short of completing its flight phase. If the project had been accomplished, X-38 would have enabled larger crews on *ISS*.

Incigul Polat-Erdogan

See also: *International Space Station*, Johnson Space Center

Bibliography

Mark Carreau, "Project's Cancellation Irks NASA," *Houston Chronicle*, 9 June 2002.

Dale R. Reed, *Wingless Flight: The Lifting Body Story* (1997).

"X-38 Fact Sheet," NASA Dryden Flight Research Center. http://www.nasa.gov/centers/dryden/news/FactSheets/FS-038-DFRC.html.

ROCKET PIONEERS AND SPACEFLIGHT VISIONARIES

Rocket pioneers and spaceflight visionaries established the conceptual and theoretical foundations that guided experimenters and inspired amateur rocketeers, particularly in the 1920s–30s. Many of these pioneers and visionaries were inspired by the technology-laden stories of Jules Verne, which provided ample fodder for active imaginations.

In the Soviet Union, Konstantin Tsiolkovsky provided the theoretical basis for the flight of rockets in space, beginning with the 1903 publication of his seminal paper on the exploration of interplanetary space using reactive devices. Much of his work was largely unknown outside the Soviet Union until the 1930s, when the Soviet Union began to take interest in internationally promoting Soviet scientific and technical achievements. Amateur rocket societies (GIRDs, or Jet Propulsion Research Groups, a translation of the Russian acronym) formed in Moscow and Leningrad, the former under the leadership of Sergei Korolev, who later became Chief Designer of the Soviet human spaceflight program after surviving imprisonment during the purges of the late 1930s and through World War II. Nikolai Rynin completed a monumental work that brought

HERMANN JULIUS OBERTH
(1894–1989)

(Courtesy NASA)

Hermann Oberth was from Transylvania in the Austro-Hungarian Empire, where as a child he was mesmerized by Jules Verne's science fiction. He translated this fascination into doctoral thesis research in Göttingen and Munich. Though rejected by his doctoral review committee as too utopian, Oberth had *Die Rakete zu den Planetenräumen* (*The Rocket into Planetary Space*), printed privately in 1923. He later proposed a three-stage rocketship design, which inspired the rocket design of the German movie *Frau im Mond*. Oberth inspired many space enthusiasts, including young Wernher von Braun, who later invited Oberth to join him at Peenemünde. In 1955, at von Braun's invitation, Oberth went to Huntsville, Alabama, where he worked as a consultant on spaceflight. Two years later he completed a study on the possibility of a lunar mission.

Guillaume de Syon

together the current state of astronautics, including the works of Tsiolkovsky, in a nine-volume encyclopedia translated as *Interplanetary Communications* (1928–32).

In Germany, Hermann Oberth, a Transylvanian by birth, wrote *Die Rakete zu den Planetenräumen* (*The Rocket into Planetary Space*, 1923), which firmly established many important concepts of human spaceflight, including an Earth-orbiting habitable space station and the use of multistage rockets to achieve Earth orbit. Oberth's book, which he published after it was rejected as the subject of a dissertation thesis, was influential in Germany and other countries around the world. Some of this impact might be attributed to the success of the VfR (Verein für Raumschiffahrt), or Society for Spaceship Travel, in building public interest in the prospects of human spaceflight and publicizing news of the rocket world in member publications and books written by VfR members. An early member of the VfR was a youthful Wernher von Braun, who later accepted an offer from the German Army to lead its rocket development program that resulted in the production of the A-4 (later more widely known as the V-2) rocket. Von Braun was taken to the United States at the end of World War II, with other captured technical specialists, and became a spokesman for the exploration and conquest of space, a voyage on which he played a leading role that culminated in the development of the Saturn V rocket that launched astronauts to the Moon.

In the United States, Robert Goddard conducted rocket experiments, including the first launch of a liquid fuel rocket (1926), and wrote an influential monograph on *A Method of Reaching Extreme Altitudes* (1919). David Lasser wrote *The Conquest of*

**ROBERT ESNAULT-PELTERIE
(1881–1957)**

Robert Esnault-Pelterie, commonly referred to as REP, was a French aviation pioneer and advocate of space travel prior to World War II. He addressed the topic of spaceflight in a paper presented to the French Physics Society in 1912 and was the first to popularize the term "astronautics." REP wrote a highly regarded and comprehensive technical assessment of astronautics in *L'Astronautique* (1930), which he followed with a supplement in 1935. He joined with André-Louis Hirsch in 1928 to establish the REP-Hirsch prize (discontinued in 1939 because of World War II) to promote international achievements in astronautics.

Michael Ciancone

Space (1931), which was the first technical book of nonfiction in English on the use of rockets for human spaceflight and which captured his vision about future prospects of spaceflight. Perhaps more important, he engaged the hopes and aspirations of other like-minded individuals to form the American Interplanetary Society (AIS), which evolved into the American Rocket Society (ARS) and later merged with the Institute of the Aeronautical Sciences to form the American Institute for Aeronautics and Astronautics. Four early members of the AIS/ARS formed Reaction Motors, Inc., in 1941, thus establishing an American lineage from early visionaries through amateur experimentation to commercial application.

In France, Robert Esnault-Pelterie, an aviation pioneer, was a prominent spaceflight promoter prior to World War II who prepared a comprehensive technical assessment of astronautics in *L'Astronautique* (1930). REP, as he was commonly known, helped to promote international achievements in astronautics through the REP-Hirsch prize, which he established with André-Louis Hirsch in 1928.

Although the successful attainment of space was the result of technical achievements worldwide, much of this success can be credited to the inspiration provided by individuals who first developed the principles of rocketry, visualized the use of rockets for human spaceflight, and promoted the notion of humans in space. Together they sparked the embers of an idea and blazed the trail long before such ideas were common or accepted.

Michael L. Ciancone

See also: American Rocket Society, Literary Foundation of the Space Age, Russian Early Rocketry, Verein für Raumschiffahrt

Bibliography
James Harford, *Korolev* (1997).
Michael Neufeld, *Von Braun: Dreamer of Space, Engineer of War* (2007).

Korolev, Sergei

Korolev, Sergei Pavlovich (1907–1966), was a Soviet Union aerospace engineer and manager, one of the founders of the Soviet space program, and Chief Designer of

OKB-1 (Experimental Design Bureau). Due to the introduction of a modern calendar in Russia in 1918 with two weeks' difference in dates, his date of birth was often quoted as 1906.

After graduation from the Moscow Bauman Technical University in 1929, Korolev worked in the aviation industry designing gliders and light airplanes. To test his creations in flight, he became a certified pilot. Korolev promoted the idea of installing a rocket engine on a glider and in 1931 joined the Moscow Jet Propulsion Research Group (MosGIRD), initially a hobby club of rocket enthusiasts. A year later, after the group became an official research organization with military support, he was appointed MosGIRD director. Korolev became an avid follower of Konstantin Tsiolkovsky, who for decades had been promoting the ideas of space exploration. He supervised the development and test flights of the first Soviet liquid-propellant rockets, GIRD-09 and GIRD-X, designed by Mikhail Tikhonravov and Friedrich Tsander in 1933. From 1934 through 1938 he worked at the Jet Propulsion Research Institute (RNII), which combined MosGIRD with the military Gas Dynamics Laboratory. Korolev developed several guided ballistic and cruise missiles in addition to a rocket-propelled glider that was tested in flight in 1940. He was severely injured during a rocket engine test in 1938.

From 1938 to 1944, Korolev was imprisoned on false charges of subversion. After barely surviving four months in a gold-mining prison camp in northeastern Siberia, he worked at the special design bureaus for imprisoned specialists, developing aircraft (1940–42) and rocket engines (1942–44). Korolev personally tested rocket engines (designed under his long-time RNII colleague, Valentin Glushko) in flights of specially equipped airplanes. In 1945 he suffered another injury and almost lost his eyesight during an engine malfunction.

Korolev played an important role in Soviet efforts to recover and reconstruct German missile weapons, working as chief engineer of the Soviet-German joint missile research enterprise, called Institute Nordhausen, in 1945–46.

In 1946 Korolev was appointed Chief Designer of a design office, later known as OKB-1, for recreation of the German A-4 (V-2) ballistic missile. He directed the early Soviet ballistic missile projects: R-1, R-2, R-3, R-11, and R-5. At that time, he organized the Council of Chief Designers, an unprecedented shortcut in traditional interdepartmental bureaucracy, which allowed him to deal directly with subcontractors instead of their bosses. From 1953 to 1957, he led the development of the first Soviet intercontinental ballistic missile (ICBM), the R-7. From 1960 to 1966, Korolev supervised the R-9 and GR-1 ICBM projects and early Soviet experiments with solid-propellant ballistic missiles.

Korolev vigorously promoted the idea of space exploration. From 1949 to 1960, with support from the Soviet Academy of Science, he organized a broad program of geophysical and biological experiments using modifications of the R-1 and R-2 rockets. At the time of the R-7 final approval, Korolev, Tikhonravov, Mstislav Keldysh, and other influential space proponents proposed artificial satellites to the Soviet government, which in 1954 authorized preliminary research in satellites and the parallel development of the R-7 as a ballistic missile and space launch vehicle. After official approval of the satellite program in 1956, Korolev organized a new space

MSTISLAV VSEVOLODOVICH KELDYSH (1911–1978)

Soviet Union mathematician Mstislav V. Keldysh came to prominence in the 1940s because of his brilliant application of mathematics to complex aeronautical problems. After World War II, he was elected a member of the Soviet Academy of Science and managed mathematical support to the development of nuclear weapons and ballistic and cruise missiles. His endorsement of early satellite proposals (1954) was crucial for the beginning of the Soviet space program. As president of the Academy of Science (1961–75), Keldysh promoted scientific space projects and was directly involved in the research, development, and implementation of all Soviet space missions.

Peter A. Gorin

systems department at OKB-1 under Tikhonravov. Because of the efforts of that department, the Soviet Union started the space era, successfully launching the world's first artificial satellite, *Sputnik*, on 4 October 1957. After *Sputnik*'s unprecedented success, other Soviet rocket designers followed Korolev's example and eagerly participated in space projects.

From 1957 to 1966, Korolev supervised development of all early Soviet artificial satellites; automated probes to the Moon, Mars, and Venus; and the Vostok, Voskhod, and Soyuz piloted spacecraft. Simultaneously he directed the first Soviet military space projects, the photoreconnaissance satellites *OD-1* and *Zenit-2*. He was constantly initiating new space projects, which eventually were combined into a comprehensive space program. That program, authorized by the government in 1960, envisioned development of powerful launch vehicles, space stations, piloted spacecraft to the Moon and Mars, and automatic probes to other planets. A year later, however, the government practically abandoned that program by redirecting resources of the rocket/space industry to the speedy development of new military missiles to catch up with the United States in the strategic arms race. Korolev supported international cooperation in space and in 1960 proposed creation of the International Space Research Institute based on OKB-1. That idea was rejected due to the extreme secrecy imposed on OKB-1.

Korolev considered the development of a universal super-heavy space launch vehicle, the N-1, and its smaller derivatives the centerpiece of the future Soviet space program. He consistently defended the N-1 project against numerous critics, especially against the desire of some specialists to use toxic storable propellants. Acknowledging the limited use of liquid oxygen for military missiles, Korolev nevertheless argued that the cryogenic propellants were essential for the space launchers due to their superior energy output. This dispute about propellant alienated Korolev from many of his patrons in the government and colleagues, including Glushko, the leading Soviet rocket engine designer. Inconsistent government support and insufficient funding forced Korolev to adopt risky design decisions in the N-1 project, causing major problems and delays. Originally conceived in 1960, the final N-1 configuration was approved by the government only after Korolev's death in 1966. His longtime first deputy and successor as OKB-1 Chief Designer, Vasili Mishin, put his career on the line trying to complete the N-1 and implement the piloted lunar mission, but the success of the U.S. Apollo program and numerous technical problems doomed Korolev's final creation. The N-1 and all other projects associated with it (including the piloted lunar program) were eventually canceled in 1976.

Korolev died unexpectedly during a surgical operation at the age of 59, creating a myth that his death was the result of the doctors' incompetence. According to Korolev's daughter Natalia, a qualified surgeon herself, the surgery was done correctly and the doctors were not to blame.

Due to the secrecy of his work, Korolev's enormous contributions to space exploration were publicly acknowledged only after his death. An asteroid, a lunar crater, a city in the Moscow Region and the Rocket and Space Corporation Energia (the former OKB-1) were named for him.

Peter A. Gorin

See also: Early Soviet Venus Program, Rocket and Space Corporation Energia, Russian Early Rocketry, Russian Launch Vehicles, Soviet Manned Lunar Program, Soviet Mars Program, Soyuz, Sputnik, Vostok, Voskhod

Bibliography
James Harford, *Korolev* (1997).
Asif A. Siddiqi, *Challenge to Apollo: The Soviet Union and the Space Race, 1945–1974* (2000).

Von Braun, Wernher

Von Braun, Wernher (1912–1977), German rocket engineer who immigrated to the United States, was one of the most influential spaceflight figures of the twentieth century. His historical role rests on four fundamental achievements: (1) as technical director of the German V-2 missile project, he led the design and construction of the world's first large rocket; (2) as a leading advocate for space travel during the 1950s, he helped sell the U.S. public on its feasibility; (3) as technical director of the U.S. Army's ballistic missile facility in Huntsville, Alabama, 1950–60, he was instrumental in launching the first American satellite in 1958; and (4) after 1960, as director of NASA's new Huntsville center, he was the consummate manager of the Saturn booster project that sent two dozen Apollo astronauts to the Moon between 1968 and 1972.

Born Wernher Maximilian Magnus Freiherr (Baron) von Braun on 23 March 1912 in Wirsitz, Germany, he was the second son of a middling Prussian aristocrat. From 1920 to 1934, his father was a wealthy banker in Berlin. Soon after Wernher began boarding school in fall 1925, the future rocket engineer discovered the works of Hermann Oberth, the German–Romanian space visionary, and became a convert to the idea of interplanetary travel, dreaming particularly of flying to the Moon. He joined the VfR (Verein für Raumschiffahrt or Society for Space Travel) in 1928 and, after graduating from high school two years later, became a member of its rocket

An image of Wernher von Braun in 1962. Von Braun developed the A-4 (V-2) ballistic missile for Nazi Germany and used in the last two years of World War II. After the war, von Braun emigrated to the United States, developing the Redstone ballistic missile for the U.S. Army, and the Jupiter-C launch vehicle used to place the first U.S. satellite into orbit in 1958. Transferring to NASA in 1960, von Braun directed the Marshall Space Flight Center and led the development of the Saturn V Moon rocket. (Courtesy Library of Congress)

experiment group, the Raketenflugplatz (Rocketport) Berlin. From 1930 to 1932, he was a mechanical engineering student at the Berlin Institute of Technology, except for one semester in Zurich.

Fall 1932 was a turning point. German Army Ordnance took an interest in von Braun's work, financing his physics dissertation on liquid-propellant rocketry at the University of Berlin. Shortly thereafter, when Adolf Hitler came to power, von Braun's success and charismatic personality quickly led him to a regular position in a small, highly secret rocket project that became better funded as rearmament accelerated. In July 1934 von Braun received his doctorate in physics; in December he launched his first Army rockets. In 1935–36 the project grew dramatically after the Army allied with the Air Force to build a secret rocket center on the Baltic coast. When the center at Peenemünde opened in May 1937, von Braun, at age 25, was technical director of the Army side and leader of 350 people; by age 30 he would lead several thousand. His genius was the management of huge engineering projects. During World War II his primary job was to bring the V-2 ballistic missile through its difficult development phase into production as a weapon for bombing Allied cities. After 1942 he also led the development of the Wasserfall antiaircraft missile.

Von Braun's Faustian bargain with Hitler's regime led him deeper into Nazi institutions and crimes. Asked to join the Nazi Party in 1937, he was pushed into becoming an SS (security forces) officer in 1940 and was highly decorated by Hitler. While he was not an ideological Nazi, there is little doubt about von Braun's loyal service to the regime. After a decision to use concentration-camp labor, he became mired in the murderous exploitation of prisoners in V-2 production. None of this protected him, however, when the Gestapo arrested him in March 1944 as the result of careless remarks at a party that he was more interested in spaceships than weapons. Those remarks became an excuse for SS leader Heinrich Himmler to punish him for not cooperating in the SS move to take over the program. His superiors' argument that he was indispensable to the V-2 project saved him.

On 2 May 1945 von Braun surrendered to the U.S. Army in Austria after evacuation from Peenemünde. By September he was already in the United States as part of a secret military program, soon called Paperclip, to exploit German scientists and engineers. The Army sent him to El Paso, Texas, where he led about 120 Germans from his program. Their tasks were to train American personnel to launch captured V-2s at the White Sands range in nearby New Mexico, to facilitate transfer of German technology, and to begin development of new missiles for the Army. From 1947 to 1950, their primary project was the Hermes II, an experimental ramjet missile to be boosted by a V-2. In 1950 the Army transferred von Braun's group, along with several hundred Americans, to Redstone Arsenal in Huntsville, Alabama. As a result of the Korean War, Hermes II was supplanted by Redstone, a short-range, nuclear-tipped ballistic missile derived from the V-2. Meanwhile, in 1947, von Braun had married a cousin, Maria von Quistorp; they had two daughters and a son.

After trying for several years to find a way to sell spaceflight to the American public, von Braun made a breakthrough in 1952 when *Collier's* began an influential series of articles on the topic, many of which he wrote. These articles persuaded many people that spaceflight was not ridiculous and launched what a later observer has called "the von Braun paradigm" for human space exploration: a winged reusable shuttle, a space station, a Moon landing, and finally a Mars expedition, all imagined on a grand scale. This picture of what a "logical" space program should be repeatedly influenced American space policymakers after the space race began. Von Braun's appearance in two 1955 Disney programs about space reinforced his fame and his influence on the public imagination of spaceflight. That same year he became a U.S. citizen.

In August 1955 von Braun's Orbiter proposal lost the competition to launch the first U.S. scientific satellite to a more advanced Navy concept called Vanguard. In November the Department of Defense tasked his group with the development of Jupiter, a nuclear-armed, intermediate-range ballistic missile. To test new materials to protect warheads during reentry into the atmosphere, the von Braun group, now the core of the Army Ballistic Missile Agency (ABMA), redesigned its Redstone-based Orbiter booster to launch nose cones, preserving its satellite launch capability. After the Soviet *Sputnik* launch of 4 October 1957, the Dwight Eisenhower presidential administration gave ABMA permission to launch satellites to back up the faltering Vanguard. When the Navy failed to orbit one prior to von Braun's group, *Explorer 1* became the first U.S. satellite on 31 January 1958.

Sputnik and *Explorer 1* vaulted von Braun to world fame and made him a leading advocate of a vigorous space competition with the Soviet Union. ABMA launched several more Explorers and two Moon probes, but for more than a year the future disposition of his group was uncertain, as the military services and the newly created NASA struggled over space policy. Finally, in October 1959, NASA got the heart of ABMA—the 4,000 people of von Braun's division—as a booster development facility. After the transfer was complete in July 1960, this group became the Marshall Space Flight Center (MSFC), with Wernher von Braun as director. With it came Saturn, a big booster with a first-stage thrust of 1.5 million lb, begun with military money in 1958.

Marshall provided Redstones to launch one-man Mercury capsules on suborbital missions, but NASA's Apollo project for a human mission to the Moon quickly came to dominate the center's agenda. Von Braun's assessment of the feasibility of the huge launch vehicle needed to outdo the Soviet Union influenced President John Kennedy's spring 1961 decision to land astronauts on the lunar surface by the end of the decade. By early 1962, NASA approved Saturn V, with 7.5 million lb thrust in its first stage. Von Braun's MSFC became the biggest NASA center, with more than 7,000 employees. Von Braun also played a crucial role in the mid-1962 decision to use lunar orbit rendezvous to land on the Moon, although MSFC had been committed to a competing concept. The Saturn series turned out to be astonishingly successful, never having a catastrophic failure, a record that must be attributed in part to von Braun's management skills.

In 1970 after two Moon landings, NASA Administrator Thomas Paine brought von Braun to Washington, DC, as deputy associate administrator for planning. It did not turn out well. Because the public had lost interest in expensive space programs, there was not much of a future to plan. Paine soon resigned. Von Braun followed two years later in June 1972, after the Space Shuttle was approved, feeling there was little more he could contribute. He became vice president for engineering at Fairchild Industries in Germantown, Maryland, but his life was soon cut short by cancer. After surviving one bout of the disease in 1973, he relapsed two years later and died on 16 June 1977. He will be remembered for his profound impact on rocketry and spaceflight in the middle decades of the twentieth century, but also for his compromises with the Nazi regime.

Michael J. Neufeld

See also: Future Studies, Redstone, Saturn Launch Vehicles, V-2, Verein für Raumschiffahrt

Bibliography
Roger Bilstein, *Stages to Saturn* (1980).
Howard McCurdy, *Space and the American Imagination* (1997).
Michael Neufeld, *The Rocket and the Reich* (1995).
———, *Von Braun: Dreamer of Space, Engineer of War* (2008).
Ernst Stuhlinger and Frederick I. Ordway III, *Wernher von Braun* (1994).

SPACECRAFT

Spacecraft are payloads put into space by space launch vehicles. There are several different types, including human-occupied space stations and transport vehicles, Earth-orbiting robotic spacecraft, deep space probes that fly by or orbit other solar system bodies, landers, and rovers. Spacecraft ultimately serve some function, and the equipment and living beings that perform these functions are themselves the payloads of the spacecraft. The equipment necessary to support the payload is often called the "bus" or the spacecraft's engineering subsystems, such as life support subsystems or electrical power subsystems.

The first spacecraft of the late 1950s were Earth-orbiting robotic craft, which performed scientific or military experiments and tested technologies. Deep space probes to the Moon, Venus, and Mars followed by the mid-1960s, along with the first human-flight vehicles. The first landers flew in the mid-1960s to the Moon, first robotic and then the Apollo human flight landers, known as Lunar Excursion Modules. Rovers, for the Apollo astronauts, and robotic rovers, by the Soviet Union, flew in the 1970s. Space stations were first orbited in the 1970s after the end of the Moon race.

Since the 1950s, electrical and computing technologies have grown more sophisticated and lighter, while major structural components have also become lighter. Spacecraft have therefore evolved to perform more sophisticated and varied functions from the late 1950s to the 2000s, often with smaller masses and volumes.

Stephen B. Johnson

See also: Subsystems and Techniques

Bibliography
Mike Gruntman, *Blazing the Trail: The Early History of Spacecraft and Rocketry* (2004).
Mark Williamson, *Spacecraft Technology: The Early Years* (2006).

Cosmos. *See* Kosmos.

Deep Space Probes

Deep space probes have provided information from across the solar system, but before the solar system could be explored from beyond lunar orbit many techniques and technologies had to be developed to ensure that spacecraft successfully reached their targets and survived long enough to complete their missions.

To reach the planets, moons, and other solar system bodies with minimum propellant, mission planners often utilize Hohmann Transfer orbits, discovered by German architect Walter Hohmann in 1925. However, a spacecraft using a Hohmann Transfer alone requires a large launch vehicle to reach the outermost planets. In the early 1960s, Michael Minovitch, a University of California at Los Angeles graduate student working at NASA's Jet Propulsion Laboratory advocated the use of gravity-assist trajectories to allow less powerful rockets to propel spacecraft to the edges of the solar system. The gravity-assist technique uses the angular momentum of planets to provide spacecraft with an increase or decrease in velocity.

Most spacecraft have used chemical rockets that provide the majority of the thrust for a spacecraft at the beginning of its flight. Ion engines, like those demonstrated on the *Deep Space 1* spacecraft, have much higher specific impulse, but provide very small thrust compared to chemical rockets. They are fired throughout much of a spacecraft's flight, but require smaller launch vehicles and fewer gravity-assist maneuvers.

To ensure a stable pointing direction, the first deep space probes used spin stabilization, which rotates a spacecraft around its major axis. This is a simple method of stabilization, but it does not allow for high quality imagery. Spin-stabilized spacecraft included: *Pioneer 10* and *11*, *Lunar Prospector*, and *Galileo*. More advanced three-axis stabilization uses either small thrusters or spinning reaction wheels that can speed up or slow down to maintain orientation, but cost and weigh more than spin stabilization. Three-axis stabilized spacecraft included *Voyager 1* and *2* and *Cassini*.

Photovoltaics (PVs) or radioisotope thermoelectric generators (RTGs) provide the electrical power necessary for operations. PVs, which convert sunlight into electricity, are used for spacecraft operating as far from the Sun as Mars and are typically constructed from crystalline silicon or gallium arsenide. Solar cells are attached directly to spinning spacecraft, while three-axis stabilized craft utilize solar arrays. Spacecraft equipped with photovoltaic technology included *Magellan*, *Deep Space 1*, *Mars Global Surveyor*, and *Lunar Prospector*. RTGs are employed beyond Mars by converting the heat generated by the natural decay of radioactive plutonium-238 to electricity. The risk of each RTG launch in the United States is evaluated and must receive presidential approval. RTGs must also be shielded from instruments to ensure that the radiation does not cause interference. Spacecraft that have used RTGs include *Pioneer 10* and *11*, *Voyager 1* and *2*, *Cassini*, and *New Horizons*. Performance of RTG and PV systems degrade at about 1–2 percent per year.

Spacecraft sent into deep space must also communicate the data they collect to Earth. To this end a series of deep space tracking and communications stations were established, beginning with Deep Space Station 11 (DSS 11), a 26 m antenna built in 1958 for tracking *Pioneer 3*. DSS 11 served as the first element of NASA's Deep Space Network (DSN), which evolved into an international network of three complexes in California, Spain, and Australia to provide continuous contact with monitored spacecraft. DSN provides tracking, telemetry, command, digital signal processing, and deep space navigation. In addition to communicating with Earth, the ability to transmit data has dramatically increased. Early spacecraft data transmission rates were in the ranges of thousands of bits per second, while by the early twenty-first century, rates reached millions of bits per second. NASA sometimes sent two probes on the same mission to ensure success if one failed or to collect additional data if both succeeded. Dual spacecraft missions included the Pioneers, Voyagers, Vikings, and Mars Exploration Rovers.

Exploration of the solar system and beyond progressed dramatically during the first 50 years of spaceflight. Improvements in propulsion, communications, power, and scientific instruments expanded the capability of deep space probes to gather more data and more sophisticated information.

Howard Trace

See also: Jet Propulsion Laboratory, Johns Hopkins University Applied Physics Laboratory, Planetary Science

Bibliography
William E. Burrows, *Exploring Space: Voyages in the Solar System and Beyond* (1991).
Peter Fortescue et al., eds., *Spacecraft Systems Engineering* (2003).

Douglas J. Mudgway, *Uplink-Downlink: A History of the Deep Space Network, 1957–1997* (2001).

Dong Fang Hong. *See* China.

Earth Orbiting Robotic Spacecraft

Earth orbiting robotic spacecraft perform a variety of functional duties, including weather prediction, observation of changes on Earth's surface, global communications, and provision of navigation signals worldwide. Due to the much lower cost and relative simplicity of robotic spacecraft, humans have launched far more robotic spacecraft (by 2009 more than 5,000) with a far greater variety of functions than crewed systems. With the faster performance, larger memories, and smaller size of computers since the late 1950s, robotic spacecraft have become capable of performing more sophisticated missions.

The first spacecraft developed by the Soviet Union and the United States demonstrated the capability of robotic spacecraft to function in orbit and to perform scientific observations. In 1957 *Sputnik* sent a simple signal to demonstrate that it functioned in orbit, and *Sputnik 2* placed the first dog, Laika, in orbit to determine her ability to survive in space. The U.S. *Explorer 1* detected extensive belts of ionized particles in orbit, soon called the Van Allen belts after the scientist who built the instruments. Subsequent launches by the early 1960s expanded the repertoire of functions that could be performed in space. The *Explorer 6* satellite provided the first fuzzy images of Earth from space. The weather satellite *Tiros 1* expanded this capability in 1960 with the first television image of the Earth from orbit. Along with weather and images, early satellites also explored enhancing communication from space, including *Telstar*, the first privately funded communications satellite. Satellite developers also examined potential military uses of space. The United States launched the first Corona (under the cover name Discoverer) satellites to collect reconnaissance imagery from space returned via a film capsule dropped back to the Earth's surface. During this period, the United States also launched the Transit project, which provided navigational capabilities to the Navy. Engineers also designed and tested new technologies to enhance the capability of these satellites. For example, in 1958 *Vanguard 1* explored alternate sources of power to the batteries used on *Explorer 1* and *Sputnik*, which severely limited mission length. It demonstrated the use of solar cells through its comparison of a battery powered transmitter, which lasted for only three months compared to its solar-powered transmitter, which broadcast for six years.

The engineering architecture of these early explorers proved remarkably successful and remained the basis for spacecraft design thereafter. A typical spacecraft consists of two major design elements: a spacecraft bus and a payload. The spacecraft bus consists of the various subsystems used for operating the spacecraft, such as the structure, power subsystem, communications, and pointing control (attitude control), while the instrument suite in the payload performs the satellite's functional mission. Early spacecraft performed simple repetitive functions in which mode changes (different functions) could be commanded from the ground through the communications system or could be preprogrammed using simple electrical or mechanical sequencing mechanisms. As computers on Earth became smaller and more capable, these simple sequencers were replaced by onboard computer systems with increasingly capable software. These improvements in computing greatly enhanced the number and complexity of the functions that the spacecraft and its payload could perform.

Spacecraft had to endure the extreme and unusual space environment. Moving in and out of Earth's shadow, the temperature swiftly changed several hundred degrees. With no air to carry heat away, heating and cooling had to depend on the insulating and cooling properties of the spacecraft's components and radiation of excess heat into space. If the spacecraft used solar power, which became typical, it would also still require batteries for use when the spacecraft was in Earth's shadow. Some spacecraft used nuclear power systems, such as Radioisotope Thermoelectric Generators first launched in 1961 onboard the *Transit 4A* spacecraft, though this was more common for deep space missions than those in Earth orbit. The zero-gravity environment meant that loose particles floated inside and around the spacecraft, which could cause unexpected damage, such as metal fragments shorting out electrical equipment. The close proximity of power and electronic equipment with radio frequency equipment to send and receive data required close attention to how and where electromagnetic signals traveled inside and external to the spacecraft.

While some scientific applications did not require any particular pointing orientation for the spacecraft, many required a fixed orientation that could be attained by spinning the spacecraft (spin stabilization). Other missions required that the spacecraft always have one part of the vehicle pointed toward Earth. This could be achieved by using a gravity boom extended toward Earth, allowing the gravity variations in space to move the craft. However by the mid-1960s many applications required that the spacecraft change its pointing direction, so as to gather images of certain locations on Earth, or for space telescopes pointing to a particular direction in space. These required three-axis stabilization using small thrusters or reaction wheels. Three-axis stabilization became the predominant type of control by the 1970s.

In the early twenty-first century, robotic spacecraft continue to perform the functions that were pioneered in the late 1950s and early 1960s, but with systematic improvements to enable more capabilities. These included more accurate pointing, lighter structures, more capable computers and autonomy, and higher power levels to enable more capable payloads and more powerful or greater number of signals to Earth.

Stephen B. Johnson and Chris Krupiarz

See also: Subsystems and Techniques

Bibliography

Helen Gavaghan, *Something New under the Sun: Satellites and the Beginning of the Space Age* (1998).

Mark Williamson, *Spacecraft Technology: The Early Years* (2006).

Kosmos

Kosmos, a publicly announced cover name for various, mostly military, Soviet Union spacecraft to conceal their real purposes and designations while complying with international agreements on mandatory declaration of every space launch. The name originated from a series of small research satellites first launched in March 1962. Beginning with *Kosmos 4* on 26 April 1962, it was assigned to all military and some civil scientific satellites, such as *Kosmos 110* in 1966. Several lunar and planetary probes that failed to leave Earth orbit were belatedly given Kosmos designations—for example, *Kosmos 21, 27,* and *60.* Military applications of Kosmos satellites included navigation, communications, photoreconnaissance, early warning, ocean surveillance, electronic intelligence, and radar calibration. Some military Kosmos satellites were observed to release multiple sub-satellites during their missions. More than 2,400 Kosmos satellites have been launched since 1962.

Peter A. Gorin

See also: Russia (Formerly the Soviet Union)

Bibliography
Nicholas L. Johnson and David M. Rodvold, *Europe and Asia in Space, 1993–1994* (1995).

Landing Craft

Landing craft are vehicles designed to perform soft landings on the surface of another planetary body, a feat successfully accomplished on the Moon, Venus, Mars, and Saturn's moon, Titan. Controlling descent until safe touchdown is the purpose of every landing craft, and one of the biggest determining technical factors is the landing destination.

The Moon was the first obvious destination. Both the United States and the Soviet Union pushed to put a human on the Moon, but before that could be accomplished, landing technologies needed to be tested by robotic lunar landing craft. After several failures beginning in 1963, on 3 February 1966, the Soviet Union *Luna 9* became the first landing craft to land successfully, using four retrorocket engines. It broadcast the first pictures of the Moon's Oceanus Procellarum. A few months later, the U.S. *Surveyor 1* successfully landed. The Surveyor Program attempted six more missions through 1968, three successfully. Surveyor's design accounted for the Moon's low gravity and lack of atmosphere, which precluded parachutes. As *Surveyor 1* reached an altitude of 75.3 km above the lunar surface, a retrorocket fired until reaching 11 km altitude, where it was discarded, having slowed the craft from 2,612 m/s to 110 m/s. Altimeter and Doppler radars controlled the remaining descent by firing vernier engines, resulting in landing speed of 3 m/s. Similar systems were used for subsequent Surveyors. *Surveyor 6* fired its vernier engines after landing and landed at a different location, demonstrating the first ascent from the Moon. The stage was set for Apollo's Lunar Modules (LM).

The heart of the Apollo program was the LM (comprised of descent/ascent modules) that first carried Neil Armstrong and Buzz Aldrin to their *Apollo 11* Moon landing on 20 July 1969. The Apollo series LM used a single deep-throttling ablative rocket, controlled by autopilot or manual controls, as its descent engine. *Apollo 12* and *14* through *17* recorded five more successful crewed landings, the last landing at the Sea of Serenity on 11 December 1972. Apollo LM pilots trained for these landings on Earth using the Lunar Lander Training Vehicle, which mimicked the dynamics

Landing Missions (Part 1 of 2)

Name of Landing Craft	Country of Origin	Destination	Landing Date(s)	Landing Mechanism	Comments
Sputnik 25, Luna 4–9, 13, 16, 17, 18, 20, 21, 23, 24	Soviet Union	Moon	Feb 1966–Aug 1976	Retrorocket engines	*Sputnik 25, Luna 4–8, 18, 23* failed
Surveyor 1–7	United States (US)	Moon	2 Jun 1966–10 Jan 1968	Retrorocket and vernier engines	*Surveyor 2, 4* failed
Apollo 11, 12, 14–17	US	Moon	20 Jul 1969–11 Dec 1972	Single deep-throttling, ablative rocket engine	*Apollo 13* landing never attempted
Venera 7–14	Soviet Union	Venus	15 Dec 1970–5 Mar 1982	Aero-braking, multiple para-chutes, drag-brake	*Kosmos 359, 482* failed
Mars 2, 3, 6, 7	Soviet Union	Mars	2 Dec 1971–Jul 1973	Aero-braking, parachutes, retrorockets	All failed
Viking 1, 2	US	Mars	20 Jul 1976, 3 Sep 1976	Aero-braking, parachutes, retrorockets	
Pioneer Venus Multiprobe	US	Venus	19 Dec 1978	Aerobraking	Not designed for soft landing, one probe survived landing
Vega 1, 2	Soviet Union	Venus	11 Jun 1985, 15 Jun 1985	Aero-braking, multiple para-chutes, drag-brake	
Phobos 2	Soviet Union	Mars	Mar 1989	Thrusters	Hopper lander, *Phobos 2* failed before landing attempt
Mars 96	Russia	Mars	Launch attempt 16 Nov 1996	Aeroshell, parachutes, cushion airbag	Failed at launch

(Continued)

Landing Missions (Part 2 of 2)

Name of Landing Craft	Country of Origin	Destination	Landing Date(s)	Landing Mechanism	Comments
Mars Pathfinder	US	Mars	4 Jul 1997	Aeroshell, parachutes, cushion airbag	
Mars Polar Lander	US	Mars	3 Dec 1999	Aeroshell, parachutes, descent engine	Failed
Deep Space 2	US	Mars	3 Dec 1999	Aeroshell, hard impactors	Piggyback with *Mars Polar Lander*, failed
Near Earth Asteroid Rendezvous	US	433 Eros	12 Feb 2001	Thrusters	Landed with spacecraft
Beagle 2	United Kingdom	Mars	25 Dec 2003	Aeroshell, parachutes, cushion airbag	Failed
Mars Exploration Rovers (*Spirit* and *Opportunity*)	US	Mars	3 Jan 2004, 24 Jan 2004	Aeroshell, parachutes, cushion airbag	
Huygens	US/Europe	Titan	14 Jan 2005	Aeroshell, parachutes	
Hayabusa/ MINERVA	Japan	25143 Itokawa	Nov 2005	Thrusters	*MINERVA* hopper failed, landing with *Hayabusa* spacecraft
Mars Phoenix	US	Mars	25 May 2008	Aeroshell, parachutes, descent engine	

and control system of the LM. The Soviet Union had seven successful robotic touch-downs with *Lunas 13*, *16*, *17*, *20*, *21*, *23*, and *24*, using one main descent engine to slow the craft until the control systems activated a bank of lower-thrust jets for final approach. The Soviet Union also had several failures.

The next series of landers were aimed at Venus. Its thick atmospheric conditions made possible a more mass-efficient landing system from lunar landers—without rockets to slow descent. The Soviet Venus program achieved a major triumph on 15 December 1970 when *Venera 7* aerobraked and parachuted through the atmosphere of Venus and transmitted the first signal from another planet. This and previous orbital and atmospheric Venera missions provided enough data for the next generation of better protected landers, *Venera 8* through *14*, protected by spherical shell construction. They entered Venus's atmosphere at 10.7 km/s. After initial aerobraking, the first parachute deployed and activated a second parachute, which separated with the top portion of the spherical shell at 64 km altitude and 250 m/s velocity. A third parachute brought the craft's velocity to 50 m/s, and the final parachute system, consisting of three parachutes, allowed the discharge of the bottom portion of the spherical shell protecting the lander. With 50 km remaining, the parachutes were detached, followed by freefall (controlled by a disk-shaped drag brake) culminating in impact speed of 7 m/s, absorbed by a landing cushion. Landers of similar design were used on the Soviet *Vega 1* and *2* missions. NASA briefly received data from the Venusian surface from one of the *Pioneer-Venus 2* probes in December 1978, though these were not designed as landers.

Engineers test the multi-lobe air bag system used to protect the Mars Pathfinder spacecraft when it landed on Mars on 4 July 1997. (Courtesy NASA/Jet Propulsion Laboratory)

The first Mars successful landing was accomplished by the Soviet Union's *Mars 3* lander, which touched down in 1971, but it survived only 20 seconds on the surface. The Soviet Union tried again in 1973, and Russia in 1996, but all of these attempts failed. U.S. *Viking 1* and *2* landers, which reached the surface of the Red Planet in 1976, were far more successful. Landing conditions on Mars determined the technological design factors. The Viking missions used a combination of aerobraking, parachutes, and retrorockets to safely land both craft on Mars with touchdown speed of 2.4 m/s. After Viking, it would be more than two decades before the *Mars Pathfinder* landed on Mars, using a major technological change in the landing mechanism of the craft. Instead of bulky and expensive retrorockets, it substituted an inflatable cushioned airbag system that landed the craft with final descent speed of 18 m/s and safely bounced the craft to rest (solid rockets were also used in the final stages of descent). A similar, but more advanced, design was used with the *Spirit* and *Opportunity* rovers that landed on Mars in 2004. The United Kingdom *Beagle 2* lander also used a parachute and airbag design, but it never contacted Earth after its landing on Mars on Christmas 2003. The European Space Agency *Huygens* probe of the Cassini-Huygens mission successfully landed on Saturn's moon, Titan, on 14 January 2005. It relied on an aeroshell and parachutes. NASA's *Mars Polar Lander* and *Mars Phoenix* both relied on descent engines instead of airbags. The Deep Space 2 probes attempted using an aeroshell and hard impact landing, but these failed along with the *Mars Polar Lander* in December 1999.

Two asteroid landings have also occurred, the U.S. *Near Earth Asteroid Rendezvous* (*NEAR*) spacecraft on 433 Eros in February 2001 and the Japanese *Hayabusa* on 25143 Itokawa in November 2005. *NEAR* was not designed as a lander, while *Hayabusa* carried a lander called *MINERVA* (*Micro/Nano Experimental Robot Vehicle for Asteroid*) designed to hop on Itokawa's surface. Unfortunately *MINERVA* was released at the wrong time and escaped into space, never performing its mission. Instead the *Hayabusa* spacecraft attempted to acquire samples. The Soviet *Phobos 2* spacecraft also carried a "hopper" lander to land on the Martian moon Phobos, but the spacecraft failed in March 1989 before the hopper had been released.

By the early 2000s, landing on other solar system bodies remained risky, but the technologies had stabilized. For landings on bodies with atmospheres, combinations of aeroshells and parachutes with either descent engines or airbags were standard. Descent engines and thrusters remained the only options for landings on bodies without atmospheres.

George Sarkisov and Stephen B. Johnson

See also: Apollo, Lunar Science, Mars, Saturn, Venus

Bibliography
Michael Hanlon, *The Real Mars* (2004).
Mikhail Ya. Marov, *The Planet Venus* (1998).
National Space Science Data Center. http://nssdc.gsfc.nasa.gov/.

LDEF, Long Duration Exposure Facility. *See* Thermal Control.

Rovers

Rovers have been deployed to planetary bodies beyond Earth. The first was the Soviet Union's *Lunokhod 1* rover, which reached the Moon in 1970. The next two years, NASA's Lunar Roving Vehicle (LRV) accompanied astronauts on the *Apollo 15*, *16*, and *17* lunar missions. Two additional NASA rover designs came more than two decades later: *Sojourner* and the twin rovers, *Spirit* and *Opportunity*, which were sent to Mars.

The Soviet Union launched two mini-rovers called PrOP-M (Pribori Otchenki Prokhodimostikhodik—Instrument for Cross Country Characteristics Evaluation on Mars) with the *Mars 2* and *3* missions of 1971, but both missions failed before they were used. Soviet lunar missions were far more successful, with *Lunokhod 1* deployed as part of the *Luna 17* lunar mission and *Lunokhod 2* during the *Luna 21* mission in 1973. The Soviet lunar rovers were telerobotic, driven by ground controllers in real time, and like other Soviet technologies of the time, quite heavy compared to similar U.S. systems. For example, *Lunokhod 1* weighed 756 kg. They were tub-shaped with eight wheels, each with its own suspension, motor, and brake. The rovers carried three television cameras, one of which was used for navigation. Solar panels charged the rovers' batteries and provided power for the polonium-210 heat source that warmed the rover during the lunar night.

During the final three Apollo missions in 1971 and 1972, the LRV carried astronauts from the landing sites to relatively distant locations across the lunar surface. Designed to operate for 78 hours, the LRV was an open-cockpit vehicle about the size of a small car. During transport to the Moon, the LRV had to fit into the Lunar Module's Quad 1 (a pie-shaped compartment) in a folded configuration. When the astronauts opened the door to Quad 1, the LRV unfolded into its deployed configuration and settled onto the lunar surface. Two 36 V batteries powered the LRV, although if one failed, the other could still power the vehicle. The LRV had front and rear independent steering with a 0.25 hp motor powering each wheel. These systems allowed for some redundancy, which was demonstrated during *Apollo 15*, when the front steering system failed. The LRV had a mass of about 209 kg and could carry about 490 kg, including the astronauts, their life-support systems, and communications and scientific equipment, in addition to site samples.

The next extraterrestrial rover mission did not occur until nearly 25 years later, when NASA landed *Mars Pathfinder*, with the attached *Sojourner* rover, on Mars. The deployment ramp unfolded, the six-wheeled *Sojourner* rover rose to its extended height and rolled down to the surface in July 1997. Roughly box-shaped, *Sojourner* was about the size of a microwave oven. It was steered with its outer four wheels and had a rocker-bogie suspension system. This suspension system, developed by NASA's Jet Propulsion Laboratory (JPL), did not have axles or springs, and it allowed the rover to traverse varied terrain, including climbing over rocks twice the size of its wheels. Powered by solar cells that provided energy (with battery backup) for an onboard processor, ground controllers drove the rover, guided by an onboard camera. During *Sojourner*'s several-meter drive across the Martian surface, all communications between it and Earth went through the *Pathfinder* lander.

In January 2004 JPL's twin Mars Exploration Rovers, *Spirit* and *Opportunity*, landed on Mars. Though they resembled large versions of the *Sojourner* rover, they incorporated many advances, including an improved rocker-bogie suspension system, and they were built to travel farther and operate longer. Unlike *Sojourner*, which was an extension of the Pathfinder mission, *Spirit* and *Opportunity* contained all vehicle and scientific instrumentation and served as mobile laboratories. The rovers communicated directly with Earth through high- and low-gain antennas and used an ultrahigh frequency antenna to communicate with Mars-orbiting satellites for image and data uplinks. Solar panels charged the batteries that powered the vehicles. Built to travel up to 100 m per day with a nominal operating mission of 90 days, both rovers were still operating four years after landing. Designed with the ability to climb inclines of 30 degrees, they traversed a variety of terrains on opposite sides of the planet. *Spirit* rolled across the rocky floor of Gusev Crater and climbed to the top of the Columbia Hills, while *Opportunity* drove across the plain of Meridiani, exploring the floors and inner slopes of three rock outcrop–strewn craters. Hazard avoidance cameras on the front and rear enabled the rovers to navigate around obstacles and travel to targets predetermined by ground controllers. A warm electronic box using gold paint, aerogel insulation, heaters, thermostats, and radiators protected computer equipment and batteries during the cold nights. The rovers carried a suite of spectro-meters, a panoramic camera, and a rock abrasion tool for examining geologic features.

Additional lunar rover designs were developed as part the Lunar Rover Initiative at the Robotics Institute at Carnegie Mellon University. One design was demonstrated in Antarctica and in simulated lunar terrain. Before 2003 the private company LunaCorp collaborated with Carnegie Mellon University, exploring the idea of deploying rovers on the Moon that would be controlled by visitors at science centers.

From the 1960s to the early twenty-first century, rovers have decreased in size and increased in capability. This has largely been due to the miniaturization of computer and communications systems, along with their vastly improved processing and data transmission capabilities. These have allowed for increasingly sophisticated software algorithms to control the rovers remotely from Earth. Sensors have also become smaller and more capable, reinforcing the trend.

As of 2008 several programs in development and planning included rovers. JPL's *Mars Science Laboratory* was scheduled to be launched in September 2009 with a planned Martian touchdown in October 2010, and NASA was developing rover proto-types to carry humans across the lunar surface for the planned missions to the Moon beginning in 2020.

Eric T. Reynolds and Stephen B. Johnson

See also: Apollo, Mars Exploration Rovers, *Mars Pathfinder*, Soviet Manned Lunar Program

Bibliography
Arthur Smith, *Planetary Exploration: Thirty Years of Unmanned Space Probes* (1988).
"Lunar Rover Initiative," Carnegie Mellon University, Robotics Institute. http://www.frc.ri.cmu.edu/project/lri/.
Mark Williamson, *Spacecraft Technology: The Early Years* (2006).

Anthony Young, *Lunar and Planetary Rovers—The Wheels of Apollo and the Quest for Mars* (2007).

SCATHA, Spacecraft Charging at High Altitude. *See* Spacecraft Charging.

Space Stations

Space stations were postulated from the early twentieth century as central to the human exploration of space by serving as permanent habitations and jumping-off points to the Moon and planets. In 1903 Soviet schoolteacher Konstantin Tsiolkovsky studied this possibility and argued for the creation of a wheeled space station that rotated to approximate gravity with centrifugal force. During the 1920s Romanian-German spaceflight theorist Hermann Oberth and Austrian engineer Hermann Noordung elaborated on the concept of the orbital space station as a base for voyages into space. In the 1950s, Wernher von Braun emphasized the role of an orbital space station as a laboratory, observatory, industrial plant, launching platform, drydock, and military facility.

Since the competition between the Soviet Union and the United States initially took shape as a race to the Moon, space stations did not figure prominently into the civilian space program of either superpower in the 1960s. Instead military leaders pursued the potential value of human-occupied, Earth-orbiting stations for reconnaissance. The U.S. Air Force began development of the Manned Orbiting Laboratory (MOL) in 1963, but the program was canceled in 1969 due to military priorities emphasizing near-term support to the Vietnam War, the success of the robotic Corona program and its successors then in development, and the prospect of NASA's civilian space station. Soviet leaders responded to MOL by authorizing Vladimir Chelomey's special design bureau OKB-52 to develop the Almaz military space station. Almaz reached orbit, though its military nature was kept secret by identifying both Almaz and the civilian Long Duration Orbital Station (DOS) as Salyut stations. In the 1970s military experiments on Almaz showed that crewed military stations were not effective in terms of cost or performance compared to robotic reconnaissance satellites, thus ending military interest in crewed space stations. Both MOL and Almaz differed from civilian designs by having large reconnaissance cameras.

As Apollo's success became more certain, NASA started plans to use leftover Apollo technology to build an experimental space station, ultimately called *Skylab*. *Skylab* utilized the structure of the Saturn V third stage, which was retrofitted on the ground for crew quarters, galley, showers, and experiments. Orbited in 1973 with almost all the supplies needed for its planned missions, NASA sent nine astronauts in three crews to the workshop, occupying it for a total of 171 days, 13 hours. Both the total hours in space and the total hours spent in extravehicular activity under microgravity conditions exceeded the combined totals of all the world's spaceflights up to that time. On 11 July 1979 *Skylab* reentered Earth's atmosphere, its debris scattered from the southeastern Indian Ocean across western Australia.

Upon losing the race to land a man on the Moon, Soviet leadership decided to emphasize space stations. They publicly claimed that they had never planned to send men to the Moon, but were instead planning much more useful crewed space stations.

To put a civilian stamp on the new race to develop a space station, the Soviet leadership forced Chelomey to give some of his Almaz structures to Vasily Mishin's design bureau (the successor to Sergei Korolev's OKB-1), which would install Soyuz-based electronics into them and launch them as DOS civilian stations. *Salyut 1* reached orbit in April 1971, though its first crew failed to enter the station. The second crew, sent on *Soyuz 11*, successfully occupied the station for 24 days, but died on reentry. The next three stations failed (two of which were not publicly acknowledged as Salyuts), but the next three, *Salyuts 3*, *4*, and *5*, succeeded, which included both DOS and Almaz stations. *Salyut 6*, launched in 1977, featured two docking ports instead of the single docking port in the earlier designs. This allowed resupply using automated Progress vehicles and hence a much longer service life and crew occupation visits. In 1986 the Soviet Union launched *Mir*, which used a modular design further extending the multiple-port design. *Mir* consisted of several components, each launched separately and attached in space to multiple-port modules. Except for a few months during the collapse of the Soviet empire and its replacement by a coalition of market economy states in 1989, from 1986 to 2000 *Mir* was occupied. While it was plagued by recurrent problems starting in early 1997, its scientific experiments added to understanding the effects of space on humans. Russia finally abandoned *Mir* in 2000 and deorbited it in 2001.

With the Space Shuttle in 1981, NASA returned to its quest for a space station as a site of orbital research and a jumping-off point to planets. NASA Administrator James Beggs persuaded President Ronald Reagan, against the advice of many of his other

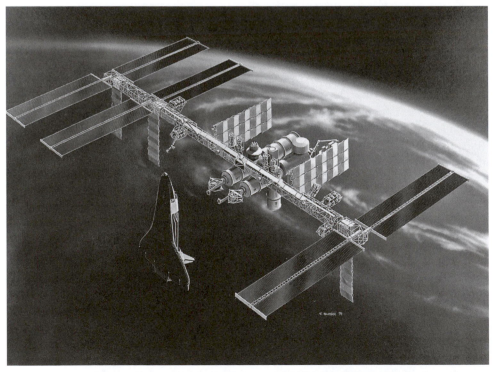

Artist's concept of Space Station Freedom. (Courtesy NASA/Marshall Space Flight Center)

advisors, to endorse building a permanently occupied space station. Despite bringing in more than a dozen international partners on the project, from the outset the Space Station Freedom program was controversial. Most of the debate centered on its costs versus its benefits. NASA designed the project to fit an $8 billion research-and-development funding profile. For both political and technical reasons, within five years the projected costs had more than tripled. Redesigns followed in 1990–93, with each iteration yielding a smaller and less capable system. As capabilities diminished, even political leaders who had once supported the program questioned its viability. An entirely new goal—helping a newly democratic Russia—saved the program, but with a radically changed design. On 7 November 1993, the United States invited Russia to join in the building of the *International Space Station* (*ISS*).

Like *Mir*, *ISS* used a modular design. The first two station components, the *Zarya* and *Unity* modules, were launched and joined together in orbit in late 1998. Several other components were at that time nearing completion in the United States and elsewhere. Orbital assembly of *ISS* began a new era of hands-on work in space, involving more spacewalks and new space robotics. The Space Shuttle and two types of Russian launch vehicles were intended to launch components of the station for orbital assembly. The *ISS*, like all space stations before it, must be supported with supplies from Earth, and refuse loaded onto spacecraft to return to Earth or burn up in the atmosphere. By the early twenty-first century the *ISS* had some capabilities to recycle air and water. The first crew of three occupied the station on 31 October 2000. To fund their ailing space program, the Russians flew the first paying tourist, U.S. citizen Dennis Tito, to the *ISS* in 2001.

Through the *Columbia* accident on 1 February 2003, 16 flights, which included 12 Space Shuttle missions to build the *ISS*, had occurred. Throughout, *ISS* continued to be hampered by problems of long duration, especially cost and schedule issues. When the Space Station Freedom program became *ISS*, NASA believed it could build the station for $17.4 billion. However, in late 2001 NASA conceded that it would require at least $23 billion. Accordingly the *ISS* devolved into a debate about questions of mismanagement and cost overruns. These issues were important factors in the announcement of the new Constellation program in 2004, which would shift U.S. human spaceflight efforts away from the *ISS* and return to the Moon.

To date, space stations have been used as vehicles for national prestige, technology development, Earth monitoring, and microgravity science. Contrary to the expectations of the early spaceflight visionaries, they have not become way-points to destinations beyond Earth orbit using rotating wheels to create artificial gravity. Rather they have remained as zero-gravitation experimental stations. By the twenty-first century, space stations as tourist destinations became a reality with Russian tourist flights to *ISS* aboard Soyuz. In addition, private ventures are beginning to experiment with potential orbital hotels. Bigelow Aerospace orbited experimental Genesis modules in 2006 and 2007 to test inflatable space hotel habitation concepts.

Stephen B. Johnson

See also: Almaz, *International Space Station*, Mir, Salyut, Skylab

Bibliography
Giovanni Caprara, *Living in Space: From Science Fiction to the International Space Station* (2000).
David M. Harland and John E. Catchpole, *Creating the International Space Station* (2002).
Roger D. Launius, *Space Stations: Base Camps to the Stars* (2003).
John M. Robert Zimmerman, *Leaving Earth: Space Stations, Rival Superpowers, and the Quest for Interplanetary Travel* (2003).

Sputnik

Sputnik, the first artificial satellite, opened the space era and sparked competition between the Soviet Union and the United States. In 1948 a group of Soviet military scientists under a prominent space pioneer, Mikhail Tikhonravov, began early studies of artificial satellites. In August 1954 the government approved its leading missile design organization, OKB-1 (Experimental Design Bureau), led by Sergei Korolev, to begin preliminary research on satellites and launch vehicles based on the R-7 ballistic missile. On 30 January 1956 another government decree authorized the development of a scientific satellite, *Object D*, to be launched in response to the U.S. Vanguard satellite program during the International Geophysical Year (July 1957–December 1958). Tikhonravov initially participated in the satellite project as a consultant and later as a head of the newly established spacecraft development department.

Object D was an ambitious and complex design: it contained 12 different sensor systems to measure solar radiation, ionosphere particles, and detect micrometeorites. The project quickly ran into unexpected impediments, and its mission was delayed until April 1958. Thus, in February 1957, OKB-1 obtained government authorization to launch a simpler satellite first.

That new project, originally designated Primitive Satellite (PS), became the first Sputnik and made history as the world's first artificial satellite after its successful launch on 4 October 1957. It consisted of a sealed aluminum sphere, 580 mm in diameter and 83.6 kg in mass, containing a radio transmitter, temperature control system, and batteries. Two pairs of antennae, 2.4 m and 2.9 m long, were attached to the surface of the satellite. *Sputnik* was launched to a 947 km × 228 km orbit by a two-stage R-7 missile prototype and remained active for three weeks. It reentered on 4 January 1958.

Realizing the important propaganda effect of the first Sputnik, the Soviet authorities urged OKB-1 to repeat the feat quickly. Hastily constructed of available components, yet intended to be more spectacular, *Sputnik 2* (*PS 2*) consisted of three containers with a compound mass of 508.3 kg, positioned in a conical frame permanently attached to the second stage of the booster rocket. The prime project task was to sustain a dog, Laika, for seven days in space and transmit her vital parameters to Earth. *PS 2* was equipped with radiation sensors. Although successfully orbited on 3 November 1957, *Sputnik 2* did not achieve its prime objective: Laika died soon after launch due to overheating caused by failed aerodynamic shroud separation. *Sputnik 2* detected radiation belts around Earth, but Soviet scientists did not immediately publish their finding, losing the discovery to the first U.S. satellite, *Explorer 1*, and James Van Allen.

MIKHAIL KLAVDIEVICH TIKHONRAVOV (1900–1974)

A Soviet military aerospace engineer and retired Air Force colonel, Mikhail Tikhonravov was one of the founders of the Soviet space program. Working at the early Soviet missile research organizations, Jet Propulsion Research Group (GIRD), and Jet Propulsion Research Institute (RNII) from 1932 to 1946, he designed the first Soviet liquid-propellant rocket (1933) and was actively involved in other projects, including experimental rockets, military missiles, and rocketplanes. Tikhonravov proposed the first Soviet piloted space project, VR-190 (1945). He performed early research on artificial satellites and introduced an unusual "rocket packet" design, realized in the R-7 ballistic missile. In the 1950s Tikhonravov publicly promoted the ideas of space exploration. From 1957 to 1966, he led the design and development of Sputnik, Vostok, and other early Soviet spacecraft at OKB-1 (Experimental Design Bureau).

Peter A. Gorin

The launch of *Object D*, on 27 April 1958, ended in a booster failure. The reserved satellite sample, now designated *Sputnik 3*, went into space on 15 May 1958. The conical satellite was 3.57 m long and 1.73 m in diameter with an orbital mass of 1,327 kg including 968 kg of scientific equipment. *Sputnik 3* was the first satellite powered by solar batteries. Unlike the previous Sputniks, that one was launched by a dedicated version of the R-7 missile.

Sputnik 3 confirmed the existence of Earth's radiation belts and determined their borders, but much of the satellite's scientific data was lost due to a failed recording device.

Peter A. Gorin

See also: Effect of Sputnik; Korolev, Sergei; Russian Space Launch Centers

Bibliography

RKK Energia, *S. P. Korolev Space Corporation Energia* (1994).
Asif A. Siddiqi, *Challenge to Apollo: The Soviet Union and the Space Race, 1945–1974* (2000).

Tethered Satellite System. *See* Italy.

SUBSYSTEMS AND TECHNIQUES

Subsystems and techniques are the building blocks and methods on which spacecraft and launch vehicles are made.

Spacecraft and launch vehicles consist of very similar subsystems, because both types of systems have similar functions that require similar, though not identical, technologies. The vehicles themselves are composed of structural components that must be strong enough to withstand the rigors of high accelerations and large random vibrations due to the dynamics of rocket engines that propel the launchers and their payloads.

The structures must also be very lightweight, as it is very expensive to place objects into space. Guidance, navigation, and control subsystems control the pointing direction and the trajectories and velocities of the space vehicles. Communications subsystems relay information to and from Earth through electromagnetic waves. Command and data-handling systems enable the complex computations required to control the vehicle engineering subsystems and the payload. These are enabled by electrical power subsystems that use batteries, solar panels, fuel cells, or nuclear power sources to create and distribute electrical energy to operate the vehicle. Propulsion subsystems contain the rocket engines that provide the thrust to place the spacecraft into orbit and change its velocity. Thermal control subsystems manage the vehicle's heat, to ensure its engineering subsystems and payloads neither freeze nor melt in the harsh vacuum of space. Space vehicles are controlled with sophisticated computing and communications systems on the ground that send commands and receive and analyze telemetry.

Integrating and testing space systems required the development of sophisticated methods, known as systems engineering, to ensure the many different subsystems and functions operate properly together. Managers developed corresponding schedule and cost control methods to provide some measure of cost and schedule predictability and stability to space system design and operations. New methods of determining and controlling trajectories required the development of orbital mechanics to new levels of sophistication. Determination of a spacecraft's location and velocity millions and even billions of miles from Earth required new technologies and precision measurements. Altogether spacecraft and launch vehicles required the development of older technologies to new levels of precision and miniaturization, new technologies such as rocket propulsion, and new methods of management and systems engineering to make spaceflight possible.

Stephen B. Johnson

See also: Technology Infrastructure and Institutions

Bibliography

Mike Gruntman, *Blazing the Trail: The Early History of Spacecraft and Rocketry* (2004).
Stephen B. Johnson, *The Secret of Apollo: Systems Management in American and European Space Programs* (2002).
Mark Williamson, *Spacecraft Technology: The Early Years* (2006).

Attitude Control

Attitude control is the function of a vehicle that establishes and maintains a position or angular orientation, known as the vehicle's attitude. Angular orientation is often referred to in terms of yaw, pitch, and roll, terms (inherited from aircraft flight dynamics terminology) that refer to the vehicle's rotation about each axis of a Cartesian coordinate system centered on the craft. A launcher's attitude control system maintains the vehicle's orientation, while its rocket engine operates during ascent. Typically a spacecraft attitude control system must be able to maintain a fixed attitude,

ATTITUDE CONTROL

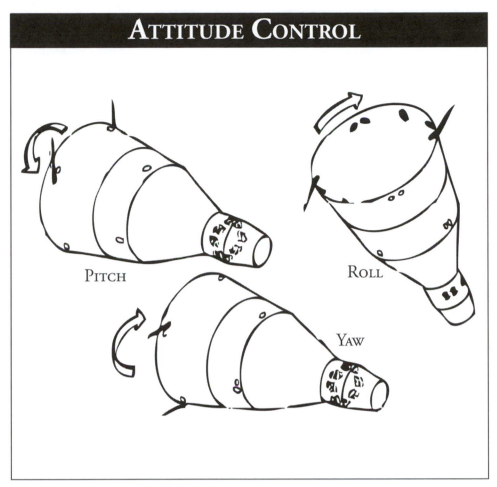

PITCH

ROLL

YAW

Diagram shows Gemini Spacecraft Responses to Orbital Attitude System's thrusters. Firing of appropriate combination of the thrusters cause pitch, roll, and yaw. (Courtesy NASA/ABC-CLIO)

while its engines fire during trajectory correction maneuvers, and must be able to maneuver the spacecraft's attitude to focus on a target to collect data and charge solar cells. To achieve these goals, an attitude control system typically consists of sensors, software, and actuators (devices that move the vehicle).

Early attitude control of rockets largely employed gyroscopes. Through conservation of angular momentum, a gyroscope will resist changes in its orientation. The gyroscope detects the rocket's orientation, which is then compared through electronic or mechanical means with the desired attitude, and the actuators provide the means to adjust the orientation. Russian schoolteacher Konstantin E. Tsiolkovsky was the first to develop the theory of gyroscopic attitude control in his 1903 article "Investigating Space with Reaction Devices." U.S. physicist Robert Goddard achieved the first use of gyroscopic attitude control for rockets in a successful rocket test near Roswell, New Mexico, in

KONSTANTIN EDUARDOVICH TSIOLKOVSKY (1857–1935)

A physics and mathematics teacher, Russian space research pioneer Konstantin Tsiolkovsky dedicated his life to studying various aspects of future space travel, which he viewed as an inevitable stage of human progress. Tsiolkovsky showed mathematically that rocket propulsion could allow escape from Earth's gravity and paved the way to liquid rocket propulsion by proposing liquefied oxygen and hydrogen as a propellant combination. He wrote several science fiction and popular books on space travel, and his main monograph, *Exploration of Space by Reactive Devices*, was published in 1903. Tsiolkovsky foresaw many features of modern astronautics, including artificial satellites, space stations, piloted spacecraft, and multistage rocket boosters. His ideas inspired numerous enthusiasts, in the Soviet Union and abroad, who later started practical explorations of space.

Peter A. Gorin

April 1932. The German rocket team under Wernher von Braun also developed gyroscopic stabilization for its unsuccessful A-3 rockets, tested in December 1937, and then improved versions for its A-4 rockets, first flown in 1942. These early systems fed the control signal to swiveling fins placed in the rocket exhaust to turn the rocket, which reduced rocket performance. Convair Corporation's MX-774 ballistic missile, first tested in 1947, was the first to use swiveling rocket nozzles instead of control fins, which became the standard for rockets during the next decade.

The simplest methods of spacecraft attitude control are passive methods, such as gravity booms and passive magnetic stabilization. These techniques are primarily used to maintain a spacecraft's orientation and have the advantage of requiring no moving parts and little or no fuel. The gravity boom consists of a rod or lever, typically a few meters long, with a mass at one end and attached to the satellite at the other. Gravitational forces ensure the satellite points toward Earth, ideal for a communications dish or similar device. Passive magnetic stabilization takes advantage of Earth's gravitational field. A simple design used on the Australian Orbiting Satellite Carrying Amateur Radio 5 satellite, launched in January 1970, consisted of a pair of bar magnets that interacted with Earth's magnetic field to maintain a known orientation. Magnetotorquers used electromagnets to provide adjustable forces. Probably the first successful on-orbit use of magnetotorquing devices was on the second Defense Meteorological Satellite Program satellite, launched in August 1962.

Another form of semi-passive stabilization is spin stabilization. The U.S. *Explorer 1* satellite was the first to attempt spin stabilization, as it was spun up during launch to maintain a fixed orientation when separated from the launcher. *Explorer 1*'s whip-like antennas were flexible, however, which dissipated the momentum of the craft, and it was unable to maintain its orientation. Engineers soon solved this problem by using more rigid structures. In the first years of the space age, most robotic spacecraft that required stabilization were spin stabilized.

However, some missions required the spacecraft to hold a fixed orientation. The *Orbiting Solar Observatory 1* spacecraft (launched March 1962), which needed to point its instruments constantly at the Sun, solved this problem by creating a dual-spin system, where one part of the spacecraft spun to keep the spacecraft pointed in one direction, and the instruments were on a despun section that exactly canceled the rotation rate of the spun section, allowing it to remain fixed in space. Crewed spacecraft, such as Vostok and Mercury, and scientific missions, such as Ranger to the Moon and Mariner to Venus, required full three-axis stabilization, where the entire vehicle was held at a fixed orientation. These required fully active attitude control systems.

Active spacecraft attitude control systems use sensors that fall into two broad categories: those (such as Sun sensors, star trackers, and horizon sensors) that detect the position of the satellite or rocket with respect to some known object and those (such as gyroscopes and accelerometers) that detect information about the motion of the satellite. To reduce the energy needed to keep gyroscopes spinning and eliminate moving parts, gyroscopic devices by the late twentieth century frequently used optical techniques, such as ring laser gyroscopes and the fiber optic gyroscopes.

Using input from the sensors, the algorithmic part of an attitude control system determines needed corrections and sends signals to the craft's actuators to achieve the desired orientation. The most common actuators are thrusters and reaction wheels, both developed in the late 1950s–60s. A thruster consists of a small rocket engine or simply a cold gas nozzle. When the engine or nozzle fires, conservation of momentum forces the craft in the opposite direction. Thrusters require propellant, which must be rigorously conserved because propellant depletion commonly marks the end of a mission. The reaction or momentum wheel provides fine attitude control by speeding up or slowing down a flywheel. As the wheel changes its motion, the craft will respond by moving in the opposition direction. Additionally the wheel can provide spin stabilization for the entire craft. When a wheel reaches its maximum or minimum spinning velocity, it is said to be saturated and must be unsaturated using other actuators, usually thrusters or magnetotorquers.

Since the 1960s steering and control algorithms have been improved with increased computing power and better computational techniques. Sensors have become smaller, more accurate, and durable. Actuators benefited from the same advances as sensors, but interesting new techniques, such as conducting tethers, piezoelectric actuators, and smart material actuators, were applied. There has been a trend over time for use of spin stabilization to decrease and for three-axis stabilization to become more common, as propellant conservation has improved with more efficient thrusters and algorithms.

R. Dwayne Ramey

See also: Rocket Pioneers and Spaceflight Visionaries

Bibliography
Mike Gruntman, *Blazing the Trail: The Early History of Spacecraft and Rocketry* (2004).
Marshall H. Kaplan, *Modern Spacecraft Dynamics and Control* (1976).
Friedrich P. J. Rimrott, *Introductory Attitude Dynamics* (1970).
Mark Williamson, *Spacecraft Technology: The Early Years* (2006).

Command and Control

Command and control, with respect to satellites, refers to the use of personnel, equipment, communications, facilities, and procedures in planning, directing, coordinating, and controlling forces and operations in the accomplishment of a satellite system's mission. Radio signals used to transfer data (also called telemetry and consisting of anything on a satellite that can be measured) require a direct line of sight, so a remote tracking station (RTS) can communicate with a satellite only while in view of a ground antenna. For low-orbiting satellites, line-of-sight periods are sometimes as brief as five minutes. Consequently RTSs must be widely scattered on land or at sea on ships, yet also linked to a central control center.

Beginning with the earliest satellite developments, engineers understood the need for a satellite command-and-control ground segment to retrieve data from and send commands to satellites. In the 1950s the U.S. government selected the Vanguard program as the first civilian space project, in part because it included plans for the first dedicated satellite control network, known as Minitrack. Vanguard's electronic tracking system used radio interferometry. Two ground receiving stations tracked the satellite broadcast signal, and by comparing the separately received signal phases, scientists could accurately calculate angles to the spacecraft and calculate its orbit. Minitrack consisted of a quartz-crystal controlled and fully transistorized oscillator, whose 10 mW output operated on a fixed frequency and had a predicted lifetime of 10–14 days. In addition to its tracking function, Minitrack included antennas and receivers to read out the data transmitted by the scientific satellites—in other words, ground telemetry stations. By the early twenty-first century two NASA organizations were performing the satellite command-and-control function: the Deep Space Network operated by the Jet Propulsion Laboratory, and the Spaceflight Tracking Network used for near-Earth spacecraft operated by Goddard Space Flight Center.

The earliest U.S. military satellite programs, including the Corona reconnaissance satellite program, included their own ground segments. The earliest configuration of the Kodiak Tracking Station is a good example of the system design. Two antennas, a Very Long Range Tracking three-pulse tracking and commanding radar, and a tri-helix telemetry-receiving Ultra High Frequency antenna, served the station. The station communicated off-site by teletype or telephone. Encrypted teletype messages told Kodiak controllers the launch and satellite support operation schedules, and then, after each satellite support operation, they sent the logs and telemetry readouts back by teletype, telephone, or mail. Controllers used unencrypted real-time voice for every other operation. By 1970, in support of its own and other agencies' satellites, the U.S. Air Force (USAF) performed satellite command and control 24 hours every day through the USAF Satellite Control Network.

The single Russian network, which combined tracking for ballistic missiles and spaceflight (both crewed and robotic) was called the Command and Measurement Complex (KIK) and dates to the 1950s. The complex included seven stations spread across the Soviet Union, and later Russian-controlled Eurasian landmass, called Scientific-Measurement Points, which relayed tracking and telemetry data to the

Coordination-Computation Center in Moscow. Soviet and Russian computer facilities determined the initial trajectory of a space launch with a high degree of accuracy. They processed orbital calculations as early as 1961 using an advanced digital computer capable of 20,000 operations per second, with another probably capable of around 50,000 operations per second. These computer capabilities rivaled their counterparts in the U.S. space program. The Soviet Union stretched two-and-a-half times the width of the United States, but satellites passed over the ground-based tracking network only 9 of every 24 hours of a satellite's orbit. When tracking requirements became more stringent for the piloted program—as they did in the American Mercury program—the lack of a global tracking network capable of continuous observation and communications with satellites became the chief limitation on Soviet capabilities for satellite command and control. Primarily the Soviet scientists filled in the gaps in their worldwide tracking coverage with tracking stations on ships, positioning them at strategic points around the world.

Other nations developed similar ground networks. The European Space Agency's worldwide network of ground stations (European Space Tracking—ESTRACK) provided links among satellites in orbit and the central Operations Control Centre at the European Space Operations Centre in Germany. The core network comprised 13 terminals sited at nine stations in six countries and was augmented by other stations as required. The Chinese telemetry, tracking, and control network was composed of the Xi'an Satellite Control Center (XSCC), several tracking stations, and oceangoing instrumentation ships. The XSCC was the communications hub, the command-and-control center, and the data processing center. Japan Aerospace Exploration Agency's Consolidated Space Tracking and Data Acquisition Department consisted of a network of six ground stations across Japan and four overseas stations to receive satellite data, monitor the status of spacecraft, transmit commands, and predict and determine the orbit of the spacecraft. The Indian agency for this important need was called the Indian Space Research Organisation Telemetry, Tracking, and Command Network (ISTRAC) and consisted of a network of stations across south Asia linked to a multimission Spacecraft Control Centre located at Bangalore.

David C. Arnold

See also: Corona, Deep Space Network, National Reconnaissance Office, United States Air Force Satellite Control Facility, Vanguard

Bibliography
David Christopher Arnold, *Spying from Space: Constructing America's Satellite Command and Control Networks* (2005).
Sunny Tsiao, *Read You Loud and Clear: The Story of NASA's Spaceflight Tracking Networks* (2006).

Communications

The communications subsystem provides the only link between the vehicle and ground stations or other spacecraft. For some satellites a communications system is the primary payload or reason the satellite is in orbit. Data passed through the communications system include telemetry (data from the spacecraft), such as space-craft health and status data, payload results, or relayed information from the ground or other spacecraft. The power that the communications system requires is a major design driver. Power-saving techniques can include low-frequency transmissions, store-and-forward data, and encryption. Some of the major components include anten-nae, transponders, amplifiers, modulating/demodulating equipment, and occasionally encryption/decryption equipment.

Antennae convert electronic carrier signals (radio waves) to polarized electromag-netic fields (data bits) and vice versa. The three main types of antennae are linear dipole (wire), parabolic reflector, and phased array. Wire antennae are used in the very high fre-quency (VHF), ultrahigh frequency (UHF) radio ranges and can provide omnidirectional coverage for operations when the main antennae are not operational. A parabolic antenna, the shape most often used for spacecraft and ground antennae, is actually a small feedhorn placed at the center of a parabolic radiofrequency reflector. Electronic steering of antennae, either through multiple narrow beams or phased arrays of separate wide-beam modules, produces multiple beams in arbitrary directions. Defense Satellite Communications System III introduced operational multiple-beam antennae in 1982, while phased array antennae were introduced onboard spacecraft in the 1990s. The Mercury Surface, Space Environment, Geochemistry, and Ranging (MESSENGER) mission to the planet Mercury (arrival in 2011) is the first deep-space mission to use a phased-array antenna for communications with Earth. Over time, communications satel-lites became more complex, using a combination of antenna types in addition to multiple antennae of a single type. For example, a communications satellite may use a phased array antenna for transmitting and receiving hundreds of communications signals—its payload—but may use a lower-power omnidirectional antenna for low data rate health and status data.

Spacecraft commands and telemetry are sent and received using electromagnetic waves through the process of modulation. Typically the information carried on the signal is either analog modulated, as in, for example, amplitude modulation (AM) or frequency modulation (FM); or digitally modulated, as in phase-shift keying. AM was the first technique used for audio radio transmissions at the turn of the twentieth century and was introduced immediately for satellites because of its simplicity. How-ever AM is inefficient because at least two-thirds of the power is concentrated in the carrier signal, whose only purpose is to indicate that a signal is present. For space signals, FM quickly replaced AM, because FM is less susceptible than AM to interfer-ence caused by thunderstorms or electronically generated noise, such as television transmissions. FM modulation affects the amplitude of a radio wave but not its frequency, so an FM signal remains virtually unchanged. Phase-shift keying is a digi-tal modulation technique, modifying the analog carrier signal by digital bits. More information can be conveyed in a given amount of time by dividing the bandwidth of a signal carrier, so that more than one signal is sent on the same carrier, known as multiplexing. Over time, spacecraft communications has shifted to digital modulation techniques so as to send more information over the same signal frequency. Jet Propulsion Laboratory's Deep Space Network (DSN) pioneered methods to compress data (such as identifying locations in images that were devoid of features) and provide error-correcting codes, so as to reduce data errors and send more information over the same frequencies. These techniques, along with increasing the size and sensitivity of DSN receiving antennae and stations, made possible missions to the outer planets.

When a transponder receives a communications signal, it isolates the neighboring radiofrequency signals and amplifies frequencies for retransmission. Typically uplink and downlink frequencies are separated on the electromagnetic spectrum to minimize interference between transmitted and received signals. Usually a lower frequency is chosen for downlink, because it requires a lower power output on the satellite that can be compensated for with a bigger parabolic antenna on the ground. Traveling wave tube amplifiers (TWTA), sophisticated vacuum tube amplifiers that use a narrow electron beam guided by a magnetic field, were an important technology for communications satellites, because they produced a high power output from a weak input signal. The first practical TWTA was produced at Bell Telephone Laboratories in 1945, and these were developed at Bell Labs and at Hughes Aircraft for space applications. Most early communications satellites, such as *Telstar 1* in 1962, used TWTAs. With the development of solid-state technology, first tested in space on the Lincoln Experimental Satellites in the 1960s, many satellites by the 1980s switched to use of lighter-weight and less expensive solid-state power amplifiers (SSPA). However because SSPAs generate less power, are half as efficient, and less capable at higher frequencies, many satellites used a combination of SSPAs and TWTAs.

Another communications issue is the selection and assignment of the frequencies to be used for commands, telemetry, and, for communications satellites, the communications payload. Spacecraft communications have used VHF (30–300 MHz), UHF (0.3–3 GHz), superhigh frequency (SHF, 3–30 GHz), extremely high frequency (EHF, 30–300 GHz), and have been typically assigned frequencies in these ranges by the International Telecommunications Union. At higher frequencies, such as SHF and EHF, signal attenuation through the atmosphere and in rainy weather increases, but these frequencies can carry far more information than lower frequencies. With the increasing need to send large quantities of data, such as imagery and data files, by satellite and over-crowding at popular lower frequencies, communications satellites have shifted over time to use higher frequency ranges in SHF and EHF for their payloads. This requires higher satellite power output to ensure transmission through the atmosphere. Commands to spacecraft and health and status data from spacecraft, which tend to require much lower data rates, have generally remained at lower frequencies (typically 2–4 GHz). Military communications, with great need to ensure communications security, have also used higher frequencies, so as to provide frequency-hopping capability to prevent jamming, spying, and spoofing of data.

David C. Arnold and Stephen B. Johnson

See also: Civilian Space Communications, Commercial Satellite Communications

Bibliography
Donald Martin, *Communications Satellites*, 5th ed. (2006).
Douglas J. Mudgway, *Uplink-Downlink: A History of the Deep Space Network, 1957–1997* (2001).
Mark Williamson, *The Cambridge Dictionary of Space* (2001).

Electrical Power

Electrical power is used to operate various spacecraft systems, including sensors, computers, thrusters, and communications systems. The electrical power subsystems of most spacecraft have typically consisted of three major components: generation, storage, and management. Chemical batteries, photovoltaic (PV) cells, and nuclear radioisotope thermoelectric generators (RTGs) frequently provided power generation, while rechargeable batteries provided electrical power storage. Typical power management functions included power distribution, regulation, and conversion onboard spacecraft.

Single-use or primary batteries powered the first several spacecraft leaving Earth's surface. The electrical power systems, like the rest of the early spacecraft, were simple, comprised mostly of nonrechargeable primary batteries capable of producing between 10–30 W for several days or weeks. In 1957 the Soviet Union launched *Sputnik*, the first artificial satellite, which broadcast radio signals throughout the duration of its three-week mission until its silver-zinc primary batteries failed. The first U.S. spacecraft, *Explorer 1*, used nickel-cadmium primary batteries, which accounted for 40 percent of the payload weight. The *Explorer 1* mission lasted approximately 111 days until the batteries failed.

To increase mission lifetimes, designers installed arrays of PV cells onboard spacecraft. These arrays generated an electrical current by collecting light photons from the Sun. The U.S. *Vanguard* satellite first used solar power in 1958, employing a small array of PV cells generating less than 1 W, for an onboard radio transmitter. In 1964 the U.S. Nimbus weather satellites contained a large solar array that generated 470 W. PV cells during this time were less than 10 percent efficient—the cells would convert only 10 percent of the solar energy striking the PV cells into usable electricity— and the efficiency of PV cells would decrease by as much as 30 percent during 10 operational years, due to high-energy particles such as electrons damaging the PV cells. Space applications played a critical role in the advancement of PV technology, and by 2005 manufacturers had increased PV efficiency to approximately 35 percent for space and terrestrial applications.

The introduction of solar power generation for space applications created the need for onboard power management. To regulate the entire power system, the management subsystem performed functions such as battery charging, power conditioning, and power distribution to other spacecraft systems, such as propulsion, navigation, and instrumentation. Rechargeable or secondary batteries became common on board spacecraft to extend mission life and to provide power during solar eclipses, when the Earth or other planetary bodies blocked sunlight from reaching the solar arrays. Common secondary batteries included lead-acid, nickel-cadmium, nickel-hydrogen, and lithium-ion varieties. The onboard power management systems diverted power from the solar arrays when needed to charge the secondary batteries, and when the solar arrays were shaded due to eclipse, the management system would switch to the batteries for spacecraft operation. Often components such as sensor and communications equipment on the spacecraft had different power requirements. The management system would distribute and condition the power for each of the components, permitting the spacecraft to support a wide range of features.

Fuel cells provided an alternative to batteries for short missions, especially for crewed missions. Fuel cells combined gaseous hydrogen and gaseous oxygen, which produced electrical power for missions. The by-product of the reaction was water, also useful for human consumption. All U.S. crewed missions from the 1965 *Gemini 5* mission to the Space Shuttle missions used fuel cells. Fuel cell technology advanced significantly in the early 2000s as an alternative to chemical batteries, but as of 2007, space missions used little of this new technology.

Deep space missions, particularly when spacecraft probes would travel farther away from the Sun, utilized radioactive power sources instead of solar cells. RTGs converted the heat generated by isotopic decay into electricity. The heat by-product generated from the radioactive decay generated several hundred watts of electrical power, using thermocouple converters consisting of two different metals that generated an electric current when heated. As the radioactive material decayed, the heat and power generated from the RTG decreased. In 1961 the U.S. *Transit 4A* and *4B* navigation satellites used RTGs, and deep space missions (*Pioneer 10* and *11*, *Voyager*, *Galileo*, *Cassini*) and planetary probes (*Mariner*) used RTGs. As of 2005 the RTGs on board *Voyager 1* and *2* continued to generate power 28 years after launch. Since their inception, RTGs have increased in efficiency and power output, but the basic operating principle remained the same into the 2000s.

For various missions that required larger amounts of power (kilowatts), the United States and the Soviet Union developed thermionic reactors. These reactors were more complex than RTGs and generated larger amounts of heat (up to 2,000 K) that boiled electrons, forcing them to flow across a cathode-anode junction. Thermionic reactors required cooling and heat regulation, which was not required for RTGs. The United States investigated these reactors on the SNAP-10 research mission, which launched a reactor on an Agena booster in 1965. The Soviet Union fielded several dozen operational reactors into space on board the Radar Ocean Reconnaissance Satellites (RORSAT) and Electronic Intelligence Ocean Reconnaissance Satellites (EORSAT), each capable of producing several kilowatts of electrical power to eliminate the atmospheric drag created by solar panels in low Earth orbit, permitting lower orbital altitudes and subsequently minimizing radar distances. One such spacecraft, *Kosmos 954*, experienced orbital degradation and crashed in northern Canada in 1978, spreading Uranium-235 throughout an area of several square miles. U.S. and Canadian authorities recovered the spacecraft remains and attempted to decontaminate the area. Several hundred pounds of sodium-potassium (NaK) reactor coolant also escaped from Soviet satellites on several occasions, creating a debris hazard in low Earth orbit. NASA estimated that approximately 115,000 NaK spheres between 2 and 3 inches in diameter were orbiting at approximately 560 miles.

The use of power-generating radioactive materials in space raised concern among the public throughout their use. Designers took measures to ensure that RTGs would not release nuclear contaminants into the environment, should a launch failure or accidental orbital reentry occur. For example, the failed *Apollo 13* Lunar Module and associated RTG eventually reentered Earth's atmosphere and the Pacific Ocean, and no evidence exists that nuclear material escaped into the environment.

For insurance and reliability reasons, satellite manufacturers and operators often preferred to use tested and proven technology, especially in critical areas such as power subsystems and command-control subsystems. This produced a culture of technology evolution versus revolution in the industry. Government research and development programs investigated and tested new potential spacecraft technologies. *Deep Space 1*, a deep space probe launched in 1998, was a NASA technology validation mission. Advanced solar arrays known as Solar Concentrator Arrays with Refractive Linear Element Technology (SCARLET) were installed that collected and focused sunlight onto narrow rows of PV cells, producing as much as 2,500 W for the mission. SCARLET employed 720 lenses to focus solar energy onto 3,600 PV cells to power the spacecraft's experimental electric ion propulsion system. Another NASA technology demonstration mission, *Earth Observing 1*, was designed primarily to validate new Earth-monitoring technology and tested lightweight, flexible solar arrays that offered a higher power-to-weight ratio when compared to conventional solar arrays at the time.

By the start of the twenty-first century, electrical power requirements had risen significantly. Demand from high-bandwidth telecommunications satellite operators was in the kilowatt range, and new high-performance electric ion thrusters became favored over conventional chemical systems for propulsion and maneuvering. Solar cell efficiency approached 40 percent, allowing construction of larger, higher-power arrays. Battery arrays became larger to meet power requirements during times of eclipse, but newer, more efficient battery designs evolved to meet demand. The solar arrays offered commercially on the Boeing 702 satellite provided up to 18 kW, using gallium arsenide PV technology. Space Systems/Loral offered up to 30 kW on its model *Geostationary Earth Orbit 1300* communications satellite.

Electrical power sources were often the limiting factor for mission lifetime and capabilities. Sophisticated, lengthy deep-space missions required large high-fidelity instrument arrays and efficient propulsion, in addition to high-performance computing and communications systems to process and transmit data. Missions to Jupiter's icy moons or to the edge of the solar system would eliminate the Sun as a power source, not only because of the distances involved, but also because of power demands from onboard electrical components and electric propulsion. As with RTGs, space nuclear reactors raised concern with the public about the use of nuclear power in space and the potential radioactive contamination effects of a launch or orbit failure. Public concern regarding the launch and use of nuclear materials led to several protests by activist groups, specifically prior to the launches and Earth flyby maneuvers of the *Galileo*, *Cassini*, and *New Horizons* deep space missions.

David Hartzell

See also: *Cassini-Huygens*, Deep Space Probes, Glenn Research Center, Russian Naval Reconnaissance Satellites

Bibliography
Gary L. Bennett, "A Look at the Soviet Space Nuclear Power Program," *Proceedings of the 24th Energy Conversion Engineering Conference* (1989).
———, "Space Nuclear Power: Opening the Final Frontier," *American Institute of Aeronautics and Astronautics* (2006).

Michael D. Griffin and James R. French, *Space Vehicle Design*, 2nd ed. (2004).
Jerry Jon Sellers, *Understanding Space: An Introduction to Astronautics* (2000).

Flight Software

Flight software is used on board a spacecraft to control the vehicle's computer. Early spacecraft used command and control from the ground or from preprogrammed hardware operations. As missions grew more complex, software-controlled computers added to spacecraft provided increased flexibility for mission design and enabled the creation of advanced architectures. Flight software is instrumental to the objectives of both crewed and robotic missions.

The first crewed spacecraft with onboard software was Gemini. Its software, which International Business Machines (IBM) developed along with the associated hardware, was involved in various flight activities, including ascent, rendezvous, and reentry. The programs for the Gemini computer were written in assembly language, a human-readable form of the machine code the computer used to conduct operations. NASA concurrently assigned the Massachusetts Institute of Technology (MIT) to develop the software and the hardware for the Apollo Guidance Computer (AGC). The AGC was used on the Command Module and the Lunar Module, with the same hardware on both spacecraft but differing software. The AGC was of critical importance during the lunar landing, because the distance to the Moon made real-time control of the landing from Earth impossible. A counterpart to the AGC during this era was the Argon-11c computer that the Soviet Union developed for its Zond Moon program. The Argon-11c system would have been the first digital computer used by Soviet cosmonauts, but the Zond was ultimately only used for uncrewed flights. IBM again developed the flight software for the subsequent U.S. Skylab and Space Shuttle missions. The Space Shuttle mission architecture was unique compared to previous crewed missions, in that it used modified general computers as opposed to custom machines, and the software was written in a higher level language called HAL/S. The Spacelab program was a major step in spacecraft software development for the European Space Agency (ESA), as NASA required that ESA meet NASA's human-rated flight software programming techniques and processes.

Robotic spacecraft have also made extensive use of computer technology. In particular, deep space probes need programmable computers for autonomous operations, because the long light-time distances at which they operate make real-time remote control impractical. Early probes, such as the Ranger and Surveyor missions to the Moon, used fixed sequencers on board the spacecraft that allowed for the spacecraft to be preprogrammed to execute a series of events at a given moment and to receive commands. The first deep space vehicles to use programmable onboard computers were the Mariner Mars probes launched in 1969. The computer and software on this spacecraft, developed by NASA Jet Propulsion Laboratory, provided the ability to reprogram spacecraft operation during the mission. Similar to the human spaceflight program, these early craft used assembly language within a custom computer. The advent of the Intel 4004 microprocessor on *Pioneer 10* enabled more functionality and flexibility. Deep space probes led the way in the development of autonomous software, since they had to make critical decisions at distances that made real-time human intervention impossible. Autonomous fault protection software was particularly important to ensure the safety of the spacecraft by responding to failures. It grew in capability and complexity from the two Voyager spacecraft through *Galileo*, *Magellan*, and *Cassini*. *Galileo* was the first deep space probe to use the HAL/S programming language.

As onboard spacecraft flight computers and microprocessors became faster and provided much greater memory storage, spacecraft flight software grew to fill the memory and to provide greater capabilities. An example of a microprocessor in the early 2000s was British Aerospace's PowerPC-based and radiation-hardened RAD6000, which had been chosen for a large number of Earth orbiters, in addition to missions to Mars, Mercury, and other solar system bodies. Typical software on these spacecraft and those closer to Earth consisted of functions, such as command processing, telemetry production, and guidance and control, with most of the code written in high-level languages such as C and C++. Deep space probes, such as *Deep Space One*, extended autonomous capabilities implemented through flight software from fault protection software that detected and responded to faults, to an autonomous onboard planner that was responsible for determining the science operations of the spacecraft. To improve capabilities and decrease operational costs, in the early twenty-first century NASA's Constellation program to return humans to the Moon began development of a software architecture that used interoperable command and telemetry protocols based on Internet Protocol.

With an increase in lines of code from a few hundred to many thousands on twenty-first-century spacecraft, the flight software development on spacecraft evolved into a stringent engineering discipline from the 1970s to the 1990s. The typical process for creating spacecraft software involved first determining what the software needed to perform in the form of requirements. Software developers designed an architecture that could meet these needs defined during the preceding exercise. Only after the requirements and design were complete did the actual production of source code begin. A rigorous testing process followed during which the software was stressed as an integrated suite under flight conditions.

Flight software continued to evolve as missions became more complex. Spacecraft architectures have advanced from the early era of exploration, where computers consisted only of preprogrammed hardware operations, to the point where software became an integral part of mission operations.

Christopher Krupiarz

See also: Human Spaceflight Programs, Jet Propulsion Laboratory, Spacecraft

Bibliography
David A. Mindell, *Digital Apollo: Human and Machine in Spaceflight* (2008).
James E. Tomayko, *Computers in Space* (1994).

Guidance and Navigation

Guidance and navigation is the subsystem of a satellite, spacecraft, or rocket responsible for determining, predicting, and controlling position along a flight trajectory or during entry into orbit. For space systems, guidance and navigation technologies evolved from ballistic missiles. Probably the first true ballistic missile guidance system was in the German A4 (V-2) rocket of World War II (WWII). The V-2 used an internal gyroscope to steer the rocket by maintaining its pitch. Some of the V-2 launches also were guided by a radio beam. The engine shut down when the rocket reached a specified velocity, as measured by integrating accelerometers. After engine shutdown, the missile followed an unguided ballistic trajectory to the target. At the end of WWII, V-2 technology found its way to the United States and the Soviet Union, where it became the basis for further development.

The superpower conflict between the United States and the Soviet Union fed the mutual development of ballistic missiles that could reach ever greater distances, by 1957 reaching intercontinental range. As ranges increased, improved navigation accuracy became critical. Although radio guidance systems were possible, military leaders wanted a system that could not be jammed, so ballistic missile guidance emphasized the development of inertial measurement systems, sophisticated gyroscopes that did not rely on external measurements after launch. Submarine-launched ballistic missiles also needed a precise initial position measurement, for which the U.S. Navy developed the Transit satellite system (first launched in 1960), whose radio signals provided the necessary initial position fix. In 1991 the Department of Defense moved from the Transit system to the Global Positioning System (GPS), which produced more frequent and accurate positioning information and also provided accurate altitude estimates, which Transit was not suited for. While the V-2's accuracy was such that it could barely hit London at 150 miles, by the 1980s U.S. and Soviet ballistic missiles could attain accuracies of less than 1/10 of a mile at intercontinental distances. In the early twenty-first century, missiles could successfully target an object the size of a truck at a distance of several thousand kilometers.

A typical sequence of events for a spacecraft is slightly different from that of a missile and consists of a launch phase in which the craft is lifted by a booster rocket to a point above Earth. Guidance during this part of the craft's flight is similar to that of a missile. After separation from the booster, the spacecraft uses its attitude control system to orient itself and prepare for firing its engine to achieve the desired trajectory away from Earth. Once the prescribed burn has been made, the craft coasts to its destination orbit. During this period, navigation consists of ground-based monitoring of the spacecraft's trajectory, and guidance is concerned with making any necessary corrections to that trajectory, again using ground-generated commanding. Finally, on reaching its destination, the craft must complete its flight by achieving orbit or some other goal. Typical orbital insertions require both the navigation and guidance system and the attitude control system.

Spacecraft guidance and navigation has usually been ground-based, depending on measurements of the spacecraft's position using radio signals. The craft broadcasts a recorded signal at a known frequency toward Earth. The relative velocity between the spacecraft and Earth causes a frequency shift in the received signal. The frequency difference between the received signal and the known frequency allows the calculation of the velocity of the spacecraft toward or away from the ground station. The distance between the ground station and the spacecraft can be determined by measuring the time between the signal being sent from a transmitter, reaching the spacecraft, and the time the retransmitted signal reaches the receiver. Because radio waves travel at the speed of light, the round-trip signal time determines the distance. Finally engineers can estimate the spacecraft's position in Earth's sky through various techniques. One of the most common and accurate is called Very Long Baseline Interferometry (VLBI). VLBI requires two receivers separated by a large distance, perhaps on two different continents. Each receiver alternately observes the spacecraft and then a remote object, such as a distant star, quasar, or pulsar. Each receiver records its data along with a highly accurate time stamp provided by an atomic clock. The two data sets are synchronized using the time stamp, and interference fringes are used to determine the location and velocity of the spacecraft. After determining the spacecraft's position and trajectory, engineers calculate needed trajectory changes and send commands for the spacecraft to fire its thrusters or engines for the necessary amount of time in a specified direction to change the trajectory.

Performing these types of measurements, especially VLBI, requires sensitive arrays of radio telescopes. One such array was NASA's Deep Space Network (DSN), begun in 1958 and first used with *Pioneer 3* and *4* in 1958 and 1959. NASA established its primary radio telescopes at DSN sites in Australia, Spain, and the United States. Other spacefaring nations established similar, if less powerful, networks of ground stations to control their spacecraft. Over time the DSN and other ground station networks improved the accuracy of their clock signals, increased the size of their antennas, and tied multiple antennas together to increase the effective antenna size, all of which significantly improved navigation accuracy.

While most spacecraft guidance has been ground-based, a few autonomous systems were tested and used by the early twenty-first century. Autonomous spacecraft navigation systems began with the Soviet Soyuz and Progress spacecraft, which used the Igla autonomous navigation system developed in the 1960s for final approach and docking to a target, usually a space station docking port. The Japanese *Hayabusa* asteroid sample return mission, launched in May 2003, utilized autonomous navigational systems for its asteroid descent with mixed success. NASA's *Demonstration for Autonomous Rendezvous Technology* mission, launched in April 2005, was a partial success, autonomously acquiring and navigating to a target satellite using GPS signals. Its final approach failed, as it made a soft collision with the target craft. The U.S. Defense Advanced Research Project Agency's Orbital Express mission, launched in March 2007, successfully tested autonomous rendezvous technologies, once the active spacecraft, *Autonomous Space Transfer and Robotic Orbiter*, was manually positioned within 3 km of the target craft. The European Space Agency's Jules Verne Automated Transfer Vehicle successfully performed an autonomous docking with the *International Space Station* in March 2008.

R. Dwayne Ramey

See also: *Hayabusa*, Navigation, Soyuz

Bibliography
Wiley J. Larson and Linda K. Pranke, *Human Spaceflight: Mission Analysis and Design* (1999).
Donald Mackenzie, *Inventing Accuracy: A Historical Sociology of Nuclear Missile Guidance* (1990).
Douglas J. Mudgway, *Uplink-Downlink: A History of the Deep Space Network, 1957–1997* (2001).

Liquid Propulsion

Liquid propulsion refers to rockets that carry one or more propellants as a liquid. Liquid-propellant rocket engines (LPREs) generally provide more thrust than comparable solid-propellant rockets, but are usually more complex and dangerous. An important LPRE characteristic is that thrust can be adjusted or terminated during flight and thus provide better trajectory and velocity control. In an LPRE the fluid reactants (fuel and oxidizer) are atomized as they are forced through an injection head, somewhat like a spray nozzle, at the top of the combustion chamber. A spark or other ignition source starts combustion, which self-sustains as long as reactants are delivered.

LPREs were initially developed, from the 1920s through the 1950s in Germany, the United States, and the Soviet Union, to act as engines for sounding rockets, ballistic missiles, and to provide extra thrust to aircraft, usually for takeoff. During this period, solid propellants could not generally deliver the performance needed for these purposes. However, by the late 1950s in the United States, solids had matured sufficiently to perform the ballistic missile task better than LPREs and by the mid-1960s had replaced most LPREs in this role for U.S. missiles. The ability of solid propellant rockets to be stored almost indefinitely and launched in seconds outweighed the better performance capability of liquid-propelled systems. LPREs remained the engine of choice for most space launch applications.

Early LPRE designers had to overcome several technical hurdles. One was to ensure that fuel and oxidizer were delivered under greater pressure than is generated in the combustion chamber. The first LPREs used an inert gas, such as nitrogen or helium, under high pressure in the propellant tanks to provide this pressure. Larger and higher-performance LPREs, such as for the German V-2 rocket of World War II, needed higher pressure, which designers provided through high-speed pumps. During the next several decades designers developed a number of schemes to power the pumps, generally using a fraction of the fuel and oxidizer feeding the combustion chamber.

Another major problem of LPREs is cooling of the engine, since an LPRE sustains and channels a controlled explosion. Early rocketeers, including U.S. physicist Robert Goddard and Soviet engineer Valentin Glushko, realized that they could pump the liquids through small tubes lining the combustion chamber and exhaust nozzle to carry away heat that otherwise would melt the engine. After 1957 small LPREs used to control the pointing direction of spacecraft in space could simply fire for short periods and radiate heat into space for cooling.

Fuel selection was another technical issue. By the 1930s rocket pioneers understood that liquid hydrogen (LH2)–liquid oxygen (LOX) engines would be the most efficient,

The launch of a U.S. Air Force Navaho cruise missile in 1957. It used an improved V-2 engine that was influential for later rockets, including Redstone and Jupiter. (Courtesy NASA/Marshall Space Flight Center)

but that because of its extremely cold temperature, liquid hydrogen, in particular, would be difficult to produce and store and dangerous to use. LOX was not quite so difficult, and a combination of LOX with alcohol or kerosene fuels was a frequent early choice, such as for early ballistic missiles, such as the German V-2 (LOX-alcohol), the MA-1 engines of the U.S. Atlas (LOX-kerosene) or the Soviet OKB 456 (Experimental Design Bureau) RD 107 and 108 engines for the R-7 (LOX-kerosene). The RL 10 engine, developed in the early 1960s by Pratt & Whitney Corporation for the U.S. Centaur upper stage, was the first to use LOX-LH2. Several other organizations and nations eventually followed suit. Hypergolic propellants, which ignite on contact and thus do not require igniters, were another possible choice. These were particularly suited to ballistic missile applications because they could be indefinitely stored at room temperature. Typical hypergolic propellants are nitrogen tetroxide (N_2O_4) oxidizer and the hydrazine family of chemicals as fuels. The U.S. Titan missiles and many Soviet ballistic missiles developed in the 1950s–60s used hypergolic propellants. In the United States, Aerojet Corporation developed a series of hypergolic engines for the Titans, while in the Soviet Union, Valentin Glushko's OKB 456 (later NPO Energomash) was the primary developer. However hypergolic propellants were toxic and corrosive, leading to some major catastrophes, such as the explosion of a Soviet R-16 ballistic missile on the launch pad in 1960 that killed dozens of workers and managers. A valuable offshoot of hypergolic engine development was the development of hypergolic thrusters, which served as fast-acting, reliable engines for maneuvering spacecraft.

Liquid monopropellant rockets employ a single energetic fluid that decomposes into hot gases on contact with a catalyst. Their thrust generally is lower than that possible with bipropellant systems. However, they are simple and reliable. Monopropellants normally are used as attitude control thrusters on spacecraft where small, brief impulses are needed. Hydrazine, which was first used in the late 1950s for U.S. spacecraft, became the most commonly used monopropellant. Hydrogen peroxide had been used on some early craft, such as the Mercury capsule.

Hybrid propulsion refers to rockets that employ a liquid oxidizer and a solid fuel, providing some advantages of both solid and liquid rockets. A typical hybrid may use LOX or nitrous oxide oxidizer and a heavy hydrocarbon-based fuel cast as a large tube (often this is a rubber-based compound). The oxidizer is pumped through a spray nozzle at the head end of the tube and burns the fuel from the inside out. The American Rocket Company developed a hybrid engine in the late 1980s. It did not fly, but this design was later updated by SpaceDev Corporation and used on the *SpaceShipOne* private suborbital spaceplane in 2004.

Another major type of liquid propulsion is the tripropellant rocket. This refers to two different types of engine. One type uses small quantities of fluorine added to increase specific impulse. This is basically a two-oxidizer system because that is fluorine's chemical regime. Tripropellants are experimental and highly dangerous because fluorine reacts with virtually all chemicals, and the exhaust is toxic. The second type refers to "dual-mode" engines that use two different fuels in different flight regimes. While experimental tripropellant systems were investigated in the United States and the Soviet Union, neither type of system moved into full development.

A final entry in the liquid rocket family is nuclear thermal rocket engines, which pump liquid hydrogen through a fission reactor to produce hot hydrogen gas. No chemical combustion is involved. Nuclear rockets are among the most efficient possible because the exhaust is pure hydrogen, the lightest of the elements. They have also been problematic because of the environmental concerns about nuclear energy.

After a period of rapid development from the 1930s to the 1970s, LPRE designs around the world largely stabilized. Several historical trends can be discerned. First, LPRE research and development spread to many nations after their start in Germany, the United States, and the Soviet Union; by the early 2000s, some 30 nations had professional or amateur LPRE developments. Through the 1970s larger and higher-performing engines came into being; after that time there was an increased emphasis on reliability. There were some improvements in production efficiency, with fewer people needed to develop and build engines as knowledge of the technology and production methods matured. By the early 2000s, some 1,500 different types of LPREs had been developed and perhaps more than 2,500 different propellant combinations considered, but a very few of these were widely used.

Dave Dooling and Stephen B. Johnson

See also: Ballistic Missiles, Expendable Launch Vehicles and Upper Stages, Hypersonic and Reusable Vehicles

Bibliography

J. D. Hunley, *The Development of Propulsion Technology for U.S. Space-Launch Vehicles, 1926–1991* (2007).

George Koopman, "Getting into the Launch Business: The American Rocket Company Story," *Quest* 12, no. 2 (2005): 30–39.

George P. Sutton, *History of Liquid Propellant Rocket Engines* (2006).

Nuclear Power and Propulsion

Nuclear power and propulsion have long been considered essential by space experts for the long-term exploration and colonization of space. Many modern spacecraft have successfully used small nuclear power units. Research conducted in the 1960s–70s demonstrated that space nuclear power devices were more robust than solar cells and less susceptible to damage from cosmic radiation, solar heat, and orbital collisions with spacecraft and debris.

In the 1950s the U.S. Air Force (USAF) and the RAND Corporation studied the feasibility of powering reconnaissance satellites under the code name of Project Feedback. Feedback officials called for the immediate development of fissioning uranium reactors and radioisotope batteries or radioisotope thermal generators (RTGs) for use in space.

Beginning in 1955 the U.S. Atomic Energy Commission (AEC) developed Feedback's concepts in a series of compact nuclear-power generators called System for Nuclear Auxiliary Power (SNAP). The idea behind SNAP was to turn a decaying radioactive element into electricity. SNAP-3 was a grapefruit-sized demonstration unit using polonium-210 as a fuel that converted the heat thermoelectrically, that is, through the use of thermocouples. Thermocouples create electricity by having different temperatures at the junction of different metals at their interface. The first public display of SNAP was held at the White House in 1959. Activated by President Dwight Eisenhower, the Earthbound unit operated for 90 days. SNAP-3A, the first space-based atomic device, powered the U.S. Navy's *Transit 3A* navigation satellite in 1960. While SNAP-3A was not entirely successful, SNAP-3B7 performed flawlessly for 15 years on board the *Transit 4A* satellite beginning in 1961.

Various U.S. satellites and Apollo lunar surface experiments employed SNAP devices to generate electricity. Notably, SNAP devices powered the *Pioneer 10* and *11* spacecraft to Jupiter and the *Viking 1* and *2* Mars landers. Similar RTGs were used to power the *Voyager 1* and *2*, *Galileo*, *Cassini*, and *New Horizons* Pluto probes.

Beginning in the early 1960s the Soviet Union experimented with SNAP-like power sources, but favored its BUK space nuclear reactor series to power satellites. Nuclear reactors actively create fission reactions, the heat from which is converted into electricity. The first reported launch of a BUK reactor into space was in 1967. An early BUK produced about 3 kWe from 100 kWt of thermal power using multiple thermoelectric elements. More than 30 BUK-type reactors were flown in space. By comparison, the United States launched only one reactor, SNAP-10A, flown on the *Snapshot* spacecraft in 1965.

In 1987 the Soviet Union flew two experimental TOPAZ thermionic reactors (reactors that convert heat into electricity by collecting ions from the hot source) that reportedly produced about 5 kWe. In the early 1990s, the U.S. Department of Defense purchased several Russian thermionic reactors (ENISEY) from the former Soviet Union for nonnuclear tests after its SP-100 (Space Power–100 kWt) space power reactor program was canceled in 1993. Russian, British, and French participants contributed to these tests at the USAF Philips Laboratory in New Mexico through 1995.

Atomic-powered rockets first appeared in science fiction literature in the 1930s. One of the first novels to describe, in detail, the inner workings of a large nuclear rocket was Robert A. Heinlein's *Rocket Ship Galileo*, written in the 1940s. U.S. physicists Stanislaw Ulam and Frederick de Hoffman conducted the first investigation of nuclear rocket propulsion in 1944 during work on the atomic bomb. Ulam first proposed the idea of "nuclear pulse propulsion" for rockets. This concept postulated a series of rapid nuclear fission explosions to propel a spacecraft to distant planets. Ted Taylor and Freeman Dyson adopted Ulam's idea as Project Orion in 1958.

NASA had a vigorous nuclear thermal rocket (NTR) engine development program called Nuclear Engine for Rocket Vehicle Application (NERVA) that followed the successful AEC Project Rover space reactor tests starting in 1959. The NERVA engine used uranium fuel elements and was ground-tested at the Nevada Test Site. NERVA used hydrogen gas channeled through a compact reactor core. The engine produced 55,000 lb of thrust and was considered as a replacement for the Saturn V rocket's upper stage J-2 engine. With no crewed Mars expedition in NASA's 1972 program budget, the NERVA project was canceled.

Following NERVA, nuclear engine designs were studied through the U.S. government's Strategic Defense Initiative in the 1980s as Project Timberwind. NASA's Project Prometheus, in close association with the U.S. Department of Energy, was established in 2003 to develop advanced nuclear power systems for space applications, including an NTR propulsion system for robotic deep-space probes. However, budget issues led to its cancellation in 2005.

Since the 1960s there have been several space nuclear power accidents by the United States and the former Soviet Union. Perhaps the most serious U.S. accident was in May 1968 when the *Nimbus B1* weather satellite fell into the Pacific Ocean off Vandenberg Air Force Base, California, after a launch abort. Its SNAP-19B2 RTG was recovered intact and used later in another spacecraft. The most famous Soviet accident was the reentry of the *Kosmos 954* spacecraft over Canada in 1978.

It was a US-A (Controlled Satellite-Active)/RORSAT (Radar Ocean Reconnaissance Satellite) with a BUK nuclear reactor. In the resulting liability dispute, the Soviet Union paid Canada $6 million for nuclear cleanup. These incidents fueled nuclear fears in some segments of the international community.

Despite the engineering rationale for using peaceful nuclear power in space, an influential and international antinuclear lobby made its development and deployment more difficult. The 1997 launch of the *Cassini* spacecraft to Saturn, using proven hardware, was almost canceled after an orchestrated protest by various environmental and leftist political groups. Hand in hand with the efforts of some political parties in the West—especially in the United States—these protests, coupled with high development costs, slowed or stymied space nuclear-power projects. This has occurred despite the fact that these systems are designed with very large margins against breakup and accidental nuclear release. In early 2009 NASA was continuing to assess nuclear power options in its assessments of lunar habitation as part of its Constellation program.

Lou Varricchio

See also: Apollo, Deep Space Probes, Russian Naval Reconnaissance Satellites

Bibliography
Gary L. Bennett, *"Space Nuclear Power: Opening the Final Frontier,"* 4th International Energy Conversion Engineering Conference (26–29 June 2006).
James A. Dewar, *To the End of the Solar System: The Story of the Nuclear Rocket* (2004).
Louis Varricchio, "U.S. Nuclear Power Systems: Past, Present, and Future," *Quest: The History of Spaceflight Quarterly* 3, no. 6 (1998): 44–50.

Orbital Mechanics

Orbital mechanics is the study of how both natural and human-made objects move under the influence of the gravitational forces of massive bodies. The beginnings of orbital mechanics can be traced to early attempts to describe the motions of the planets known to the ancients. The idea that circular motion was perfect pervaded much of the early work in the subject. The first major advance came from the work of Johannes Kepler. Using observations of Mars obtained by Tycho Brahe, Kepler was able to develop his three laws of planetary motion: (1) the orbits of the planets are ellipses with the Sun at one focus, (2) the line joining the planets to the Sun sweeps out equal areas in equal times, and (3) the square of the period of an orbit is directly proportional to the cube of the average distance from the Sun. Kepler published his first two laws in 1609 and the third in 1619.

Kepler's laws of planetary motion were a major step forward in the description of planetary motion, but they did not explain the mechanics that led to that motion. Isaac Newton's discovery (building on the work of Kepler and Galileo) of the laws of motion and gravitation provided a firm theoretical background that told why the orbits of the planets obeyed Kepler's laws. Newtonian mechanics also showed that any body orbiting a much larger body would obey Kepler's laws of motion, and that, in general, motions in space could be any conic section, including an ellipse or, if the body was not bound by the gravitational effects of the central body, a parabola or hyperbola.

Newton's laws of motion, combined with advances in observational astronomy, led to the problem of determining the orbits of one body about another. This problem was initially applied to the orbits of the planets, comets, and asteroids. Later the same techniques were improved and expanded for use in orbit determination for artificial satellites. This problem can be divided into two main areas: developing an initial estimate of an orbit using a minimum number of observations, and determining an optimal estimate of an orbit given more observations than are necessary to uniquely determine the orbit. The first, known as initial determination, was originally applied to the orbits of comets. Initial orbit determination requires at least three independent angular observations using the telescopic techniques available when the techniques were developed. Newton published the first practical method for initial orbit determination. Edmund Halley used Newton's technique to determine the orbit of the comet that came to be known as Halley's Comet. Newton's technique was extremely complex; Karl Friedrich Gauss developed the first widely used technique. Gauss's method became the basis of the techniques used for initial orbit determination.

The second area of orbit determination, called optimal estimation, was widely attacked by the mathematical community. In 1795 Gauss developed the method of weighted least squares, where the optimal solution is the solution that minimizes the square of the errors of the estimate compared to the observations. This method was used to estimate the orbits of comets and, later, the orbits of asteroids. Weighted least squares means that some weighting criteria are given when some observations are known to be more accurate than others. Although Gauss used the method, he did not publish his results until 1809. In the meantime Adrien-Marie Legendre independently developed the method and published his results in 1806. Both Legendre and Gauss developed the method to use all observations to obtain the best possible estimate of the orbits of comets. No major advances were made in optimal estimation theory until the twentieth century, when the availability of computers led to the development of the Kalman, or sequential, filter. The major advantage of the Kalman filter over the traditional weighted least squares was that it allowed individual observations to be incorporated into the estimate as they were collected, instead of waiting for a batch of observations to be available and then processing them together. The development of the Kalman filter and its application to orbit determination involves some historical controversy. In 1958 Peter Swerling published a report that used a filter similar to the Kalman filter for orbit determination. However Kalman's work, which did not appear until 1960 and was applied to the general estimation problem, became better known. Variations of weighted least squares and the Kalman filter remained in use for orbit determination of both artificial and natural satellites into the twenty-first century, by which time the observation techniques included radars, radio transponders, laser ranging, and the Global Positioning System, in addition to traditional telescopic observations.

The orbital mechanics of the three-body and n-body problems have received significant attention. The n-body problem is determining the trajectory of an arbitrary number of bodies, each interacting with the gravitational fields of all the other bodies.

The three-body problem is where n is equal to three. Mathematician Joseph-Louis Lagrange discovered the complete solution to the circular restricted three-body problem. The restricted three-body problem assumes that the mass of one body is negligible compared to the other two masses, and, therefore, the trajectory of this body is completely determined by the gravitational fields of the other two. The circular restricted three-body problem restricts the two massive bodies to move in circular orbits around their mutual center of mass. Lagrange found the five libration (or Lagrange) points as a solution to this problem. These are equilibrium points where the rotational and gravitational accelerations cancel. There are three equilibrium points in a line that connect the two major bodies in the system (L1 through L3). All three collinear points represent unstable equilibrium points, meaning that objects placed at the points will drift away over time if something moves them away from the point. The other two libration points (L4 and L5) are the equilateral or triangular libration points. They are located in the same orbit as the secondary body, but either 60 degrees ahead or 60 degrees behind the secondary body. These points represent stable equilibrium points. The numbering of the libration points is not uniform in all sources.

Henri Poincaré showed that there are no closed-form solutions to the unrestricted three-body problem. In addition, Poincaré's work on nonlinear dynamical systems laid the foundation for the examination of chaos theory. The advent of modern computers made possible new methods of analysis for nonlinear systems, including numerical solutions of various n-body problems.

Modern computers also enabled the analysis of complex trajectories that led to the concept of gravity assist maneuvers, discovered in the early 1960s at the Jet Propulsion Laboratory. Gravity assist maneuvers use the mass of a planet to change the velocity of a spacecraft relative to the Sun. The planet's velocity is also changed by a negligible amount in the maneuver because of conservation of momentum. The first mission to use a gravity assist was *Mariner 10*, which used a gravity assist from Venus to perform several flybys of Mercury. The most famous missions to use gravity assist were *Voyager 1* and *2*. Both missions used a gravity assist from Jupiter to reach Saturn. In addition, *Voyager 2* used a gravity assist from Saturn to reach Uranus and Neptune. *Galileo* and *Cassini* used gravity assists from Venus and Earth to reach the outer solar system, and *Ulysses* used a gravity assist from Jupiter to go into a solar orbit out of the plane of the solar system.

There are two main focus areas in modern orbital mechanics. One is the motion of natural objects, including the long-term stability of the solar system and other planetary systems, and the determination of the orbits of comets and asteroids and the possibility they will collide with Earth. The second focus is motion of artificial satellites, including complex trajectories that take advantage of perturbations to produce desired trajectories with a minimum amount of fuel. This also includes better description of the forces that affect a satellite in motion, including more accurately modeling the gravity fields and atmospheric density and radiation environment of Earth and other celestial bodies.

Craig A. McLaughlin

See also: *Mariner 10*, Voyager

Bibliography
Anton Pannekoek, *A History of Astronomy* (1961).
Harold W. Sorenson, "Least-Squares Estimation: From Gauss to Kalman," *Institute of Electrical and Electronics Engineers* (IEEE) Spectrum 7, no. 7 (1970): 63–68.
David A. Vallado, *Fundamentals of Astrodynamics and Applications* (2001).

Solid Propulsion

Solid propulsion is a method of providing thrust for rockets using solid fuels and oxidizers. In the form of black powder (charcoal, sulphur, and potassium nitrate), solid propellants (evidently invented by the Chinese) have existed for centuries. Military rockets have used them since at least 1232. Despite improvements over the years—including those by William Congreve in British rockets used in the Napoleonic wars—black powder had too little performance for use in the sophisticated missiles and rockets developed after World War II.

Since that time, a series of technological improvements has enabled solid-propellant rockets to serve as boosters and upper stages in a large number of strategic missiles and launch vehicles, in addition to motors for tactical missiles, vernier rockets, ullage motors, satellite apogee and perigee kick motors, stage-separation motors, and sounding rockets. Many entire missiles and launch vehicles, such as Polaris, Minuteman, and the Scout launch vehicle, have used only solid propulsion, which is simpler and more reliable than liquid-propellant systems. The exhaust velocity and performance of solid propellants are lower than many liquid propellants, and liquids offer the advantages of throttleability, easy shutdown, and reignition. But liquids entail extra weight in the form of tanks, plumbing, and complicated valves. The absence of these systems in solid-propellant rockets results in a higher thrust-to-weight ratio, enabling a more rapid liftoff when solids are used in boosters on liquid-propellant launchers. This was why the U.S. Space Shuttle and the European Ariane launch vehicles used large solid-propellant boosters.

Solid propellants advanced from the low-performing versions used before World War II to the twenty-first century in complicated and different ways from one country to another. The Soviet Union/Russia relied more heavily than the United States on liquid propellants, with its large Energia launch vehicle using big liquid-propellant boosters, but in the late twentieth century, Russian light- and medium-lift launch vehicles such as Start and Start-1 began using solid-propellant stages.

In the United States several organizations contributed to significant improvements in solid-propellant technology during World War II. By this time most rockets for military uses employed double-base (mostly nitroglycerine and nitrocellulose) propellants, which had higher performance than black powder. But the propellants had to be extruded (forced) through dies to form grains (masses of propellant). The process of extrusion limited the size of the grain. A California Institute of Technology (Caltech) program developed dry-extruded double-base rockets, mostly for the U.S. Navy. To test and evaluate these rockets, the Navy established the Naval Ordnance Test Station

(NOTS) in the desert north of the San Gabriel Mountains. NOTS would continue to play a prominent role in rocket development after the war. Caltech and the Navy pioneered U.S. use of a star-shaped, internal-burning cavity in the middle of the grain. This design, used earlier in the United Kingdom, became common in later solid-propellant rockets. It, and other internal-burning designs, allowed the grain to insulate the rocket case from the heat of combustion, permitting much thinner, lighter cases.

Research during World War II at the National Defense Research Committee's Explosives Research Laboratory at Bruceton, Pennsylvania, led to a process of casting double-based propellants (curing them in a case), apparently first used by the Hercules Powder Company after the war at the U.S. government-owned Allegheny Ballistics Laboratory (ABL), which, together with Hercules, played an important role in solid-propellant rocketry. Casting yielded a larger grain than could be produced by extrusion, making the technology useful for upper stages in launch vehicles, beginning with Vanguard on its last launch (18 September 1959), which used an ABL third stage to launch a 52 lb satellite into orbit.

The Guggenheim Aeronautical Laboratory at the California Institute of Technology (GALCIT) began studying jet-assisted takeoff (JATO) devices for aircraft in 1939. It found that compressed powder exploded, so, in 1942, chemist John Parsons at GALCIT combined asphalt, as a binder, and fuel with potassium perchlorate, as an oxidizer, to create the first castable composite solid propellant. Asphalt was not a satisfactory binder, and potassium perchlorate created smoke as it burned. The two ingredients did not have the high performance of double-base propellants, but they provided the basis for many subsequent castable composite propellants later used in large boosters. Many of them had lower performance than castable double-base propellants, especially after the latter began to incorporate additional fuels and oxidizers, such as aluminum and ammonium perchlorate in what were called composite modified double-base propellants. But the non–double-base composite propellants were less inclined to detonate than double-base ones, making them safer to use in lower stages (especially important on the Space Shuttle with astronauts aboard).

Aluminum's use as a principal fuel in high-performing solid compositions marked another key discovery. In the United States, credit for this development belonged to Keith Rumbel and Charles B. Henderson, chemical engineers working for Atlantic Research Corporation (ARC). The Navy Bureau of Ordnance had contracted with ARC to improve the performance of solid propellants. Rumbel and Henderson began theoretical studies in 1954. Defying existing theory, they discovered that adding significant amounts of aluminum to solid propellants boosted performance. This contribution, among others, made it possible to develop the Polaris and Minuteman missiles in the late 1950s and early 1960s, following which all subsequent U.S. strategic missiles shifted from liquid to solid propellants. The propellant used in Minuteman, polybutadiene acrylic acid acrylonitrile, formed the basis for those in the large Titan 3 and 4 solid rocket motors and the even larger solid rocket boosters used on the Space Shuttle. The Ariane 5 boosters used a similar propellant and a polymeric binder compound that retained its shape, resisted cracking, and yet permitted fuel and oxidizer to be loaded in it when it was in a liquid state, before curing into a rubbery substance.

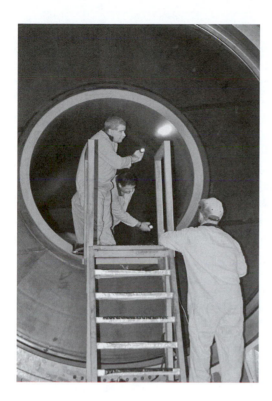

In the Rotation, Processing, and Surge Facility workers conduct propellant grain inspection of the solid rocket booster segment required as part of safety analysis, 2007. (Courtesy NASA/Kennedy Space Center)

Another important solid-propellant technology was the development in the late 1960s of an adhesive, sealant, and coating known as hydroxy-terminated polybutadiene (HTPB) for use as a propellant binder. Among other things, this required discovery of suitable bonding agents to link the HTPB polymer to oxidizers, such as ammonium perchlorate, and fuels, such as aluminum. Tight links were necessary to withstand temperature cycling, ignition pressure, and other forces that could cause the solid particles to separate from the binder network, producing voids in the grain that might result in cracks and structural failure. The U.S. Air Force Rocket Propulsion Laboratory at Edwards Air Force Base, California; Thiokol; Aerojet; the U.S. Army Redstone Arsenal; ARC; Hercules; and the Navy were involved in developing HTPB-based propellant. HTPB found uses in upper-stage motors such as the Payload Assist Module and upper stages for the Scout launch vehicle. But it did not replace the polybutadiene acrylic acid acrylonitrile propellant in the large solid rocket motors on Titan 3 and 4, the Space Shuttle, and the Ariane 5, because the older propellant was less expensive to produce.

Other technologies developed for the various uses of solid propellants included sophisticated grain designs to adjust the thrust profile and reduce erosive burning; various measures to ameliorate combustion instability (still not fully understood); lighter case materials and substances to permit exhaust nozzles (especially their throats) to withstand the heat and high speeds of exhaust gases; methods of rotating the thrust vector (direction) for steering, ranging from liquids injected into the exhaust stream to moveable nozzles; joints for large boosters, which needed to be segmented for ease of transportation from manufacturing locations to launch sites; and extendable nozzles to adjust the expansion ratio for maximum efficiency at different altitudes.

Developing these technologies required extensive cooperation among a large number of different organizations. The United States and the Soviet Union, especially, developed many technologies first for military missiles and then transferred them to other military and civilian uses. France and China relied less on their own missiles to develop space-launch capabilities.

These and other countries have relied in varying degrees on solid propellants in their launch vehicles and other space-related uses. Japan, for example, made use of U.S. technology and included solid-propellant stages on most of their launch vehicles, including two large solid boosters on versions of the H2. India based its Satellite Launch Vehicle (SLV) 3 on sounding rockets. With four solid-propellant stages, the SLV 3 resembled the U.S. Scout launch vehicle. Its successor, the Augmented SLV was more powerful, with solid boosters and four solid stages. Israel and Brazil developed smaller all-solid launch vehicles.

Solid propellants have made important contributions to overall space technology. Their uses have evolved throughout time and will continue to do so.

J. D. Hunley

See also: Ariane, Ballistic Missiles, Jet Propulsion Laboratory, Space Shuttle, Titan

Bibliography
J. D. Hunley, *The Development of Propulsion Technology for U.S. Space Launch Vehicles 1926–1991* (2007).
———, "The Evolution of Large Solid Propellant Rocketry in the United States," *Quest* 6 (Spring 1998): 22–38.
Steven J. Isakowitz et al., *International Reference Guide to Space Launch Systems* (2004). Martin J. L. Turner, "Solid Propellant Rocket Motors," in *Rocket and Spacecraft Propulsion: Principles, Practice, and New Developments* (2000), 87–104.

Space Elevator

Space elevator, a technological concept for routinely sending payloads into orbit without using rocket-powered boosters. Konstantin Tsiolkovsky originated the idea in 1895, but scientists and engineers did not reconsider it seriously until the 1960s. Further scientific discussion of the concept appeared in the mid-1970s, with science fiction writers, such as Arthur C. Clarke and Robert A. Heinlein, introducing it to a wider audience shortly thereafter. Also known as an orbital skyhook, geosynchronous orbital tether, or beanstalk, the space elevator became more feasible after the 1990s because of the development of carbon nanotubes and laser power beaming.

At the dawn of the twenty-first century, U.S. entrepreneurs joined scientists to promote building a space elevator. Scientist Bradley C. Edwards and businessman Michael Laine formed HighLift Systems, a company that received funds from the NASA Institute for Advanced Concepts to explore design and deployment of such a system. In April 2003 Laine founded LiftPort Incorporated, which soon expanded into a group of companies to accelerate commercial development of the space elevator. The LiftPort Group announced in February 2006 that it had tested a mile-long segment of space-elevator tether consisting of carbon-fiber composite strings wrapped with fiberglass tape. Meanwhile, advocates of the space elevator began holding annual conferences in 2004. Organized competitions arose, with monetary prizes from government and private sources, for demonstration of relevant technologies.

Although NASA officials predicted it would take another century to complete a working version of the space elevator, Japan projected one by 2030. Most experts agreed the estimated $7–10 billion cost represented a major roadblock for near-term construction of a space elevator. Technological hurdles also persisted, especially designing an elevator that could survive collisions with orbital debris. Despite widespread skepticism among professionals in space-related fields, true believers in the validity of the space-elevator concept anticipated less expensive access to outer space and even envisioned construction of space elevators for operations on the Moon and planets.

Rick W. Sturdevant

See also: Future Studies, Space Advocacy Groups

Bibliography

Bradley C. Edwards and Eric A. Westling, *The Space Elevator: A Revolutionary Earth-to-Space Transportation System* (2003).
Bill Fawcett et al., eds., *Liftport–The Space Elevator: Opening Space to Everyone* (2006).

Spacecraft Charging

Spacecraft charging is a phenomenon that occurs to spacecraft, whereby its components or the entire vehicle become electrically charged due to interaction with the surrounding space environment, possibly resulting in damage. The space environment contains high-energy particles, including negatively charged electrons and positively charged ions. All spacecraft, whether in Earth orbit or in interplanetary space, encounter these particles, and over time a charge relative to the surrounding space environment accumulates. Photoelectron current represents another cause of charging, whereby a spacecraft can accumulate a positive charge when solar radiation frees electrons from the surface. Accumulation of charge on a spacecraft can be absolute, when compared to the surrounding space, or differential, when compared to various sections of the spacecraft.

Whether differential or absolute, detrimental effects of charging can occur, including electrical discharge arcing and electromagnetic interference (EMI). Low-Earth-orbit satellites frequently accumulate a negative charge relative to the surrounding space, due to higher rates of low-energy particles. This negative charge can attract positively charged ions, which have sufficient mass, resulting in physical damage to the surface of the satellite, changing thermal properties, and creating a debris field in the vicinity of the satellite. Because of higher energy particles with energies up to 50,000 eV, occurrences of differential charge occur with high-Earth-orbit satellites, resulting in large EMI discharges between components, such as the chassis and internal electronics. A satellite in geosynchronous orbit may experience charge levels in excess of 10,000 V relative to the surrounding plasma. Discharging events can cause various failures and anomalies, possibly resulting in complete failure of satellites or missions.

Spacecraft designs often incorporate features to eliminate or mitigate charging effects. Knowledge of spacecraft charging and its effects were apparent and have been researched since the dawn of the space age. *Sputnik 3* measured electrical potential between the spacecraft and the surrounding environment. In 1973 the U.S. Air Force (USAF) *Defense Satellite Communications Satellite 2-01* (*DSCS 2-01*) experienced a destructive discharge event that rendered the communications system useless. The failure of the TRW-built *DSCS 2-01* less than three years after launch resulted in research and development into mitigating the risks of spacecraft charging. On-orbit anomalies with the *Meteosat 1* weather satellite prompted the European Space Agency (ESA) to investigate the impact of charging events in 1978. ESA concluded that the anomalies were charging events roughly correlating with higher levels of geomagnetic activity. Laboratory testing resulted in improved grounding techniques and arc-detection systems for *Meteosat 2*. In 1979, NASA and the USAF launched the Martin-built *Spacecraft Charging at High Altitude* (*SCATHA*) satellite into a highly elliptical orbit (185 × 43,000 km), with the goal of obtaining information about charging and its effects. *SCATHA* experienced and recorded several discharge events during its lifetime, returning significant information on charging and discharging events. Data collected from *SCATHA* resulted in better engineering and materials selection for spacecraft builders. As spacecraft became larger and incorporated more sophisticated digital control systems, mitigating the effects of spacecraft charging became important. NASA and the U.S. Office of Naval Research launched, in 1990 the *Combined Release and Radiation Effects Satellite*, which mapped Earth's radiation belts and studied the radiation environment in space and measured the impact the space environment has on microelectronic devices. Charging events also remained an issue with human spaceflight vehicles such as the *International Space Station*, requiring designers to implement a variety of techniques to minimize the impact of charging on these vehicles, to protect systems against catastrophic discharge events that might endanger human lives. For example, large solar arrays may build a differential charge relative to the surrounding plasma or the station's modules. If not mitigated, this charge could result in arcing, possibly destroying power junctions and subsequently reducing the station's ability to produce power.

Since the late 1970s, the international Spacecraft Charging Technology Conference has met to discuss the topic of spacecraft charging. Topics discussed at the conference included spacecraft materials, design techniques, and the physics of spacecraft charging.

David Hartzell

See also: Defense Satellite Communications System, *International Space Station*, Meteosat, Space Shuttle

Bibliography

Michael D. Griffin and James R. French, *Space Vehicle Design* (2004).

D. Hoge and D. Leverington, "Investigation of Electrostatic Discharge Phenomena on the Meteosat Spacecraft," *ESA Journal* 3 (1979): 101–13.

R. D. Leach and M. D. Alexander, *Failures and Anomalies Attributed to Spacecraft Charging* (1996).

Systems Engineering

Systems engineering is the set of organizational methods used by engineers to coordinate large-scale engineering projects. Developed primarily in the U.S. aviation, space, and computing industries from the mid-1940s through the 1960s, systems engineering then spread to other nations and engineering sectors. Systems engineering consists of early analyses of potential future systems; development of requirements and specifications; progressive design and development of the hardware, software, and operations concepts; integration of the components; testing and verification of the components and the integrated prototypes; and deployment of the design into manufacturing and operations.

Several industries and programs contributed elements to systems engineering. American Telephone and Telegraph (AT&T) Company began to conceive of its network of telephone exchanges, wires, and operations as a "system" by the mid-1930s. Bell Telephone Laboratories, the research arm of AT&T, developed formal specifications and drawings for new technologies that it gave to Western Electric, AT&T's manufacturing arm, to build. Because many engineers from Bell later worked on or led major aerospace projects, this formal means of communication came with them. Another source of ideas in the 1930s was the International Business Machines (IBM) Corporation of the punch card machine business, the precursor to the computer industry. IBM's tabulators, printers, and sorting machines came to be called "systems" because all had to be tailored together to meet the needs of particular organizations. During World War II, physicist Ivan Getting of the Massachusetts Institute of Technology (MIT) Radiation Laboratory led efforts to connect radar systems, which detected aircraft, with analog computers that pointed anti-aircraft guns. The complexity of this task led him to create the method of systems integration, an important element of systems engineering. The U.S. B-29 and German V-2 projects were also important examples of systems engineering, as both these weapons integrated diverse new technologies in ways never before accomplished.

However, these methods did not come to prominence until the mid-1950s, particularly on U.S. missile, submarine, command-and-control, and aircraft projects. The Jet Propulsion Laboratory (JPL) of the California Institute of Technology developed the method of the "design freeze" coupled with "change control" to coordinate design changes on its Sergeant ballistic missile project. JPL also created vibration tables and later thermal vacuum chambers to test missile and spacecraft components in the launch and space environments, while closely tracing and assessing the quality of individual components. By the early 1960s, JPL instituted systems engineering as a division within its organization as part of its reforms to institute a matrix structure to handle several projects, and each space project had its chief systems engineer. The U.S. Air Force (USAF) hired the Ramo-Wooldridge Corporation (later TRW) to create similar methods and test equipment on the Atlas, Thor, and Titan ballistic missile programs. Ramo-Wooldridge also developed new methods to document and trace specifications. When The Aerospace Corporation, under Ivan Getting, became the USAF's technical direction contractor for missiles and space in 1960, it inherited these methods from TRW. Lockheed, working on the U.S. Navy's Polaris nuclear submarine program, developed similar methods. Computers and large-scale software programs were another spur to formal methods to trace specifications to design, in the Semi-Automatic Ground Environment air defense project of the 1950s. MIT's Lincoln Laboratory ran these efforts, and then passed them on to the MITRE Corporation, which became the USAF's technical direction contractor for electronics and command-and-control systems. Finally, complex new aircraft such as the B-52 and F-104 spurred similar formal coordination efforts in its contractors and government managers. Low reliability in the early ballistic missile and command and control systems led to improvements in the estimation of reliability, the assessment of failure modes and effects, and in tracking engineering changes into the manufacturing process. The military's systems engineering methods crystallized into formal procedures by the early 1960s.

Engineering theorists began to catch up with these new practices with the writing of new textbooks, the first of which was H. Goode and R. Machol's *Systems Engineering*, published in 1957. Machol wrote a later text in 1965 that became the aerospace industry standard for many years. Universities such as MIT and Stanford began to teach systems engineering courses in the 1950s and 1960s. However, not until 1990 did systems engineers create their own professional society, the International Council on Systems Engineering.

When NASA was created in 1958, its engineers, who were primarily researchers from National Advisory Committee for Aeronautics laboratories, were mostly unfamiliar with systems engineering or the management and coordination of large-scale programs. NASA's acquisition of JPL in 1959 brought JPL's systems engineering methods, and its acquisition of Naval Research Laboratory and Army Ballistic Missile Agency personnel brought in more large-scale program management skills. In the 1960s on large programs such as Mercury, Gemini, and Apollo in the human flight program, and various robotic Earth-orbiting satellites and deep space probes, NASA personnel further refined and developed systems engineering techniques into standard procedures.

European nations and space organizations, which were trying to collaborate with one another on launcher and satellite programs, had a difficult time learning systems engineering due largely to the political, cultural, and communication problems of cooperating across many languages and nations. The European Space Research Organisation was much more successful in its efforts than its launcher counterpart, the European Space Vehicle Launcher Development Organisation. With assistance from the United States on NASA-European programs such as *Spacelab*, Europeans had mastered systems engineering by the mid-to-late 1970s.

Soviet engineers developed their own methods of coordinating large space projects in the 1950s and 1960s, paralleling efforts in the United States. While the history of Soviet systems engineering methods has not yet been researched, interactions between

U.S., European, and Russian engineers in the 1990s and 2000s have revealed that Russian systems engineering methods are more personalized and less formal than those used in the United States and Europe.

Systems engineering remains the standard engineering coordination process for space projects in the early twenty-first century.

Stephen B. Johnson

See also: The Aerospace Corporation, Jet Propulsion Laboratory, MITRE Corporation, Systems Management, TRW Corporation

Bibliography

Stephen B. Johnson, *The Secret of Apollo: Systems Management in American and European Space Programs* (2002).

Howard E. McCurdy, *Inside NASA: High Technology and Organizational Change in the U.S. Space Program* (1993).

Systems Management

Systems management is the set of organizational methods used to manage large-scale space development projects. The essential elements of systems management are project management, systems analysis, systems engineering, configuration management, and phased planning. These techniques originated in the 1940–50s on ballistic missile, command-and-control, and space programs, primarily in the U.S. armed forces and their contractors.

Project management, which structures an organization around the functions of the end product, came into the aviation industry through complex World War II projects, such as the B-29 bomber, P-61 night fighter, the Manhattan project, and the German V-2 program. In each case the military organizations realized that they had to break up disciplinary and hierarchical groupings to facilitate problem solving and communication among diverse teams of experts. By the early 1950s the Jet Propulsion Laboratory (JPL) of the California Institute of Technology, funded by U.S. Army Ordnance and the U.S. Air Force (USAF) and its contractors, had implemented project organizations for its largest technical projects: ballistic missiles and command-and-control systems. As corporations came to manage several projects by the late 1950s and early 1960s, several implemented a matrix structure to handle the movement of personnel between projects and other cross-project disciplinary issues. Personnel reported to their project managers on their project assignments and to functional managers for technical standards and for their assignment to various projects. Project management methods came to include scheduling tools such as the Program Evaluation and Review Technique.

Systems analysis, originally created by analysts at the RAND Corporation in the late 1940s, was a derivative of operations research methods developed in World War II. During the war, operations researchers used mathematical and physical theories to reorganize military operations, such as determining the optimal geometry for a flight of bombers to defend against German fighters, or the best aircraft search patterns to detect German submarines. RAND analysts used these methods to investigate future conflicts and systems, such as technologies and strategies to fight nuclear war. Colonel Bernard Schriever of the USAF Development Planning Office imported these methods into USAF planning in 1951–52.

Systems engineering stemmed from government-industry interactions among the Massachusetts Institute of Technology, Bell Telephone Laboratories, and the U.S. Army during this period, and it was instituted in the USAF through the hiring of the Ramo-Wooldridge Corporation in 1953, and later the creation of the Aerospace Corporation in 1960. Initially systems engineering involved coordinating engineering and manufacturing design groups, controlling the hardware connections (interfaces) between components, technical oversight of contractors, and managing design specifications.

These ideas came together between 1953 and 1962 for the Atlas intercontinental ballistic missile (ICBM) program and later throughout Air Research and Development Command, led in the late 1950s by Schriever. Schriever circumvented the USAF's usual development processes, and his team combined project management, systems analysis, and systems engineering into a new set of methods better suited to large, complex systems with high reliability requirements. These resulted in the 375 series of USAF regulations for systems management, published between 1959 and 1961. In 1961 these procedures became the basis for the new Air Force Systems Command, which had development responsibility for large USAF systems. Logistical considerations were separated into a new USAF Logistics Command. During the same time period, JPL was creating a parallel set of processes on its Sergeant ballistic missile program, the U.S. Navy was creating another similar set on its Polaris submarine program, and Lincoln Laboratory of the Massachusetts Institute of Technology was creating its own version on its Semi-Automatic Ground Environment (SAGE) computer-based air defense project.

Systems management evolved further in the 1960s in response to high ballistic missile and space system failure rates. A key development was the implementation of configuration management, which tied systems management to a system for estimating costs and schedules by requiring engineers to give cost and schedule estimates with every design change. Originally developed by the Boeing Company in the late 1950s for aircraft manufacturing, this process came to the USAF through Boeing's involvement as the integrating contractor on the Minuteman ICBM project. Another important change was the implementation of phased planning, which divided the development cycle into several phases, each of which a manager had to approve before the program continued. Secretary of Defense Robert McNamara implemented phased planning in the Department of Defense (DoD) in 1961–62. McNamara made the USAF's version of systems management the DoD standard in 1965. Finally, systems management also came to incorporate various quality process measures in the 1950s–60s, such as environmental testing, parts tracking and inspection, and software quality.

Since the mid-1960s systems management has become the standard for space programs throughout the Western world. NASA adopted systems management in the 1960s, primarily through implementation of the Apollo program and through JPL's robotic probe programs. Systems management came to the European space programs through the European Space Research Organisation's (ESRO) early satellite programs such as *ESRO 1* and the *Highly Eccentric Orbiting Satellite*, often from U.S. contractors working as consultants for European companies, and NASA working with ESRO personnel. On the Spacelab project, part of NASA's Space Shuttle program, NASA imposed further systems management methods on ESA, which became ESA standards. Many space organizations came to use it through interactions with the U.S. space agency. The Space Shuttle program, along with the later Space Station Freedom and International Space Station (ISS) programs, decentralized systems management by dividing top-level responsibilities among different NASA field centers. Many experts, particularly after the *Challenger* accident of 1986, considered this a major problem with NASA's post-Apollo management system.

Unhappy with continuing program cost and schedule issues, from the late 1980s to the mid-1990s, NASA, the DoD, and their contractors experimented with a few managerial theories and ideas. Most prominent among them were Total Quality Management (TQM) and reengineering. Implementation of TQM improved long-term logistics operations in the DoD, but elsewhere was judged to be a failure. Reengineering was applied at JPL in the 1990s, with some success, but with relatively little long-term impact. After the *Challenger* accident dramatically reduced the number and frequency of flight opportunities, Goddard Space Flight Center developed the Small Explorer mission concept to provide more frequent flight opportunities for instruments using small, low-cost, and rapidly developed spacecraft launched on small rockets. The DoD's Strategic Defense Initiative Office also began low-cost, rapidly scheduled space projects so as to perform missile defense technology experiments.

By the early 1990s NASA Administrator Dan Goldin, who was under pressure to reduce costs in its robotic space programs so as to fund the ISS program and keep the Space Shuttle operating, began to implement so-called "faster, better, cheaper" approaches, which scaled back systems management techniques in favor of relatively informal but tight-knit teams. This modification to systems management, which derived in part from small-scale DoD programs, had some success, with several highly publicized successes such as Mars Pathfinder, but spacecraft failure rates increased significantly in the late 1990s. After the much-publicized failures of the *Mars Climate Orbiter* and *Mars Polar Lander* in 1999, the managerial pendulum swung back toward classical systems management. In the human flight programs, NASA modestly recentralized management of the International Space Station program, when in 1993 NASA hired Boeing to be the single systems integration contractor.

Within the USAF, continued cost pressures led in the early 1970s to the implementation of the Initial Operational Test and Evaluation phase, further testing prior to production of major aircraft and satellite systems. Deputy Secretary of Defense David Packard, cofounder of Hewlett-Packard Corporation, led these and later USAF initiatives to improve USAF acquisition processes. In the mid-1980s Packard led a commission that called for further changes, essentially creating corporation-like organizational structures with strong project managers for large programs. During the 1960s–80s Air Force Systems Command (AFSC) was the major organizational group involved with these structural modifications. In 1992, with the deactivation of AFSC and the creation of Air Force Materiel Command, the reintegration of development with logistics considerations was complete. Throughout these changes within NASA and the USAF, which aimed to run programs in a more business-like manner, most of the underlying systems management processes have remained quite stable.

Stephen B. Johnson

See also: Jet Propulsion Laboratory, National Aeronautics and Space Administration, Systems Engineering, United States Air Force

Bibliography

Lawrence R. Benson, *Acquisition Management in the United States Air Force and Its Predecessors* (1997).

Stephen B. Johnson, *The Secret of Apollo: Systems Management in American and European Space Programs* (2002).

Howard E. McCurdy, *Faster, Better, Cheaper: Low-Cost Innovation in the U.S. Space Program* (2001).

Peter J. Westwick, *Into the Black: JPL and the American Space Program, 1976–2004* (2007).

Telerobotics

Telerobotics is the capability of controlling robots from a remote location using visual "telepresence" to aid operators in assessing the external environment in which the robot is functioning. Controlling telerobots in real time is through teleoperation, often through hand controllers similar in function to video game controls. The feasibility of teleoperation is determined by the distance, and hence the communications time delay, between the operator and the robot. For greater distances, such as between Earth and Mars, teleoperation is not practical. Telerobotics evolved in the late twentieth century and was used in space because it complemented the capabilities of humans. Robots performed tasks too dangerous, difficult, or expensive for people to undertake. They also directly assisted astronauts in space operations and relieved them of routine or repetitive tasks so that they could pursue more productive activities.

The earliest teleoperated space robots were manipulators. The Surveyor lunar landers of the 1960s were equipped with robotic arms, operated from the Jet Propulsion Laboratory (JPL), to scoop samples of the lunar soil for analysis. Similar manipulators were on the two Viking Landers that touched down on Mars in 1976, although light-time delays required that manipulator arm commands be preprogrammed and sent to the landers. Among the best-known robotic manipulator systems is the Shuttle Remote Manipulator System, or Canadarm robotic arm, on the U.S. Space Shuttle. Since its maiden voyage aboard the Shuttle *Columbia* in 1981, the Canadarm has been used to deploy, handle, and retrieve payloads in orbit and to assist spacewalking astronauts. In 2001 the next-generation Canadarm2 was delivered to the *International Space Station (ISS)*. The Canadarm2 was a seven-jointed manipulator that was capable of handling large payloads for station assembly and maintenance. Both of these manipulator systems were operated in real time by astronauts aboard the Shuttle or the *ISS*.

The Canadarm2 aboard the International Space Station has multiple joints and is capable of maneuvering payloads as massive as 116,000 kilograms, equivalent to a fully loaded bus. (Courtesy NASA/Goddard Space Flight Center)

Robotic rovers have been successfully employed for the surface exploration of the Moon and Mars. In the early 1970s the Soviet Union landed two robotic Lunokhod rovers on the Moon. Five-person teams of controllers on Earth operated the Lunokhods under remote control, using images from the rovers' television cameras. These vehicles were the first roving teleoperated robots to land on another world.

Mars rovers, required more autonomy, with JPL's human operators specifying target destinations based on visual cues (telepresence) from onboard cameras, and the onboard control system self-navigating a safe route to the targets. The *Sojourner* rover, deployed in 1997 as part of the Mars Pathfinder spacecraft, and *Spirit* and *Opportunity* rovers, part of the Mars Exploration Rover (MER) missions, landed in 2003. The MER rovers were equipped with small manipulators used to deploy scientific instruments against rocks.

Several advanced manipulator systems and planetary rovers were under development in the first decade of the twenty-first century, including Dextre, a two-armed robot capable of performing routine maintenance tasks on the *ISS* exterior, and a manipulator arm to be flown on the Orbital Express mission to demonstrate the feasibility of robotically repairing satellites in Earth orbit. NASA and the European Space Agency planned various Mars rovers with greater autonomy and capabilities. Microrovers have been proposed for the future surface exploration of Mars, Mercury, and near-Earth asteroids.

Eric Choi

See also: Deep Space Probes, *International Space Station*, Rovers, Space Shuttle

Bibliography

Savan Becker, "Astro Projection: Virtual Reality, Telepresence, and the Evolving Human Space Experience," *Quest* 12, no. 3 (2005): 34–55.

———, "The Rise of the Machines: Telerobotic Operations in the U.S. Space Program," *Quest* 11, no. 4 (2004): 14–39.

Thermal Control

Thermal control and protection of space vehicles relies on a wide variety of technical disciplines, including thermomechanics, thermodynamics, mechanical and electrical engineering, knowledge of the space environment, and materials engineering. Ultimately spacecraft thermal control depends on the materials used to construct the vehicle, how well they perform, and how resistant they are to the space environment. These include the extremely severe temperature swings, between being in the Sun's light or in shadow, and the lack of air, which means that convective air currents that exist on Earth cannot be used. Instead thermal conduction and insulation and thermal radiation generally become the primary means to control temperature in space. Getting to and returning from space are equally difficult, with very high heating due to extremely high speeds in the atmosphere.

In the 1950s both the United States and the Soviet Union recognized that the severe heating of intercontinental ballistic missile (ICBM) reentry systems was a major hurdle to the success of these weapons. While many engineers' intuition was that a strongly pointed surface would best dissipate heat, H. Julian Allen at the National Advisory Committee for Aeronautics Ames Research Center showed that in fact a blunt surface worked much better, which was proven by testing reentry vehicle configurations. The blunt configuration combined with ablative materials that melted during reentry, and thus removed the heated materials, provided the means by which ICBM and spacecraft (human and robotic) reentry vehicles successfully returned to Earth. These led to the development of new materials, such as the Space Shuttle's thermal tiles.

Engineers of the 1950s also considered the problem of thermal control of spacecraft in the vacuum of space. A. R. Hibbs of the California Institute of Technology Jet Propulsion Laboratory (JPL) published one of the first treatises on the subject in March 1956. The JPL *Explorer 1* spacecraft, the first satellite successfully launched by the United States, dealt with thermal effects by spinning the cylindrical spacecraft and also by painting white stripes on it to increase solar reflectivity, thus reducing the average temperature. Soviet engineers were equally concerned about overheating and rigorously polished *Sputnik*'s spherical surface to reflect the Sun's rays.

In 1961 JPL engineers asked the Illinois Institute of Technology Research Institute (IITRI) if it could develop an ultraviolet (UV), stable, white, thermal control coating that could be applied to the exteriors of the upcoming Gemini spacecraft. IITRI engineers tested many dozens of various pigments and binders and their combinations. Eventually they found that formulation Z-93 (zinc oxide pigment in a potassium silicate binder) was among the best inorganic performers. The Z-93 formulation was a white paint that was used on the Gemini space capsules, and in the early twenty-first century the Z-93P variant was still considered state of the art. Additionally organic binders were also investigated and engineers found that the formulations of S-13 and S-33 (zinc oxide pigment in a methyl silicone binder) were the best performers. Eventually the S-33 formulation lost out to the S-13 formulation, which was still used in the twenty-first century in its evolutionary form of S-13GP:6N/LO-1.

In 1965 Robert M. Van Vliet of the Air Force Materials Laboratory (AFML; over the years this organization experienced several name changes; in 2009 the name was the Air Force Research Laboratory (AFRL) Materials and Manufacturing Directorate) authored a book, *Passive Temperature Control in the Space Environment*, the first general text for the discipline of spacecraft thermal design. AFML was a critical site for the development of materials that would have a low solar absorptance and a high thermal emittance, while stable in the space environment, and engineers were experimenting by exposing various materials to UV radiation in a vacuum chamber. A dozen or so of these simulation chambers were developed and utilized across the United States.

As the Gemini and then the Apollo programs wound down, thermal control researchers began to desire spaceflight experiments to measure the performance of candidate materials on orbit by monitoring the equilibrium temperatures of specimens mounted in calorimeters, which thermally isolated them from the rest of the spacecraft. The solar absorptance could be calculated from the temperature data, because the material's thermal emittance does not change as a result of being exposed to the space environment. Thus the stability of the thermal control material can be monitored on orbit and be used to correlate ground-based testing results to the actual space environment.

In 1966 AFML began a series of three spaceflight experiments with the launching of the *OV1-4* (orbital vehicle), *OV1-10*, and *OV1-17* experiments. The specimens were suspended on cords and thermistors were used to monitor their temperatures, which was telemetered to Earth. These experiments found that even though the white coatings degraded more than expected, the ground-based combined effects chambers were providing reasonable simulations. The last experiment, *OV1-17*, was launched in March 1969 and cost AFML $10,000.

In October 1972 another AFML spaceflight experiment, *ML-101*, was launched. Designed to be in orbit for one year, it provided data for 28 months in orbit, and additional data was obtained after five years in orbit. In this experiment, as in all previous ones, the early changes in solar absorptance were considerably larger than expected, and it was hypothesized that the vehicle was contaminating the thermal control surfaces.

The first real opportunity for the return of materials to Earth from an actual spaceflight experiment came about when NASA launched *Skylab*. The AFML conducted the *Skylab* DO24 experiment on thermal control coatings (TCC) and polymeric films, which was designed to assess the performance and properties of a wide variety of TCCs and polymeric films. This experiment consisted of mounting two identical sample trays outside *Skylab* (SL) during the SL-1 mission and retrieving one of the two trays during the SL-2 mission and the second tray during the SL-3 mission. These returned, sealed specimen trays were then to be remotely opened inside a vacuum chamber on Earth, and its infrared reflectance spectra would be measured in the vacuum. This ground-based system was known as the In situ Device Infrared Optical Test (IDIOT). Unfortunately the seals did not work on some of the canisters, and on others the astronauts inserted the trays backward. The IDIOT's sample manipulation system could not remove the specimens, and they were exposed to Earth's atmosphere. Engineers discovered that the specimens were covered with contaminants; it was not possible to discern whether the increase in solar absorptance was due to the space environment or the contamination. A third specimen tray was therefore deployed outside *Skylab* in the beginning of the SL-4 mission and later returned to Earth.

The next spaceflight experiment opportunities were the U.S. Air Force (USAF)/ NASA/Navy Navigation Technology Satellite-2 TCC spaceflight experiment, launched in June 1977, which investigated new antistatic and quartz-fiber fabric materials and the TCC materials enhanced calorimeter on a USAF Defense Support Program satellite. This experiment, which lasted more than four years, was the first opportunity to directly monitor the performance of thermal control materials in geosynchronous orbit.

In January 1979 the *Spacecraft Charging at High Altitudes (SCATHA)* P78-2 satellite, managed and funded by the USAF Space Test Program, was launched. This program was designed to measure the environment that causes spacecraft charging and its effects on the spacecraft. This satellite contained about a dozen experiments, one of which was the ML12 Spacecraft Contamination Plus Thermal Control Materials Monitoring experiment. The contamination was monitored by a temperature-controlled quartz crystal microbalance and retarding potential analyzer, while the thermal control experiment consisted of a calorimeter. The combination of the calorimeter and the contamination monitors provided the ability to correlate the materials change in solar absorptance to either contamination or the space environment. The Aerospace Corporation convinced the program managers to keep the satellite and the ML12 experiment from being deactivated until after 11½ years of service, when the satellite experienced severe transmitter problems. The satellite was deactivated in December 1990.

The *Long Duration Exposure Facility (LDEF)* satellite, managed by NASA Langley Research Center, was launched in 1984 on board the Space Shuttle *Challenger*. It was a totally passive satellite that was a cylinder about the size of a school bus. It was placed in a three-axis stabilized orbit with one end of the cylinder facing Earth and the other end pointing to deep space. This satellite was covered with 86 experiment trays and had a tremendous variety of materials mounted on its exterior. The experiments involved the participation of more than 200 principal investigators from 33 private companies, 21 universities, seven NASA centers, nine Department of Defense laboratories, and eight foreign countries. It was designed to be in orbit for one year, but the *Challenger* disaster and other events caused the retrieval to be delayed for five and one-half years. It was rescued from a rapidly decaying orbit in 1990 by a Space Shuttle. Included on *LDEF* were a wide variety of TCC materials. *LDEF* once again showed that contamination is a large problem for any spacecraft, and it also showed the extreme damage that atomic oxygen can cause to some materials. However, for the white paint TCC, atomic oxygen actually improved performance. One result of the *LDEF* program was the development of the NASA Space Environmental Effects (SEE) program, which funded further research in areas of charging, ionizing radiation, materials and processes, meteoroid and debris, and contamination.

After these successes, the NASA SEE program and the materials community sought other spaceflight opportunities. They utilized the Shuttle as a platform for various short-term spaceflight experiments. Once NASA started visiting the Russian *Mir* space station, the SEE program sought an opportunity to fly an experiment on board *Mir*. One was the Space Portable Spectro-Reflectometer, which measured the solar absorptance of a material with a portable space qualified spectroreflectometer.

The program developed the Mir Environmental Exposure Program (MEEP), which provided suitcase-like experiments that held materials specimens and opened like a butterfly, which was then mounted on the outside of spacecraft. Two experiments were flown in the MEEP program, Passive Optical Sample Assemblies (POSA) 1 and POSA 2. These experiments contained numerous thermal control materials. All the *Mir* experiments documented the extreme contamination experienced on this spacecraft. In fact there was evidence on one of the POSA experiments of splattered liquid contamination that looked as if it came from a waste dump from the Shuttle, although NASA denied the possibility.

The *International Space Station (ISS)* provided further opportunities for TCC material development. Once again the AFRL Materials and Manufacturing Directorate, NASA Langley Research Center, NASA Marshall Space Flight Center, the Naval Research Laboratory, The Boeing Company, and various other commercial partners participated in a program to reuse the Passive Experiment Container suitcases from the MEEP program to perform their experiments. This program was named the Materials International Space Station Experiment (MISSE) and grew into a series of seven and possibly eight experiments. MISSE 1 and MISSE 2 were installed on the exterior of the *ISS* in August 2001. MISSE 1 and MISSE 2 each had more than 400 candidate spacecraft materials and were scheduled to be on orbit for two years. However, the *Columbia* accident delayed retrieval until July 2005. MISSE 1 and 2 showed over 100 micrometeoroid impacts with a large number of the polymeric specimens completely eroded by atomic oxygen, but some samples survived. Particulate contamination was observed as well as optical property changes in thermal control materials. However, the extent of contamination that plagued *Skylab* and *Mir* was greatly reduced, helping the performance of the thermal control materials.

Spaceflight experiments were complemented by many ground-based simulation efforts and materials development programs. These programs have looked at anodized processes for aluminum, rare earth oxides for more stable pigments, transparent conductive coatings for charge dissipation, glass fabrics, carbon-carbon composite substrates, plasma sprayed coatings, thin-film dielectric stacks for reflection, material systems with variable emittance schemes, adhesively backed metalized polymeric films, and pigment-less coatings that utilize hollow silica spheres as the pigment.

Clifford A. Cerbus, William L. Lehn, and Stephen B. Johnson

See also: *International Space Station, Mir, Skylab*

Bibliography
T. A. Heppenheimer, *Facing the Heat Barrier: A History of Hypersonics* (2007).
R. D. Karam, *Satellite Thermal Control for Systems Engineers* (1998).
NASA JPL Mission and Spacecraft Library. http://www.jpl.nasa.gov/missions/index.cfm.
NASA Langley (LaRC) LDEF Archive System. http://setas-www.larc.nasa.gov/LDEF/.
NASA LARC MISSE. http://misse1.larc.nasa.gov/.

TECHNOLOGY INFRASTRUCTURE AND INSTITUTIONS

Technology infrastructure and institutions were established in many nations to support space endeavors. The most obvious infrastructure elements for spaceflight are launch facilities, which require large, expensive launch pads and transportation facilities, and large land requirements to protect against catastrophic explosions. These facilities also generally include large buildings to integrate launch vehicles and their payloads. As more nations build and operation launch vehicles, launch sites have proliferated to match. Almost as well known are mission operations facilities, such as the Mission Control Center at NASA Johnson Space Center, or Russia's TsUP (Tsentr Upravleniye Polyotami) in Kaliningrad. With the maturation and spread of sophisticated computing capabilities, mission control centers have moved from centralized facilities to many smaller centers for a variety of spacecraft, often at corporate or university facilities.

Less obvious are organizations and facilities to develop space technologies. Rocket testing organizations and facilities are much like launch facilities, requiring remote locations and large facilities to preclude damage to other facilities and people. Other research organizations often house expensive facilities, such as vacuum chambers, wind tunnels, vibration test facilities, anechoic chambers (to test electromagnetic compatibility), clean rooms, and other specialized research and test facilities and hardware. Space organizations use these various resources to develop new technologies and to test launch vehicles and spacecraft. While many of these facilities are at government facilities run by government organizations, many private corporations have developed their own space technology research and development infrastructure.

Stephen B. Johnson

See also: Subsystems and Techniques

Bibliography
Roger E. Bilstein, *Orders of Magnitude: A History of NACA and NASA, 1915–1990* (1989).
Brian Harvey, *Russia in Space: The Failed Frontier?* (2001).

Ames Research Center

The Ames Research Center has led NASA's efforts in supercomputing and information technology, astrobiology and the space life sciences, and nanotechnology and materials science. In addition NASA Ames hosted the world's greatest collection of wind tunnels and flight simulation facilities, a legacy of its origins in the National Advisory Committee for Aeronautics (NACA).

NACA broke ground for its second laboratory on 20 December 1939, as the United States was preparing for World War II. The laboratory was located at a U.S. Navy airfield in Sunnyvale, California (later named Moffett Federal Airfield), and was named Ames Aeronautical Laboratory in tribute to Joseph Sweetman Ames, founding chairman of the NACA and an architect of U.S. aeronautical research. Smith DeFrance, director from its founding to 1965, attracted some the brightest aeronautical engineers to Ames, encouraged them to build test facilities to prove their ideas, and gave them freedom to pursue useful work.

For example, Lew Rodert directed an urgent program on icing research that won the Collier Trophy. R. T. Jones developed his theories of wing sweep and the transonic area rule. Harry Goett (later director of Goddard Space Flight Center) led a research effort in subsonic and supersonic aerodynamics that solved problems in jet flight in the 1950s. H. Julian Allen proposed his concept of a blunt body shape for reentry vehicles, and then devised a series of hypersonic and high-heat wind tunnels to create technologies so that astronauts could return safely to Earth.

When Ames became part of NASA in 1958, it continued fundamental research in new sciences and component technologies. Building on its expertise in human factors and pilot workload research, Ames became NASA's lead center in basic life sciences research, which included radiation biology, adaptability to microgravity, and exobiology. Some Ames aerodynamicists explored the complex airflows around rotorcraft and devised the first tiltrotor aircraft, while others modeled airflows using supercomputers, creating computational fluid dynamics. Ames engineers and planetary scientists managed a series of airborne science aircraft, of planetary atmosphere probes, and robotic explorers like the Pioneers and Lunar Prospector.

Into the 1990s Ames created the NASA Research Park as a place to draw university and corporate partners from nearby Silicon Valley into space exploration. Ames leaders continued to explore new ways to develop new technological capabilities—in astrobiology, robotics, communications, instrumentation, and small satellites—and apply them to NASA's evolving missions.

Glenn Bugos

See also: *Lunar Prospector, Pioneer 10* and *11*

Bibliography
Glenn E. Bugos, *Atmosphere of Freedom: Sixty Years at the NASA Ames Research Center* (2000).
Edwin P. Hartman, *Adventures in Research: A History of Ames Research Center, 1940–1965* (1970).
NASA Ames History Office. http://www.nasa.gov/centers/ames/about/history/index.html.

Deep Space Network

Deep Space Network (DSN), an international network of antennas operated by the Jet Propulsion Laboratory (JPL) for NASA, supported interplanetary spacecraft missions and radio and radar astronomy observations for the exploration of the solar system and the universe. In the early twenty-first century, it consisted of three deep-space communications facilities placed approximately 120 degrees apart around the world at Goldstone, in California's Mojave Desert; Robledo de Chavela near Madrid, Spain; and near Canberra, Australia, which permitted constant observation of spacecraft as Earth rotated. DSN was the world's largest and most sensitive scientific telecommunications system. All stations were remotely operated from a centralized signal-processing center at each complex. The centers housed the electronics that pointed and controlled the antennas, received and processed the telemetry data, transmitted commands, and generated the spacecraft navigation data.

DSN provided a vital two-way communications link that guided and controlled automated spacecraft exploring the planets and brought back images and new scientific information they collected. All DSN antennas were steerable, high-gain, parabolic reflector antennas. The antennas and data delivery systems made it possible to acquire telemetry data from spacecraft, transmit commands to spacecraft, track spacecraft position and velocity, perform very-long-baseline interferometry observations, measure variations in radio waves for radio science experiments, gather science data, and monitor and control the network's performance.

The forerunner of DSN was established in January 1958 when JPL, then under contract to the U.S. Army, deployed portable radio tracking stations in Nigeria, Singapore, and California to receive telemetry and plot the orbit of the Army-launched *Explorer 1*, the first successful U.S. satellite. On 3 December 1958 JPL was transferred from the Army to NASA and given responsibility for the design and execution of lunar and planetary exploration programs using automatically operated spacecraft. Shortly afterward NASA established DSN as a separately managed and operated communications facility to accommodate all deep space missions, avoiding the need for each program to acquire and operate its own specialized communications network. DSN was given responsibility for its own research, development, and operation in support of its users, becoming a world leader in the development of low-noise receivers, tracking, telemetry and command systems, digital signal processing, and deep space navigation.

DSN's facilities continued to evolve and, by the start of the twenty-first century, each complex consisted of at least four deep space stations equipped with ultrasensitive receiving systems and large parabolic dish antennas. They included one 34 m diameter-high efficiency antenna; one or more 34 m beam waveguide antennas (three at the Goldstone complex, two at the Robledo de Chavela complex, and one at the Canberra complex); one 26 m antenna; and one 70 m antenna. Five of the 34 m beam waveguide antennas were added to the system in the late 1990s: three were located at Goldstone and one each at Canberra and Madrid. A second 34 m beam waveguide antenna (DSN's sixth) was completed at the Madrid complex in 2004. Also in the 1990s DSN incorporated the ability to array several antennas to improve the data returned from the *Galileo* spacecraft, whose high gain antenna had failed to deploy properly and forced significant improvements to DSN to acquire enough data to salvage the mission. The array electronically linked the 70 m antenna at the DSN complex in Goldstone, California, with an identical antenna located in Australia, in addition to two 34 m antennas at the Canberra complex. The California and Australia sites were used concurrently to pick up communications with the Saturn probe *Cassini*.

David C. Arnold

See also: Command and Control, Jet Propulsion Laboratory

Bibliography
David C. Arnold, *Spying from Space: Constructing America's Satellite Command and Control Systems* (2005).
Douglas J. Mudgway, *Uplink-Downlink: A History of the Deep Space Network, 1957–1997* (2001).

Glenn Research Center

The Glenn Research Center in Cleveland, Ohio, a NASA center that initially specialized in aerospace propulsion research. Established as the Aircraft Engine Research Laboratory in 1941 by the National Advisory Committee for Aeronautics (NACA), the center was central in development of turbojet engines in the 1950s and rocket engines beginning in the 1960s. Its propulsion research continued in aeronautics with high-speed, hypersonic, and efficient engines, and in space with ion, nuclear, and electric propulsion.

As early as 1945 the center (renamed the Flight Propulsion Research Laboratory in 1947 and Lewis Flight Propulsion Laboratory in 1948 in honor of George Lewis, NACA director of aeronautical research from 1924 to 1947) had a small group of rocket researchers who concentrated on alternative, high-energy propellants such as

Glenn Research Center's Spacecraft Propulsion Research Facility is the only facility in the world capable of testing full-scale upper-stage space vehicles in a simulated space environment. The large vacuum test chamber extended 176 feet below ground. (Courtesy NASA/Glenn Research Center)

liquid hydrogen. Conceived in 1952, the Rocket Engine Test Facility tested these propellants and rocket engines such as the RL10, F-1, and J-2. The center was also a leader in electric, ion, and nuclear propulsion. Plum Brook Station was acquired in 1956 to construct NASA's only nuclear test reactor, later the world's largest vacuum chamber and the only facility capable of testing full-scale upper-stage launch vehicles.

After becoming part of NASA in 1958, the center, renamed Lewis Research Center, doubled its staff and focused on space programs. Early space contributions included the prototype Mercury capsule, astronaut training in the spinning gimbal rig, microgravity research, and the Centaur upper stage, the first liquid-hydrogen fuel rocket. The center managed Centaur missions, including Surveyor, Mariner, Viking, Pioneer, and Voyager, until 1997. The Zero Gravity Research Facility, the nation's largest drop tower, began conducting spaceflight component and basic physics studies in a weightless environment in 1966.

Having little initial involvement with the Space Shuttle, the center returned to aeronautics and instituted new Earth energy studies after the Apollo program ended in 1972. Plum Brook closed and budgets and staff levels declined throughout the decade.

The center reinvented itself in the early 1980s, pursuing major programs such as the Space Station Freedom Electrical Power System, the *Advanced Communications Technology Satellite*, and the Advanced Turboprop. Plum Brook reopened and several major facilities were renovated. The center designed solar arrays for *Mir* and the *International Space Station* and served as the nuclear thermal propulsion lead for the 1989 Space Exploration Initiative. In 1992 the world's largest slush hydrogen production plant was built at Plum Brook for the National Aero-Space Plane program. In 1999 NASA renamed the center the John H. Glenn Research Center at Lewis Field.

In addition to aerospace propulsion research, NASA Glenn branched out into space power, microgravity studies, communications satellites, aircraft icing, photovoltaics, and testing large space-bound hardware, such as the Mars Pathfinder Landing System.

Robert S. Arrighi

See also: Centaur, Electrical Power, Nuclear Power and Propulsion

Bibliography
Mark D. Bowles and Robert S. Arrighi, *NASA's Nuclear Frontier: The Plum Brook Reactor Facility* (2004).
Virginia P. Dawson, *Engines and Innovation: Lewis Laboratory and American Propulsion Technology* (1991).
———, *Ideas into Hardware: A History of the Rocket Engine Test Facility at the NASA Glenn Research Center* (2004).
Glenn Research Center History Office web page.

Khrunichev Center

Khrunichev Center, the State Research and Production Space Center (GKNPC Khrunichev), one of Russia's major aerospace organizations, responsible for the development and manufacture of space launch vehicles, space station modules, satellites and

VLADIMIR NIKOLAEVICH CHELOMEY (1914–1984)

A Soviet aerospace engineer and member of the Soviet Academy of Science, Vladimir Chelomey designed the first Soviet pulsejet engine (1942) and developed a canister-launched cruise missile with folding wings (1955). From 1955 to 1984, Chelomey headed the OKB-52 (Experimental Design Bureau) research-and-development organization. The cruise and ballistic missiles of OKB-52 became the prime weapons of the Soviet Navy and strategic forces in the 1960s through the 1980s. Chelomey started a program of maneuverable satellites for orbital interception and navy reconnaissance. He supervised development of the Proton launch vehicle, Almaz space station, and TKS (Transport Supply Ship) piloted spacecraft and proposed detailed projects, such as the UR-700 super-heavy booster, alternative human lunar expeditions, and reusable spaceplanes.

Peter A. Gorin

other spacecraft, and subsystems and components for them. Named for Mikhail Vasilyevich Khrunichev (1901–61), minister of aviation and later deputy chair of the Council of Ministers of the Soviet Union, the center started as an automobile assembly plant in 1916. In 1923 it began to produce aircraft under license from the German manufacturer Junkers. In 1924–25 it commenced production of newly designed Soviet aircraft alongside the Junkers models. In 1927 Junkers aircraft production was shut down and the contract with the German manufacturer terminated. The plant continued with production of Soviet airplanes, producing the TB-2, TB-3, PE-2, and IL-4 World War II bombers.

In 1951 the plant became OKB-23 (Experimental Design Bureau) under V. M. Myasishchev, designing and manufacturing the jet-powered M-4 strategic bomber (Bison in the West), the unsuccessful Buran cruise missile, and M-50 (Bounder) supersonic bomber. The failures prompted the transfer of the organization in 1960, becoming Branch 1 of Vladimir Chelomey's OKB-52. Chelomey quickly used the facilities for design and manufacturing of his universal rocket (UR) 100, 200, and 500 intercontinental ballistic missiles (ICBM). The UR-500 was converted to the Proton space launch vehicle, first launched in 1965. Chelomey also assigned the Almaz military space station work to the organization, which led to involvement with both the military and civilian Salyut space stations, and later to work on *Mir* and the *International Space Station*. The organization became the KB Salyut Branch of NPO (Scientific-Production Association) Energia in 1981, and then separated from NPO Energia in 1988, becoming NPO EM KB Salyut, and briefly, KB Salyut. In 1993 KB Salyut combined with the Khrunichev Machine Building Plant to become GKNPC Khrunichev.

Beginning in 1995 the center provided Proton launches marketed by International Launch Services, a joint venture with U.S. partner Lockheed Martin. Khrunichev leaders also partnered with Daimler Chrysler of Germany in Eurockot, a joint project to market launches of Rockot (Russian for thunder), its lighter launch vehicle converted from a Soviet ICBM. As of 2006 the Khrunichev launch vehicle lineup consisted of Proton K and Proton M (and Breeze upper stages for them) and Rockot, and a new generation of Angara launch vehicles (named after a river in the Russian Far East) was under development. Khrunichev also developed the small Yakhta (yacht) spacecraft bus, the basis for satellites such as *Monitor-E* and *Kazsat*.

Stephen B. Johnson

See also: Almaz, Commercial Space Launch, *Mir*, Russian Launch Vehicles, Salyut

Bibliography
Asif A. Siddiqi, *Challenge to Apollo: The Soviet Union and the Space Race, 1945–1974* (2000).
Steven J. Zaloga, *The Kremlin's Nuclear Sword: The Rise and Fall of Russia's Strategic Nuclear Forces, 1945–2000* (2002).

Langley Research Center

Langley Research Center, founded in 1917 by the National Advisory Committee for Aeronautics (NACA), was the first U.S. civilian aeronautics research facility. Originally named Langley Memorial Aeronautical Laboratory (for aviation pioneer Samuel Pierpont Langley) and renamed Langley Aeronautical Laboratory in 1948, the center designed and built the world's premier test facilities, including variable-density, propeller research, full-scale, and high-speed wind tunnels. At these facilities, Langley engineers studied almost every aspect of flight, providing vast amounts of data to the U.S. aviation industry. In 1946, at the request of the military, the center created the Pilotless Aircraft Research Division (PARD) to conduct studies on uncrewed aircraft (rockets and jet-powered models). PARD developed the basic concepts for the X-1 and X-15 high-speed research aircraft and developed, launched, and monitored missiles and rockets. In the late 1950s PARD's expertise in high-altitude supersonic flight and rockets triggered Langley engineers' interest in the problems of space exploration, leading them to develop the first concept for the one-person spacecraft used in Project Mercury.

In 1958 the U.S. government established NASA, and NACA became a part of this new organization. Langley Aeronautical Laboratory became a NASA facility, Langley Research Center (LaRC). Throughout the space age, the center made notable contributions to spaceflight, developing rockets, satellites, and crewed spacecraft. LaRC created the Space Task Group (STG), which conceived and managed Project Mercury and trained the first seven U.S. astronauts between 1959 and 1961. With the advent of the Apollo program, STG moved to Houston, Texas, becoming the nucleus of the newly established Manned Space Center (later Johnson Space Center) in 1962. LaRC made significant contributions to Apollo, including developing the Lunar Orbiters used to select Apollo landing sites, proving the feasibility of lunar-orbit rendezvous, and performing simulations and tests for rendezvous/docking and lunar landing. In the 1960s LaRC developed spacecraft for space and Earth observation, including *Echo*, *Explorers 9* and *19*, and *Passive Geodetic Satellite*. In the 1970s LaRC managed the Viking program, a mission to search for life on Mars. The center was deeply involved in the design, development, and testing of the Space Shuttle, using its experience from hypersonic planes, such as the X-15. LaRC studied concepts for Earth-orbiting space stations since the late 1950s, providing a solid base for the development of the *International Space Station*. Besides its studies in aeronautics, LaRC conducted researches in atmospheric sciences and developed technology for advanced space transportation systems, small spacecraft, and instruments.

Incigul Polat-Erdogan

See also: Lunar Orbiter, Viking, X-15

Bibliography

James R. Hansen, *Engineer in Charge: A History of the Langley Aeronautical Laboratory, 1917–1958* (1987).

―――, *Spaceflight Revolution: NASA Langley Research Center from Sputnik to Apollo* (1995).

Lewis Research Center. *See* Glenn Research Center.

Launch Facilities

Launch facilities are designated areas and complexes for spacecraft launch operations. Most launch complexes began as testing ranges for sounding and suborbital rockets and later developed to accommodate space launchers. The launch complex can be fixed or mobile; launches have been performed from sea platforms, submarines, and aircraft.

Selection of fixed launch sites involves several factors. Climate and political factors can impose limits on the number of launch dates available at a given site. The latitude of a given location determines the amount of launch assistance from Earth's rotation. Launch trajectories generally must avoid populated areas and adversary states because regular launches and accidents generate falling debris. While these factors push launch sites to unpopulated areas, launch sites need transportation access for workers and for the large and complex hardware that must be sent to the site, where launchers undergo final assembly and testing. Launch facilities generally require an array of complex facilities for assembly, processing, integration, technical support, launch operation, and tracking of the launch vehicles.

Initial testing of German V-2 rockets and the Soviet-manufactured version of the V-2, known as the R-1, began in 1947 at a remote site southeast of Volgograd known as Kapustin Yar. It continued as a launch site for Kosmos launchers. The 1954 decree to develop the R-7 intercontinental ballistic missile (ICBM) also included the directive to establish a launch site, which the next year was selected near Tyuratam in Kazakhstan. In an unsuccessful bid to confuse foreigners as to its location, Soviet leaders called it Baikonur after a town several hundred kilometers distant. The name stuck, and it became the launch site for the first Soviet space launches, including all human flight missions. After the demise of the Soviet Union, Baikonur was in the new nation of Kazakhstan, but was leased to Russia. A third major Soviet launch site was established several hundred kilometers northeast of Moscow at Plesetsk as an R-7 ICBM launch facility. It became the primary site for military satellite launches and polar trajectories.

The site of the first U.S. satellite was Cape Canaveral, Florida, which the U.S. military selected as a missile test site in 1947. The U.S. Air Force (USAF) initially

Launch Facilities (Part 1 of 2)

Name	Location	Operating Nation	Operational Dates for Space Launch	Comments
Baikonur	Near Tyuratam, Kazakhstan	Soviet Union/Russia	1957–	Soyuz, Zenit, Tsiklon, Proton; leased by Russia from Kazakhstan
Cape Canaveral	Florida	United States	1957–	NASA and United States Air Force (USAF) facilities
Vandenberg	Vandenberg Air Force Base, California	United States	1960–	USAF operation, some NASA launches
Wallops	Wallops Island, Virginia	United States	1961–1985	NASA
Kapustin Yar	Near Volgograd, Soviet Union/ Russia	Soviet Union/Russia	1962–1999	Kosmos
Hammaguir	Algeria	France	1965–1967	CNES (French Space Agency)
Plesetsk	Soviet Union/Russia	Soviet Union/Russia	1966–	Soyuz, Molniya, Tsiklon, Rokot, Start; polar orbit vehicles
San Marco	Off Kenya, Africa, coast	Italy	1967–1988	Sea launch platform
Woomera	Australia	United Kingdom	1967–1971	European Space Vehicle Launcher Development Organisation
Kourou	French Guiana	France	1970–	CNES, European Space Agency
Kagoshima	Japan	Japan	1970–	Institute of Space and Astronomical Science, Japanese Aerospace Exploration Agency (JAXA)

Launch Facilities (Part 2 of 2)

Name	Location	Operating Nation	Operational Dates for Space Launch	Comments
Jiuquan	China	China	1970–	
Tanegashima	Japan	Japan	1975–	National Space Development Agency, JAXA
Sriharikota	India	India	1980–	Indian Space Research Organisation
Xichang	China	China	1984–	
Taiyuan	China	China	1988–	
Palmahim	Israel	Israel	1988–	
Air Launch	United States	United States	1990–	Pegasus
Svobodny	Russia	Russia	1997–	Rokot and Start 1 launch vehicles (LVs)
Submarine Launch	Russia	Russia	1998	
Sea Launch	Long Beach, California	United States, Russia, Norway	1999–	Sea Launch Corporation, Odyssey platform
Musudan	North Korea	North Korea	2006–	Taepodong LVs
Dasht-e-kabir	Iran	Iran	2009–	Safir 2 rocket, Omid satellite

managed the site, but the U.S. Army also established launch facilities there, firing the first V-2 from Canaveral in 1950. The Army facilities became part of NASA Marshall Space Flight Center in 1960 with the transfer of the Army Ballistic Missile Agency to the space agency. NASA purchased more land nearby, and in 1962 the NASA facility became the Launch Operations Center, renamed Kennedy Space Center the next year. The USAF continued to operate its portion of the Cape as the Cape Canaveral Air Force Station. All U.S. geosynchronous and human spaceflight launches fly from Cape Canaveral. Vandenberg Air Force Base became the primary U.S. polar-orbit launch site, with the first space launch in 1959.

In the 1960s other countries began entering the space frontier. France launched its initial rockets from Hammaguir, Algeria, but as part of France's withdrawal from its former colony, it sought another launch site, eventually located near the equator in Kourou, French Guiana, on the northern coast of South America. The Guiana Space Center became operational at Kourou in 1968 as part of the Diamant launcher program. The European Space Vehicle Launcher Development Organisation (ELDO), was created initially in 1964 at the initiative of the United Kingdom (UK), and as the major financial contributor to ELDO the United Kingdom was able to convince other Member States to place its launch site near Woomera, Australia. However, by the mid-1960s the UK's interest and contribution to ELDO declined, and France became the driving force behind a European launcher. ELDO's launch site was moved to Kourou, and later the European Space Agency (ESA) Ariane program (first launch 1979) continued to use it. Italy operated its offshore San Marco platform off the coast of Kenya during the 1960s–80s.

China has been launching spacecraft into orbit since 1970, and its Long March family of rockets launch from three mainland complexes; Jiuquan Satellite Launch Center, Xichang Satellite Launch Center, and Taiyuan Satellite Launch Center. Shenzhou human spaceflight missions flew from Jiuquan, and Feng Yun weather satellites from Taiyuan. Because of China's high population density, debris from several launches has resulted in casualties and property damage. A fourth launch site on Hainan Island was under construction in 2009.

In 1975 Japan saw the first launch of its orbital satellite from the Tanegashima Space Center. Located on the Tanegashima Island south of Japan's main islands, Tanegashima is the major manufacturing, integration, and launch center for the Japanese Aerospace Exploration Agency. A second smaller launch complex is located at the Kagoshima Space Center and is primarily used for scientific and educational payloads. The Japanese launch facilities are restrained by a unique circumstance; the launch hazards and noise disturbances imposed on the fishing industry brought legislation that confined all launches to two launch windows around every February and September.

Independent launches have been also performed by India, Brazil, Israel, North Korea, and Iran. The Palmachim Air Force Base in Israel is used to launch the Shavit and LK-A rockets over the Mediterranean Sea to steer clear of flying over surrounding Arab countries. Indian space launch operations take place at the Satish Dhawan Space Center, Shriharikota. Direct North and South Pole launches are not possible from Shriharikota due to population densities in those directions, requiring energy-consuming in-flight yaw maneuvers. The annual monsoon season from October through December generally precludes launches during those months. Brazil's Alcantara Launch Center was favorably sited on the equatorial coast. North Korea and Iran established missile test sites used for orbital launch attempts at Musudan and Dasht-e-kabir, respectively.

Commercial endeavors brought new launch capabilities online or reactivated older facilities. The Woomera site was nearly revived with Kistler Aerospace plans to launch its K-1 vehicle from the location. Kistler was purchased by Pioneer Rocketplane, which planned to launch from Oklahoma, but this fell through with Rocketplane Kistler's bankruptcy in 2008. The Sea Launch mobile platform *Odyssey* based in Long Beach, California, provided equatorial launches in the Pacific Ocean starting in 1999. Sea Launch formed as a multinational effort involving U.S., Russian, Ukrainian, and Norwegian companies. The U.S. Office of the Associate Administrator for Commercial Space Transportation encouraged commercialization of space access, issuing nongovernmental spaceport licenses in the United States since 1996. In the late 1990s and early 2000s, commercial and state-operated spaceports were considered or under development in Alaska, Virginia, California, New Mexico, Oklahoma, Florida, Nevada, Texas, and Montana. These spaceports promised to reduce bureaucratic delays and hurdles for new commercial launch operators.

George S. Sarkisov and Stephen B. Johnson

See also: Brazil, China, Commercial Space Launch, European Space Agency, European Space Vehicle Launcher Development Organisation, France, India, Israel, Italy, Japan, Russia (Formerly the Soviet Union), Russian Space Launch Centers, United States

Bibliography

Boris Chertok, *Rockets and People*, vol. 2 (2006).

Brian Harvey, *China's Space Program: From Conception to Manned Spaceflight* (2004). Kenneth Lipartito and Orville R. Butler, *A History of Kennedy Space Center* (2007).

Federation of American Scientists website.

Russian Flight Control Centers

Russian flight control centers located near Moscow provide continuous operation and data processing of Russian spacecraft, including the *International Space Station*. The Russian Federal Space Agency (Roskosmos) oversees the civilian-run Flight Control Center (TsUP) in Podlipki, and the military-driven Russian Space Forces manage the Scientific Test Center of Space Systems (GNIITs KS) in Krasnoznamensk.

Both centers evolved from the range control infrastructure organized in 1956–57 for testing of the R-7, the first Soviet intercontinental ballistic missile. This included ballistic data processing centers in the NII-4 and NII-88 research institutes, the Applied Mathematics Institute and 13 receiving stations (NIPs) throughout the Soviet Union. In the 1960s that infrastructure was enlarged and updated with the latest computer technology to support piloted space missions, while the ballistic centers were reorganized into the military Command-Measurement Complex (KIK) and civilian Coordinate Processing Center (KVTs) under the Ministry of General Machine

Building (MOM). Until 1975 KIK controlled practically all space missions, whereas KVTs played mostly a supplemental role.

Initially NIPs supplied only raw data for KIK and KVTs, but later some of them were reequipped for more important tasks. In the mid-1960s NIP-15, located in Ussuriysk, Far Eastern Russia, tested the country's first space communications system. NIP-10 and NIP-16, located in the cities of Simferopol and Yevpatoria, Crimean Peninsula, Ukraine, provided communications for lunar and planetary probes during the 1960s–80s. NIP-16 also controlled piloted missions during 1966–75. Earlier piloted flights were operated from the Baikonur Cosmodrome in Kazakhstan. These NIPs became known as the Long Range Space Communications Centers.

Since 1959 ship-based floating NIPs were employed to ensure spacecraft communications beyond the mainland of the Soviet Union. By 2005, however, all 11 NIPs vessels were scrapped due to high operating costs and the advent of relay communications satellites.

In 1973 the Soviet Union developed TsUP, based on KVTs, in preparation for the Apollo-Soyuz Test Project in 1975. TsUP took control of piloted, lunar, and planetary missions during 1975–77 and some scientific satellites in 1999. KIK, reorganized into GNIITs KS in 1986, was given control of all other satellites.

Peter A. Gorin

See also: *International Space Station*, Russian Federal Space Agency, Russian Launch Vehicles, Russian Space Forces

Bibliography
Russia's Arms Catalog, Volume VI, Missiles and Space Technology, in Russian and English (1998).
Russian Military Historical Society website in English.

Abbreviations

Corp.   Corporation
Ltd.   Limited

Acronyms

AAF	Army Air Forces
AAP	Apollo Applications Program
AAS	American Astronautical Society
AATSR	Advanced Along Track Scanning Radiometer
ABC	American Broadcast Company
ABL	Allegheny Ballistics Laboratory
ABM	Antiballistic Missile
ABMA	Army Ballistic Missile Agency
ACE	Advanced Carrier Electronics
ACE	Advanced Composition Explorer
ACeS	Asia Cellular Satellite
ACP	Aerosol Collector and Pyrolyser
ACRIM	Active Cavity Radiometer Irradiance Monitor
ACRIMSAT	Active Cavity Radiometer Irradiance Monitor Satellite
AD	Atmospheric Density
ADCS	Advanced Data Collection System
ADEOS	Advanced Earth Observing Satellite
ADS	Astronautical Development Society
AE	Atmospheric Explorer
AEC	Atomic Energy Commission
AEM	Applications Explorer Mission
AEPI	Atmospheric Emissions Photometric Imaging
AFB	Air Force Base
AFM	Air Force Manual
AFML	Air Force Materials Laboratory
AFRL	Air Force Research Laboratory
AFS	Air Force Station
AFSA	Armed Forces Security Agency
AFSATCOM	Air Force Satellite Communications
AFSC	Air Force Systems Command
AFSCF	Air Force Satellite Control Facility
AFSPC	Air Force Space Command

AG	Aktiengesellschaft (German, corporation)
AGC	Apollo Guidance Computer
AGN	Active Galactic Nuclei
AIAA	American Institute for Aeronautics and Astronautics
AIM	Aeronomy of Ice in the Mesosphere
AIMP	Anchored IMP
AIRS	Atmospheric Infrared Sounder
AIS	American Interplanetary Society
AKM	apogee kick motor
ALAE	Atmospheric Lyman-Alpha Emissions
ALARM	Alert Locate and Report Missiles
ALERT	Attack and Launch Early Reporting to Theater
ALFE	Advanced Liquid Feed Experiment
ALMV	Air-Launched Miniature Vehicle
ALOS	Advanced Land Observation Satellite
ALSEP	Apollo Lunar Surface Experiments Package
AM	Amplitude Modulation
AMC	Americom
Americom	American Communications
AMI	Active Microwave Instrument
AMSAT	Amateur Satellite
AMSR	Advanced Microwave Scanning Radiometer
AMSR E	Advanced Microwave Scanning Radiometer for EOS
AMSU	Advanced Microwave Sounding Unit
AMU	Astronaut Maneuvering Unit
ANNA	Army, Navy, NASA, Air Force
ANS	Astronomical Netherlands Satellite (Astronomische Nederlandse Satelliet)
ANU	Australian National University
AOL	American Online
APA	Active Phased Array
APL	Applied Physics Laboratory
APM	Ascent Particle Monitor
APOD	Astronomy Picture-of-the-Day
APOS	Autonomous Piloted Orbital Station
APS	Aerosol Polarimetry Sensor
APT	Automatic Picture Transmission
Arabsat	Arab Satellite Communications Organization
ARC	Ames Research Center
ARC	Atlantic Research Corporation
ARGOS	Advanced Research and Global Observation Satellite
ARPA	Advanced Research Projects Agency
ARISS	Amateur Radio aboard the International Space Station
ARRL	American Radio Relay League
ARS	American Rocket Society

ARSPACE	Army Space Command
ARTEMIS	Advanced Relay Technology Mission
ARTS	Automated Remote Tracking Station
A/S	Apollo-Saturn
ASAR	Advanced Synthetic Aperture Radar
ASAT	Antisatellite
ASC	Agence Spatiale Canadienne (or Canadian Space Agency, CSA)
ASCA	Advanced Satellite for Cosmology and Astrophysics
ASD	Assistant Secretary of Defense
ASI	Agenzia Spaziale Italiana (or Italian Space Agency)
ASLV	Augmented Space Launch Vehicle
ASM	All-Sky Monitor
ASO	Australian Space Office
ASP	Attitude Sensor Package
ASPERA	Analyzer of Space Plasma and Energetic Atoms
ASSET	Aerothermodynamic/elastic Structural Systems Environmental Tests
AST	associate administrator for Commercial Space Transportation
ASTER	Advanced Spaceborne Thermal Emission and Reflection Radiometer
ASTP	Apollo-Soyuz Test Project
AT&T	American Telephone & Telegraph
ATDA	Augmented Target Docking Adapter
ATK	Alliant Techsystems
ATLAS	Atmospheric Laboratory for Applications and Science
ATM	Apollo Telescope Mount
ATMS	Advanced Technology Microwave Sounder
ATMOS	Atmospheric Trace Molecule Spectroscopy
ATN	Advanced Tiros N
ATO	Abort to Orbit
ATS	Applications Technology Satellite
ATSR	Along Track Scanning Radiometer
ATV	Automated Transfer Vehicle
AU	Astronomical Unit
AURA	Association of Universities for Research in Astronomy
AVCS	Advanced Vidicon Camera System
AVHRR	Advanced Very High Resolution Radiometer
AVNIR	Advanced Visible and Near Infrared Radiometer
AXAF	Advanced X-ray Astronomical Facility
BAC	British Aircraft Company
BAe	British Aerospace
BAT	Burst Alert Telescope
BATSE	Burst and Transient Source Experiment
BBRC	Ball Brothers Research Corporation
BBXRT	Broadband X-ray Telescope

BCE	Before Common Era
BD	Beidou
BE	Beacon Explorer
BE	Brilliant Eyes
BeppoSAX	Beppo Satellite per Astronomia X
BIS	British Interplanetary Society
BiSON	Birmingham Solar Oscillations Network
BMD	Ballistic Missile Defense
BMDO	Ballistic Missile Defense Organization
BMEWS	Ballistic Missile Early Warning System
BNCSR	British National Committee on Space Research
BNSC	British National Space Centre
BoB	Bureau of the Budget
BOMARC	Boeing-Michigan Aeronautics Research Center
BOMI	bomber missile
BOR	unpiloted orbital rocket (translated from Russian)
BP	Brilliant Pebbles
BremSat	University of Bremen Satellite
BSB	British Satellite Broadcasting
BSE	Broadcast Satellite Experimental
BSS	broadcast satellite service
BSTS	Boost Surveillance and Tracking System
C	Commander
C3I	Command, Control, Communications, and Intelligence
C3ISR	Command, Control, Communications, Intelligence, Surveillance, Reconnaissance
CAIB	Columbia Accident Investigation Board
CAC	China Aerospace Corporation
CALIPSO	Cloud-Aerosol Lidar and Infrared Pathfinder Satellite Observations
CALT	China Academy of Launch Vehicle Technology
Caltech	California Institute of Technology
CANDOS	Communications and Navigation Demonstration on Shuttle
CAPCOM	Capsule Communicator
CAPL	Capillary Pumped Loop
CAPS	Cassini Plasma Spectrometer
CASC	see CASTC
CASTC	China Aerospace Science and Technology Corporation (also known as CASC)
CBAS	Combined British Astronautical Societies
CBERS	China Brazil Earth Resources Satellite
CBR	Cosmic Background Radiation
CCD	Charge-Coupled Device
CCE	Charge Composition Explorer
CCDS	Centers for the Commercial Development of Space
CDA	Command and Data Acquisition

CDA	Cosmic Dust Analyzer
CDEP	Concept Development and Evaluation Program
CDR	Commander
CEPT	Conférence européenne des administrations des postes et des télécommunications (or Conference of Postal and Telecommunications Administrations)
CERN	Conseil Européen pour la Recherche Nucléaire (or European Council for Nuclear Research)
CESR	Centre d'Etude Spatiale des Rayonnements
CETS	European Conference on Satellite Communications (translated from French)
CFC	chlorofluorocarbon
CFE	Cryogenic On-orbit Long-life Active Refrigerator Flight Experiment
C&GS	Coast and Geodetic Survey
CGMS	Coordination Group for Meteorological Satellites
CGMS	Coordination of Geostationary Meteorological Satellites
CGP	CSE/GLO Payload
CGRO	Compton Gamma Ray Observatory
CHAMP	Challenging Minisatellite Payload
CGWIC	China Great Wall Industry Corporation
CHASER	Colorado Hitchhiker and Student Experiment of Solar Radiation
CHIPS	Cosmic Hot Interstellar Plasma Spectrometer
CHRIS	Compact High Resolution Imaging Spectrometer
CIA	Central Intelligence Agency
CERES	Clouds and the Earth's Radiant Energy System
CIE	collisional ionization equilibrium
CIRS	Composite Infrared Spectrometer
CLES	Closed Loop Ecological System
CLV	Crew Launch Vehicle
CM	Command Module
CMA	China Meteorological Administration
CMC	Central Military Commission
CME	Coronal Mass Ejection
CMIS	Conical Scanning Microwave Imager/Sounder
CNES	Centre National d'Études Spatiales
CNN	Cable News Network
CNR	Consiglio Nationale della Ricerche
CNRS	Centre National de la Recherche Scientifique
CNSA	Chinese National Space Administration
COB	CAPL/ODERACS/BremSat
COBE	Cosmic Background Explorer
COMEST	Commission mondiale d'éthique des connaissances scientifiques et des technologies (Commission on the Ethics of Scientific Knowledge)

COMPTEL	Compton Space Telescope
Comsat	Communications Satellite
COMINT	Communications Intelligence
CONAE	Comision Nacional de Actividades Espaciales, Argentina
CONCAP	Consortium Complex Autonomous Payload
CONUS	Continental United States
COPUOS	Committee on the Peaceful Uses of Outer Space
COROT	Convection, Rotation, and Planetary Transits
COS	Not an acronym, but derived from Cosmic Ray
COSPAR	Committee on Space Research
COSPAS	Space System for the Search of Vessels in Distress (translated from Russian)
COSTAR	Corrective Optics Space Telescope Axial Replacement
CP	Command Module Pilot
C/P	Coronagraph/Polarimeter
CPL	Capillary Pumped Loop
Cremains	Cremated remains
CrIS	Cross-track Infrared Sounder
CRISM	Compact Infrared Imaging Spectrometer for Mars
CRISTA-SPAC	Cryogenic Infrared Spectrometers and Telescopes for the Atmosphere-Shuttle Pallet Satellite
CRPL	Central Radio Propagation Laboratory
CRTC	Canadian Radio-television and Telecommunications Commission
CryoFD	Cryogenic Flexible Diode
CryoFD-R	CryoFD-Reflight
CryoHP	Cryogenic Heat Pipe
CryoTP	Cryogenic Two-Phase
CryoTSU	Cryogenic Thermal Storage Unit
CSA	Canadian Space Agency (also ASC)
CSA	Chinese Society of Astronautics
CSAGI	Special Committee for the International Geophysical Year (from French)
CSE	Cryogenic System Experiment
CSIRO	Commonwealth Scientific and Industrial Research Organisation
CSLA	Commercial Space Launch Act
CSM	Command and Service Module
CSR	Center for Space Research
CTB	Comprehensive Test Ban
CTBT	Comprehensive Test Ban Treaty
CTIO	Cerro Telolo Inter-American Observatory
CTS	Communications Technology Satellite
CVX	Critical Viscosity of Xenon
CZ	Chang Zheng (Long March in Chinese)
CZCS	Coastal Zone Color Scanner

DARA	Deutsche Agentur für Raumfahrt-Angelegenheiten (or German Agency for Space Issues)
DARPA	Defense Advanced Research Projects Agency
DASA	Deutsche Aerospace AG
DATA	Distribution and Automation Technology Advancement
DBS	Direct Broadcast Satellite
DC	Delta Clipper
DCS	Data Collection System
DDR&E	Director, Defense Research and Engineering
DE	Dynamics Explorer
DEFA	Direction des Études et des Fabrications d'armements (Armaments Studies and Construction Authority)
DEW	Distant Early Warning
DF	Dongfeng (Chinese)
DFG	Deutsche Forschungsgemeinschaft (or German Research Foundation)
DFH	Dong Fang Hong (Chinese, East is Red)
DGE	Doppler Gravity Experiment
DIDM	Digital Ion Drift Meter
DIRBE	Diffuse Infrared Background Experiment
DISCO	Dual Spectral Irradiance and Solar Constant Orbiter
DISR	Descent Imager and Spectral Radiometer
DIXI	Deep Impact Extended Investigation
DLR	Deutsche Forschungsanstalt fur Luft und Raumfahrt (or German Aerospace Research Center)
DME	Direct Measurement Explorer
DMR	Differential Microwave Radiometer
DMSP	Defense Meteorological Satellite Program
DOC	Department of Commerce
DoD	Department of Defense
DODGE	Department of Defense Gravity Experiment
DOE	Department Of Energy
DOI	Department of Interior
DORIS	Dual-Frequency Doppler Tracking System Receiver (from French)
DORIS	Doppler Orbitography and Radio positioning Integrated by Satellite
DOS	Department of State (United States)
DOS	Department of Space (India)
DOS	Long Duration Orbital Station (translated from Russian)
DOT	Department Of Transportation
DP	Docking Module Pilot
DRID	Defense Reform Initiative Directive
DS	Deep Space
DSCS	Defense Satellite Communications System
DSE	Data Systems Experiment
DSN	Deep Space Network

DSS	Deep Space Station
DSP	Defense Support Program
DTH	Direct-to-Home
DUSD	Deputy Under Secretary of Defense
DWE	Doppler Wind Experiment
DXS	Diffuse X-ray Spectrometer
E	Engineer
EAC	European Astronaut Centre
EADS	European Aeronautic Defence and Space
ECLSS	Environmental Control and Life Support System
ECS	European Communications Satellite
ECT	Emulsion Chamber Technology
EDC	EROS Data Center
EDL	Entry, Descent, and Landing
EDS	Earth Departure Stage
EELV	Evolved Expendable Launch Vehicle
EGNOS	European Geostationary Navigation Overlay System
EGRET	Energetic Gamma Ray Telescope
EGRS	Electronic and Geodetic Ranging Satellite
EHF	Extremely High Frequency
EISG	Experimental Investigation of Spacecraft Glow
ELDO	European Space Vehicle Launcher Development Organisation
ELINT	Electronic Intelligence
ELSS	EVA Life Support System
ELV	Expendable Launch Vehicle
EMI	electromagnetic interference
EMS	Electromagnetic Sciences
EMU	Extravehicular Mobility Unit
EMZ	Experimental Machine Building Factory (translated from Russian)
ENVISAT	Environmental Satellite
EO	Earth Observing
EOC	Earth Observation Center
EOR	Earth Orbit Rendezvous
EORSAT	ELINT Ocean Reconnaissance Satellite
EOS	Earth Observing System
EOS	electrophoresis operations in space
EOSAT	Earth Observation Satellite Company
EOSDIS	Earth Observing System Data and Information System
EP	Explorer Platform
EPAS	Apollo Soyuz Experimental Flight (translated from Russian)
EPE	Energetic Particle Explorer
EPIC	European Photon Imaging Camera
EPIRB	Emergency Position-Indicating Radio Beacons
EPO	Education and Public Outreach
EPOCh	Extrasolar Planet Observation and Characterization

EPOS	Experimental Passenger Orbital Aircraft (translated from Russian)
EPOXI	EPOCh and DIXI
ERB	Earth Radiation Budget
ERBE	Earth Radiation Budget Experiment
ERBS	Earth Radiation Budget Satellite
EREP	Earth Resources Experiment Package
ERNO	Entwicklungsring Nord (German, "Northern Development Circle")
EROS	Earth Resources Observation System
ERTS	Earth Resources Technology Satellite
ERS	European Remote Sensing
ESA	European Space Agency
ESCAPE	Experiment of the Sun for Complementing the Atlas Payload and for Education
ESCES	Experimental Satellite Communication Earth Station
ESE	Earth Science Enterprise
ESOC	European Space Operations Centre
ESRI	Environmental Systems Research Institute
ESRIN	European Space Research Institute
ESRO	European Space Research Organisation
ESSA	Environmental Science Services Administration
ESSP	Earth System Science Pathfinder
ESSS	Unified System of Satellite Communications (translated from Russian)
ESTEC	European Space Research and Technology Centre
ESTRACK	European Space Tracking
ET	External Tank
ETH	Eidgenössische Technische Hochschule, Zurich
ETM	Enhanced Thematic Mapper
ETS	Engineering Test Satellite
EU	European Union
EUMETSAT	Not an acronym, but derived from European Meteorological Satellites
Eutelsat	European Telecommunications Satellite
EUV	Extreme Ultraviolet
EUVE	Extreme Ultraviolet Explorer
EVA	Extravehicular Activity
EXOSAT	European X-ray Observing Satellite
FAA	Federal Aviation Administration
FASat	Fuerza Aéra de Chile Satellite
FAST	Fast Auroral Snapshot Explorer
FAUST	Far Ultraviolet Space Telescope
FCC	Federal Communications Commission
FCR	Flight Control Room
FEWS	Follow-on Early Warning System
FGB	Functional Cargo Block (translated from Russian)

FGGI	First GARP Global Experiment
FIA	Future Imagery Architecture
FIRAS	Far Infrared Absolute Spectrophotometer
FLTSATCOM	Fleet Satellite Communications
FM	Frequency Modulation
FOBS	Fractional Orbital Bombardment System
FORMOSAT	Formosa Satellite
FPR	Flat Plate Radiometer
FREESTAR	Fast-Reaction Experiments Enabling Science, Technology, and Research
FREGAT	French Gamma-Ray Telescope
FSA	Florida Space Authority
FSS	Fixed Service Satellite
FSW	Fanhui Shi Weixing (Chinese, Recoverable Test Satellite)
ftp	file transfer protocol
FUSE	Far Ultraviolet Spectroscopic Explorer
FUV	Far Ultraviolet
FY	Feng Yun (Chinese, Wind and Cloud)
FY	Fiscal Year
Gagan	GPS-aided GEO Augmented Navigation
GALCIT	Guggenheim Aeronautical Laboratory at the California Institute of Technology
GALEX	Galaxy Evolution Explorer
GANE	GPS Attitude and Navigation Experiment
GARP	Global Atmospheric Research Program
GAS	Get Away Special
GATV	Gemini Agena Target Vehicle
GBM	Gamma-ray Burst Monitor
GBS	Global Broadcast Service
GCMS	Gas Chromatograph Mass Spectrometer
GCP	GLO-CryoHP Payload
GD	General Dynamics
GDL	Gas Dynamics Laboratory
GDP	Gross Domestic Product
GE	General Electric
GEC	General Electric Company (in UK, Not same as US GE)
GEO	Geosynchronous (Geostationary) Orbit
GEODSS	Ground-based Electro-Optical Deep Space Surveillance
GEONS	GPS Enhanced Onboard Navigation System
GEOS	Geodetic Earth Observing Satellite
GEOS	Geodynamics Experimental Ocean Satellite
GEOS	Geostationary Satellite
GFO	Geosat Follow-on
GFZ	GeoforschungsZentrum (German, Geo Research Center)
GGS	Global Geospace Spacecraft

GIOVE	Galileo In-Orbit Validation Element
GIRD	Jet Propulsion Research Group (translated from Russian)
GIS	Geographic Information System
GIS	Gas Imaging Spectrometer
GKNPC	State Research and Production Space Center (translated from Russian)
GLAS	Geoscience Laser Altimeter System
GLAST	Gamma-ray Large Area Space Telescope
GLI	Global Imager
GLO	Shuttle Glow Experiment
GLONASS	Global Navigation Satellite System
GmbH	Gesellschaft mit beschranker Haftung (German, Limited Liability Company)
GMD	Ground-based Midcourse Defense
GMDSS	Global Maritime Distress and Safety System
GMES	Global Monitoring for Environment and Security
GMS	Geostationary Meteorological Satellite
GMT	Greenwich Mean Time (also called UT)
GMVK	State Interdepartmental Commission (translated from Russian)
GNIITs	Scientific Test Center of Space Systems (translated from Russian)
GOCE	Gravity Field and Steady-State Ocean Circulation Explorer
GOCNAE	Grupo de Organização da Comissão Nacional de Atividades Espacias (Portuguese, Group of Organizations of the National Commission on Space Activities)
GOES	Geostationary Operational Environmental Satellites
GOME	Global Ozone Monitoring Experiment
GOMOS	Global Ozone Monitoring by Occultation of Stars
GOMS	Geostationary Operational Meteorological Satellite
GONG	Global Oscillation Network Group
GP	Gravity Probe
GPALS	Global Protection Against Limited Strikes
GPP	GLO/PASDE Payload
GPS	Global Positioning System
GPSDR	Global Positioning System Demonstration Receiver
GR	Global Missile (translated from Russian)
GRAB	Galactic Radiation And Background
GRACE	Gravity Recovery and Climate Experiment
GRB	Gamma Ray Burst
GRC	Glenn Research Center
GRIST	Grazing Incidence Solar Telescope
GRO	Gamma Ray Observatory
GRS	Gamma Ray Spectrometer
GRS	Geodetic Reference System
GSFC	Goddard Space Flight Center
GSLV	Geosynchronous Launch Vehicle

GTE	General Telephone and Electronics
GTO	Geostationary Transfer Orbit
GTsMP	Central State Multi-Purpose Test Range (translated from Russian)
GUKOS	Chief Directorate of the Space Systems
GURVO	Chief Directorate of Rocket Armaments (translated from Russian)
HACCP	Hazard Analysis Critical Control Point
HALCA	Highly Advanced Laboratory for Communications and Astronomy
HASI	Huygens Atmospheric Structure Instrument
HCMM	Heat Capacity Mapping Mission
HD	Henry Draper (catalog)
HEAO	High-Energy Astronomy Observatory
HEAT	Hitchhiker Experiments Advancing Technology
HEO	Highly Elliptical Orbit
HEOS	Highly Eccentric Orbiting Satellite
HESSI	High Energy Solar Spectroscopic Imager
HETE	High Energy Transient Explorer
HEXTI	High Energy X-ray Timing Experiment
HF	Henry Frieden (radiometer)
HF	High Frequency
HH	Hitchhiker
HH-G	Hitchhiker-Goddard
Hipparcos	High Precision Parallax Collecting Satellite
HIRDLS	High Resolution Dynamics Limb Sounder
HIRS	High Resolution Infrared Sounder
HLV	Heavy Lift Vehicle
HOPE	H-2 Orbiting Plane Experimental
HOTOL	Horizontal Takeoff and Landing
HRIR	High Resolution Infrared Radiometer
HS	Hyper-spectral
HSB	Humidity Sounder for Brazil
HSD	Hawker Siddeley Dynamics
HST	Hubble Space Telescope
HTPB	hydroxy-terminated polybutadiene
HTV	H-2 Transfer Vehicle
HXIS	Hard X-ray Imaging Spectrometer
HXRBS	Hard X-ray Burst Spectrometer
HY	Haiyang (Chinese for Ocean)
IAA	International Academy of Astronautics
IABG	Industrieanlagen-Betriebsgesellschaft (German, Industrial Works Business Society)
IAC	International Astronautical Congress
IACG	Inter-Agency Consultive Group
IAF	International Astronautical Federation
IAS	Institute of the Aeronautical Sciences
IAU	International Astronomical Union

IBM	International Business Machines
IBP	Institute of Biophysics
ICBC	IMAX Cargo Bay Camera
ICBM	Intercontinental Ballistic Missile
ICE	International Cometary Explorer
ICESat	Ice, Cloud, and Land Elevation Satellite
ICS	International Council for Science
ICSU	International Council of Scientific Unions
IDCSP	Initial Defense Communications Satellite Program
IDIOT	In situ Device Infrared Optical Test
IDSCS	Initial Defense Satellite Communications System
IEEE	Not an acronym, but derived from Institute of Electrical and Electronic Engineers
IEH	International EUV Hitchhiker
IERS	International Earth Rotation and Reference Systems Service
IGO	Intergovernmental Organization
IGS	Information Gathering Satellite
IGY	International Geophysical Year
IIS	Institute for Industrial Science
IISL	Institute of Space Law
IITRI	Illinois Institute of Technology Research Institute
IKI	Space Research Institute (translated from Russian)
ILAS	Improved Limb Atmospheric Spectrometer
ILC	International Latex Corporation
ILS	International Launch Services
IMAGE	Imager for Magnetopause-to-Aurora Global Exploration
IMAX	Image Maximum
IMBP	Institute for Biomedical Problems (translated from Russian)
IMD	India Meteorological Department
IMEWS	Integrated Missile Early Warning Satellite
IMG	Interferometric Monitor for Greenhouse Gases
IML	International Microgravity Laboratory
IMP	Interplanetary Monitoring Platform
IMS	International Magnetospheric Study
IMSO	International Mobile Satellite Organization
INF	Intermediate-range Nuclear Forces
Inmarsat	International Maritime Satellite Organisation
INMS	Ion and Neutral Mass Spectrometer
INPE	Instituto Nacional de Pesquisas Espaciais (or Brazilian National Institute for Space Research)
INSAT	Indian National Satellite
INTEGRAL	International Gamma-Ray Astrophysics Laboratory
Intelsat	International Telecommunications Satellite Organization
IOC/UNESCO	Intergovernmental Oceanographic Commission of the United Nations Educational, Scientific, and Cultural Organization

IONDS	Integrated Operational Nuclear Detection System
IP	Internet Protocol
IPO	Integrated Program Office
IPY	International Polar Year
IR	Infrared
IRAS	Infrared Astronomical Satellite
IRBM	Intermediate-Range Ballistic Missile
IRFNA	inhibited red fuming nitric acid
IRIS	Internet Routing In Space
IRMB	Royal Meteorological Institute of Belgium (French)
IRR	Infrared Radiometer
IRS	Indian Remote Sensing
IR&D	Internal Research and Development
IS	Istrebitel Sputnikov (Russian, satellite destroyer)
ISA	Israeli Space Agency
ISAS	Institute of Space and Astronautical Science
ISDC	International Space Development Conference
ISEE	International Sun-Earth Explorer
ISF	Industrial Space Facility
ISIR	Infrared Spectral Imaging Radiometer
ISIS	Integrated Satellite Interface System
ISIS	International Satellite for Ionospheric Studies
ISM	interstellar medium
ISO	Imaging Spectrometric Observatory (ATLAS article)
ISO	Infrared Space Observatory
ISPM	International Solar Polar Mission
ISRO	Indian Space Research Organisation
ISS	Imaging Science Subsystem (Cassini-Huygens article)
ISS	International Space Station
ISTP	International Solar Terrestrial Physics
ISTRACK	Indian Space Research Organisation Telemetry, Tracking, and Command Network
ISU	International Space University
ITAR	International Traffic in Arms Regulations
ITOS	Improved Tiros Operational System
ITSA	Institute for Telecommunication Sciences and Aeronomy
ITSO	International Telecommunications Satellite Organization
ITW/AA	Integrated Tactical Warning and Attack Assessment
IUE	International Ultraviolet Explorer
IUS	Inertial Upper Stage
J-T	Joule-Thomson
JATO	Jet-Assisted Takeoff
JAWSAT	Joint Air Force Weber Satellite
JAXA	Japanese Aerospace Exploration Agency
JB	Jian Bing (Chinese for Pathfinder)

JBIS	Journal of the British Interplanetary Society
JE	Johnson Engineering
JEM	Japanese Experiment Module
JERS	Japanese Earth Resource Satellite
JLRPG	Joint Long Range Proving Ground
JMA	Japan Meteorological Agency
JOP	Jupiter Orbiter with Probe
JPL	Jet Propulsion Laboratory
JSC	Johnson Space Center
JSLC	Jiuquan Satellite Launch Center
KB	Design Bureau (translated from Russian)
KBO	Kuiper Belt Object
KBS	Salyut Design Bureau (translated from Russian)
KE	Kinetic Energy
KGB	Committee for State Security (translated from Russian, Komitet Gosudarstvennoi Bezopasnosti)
KH	Not an acronym, but derived from Keyhole
KIK	Command and Measurement Complex
KKV	Kinetic Kill Vehicle
KMA	Korea Meteorological Agency
KOMPSAT	Korea Multipurpose Satellite
KPNO	Kitt Peak National Observatory
KREEP	Potassium (element symbol K), Rare Earth Elements and Phosphorus
KSC	Kagoshima Space Center
KSC	Kennedy Space Center
KV	Space Forces (translated from Russian, Kosmicheskie Voiska)
KVT	Coordinate Processing Center (translated from Russian)
LAGEOS	Laser Geodynamics Satellite
LANL	Los Alamos National Laboratory
LaRC	Langley Research Center
LAS	Large Astronomical Satellite
LASS	Land Application Satellite System
LAT	Large Area Telescope
LDEF	Long Duration Exposure Facility
LEM	Lunar Excursion Module (also LM)
LEO	low Earth orbit
LeRC	Lewis Research Center
LES	Lincoln Experimental Satellites
LHP	Loop Heat Pipe
Lidar	Light Detection and Ranging
LINEAR	Lincoln Near Earth Asteroid Research
LISS	Linear Imaging Self-scanning Sensor
LITE	Lidar In-Space Technology Experiment
LK	Lunniy Korabl (or Lunar Cabin)

LLC	Limited Liability Corporation
LM	Lockheed Martin
LM	Lunar Module (also LEM)
LM	Long March
LMTE	Liquid Metal Thermal Experiment
LOBS	Long-duration Orbital Bombardment System
LOH	liquid hydrogen
LOI	Lunar Orbit Insertion
LOK	Lunar Orbiting Spacecraft (translated from Russian, Lunnei Orbitalnei Korabl)
LONEOS	Lowell Observatory Near Earth Object Search
LOR	Lunar Orbit Rendezvous
LORE	Limb Ozone Retrieval Experiment
LOX	Liquid Oxygen
LPAR	Large Phased Array Radar
LPL	Lunar and Planetary Laboratory
LPRE	Liquid-propellant rocket engine
LPS	Lunar Power Satellite
LPSC	Launching Programme Sub-Committee
LPT	Low-Power Transceiver
LRA	Laser Retroreflector Array
LRBA	Laboratoire de Recherches Balistiques et Aérodynamiques; Ballistics and Aerodynamics Research Laboratory
LRL	Lunar Receiving Laboratory
LRR	Laser Retroreflector
LRV	Lunar Roving Vehicle
LSAM	Lunar Surface Access Module
LSC	Legal Subcommittee
LST	Large Space Telescope
MAA	Manchester Astronautical Association
MACH	Multiple Application Customized Hitchhiker
MAD	Mutual Assured Destruction
MAG/ER	Magnetometer/Electron Reflectometer
Magsat	Magnetic Field Satellite
MAKS	Multi-purpose Aerospace System (translated from Russian)
MARECS	Maritime European Communications Satellite
Marots	Maritime OTS
MARSIS	Mars Advanced Radar Subsurface and Ionosphere Sounding
MAS	Millimeter-wave Atmospheric Sounder
MBB	Messerschmitt-Bölkow-Blohm
MCC	Mission Control Center
MCF	Master Control Facility
MDA	MacDonald Dettwiler and Associates
MDA	Missile Defense Agency
MDAC	McDonnell Douglas Aerospace Corporation

MDC	McDonnell–Douglas Corporation
MEASAT	Malaysia East Asia Satellite
MECA	Microscopy, Electrochemistry, and Conductivity Analyzer
MEEP	Mir Environmental Exposure Program
MEIDEX	Mediterranean Israeli Dust Experiment
MEO	medium Earth orbit
MER	Mars Exploration Rovers
MERIS	Medium Resolution Imaging Spectrometer
MESSENGER	Mercury Surface, Space Environment, Geochemistry, and Ranging
MESUR	Mars Environmental Survey
MET	Meteorological Station
MetOp	Meteorological Operational
MGS	Mars Global Surveyor
MHI	Mitsubishi Heavy Industries
MHT	Matra Hautes Technologies
MIDAS	Missile Defense Alarm System
MIDEX	Mid-sized Explorer
MightySat	Mighty Satellite
MIMI	Magnetosphere Imaging Instrument
MINERVA	Micro/Nano Experimental Robot Vehicle for Asteroid
MIRACL	Mid-Infrared Advanced Chemical Laser
MIPAS	Michelson Interferometric Passive Atmospheric Sounder
MIRV	Multiple Independently Targeted Reentry Vehicle
MISR	Multi-angle Imaging Spectro-Radiometer
MISS	Man-in-Space Soonest
MISSE	Materials International Space Station Experiment
MIT	Massachusetts Institute of Technology
MKRT	Naval Space Reconnaissance and Targeting System (translated from Russian)
MLS	Microwave limb Sounder
MLTI	Mesosphere and Lower Thermosphere/Ionosphere
MMC	Martin Marietta Corporation
MMH	monomethyl hydrazine
MMS	Matra Marconi Space
MMS	Multi-Mission Modular Spacecraft
MMU	Manned Maneuvering Unit
MOBS	Multiple Orbital Bombardment System
MOCR	Mission Operations Control Room
MOD	Ministry of Defence
MODIS	Moderate resolution Imaging Spectroradiometer
MODS	Military Orbital Development System
MOL	Manned Orbiting Laboratory
MOM	Ministerstvo Obshchego Mashinostroeniya (or Ministry of General Machine Building)
MOP	Meteosat Operational

MOPITT	Measurements of Pollution in the Troposphere
MOS	Modular Opto-elecronics Scanner
MosGIRD	Moscow Jet Propulsion Research Group (translated from Russian)
MOST	Microvariability and Oscillations of Stars (or Microvariabilité et Oscillations STellaire)
MOTIF	Maui Optical Tracking and Identification Facility
MPI	Max Planck Institute
MPL	Mars Polar Lander
MRIS	Medium-resolution imaging spectroradiometer
MRO	Mars Reconnaissance Orbiter
MS	Mission Specialist
MS	Multi-spectral
MSAS	Multi-transport Satellite Augmentation System
MSAT	mobile satellite system
MSC	Manned Spacecraft Center
MSFC	Marshall Space Flight Center
MSG	Meteosat Second Generation
MSL	Microgravity Science Laboratory
MSM	master in space management
MSMR	Multifrequency Scanning Microwave Radiometer
MSS	master of space studies
MSS	Multi-Spectral Scanner
MSTI	Miniature Sensor Technology Integration
MSU	Microwave Sounding Unit
MSX	Midcourse Space Experiment
MTP	Meteosat Transition Program
MTPE	Mission to Planet Earth
MTS	Meteoroid Technology Satellite
MTSAT	Multi-functional Transport Satellite
MUSES	Mu Space Engineering Spacecraft
MV	Ministry of Armaments (translated from Russian)
MVISR	Multichannel Visible and IR Scan Radiometer
MWS	Microwave Sounder
NAA	North American Aviation
NACA	National Advisory Committee for Aeronautics
NAD	North American Datum
NAL	National Aerospace Laboratory
NAR	North American Rockwell
NAS	National Academy of Sciences
NASA	National Aeronautics and Space Administration
NaSBE	Sodium-Sulfur Battery Experiment
NASCOM	NASA Communications Network
NASDA	National Space Development Agency (Japan)
NASM	National Air and Space Museum
NASP	National Aero-Space Plane

NATO	North Atlantic Treaty Organization
NAVSPASUR	Navy Space Surveillance
NAVSTAR	Not an acronym, but derived from Navigation Satellite Time and Ranging
NCST	Naval Center for Space Technology
Nd:YAG	Neodymium-doped Yttrium Aluminum Garnet
NEA	Near Earth Asteroid
NEAR	Near Earth Asteroid Rendezvous
NEAT	Near Earth Asteroid Tracking
NEC	Nippon Electric Company
NEO	Near Earth Object
NERVA	Nuclear Engine for Rocket Vehicle Application
NESDIS	National Environmental Satellite, Data, and Information Service
NESS	National Environment and Satellite Service
NExT	New Exploration of Tempel 1
NGA	National Geospatial-Intelligence Agency
NGC	New General Catalog
NHFRF	National Hypersonic Flight Research Facility
NII	Scientific Research Institute (translated from Russian)
NIIP	Scientific Research Test Range (translated from Russian)
NIMA	National Imagery and Mapping Agency
NIPs	Receiving Station (translated from Russian)
NKH	Japan Broadcasting Company (translated from Japanese)
NMD	National Missile Defense
NMP	New Millennium Program
NOAA	National Oceanic and Atmospheric Administration
NOAO	National Optical Astronomy Observatory
NORAD	North American Aerospace Defense Command
NOSS	Naval Ocean Surveillance System
NOTS	Naval Ordnance Test Station
NPIC	National Photographic Interpretation Center
NPO	Scientific-Production Association (translated from Russian)
NPOESS	National Polar Orbiting Environmental Satellite System
NPS	Naval Postgraduate School
NRA	NASA Radar Altimeter
NRC	National Research Council
NRL	Naval Research Laboratory
NRO	National Reconnaissance Office
NRP	National Reconnaissance Program
NS	Neutron Scatterometer
NS	Neutron Spectrometer
NSA	National Security Agency
NSC	National Science & Technology Council
NSCAT	NASA Scatterometer
NSF	National Science Foundation

NSI	National Space Institute
NSS	National Space Society
NSS	New Skies Satellite
NSSA	National Security Space Architect
NSTC	National Science and Technology Council
NTR	nuclear thermal rocket
NWS	National Weather Service
OAMS	Orbit Attitude and Maneuvering System
OAO	Orbiting Astronomical Observatory
OAR	Office of Oceanic and Atmospheric Research
OAST	Office of Aeronautics and Space Technology
OCM	Ocean Color Monitor
OCST	Office of Commercial Space Transportation
OCTS	Ocean Color and Temperature Scanner
ODERACS	Orbital Debris Radar Calibration Spheres
OIF	Operation Iraqi Freedom
OFK	Official Flight Kit
OGO	Orbiting Geophysical Observatory
OKB	Experimental Design Bureau, except when referring to OKB-586, or OKB-1 between 1950 and 1956, in which case it is Special Design Bureau (translated from Russian)
OM	Optical Monitor
OMB	Office of Management and Budget
OMEGA	Observatoire pour la Minéralogie, l'Eau, les Glaces, et l'Activité
OMI	Ozone Monitoring Instrument
OMPS	Ozone Mapping and Profiler Suite
OMS	Orbital Maneuvering System
OOSA	Office of Outer Space Affairs
OPF	Orbiter Processing Facility
OPP	other physical principles
OPS	Optical Sensor
OPS	Orbital Piloted Station
ORBIT Act	Open-market Reorganization for the Betterment of International Telecommunications Act
ORS	Ocean Remote Sensing
OSC	Office of Space Commercialization
OSC	Orbital Sciences Corporation
OSCAR	Orbital Satellite Carrying Amateur Radio
OSO	Orbiting Solar Observatory
OSSE	Oriented Scintillation Spectrometer Experiment
OSTA	Office of Space and Terrestrial Application
OSTM	Ocean Surface Topography Mission
OSTP	Office of Science and Technology Policy
OTA	Office of Technology Assessment
OTP	Office of Telecommunications Policy

OTS	Orbital Test Satellite
OTV	Orbital Transfer Vehicle
OV	Orbital Vehicle (or Orbiting Vehicle)
P	Pilot
PACS	Particle Analysis Cameras for Shuttle
PAGEOS	Passive Geodetic Earth Orbiting Satellite
PALSAR	Phased Array L-band Synthetic Aperture Radar
PAMS	Passive Aerodynamically Stabilized Magnetically Damped Satellite
PAN	panchromatic
PANSAT	Petite Amateur Navy Satellite
PARASOL	Polarization and Anisotropy of Reflectances for Atmospheric Science coupled with Observations from a Lidar
PARD	Pilotless Aircraft Research Division
PAS	Pan American Satellite
PASDE	Photogrammetric Appendage Structural Dynamics Experiment
PAWS	Phased Array Warning System
PCA	Proportional Counter Array
PDD	Presidential Decision Directive
PE	Perkin-Elmer
PEB	Payload Equipment Bay
PGM	Platinum Group Metals
PI	Principal Investigator
PKA	Gliding Cosmic Apparatus (translated from Russian)
PLC	Public Limited Company
PLSS	Portable Life Support System
PLT	Pilot
PNTB	Partial Nuclear Test Ban
POES	Polar-orbiting Operational Environmental Satellite
POLDER	Polarization and Directionality of the Earth's Reflectances
POSA	Passive Optical Sample Assemblies
PP	Polar Platform
PPBE	Planning, Programming, Budgeting, and Execution
PPBS	Planning, Programming and Budgeting System
PPK	Personal Preference Kit
PPS	Precise Positioning Service
PRARE	Precise Range and Range Rate Experiment
PRIME	Precision Recover Including Maneuevering Entry
PRISM	Panchromatic Remote-Sensing Instrument for Stereo Mapping
PROBA	Project for On-Board Autonomy
PrOP-M	Pribori Otchenki Prokhodimostikhodik—Instrument for Cross Country Characteristics Evaluation on Mars
PS	Payload Specialist
PSLV	Polar Satellite Launch Vehicle
PSRD	Prototype Synchrotron Radiation Detector
PSSRI	Planetary and Space Science Research Institute

PTBT	Partial Test Ban Treaty
PTT	Postal, Telegraph and Telephone
PV	photovoltaic
PVO	Pioneer Venus Orbiter
RA	Radar Altimeter
RAE	Radio Astronomy Explorer
RAL	Rutherford Appleton Laboratory
RAND	Not an acronym, but derived from Research and Development Corporation
RARC	Regional Administrative Radio Conference
RBV	Return Beam Vidicon
RCA	Radio Corporation of America
RCS	Reaction Control System (or Re-entry Control System)
RGS	Reflection Grating Spectrometer
RHESSI	Reuven Ramaty High Energy Solar Spectroscopic Imager
RIKEN	Institute for Chemistry and Physics (Japan)
RIS	Retroreflector in Space
RKA	Rossiyskoe Kosmicheskoe Agentstvo (or Russian Space Agency, RSA)
RLV	Reusable Launch Vehicle
RMI	Reaction Motors Inc.
RMI	Royal Meteorological Institute, Belgium
RMS	Radiation Meteoroid Satellite
RMS	Remote Manipulator System
RNII	Jet Propulsion Research Institute (translated from Russian)
Robo	Rocket bomber
ROCSAT	Republic of China Satellite
ROMPS	Robot Operated Processing System
RORSAT	Radar Ocean Reconnaissance Satellite
ROSAT	Röntgen Satellite
ROSCOSMOS	Russian Federal Space Agency ROSHYDROMET
Russian Federal Service for	Hydrometeorology and Environmental Monitoring
RPWS	Radio and Plasma Wave Spectrometer
RSA	Russian Space Agency (also RKA)
RSS	Radio Science Subsystem
R&D	Research and Development
RTG	Radioisotope Thermoelectric Generator
RTS	Remote Tracking Station
RTS	Research Test Series
RV	Reentry Vehicle
RVSN	Strategic Missile Forces (translated from Russian)
RXTE	Rossi X-ray Timing Explorer
SA	Selective Availability

SA	Corporation (Spanish, Sociedad Anónima, literally "anonymous society")
SA	Space Architect
SAC	Satélite de Aplicaciones Científicos (Argentina)
SAC	Strategic Air Command
SAC	Space Activities Commission
SAFER	Simplified Aid for EVA Rescue
SAGE	Semi-Automatic Ground Environment
SAGE	Stratosphere Aerosol Gas Experiment
SAINT	Satellite Interceptor (or Satellite Inspector)
SALT	Strategic Arms Limitation Treaty (or Strategic Arms Limitation Talks)
SAM	School of Aviation Medicine
SAMOS	Not an acronym, but derived from Satellite and Missile Observation Satellite
SAMPEX	Solar Anomalous and Magnetospheric Particle Explorer
SAMPIE	Solar Array Module Plasma Interaction Experiment
SAO	Smithsonian Astronomical Observatory
SAR	Synthetic Aperture Radar
SAREX	Shuttle Amateur Radio Experiment
SARSAT	Search and Rescue Satellite Aided Tracking
SAS	Small Astronomy Satellite
SAS	Space Adaptation Syndrome
SAST	Shanghai Academy of Space Technology
SATCOM	Satellite Communications
SAX	Satellite per Astronomia X
SBIRS	Space Based Infrared System
SBS	Satellite Business Systems
SBSS	Space-Based Space Surveillance
SBUV	Solar Backscattered Ultraviolet Radiometer
SCARLET	Solar Concentrator Arrays with Refractive Linear Element Technology
SCATHA	Spacecraft Charging at High Altitude
SCD	*Satélite de Coleta de Dados* (Portuguese, *Data Gathering Satellite*)
SCIAMACHY	Scanning Imaging Absorption Spectrometer for Atmospheric Chartography
SCISAT	Not an acronym, but originally derived from "Scientific Satellite"
SCO X-1	Scorpius X-1
SCORE	Signal Communication by Orbiting Relay Equipment
SCOSTEP	Scientific Committee on Solar-Terrestrial Physics
SDARS	satellite digital audio radio services
SDI	Strategic Defense Initiative
SDIO	Strategic Defense Initiative Organization
SDS	Satellite Data System
SDS	Strategic Defense System

SDV	Shuttle-derived Vehicle
SeaWIFS	Sea-Viewing Wide Field-of-View Sensor
SecDef	Secretary of Defense
SECOR	Sequential Correlation Of Range
SEDS	Students for the Exploration and Development of Space
SEE	Space Environmental Effects
SEECM	Shuttle Environmental Effect on Coated Mirror
SEH	Solar EUV Hitchhiker
SEI	Space Exploration Initiative
SELENE	Selenological and Engineering Explorer
SEM	Space Environmental Monitor
SEM	Space Experiment Module
SEPAC	Space Experiments with Particle Accelerators
SEREB	Société pour l'Étude et la Réalisation d'Engins Balistiques (Ballistic Missiles Research and Development Company)
SES	Société Européenne des Satellites
SESS	Space Environment Sensor Suite
SETI	Search for Extraterrestrial Intelligence
SFOC	Space Flight Operations Contract
SGAC	Space Generation Advisory Council
SGSR	Steering Group for Space Research
SHF	Superhigh Frequency
SHOOT	Superfluid Helium On-Orbit Transfer
SIG	Special Interagency Group
SIGINT	Signals Intelligence
SII	Space Industries, Incorporated
SIM	Spectral Irradiance Monitor
SIR	Shuttle Imaging Radar
SIR	Spaceborne Imaging Radar
SIRIO	Satellite Italiano Ricerca Industriale Operativa (Italian, Italian Satellite for Operational Industrial Research)
SIRTF	Shuttle Infrared Telescope Facility
SIRTF	Space Infrared Telescope Facility
SIS	Solid-State Imaging Spectrometer
SITE	Satellite Instructional Television Experiment
SJ	Shijian
SKIRT	Spacecraft Kinetic Infrared Test
SL	Spacelab
SLA	Shuttle Laser Altimeter
SLBD	SeaLite Beam Director
SLBM	Submarine Launched Ballistic Missile
SLEP	Service Life Enhancement Program
SLS	Spacelab Life Sciences
SLV	Satellite Launch Vehicle
SM	Service Module

SM	Standard Missile
SMART	Small Missions for Advanced Research in Technology
SMC	Space and Missile Systems Center
SMDC	Space and Missile Defense Command
SME	Solar Mesospheric Explorer
SMEX	Small Explorer
SMM	Solar Maximum Mission
SMS	Space motion sickness
SMS	Synchronous Meteorological Satellite
SMTS	Space and Missile Tracking System
SNAP	System for Nuclear Auxiliary Power
SNG	satellite news gathering
SNIAS	Société Nationale Industrielle Aérospatiale (French, National Aerospace Industrial Corporation)
SNOE	Student Nitric Oxide Explorer
SOHO	Solar and Heliospheric Observatory
SOLAS	Safety of Life at Sea Convention
SOLCON	Solar Constant
SOLRAD	Solar Radiation
SOLSE	Shuttle Ozone Limb Sounding Experiment
SOLSPEC	Solar Spectrum Measurement
SOLSTICE	Solar Stellar Irradiance Comparison Experiment
SORCE	Solar Radiation and Climate Experiment
SORT	Strategic Offensive Reductions Treaty
SOVA	Solar Constant and Variability
SPADATS	Space Detection And Tracking System
Spartan	Not an acronym, but derived from Shuttle-Pointed Autonomous Research Tool for Astronomy
SPAWAR	Space and Naval Warfare Systems Command
SPELDA	Structure Porteuse Externe pour Lancements Doubles Ariane
SPM	Solar Proton Monitor
SPO	Sacramento Peak Observatory
SPOC	Space Program Operations Contract
SPOT	Satellite Pour l'Observation de la Terre
SPS	Service Propulsion System
SPS	Solar Power Satellites
SPS	Standard Positioning Service
SR	Scanning Radiometer
SRB	Solid Rocket Booster
SRC	sample return capsule
SRD	Synchrotron Radiation Detector
SRDL	Signal Research and Development Laboratory
SRET	Satellite de Recherches et d'Études Techniques
SRL	Space Radar Laboratory
SRM	Solid Rocket Motor

SRTM	Shuttle Radar Topography Mission
SS	Schutzstaffel (German, Protective Squadron)
SS	Survivability Sensor
SSA	space situational awareness
SSALT	Single-Frequency Solid State Radar Altimeter
SSBUV	Shuttle Solar Backscatter Ultraviolet
SSETI	Student Space Exploration & Technology Initiative
SSF	Space Station Freedom
SSG	Senior Steering Group
SSI	Space Services Incorporated
SSL	Space Sciences Laboratory
SS/L	Space Systems/Loral
SSME	Space Shuttle Main Engine
SSN	Space Surveillance Network
SSP	Summer Session Program
SSP	Surface Science Package
SSPA	solid-state power amplifier
SSPAR	Solid State Phased Array Radar
SSRMS	Space Station Remote Manipulator System
SSS	Small Scientific Satellite
SSS	Space Surveillance System
SSTL	Surrey Satellite Technology Limited
SSTO	Single Stage to Orbit
SSTS	Space Surveillance and Tracking System
SSU	Stratospheric Sounding Unit
STA	Science and Technology Agency
STA	Space Transportation Association
STARLITE	Spectrograph Telescope for Astronomical Research
STARSHINE	Student-Tracked Atmospheric Research Satellite for Heuristic International Networking Experiment
START	Spacecraft Technology and Advanced Reentry Test
START	Strategic Arms Reduction Treaty
STC	Satellite Television Corporation
STEDI	Student Explorer Demonstration Initiative
STEREO	Solar Terrestrial Relations Observatory
STG	Space Task Group
STIS	Space Telescope Imaging Spectrograph
STL	Space Technology Laboratory
STP	Solar Terrestrial Probes
STP	Solar Terrestrial Program
STP	Space Test Payload
STP	Space Test Program
STS	Space Transportation System
STScI	Space Telescope Science Institute

STSS	Space Tracking and Surveillance System
STWG	Scientific and Technical Working Group
SUSIM	Solar Ultraviolet Irradiance Monitor
SWAS	Submillimeter Wave Astronomy Satellite
SXC	Soft X-ray Camera
SXT	Soft X-ray Telescope
SY	Shiyan (Chinese for Experiment)
TA	Thor-Agena
TACSAT	Tactical Communications Satellite
TAD	Thrust-Augmented Delta
TAS	Technology Applications and Science
TAT	Transatlantic
TAV	Trans-Atmospheric Vehicle
TB	Thor-Burner
TC	Tan Ce (Chinese for Explorer)
TCC	thermal control coating
TD	Thor-Delta
TDF	Télédiffusion de France
TDRS	Tracking and Data Relay Satellite
TDRSS	Tracking and Data Relay Satellite System
TEAMS	Technology Experiments Advancing Missions in Space
TEGA	Thermal and Evolved Gas Analyzer
TENCAP	Tactical Employment of National Capabilities
TERLS	Thumba Equatorial Rocket Launching Station
TES	Technology Experiment Satellite
TES	Thermal Energy Storage
TES	Tropospheric Emission Spectrometer
TEXUS	Technologische Experimente Unter Schwerelosigkeit (German, Technological Experiments Under Weightlessness)
TFSP	Teacher From Space Program
TGS	triglycine sulfate
THEMIS	Time History of Events and Macroscale Interactions during Substorms
TIM	Total Irradiance Monitor
TIMED	Thermosphere Ionosphere Mesosphere Energetics and Dynamics
TIROS	Television Infrared Observation Satellite
TISP	Teacher In Space Project
TKS	Transport-Supply Ship (translated from Russian)
TM	Thematic Mapper
TM	Transport Modification (translated from Russian)
TMD	Theater Missile Defense
TMI	TRMM Microwave Imager
TMR	TOPEX Microwave Radiometer
TNT	Trinitrotoluene
TOMS	Total Ozone Mapping Spectrometer

TOMS-EP	Total Ozone Mapping Spectrometer Earth Probe
TOPEX	Topography Experiment
TOS	Tiros Operational Satellites
TOS	Transfer Orbit Stage
TOS	Transportable Optical System
TOVS	Tiros Operational Vertical Sounder
TPF	Two-Phase Flow
TRACE	Transition Region and Coronal Explorer
TRRM	Tropical Rainfall Measuring Mission
TRW	Not an acronym, but derived from Thompson-Ramo-Wooldridge
TS	Tansuo (Chinese for Exploration)
TSIS	Total Solar Irradiance Sensor
TsKBEM	Central Construction Bureau of Experimental Machine Building (translated from Russian)
TsNII	Central Scientific-Research Institute (translated from Russian)
TsNIIMash	Central Research Institute for Machine Building (translated from Russian)
TsNIRTI	Central Scientific Research Radiotechnical Institute (translated from Russian)
TsPK	Cosmonaut Training Complex (translated from Russian)
TSS	Tethered Satellite System
TsSKB	Central Specialized Design Bureau (translated from Russian)
TSTO	Two Stage to Orbit
TsUKOS	Central Directorate of Space Systems (translated from Russian)
TsUP	Flight Control Center (translated from Russian, Tsentr Upravleniye Polyotami)
TsVNIAG	Central Military Scientific Aviation Hospital (translated from Russian)
TV	Television
TWTA	Traveling Wave Tube Amplifiers
UAL	Upper Air Laboratory
UARS	Upper Atmosphere Research Satellite
UAV	Unmanned Aerial Vehicle
UCB	University of California, Berkeley
UDMH	Unsymmetrical dimethyl hydrazine
UFO	UHF Follow-on
UFO	Unidentified Flying Object
UHF	Ultrahigh Frequency
UK	United Kingdom
UN	United Nations
UNAMSAT	National University of Mexico Satellite
UNESCO	United Nations Educational, Scientific, and Cultural Organization
UNEX	University-Class Explorer
UNKS	Directorate of the Space Systems Commander (translated from Russian)

UR	Universal Rocket (translated from Russian)
US	United States
USA	United Space Alliance
USA	United States of America
US-A	Controlled Satellite-Active (translated from Russian)
USAF	United States Air Force
USASDC	U.S. Army Strategic Defense Command
USASSDC	U.S. Army Space and Strategic Defense Command
US-A	Controlled Satellite-Active (translated from Russian)
USD	United States dollars
USDA	United States Department of Agriculture
USGS	United States Geological Survey
USML	United States Microgravity Laboratory
USMP	United States Microgravity Payload
USN	United States Navy
US-P	Controlled Satellite-Passive (translated from Russian)
USS	United States Ship
USSPACECOM	United States Space Command
USSRC	U.S. Space and Rocket Center
USWB	U.S. Weather Bureau
UT	Universal Time (also called GMT)
UTC	Universal Time Code
UV	Ultraviolet
UVIS	Ultraviolet Imaging Spectrograph
UVLIM	Ultraviolet Limb Imaging
UVOT	Ultraviolet Optical Telescope
UVSP	Ultraviolet Spectrometer and Polarimeter
UVSTAR	Ultraviolet Spectrograph Telescope for Astronomical Research
VA	Vozvrashemui Apparat (Russian, Return Apparatus)
VAB	Vehicle Assembly Building (formerly Vertical Assembly Building)
VAFB	Vandenberg Air Force Base
VAKO	All-Russian Youth Aerospace Society
VfR	Verein für Raumschiffahrt (German Society for Space Travel)
VFW	Vereinigte Flugtechnische Werke (German, "United Flight Technology Works")
VHF	Very High Frequency
VHRR	Very High Resolution Radiometers
VHRSR	Very High Resolution Scanning Radiometer
VIIRS	Visible/Infrared Imager Radiometer Suite
VIMS	Visible and Infrared Mapping Spectrometer
VIRGO	Variability of Solar Irradiance and Gravity Oscillations
VIRR	Visible and Infrared Radiometer
VIS	visible
VISSR	Visible Infrared Spin Scan Radiometer
VKS	Military Space Forces (translated from Russian)

VLBI	Very Long Baseline Interferometry
VLS	Vertical Launch System
VLS	Veículo Lançador de Satélite (Portuguese, Satellite Launch Vehicle)
VLSI	Very Large-Scale Integrated
VNIIEM	All-Union Scientific Research Institute of Electromechanics
VOIR	Venus Orbiting Imaging Radar
VSE	Vision for Space Exploration
VSOP	VLBI Space Observatory Programme
VTPR	Vertical Temperature Profile Radiometers
VTRE	Vented Tank Resupply Experiment
WAC	Without Attitude Control
WARC	World Administrative Radio Conference
WB	Weather Bureau
WCRP	World Climate Research Program
WDL	Western Development Laboratories
WFPC	Wide Field Planetary Camera
WGS	Wideband Gapfiller System
WGS	World Geodetic System
WGS	Wideband Global System
WiFS	Wide Field Sensor
WIRE	Wide Field Infrared Explorer
WMAP	Wilkinson Microwave Anisotropy Probe
WMD	Weapons of Mass Destruction
WMO	World Meteorological Organization
WRESAT	Weapons Research Establishment Satellite
WWI	World War I
WWII	World War II
WWW	World Weather Watch
WXM	Wide Field X-ray Monitor
XMM	X-ray Multi-Mirror
XPS	Extreme Ultraviolet Photometer System
XRP	X-ray Polychromator
XRT	X-ray Telescope
XSCC	Xi'an Satellite Control Center
XSS	Experimental Satellite System
XTE	X-ray Timing Explorer
ZY	Zi Yuan (Chinese for Earth Resources)

Glossary

ablation The process of removing by cutting, erosion, melting, evaporation, or vaporization. In aerospace, typically refers to the process of removing heat by vaporizing specialized materials.

absorption lines The dark lines seen in a spectrum due to light being absorbed by atoms and molecules at particular wavelengths/frequencies.

accretion disk A disk of material created by one large rotating object acquiring material from another object, such as one star of a binary pair acquiring material from its companion star.

aerobraking Using the resistance caused by the atmosphere of a planet to slow down a spacecraft.

aerodynamic braking *See* aerobraking.

aeroform The external shape of a vehicle that must move through an atmosphere.

aerogel A manufactured material formed by removal of fluid portions of a gel from the solid portions of the gel, resulting in extremely low bulk density.

aeroshell A rigid heat-shielded shell that slows a vehicle and protect it from pressure and heat created by drag during atmospheric entry.

albedo Reflectivity of an astronomical object. The units used are either on the scale 0 to 1 or 0 percent to 100 percent.

altimeter An instrument for measuring altitude.

anisotropy Exhibition of different properties with different values when measured along axes in different directions.

antipode Situated at the opposite side of sphere.

aperture The diameter of the objective lens or mirror of a telescope.

apogee The point in the orbit of the Moon or an Earth-orbiting spacecraft where it is furthest away from the Earth.

apoapsis The point in the path of an orbiting body at which it is furthest from the primary.

architecture In aerospace, the components and relationship of components in a system.

asteroseismology The technique of probing the internal structure of stars through the frequencies of their surface vibrations, measured from periodic variations in light output or spectral characteristics (most commonly Doppler shifts due to surface motions along the line of sight). The surface vibrations are caused by sound or buoyancy waves traveling through the gas, exciting resonances in the star that are sensitive to interior properties.

astronomical unit (AU) The mean distance of the Earth from the Sun.

aurora A luminous phenomenon that consists of streamers or arches of light appearing in the upper atmosphere of a planet's polar regions, caused by the emission of light from atoms excited along the planet's magnetic field lines.

avionics The electronic and electromagnetic devices used on board aircraft, missiles, and spacecraft, including computers.

azimuth The angular distance, measured in degrees from true north in a clockwise direction (compass direction), that an antenna must be pointed to receive an electronic signal.

ballistic missile A rocket-propelled projectile carrying a warhead and following an elliptical path determined by the duration of its powered flight phase, the effect of earth's gravitational field, and uncontrolled aerodynamic interactions with the atmosphere during its final, free-falling flight phase.

bandwidth The range of frequency or wavelength, expressed in cycles per second or Hertz (Hz), required for transmitting a specific signal without distortion or data loss.

basalt A volcanic igneous rock consisting mainly of pyroxene and plagioclase, with smaller quantities of ilmenite.

Be stars B-type stars with hydrogen emission lines superimposed on the normal spectrum.

Big Bang The cosmological theory that the universe expanded from an extremely hot and dense initial condition at a finite time in the past (currently estimated at 13.7 billion years ago).

bilateral Pertaining to, involving, or affecting two or both sides, factions, or parties.

binary stars Two stars that orbit a common center of mass.

binaries *See* binary stars.

biosphere The parts of the land, sea, and atmosphere in which organisms are able to live.

black hole A region in space with a gravitational field so strong it prevents light from escaping from it.

bolometer A device that detects radiation by changing its electrical resistance when heated by the radiation.

bow shock The boundary around a planet's magnetosphere where the solar wind is deflected.

broadband spectroscopy Techniques for measuring spectra across a wide band of wavelengths.

bus The portions of a spacecraft that carry and provide the functions necessary for the payload to perform its functions. These typically include the structure, electrical power subsystem, thermal control subsystem, command and data handling subsystem, radio frequency subsystem, and guidance, navigation, and control subsystem.

caldera A large volcanic crater created by the collapse of the surface and/or by an explosive eruption.

calorimeter A device to measure quantities of absorbed or evolved heat.

cardiovascular Of, or relating to the heart or blood vessels.

castable A material that can be given a shape by pouring in liquid or plastic form into a mold and letting it harden without pressure.

cataclysmic variable A star whose brightness increases dramatically as the result of a large explosion.

catalytic oxidation To chemically combine a substance with oxygen using a process that initiates the chemical reaction and enables it to proceed under different conditions than otherwise possible (such as lower temperature).

celestial Pertaining to the sky or visible heaven.

charge-coupled device (CCD) A device that detects radiation via the production of electrons within a silicon substrate. Invariably in the form of many individual detectors forming an array, the image is read-out by transferring (coupling) the charges accumulated within the individual detectors from one to another.

chemosynthetic life Life based on the conversion of one or more carbon molecules into organic matter through the oxidation of inorganic molecules or methane as a source of energy, rather than using sunlight, as in photosynthesis.

chromatography A process in which a chemical mixture carried by a liquid or gas is separated into components as a result of differential distribution of the solutes as they flow around or over a stationary liquid or solid phase.

chromosphere The region of the Sun above the photosphere. The temperature of the chromosphere falls from about 6,000 K at the top of the photosphere to 4,500 K about 500 km above the photosphere, before increasing to 20,000 K at the top of the chromosphere.

chronometer A highly accurate timekeeping instrument, especially one used by mariners for determining longitude.

cirrus A wispy cloud usually of minute ice crystals formed at altitudes of 20,000 to 40,000 feet.

coded mask telescope A telescope that records an unfocused shadow image from a coded pattern mask of selected transparent and opaque portions. The image is reconstructed from image processing algorithms that account for the specific pattern mask(s).

collectivism The political principle of centralized social and economic control, especially of all means of production.

collimate To make parallel (as in lines of light).

collimator A device to make light or particle rays parallel.

coma of a comet Otherwise called the head. It is the diffuse luminous sphere that surrounds the comet's nucleus.

composite grain solid propellant A solid propellant in which a powdered oxidizer and powdered metal fuel are intimately mixed and immobilized with a rubbery binder, which also acts as a fuel.

conjunction An alignment of three bodies. So, for example, Mercury can be in line with the Earth and Sun, and be either between the Earth and Sun, in an arrangement called inferior conjunction, or on the other side of the Sun to the Earth, when it is at superior conjunction.

consumerism The concept that an ever-expanding consumption of goods is advantageous to the economy.

constellation In the space engineering context, a set of identical or related set of satellites required to perform a function, usually in similar orbits.

continuous-wave laser A laser that produces a continuous, amplified light beam as opposed to one that produces pulses of light.

convection The circulatory motion that occurs in a fluid at a nonuniform temperature owing to the variation of density and the action of gravity.

corona the extended outer atmosphere of the Sun or stars, above their chromospheres, which have temperatures of millions of degrees Kelvin.

coronagraph An instrument containing an occulting disc that creates an artificial eclipse of a bright source like the sun, and allows observers to detect faint objects close to the bright source.

coronal hole Areas where the Sun's corona is darker, colder, and lower density than average. The fastest solar winds emanate from coronal holes.

coronal mass ejection Huge bubbles of gas threaded with magnetic field lines that are suddenly ejected from the Sun.

cosmic rays Very energetic elementary particles traveling through space. When they are above the Earth's atmosphere they are called *primary cosmic rays* or *cosmic ray primaries*. After collisions with the Earth's upper atmosphere they produce showers of *secondary cosmic rays*.

cosmology The branch of astronomy that deals with the origin, structure, and space-time structures of the universe.

cryogenic Of, or relating to the production of very low temperatures.

cryostat A device for creating and maintaining very low temperatures.

cryptanalyst Also known as a cryptologist; a person skilled in solving cryptograms or devising methods for cracking cryptographic systems; a code-breaker, or someone who reads coded texts.

cryptolinguist Someone who acquires, identifies, and translates or transcribes intercepted, foreign-language communications and examines them for key words or indicators.

dark energy Hypothetical repulsive energy proposed to explain the increasing expansion rate of the Universe.

dark matter Hypothetical matter that is undetectable by its emitted radiation, but whose presence can be inferred from gravitational effects on visible matter such as rotational speeds of galaxies and rotational speeds of galaxies in clusters.

degenerate gas A gas characterized by atoms stripped of their electrons and by very great density.

despun A section of a spinning spacecraft, which does not spin with the rest of the spacecraft.

deuterium The hydrogen isotope with a nucleus that contains one neutron in addition to one proton of normal hydrogen.

dipole A pair of equal and opposite electric charges or magnetic poles of opposite sign separated by a small distance.

dipole antenna A simple rod antenna.

direct-to-home (DTH) broadcasting Broadcast signals from satellites intended to be received by individual antennae at each customer's home.

Doppler effect A change in the observed frequency of a sound or light wave, occurring when the source and observer are in motion relative to each other, with the frequency increasing when the source and observer approach each other and decreasing when they move apart; also known as Doppler shift.

dosimeter A device for measuring doses of radiation.

drogue parachute A small parachute for stabilizing or decelerating an object, or for pulling a larger parachute out of stowage.

ecliptic The mean plane of the Earth's orbit around the Sun.

ejecta Material thrown out (as from a volcano or from a meteor strike).

electrocardiogram The tracing of the changes of electrical potential occurring during the heartbeat.

electromagnetic pulse A high-intensity, short-duration burst of energy or radiation, which results in electric and magnetic fields that can produce damaging current and voltage surges in electrical systems or electronic devices.

electron avalanche A large number of electrons produced by the ion of some of the atoms within a gas. In the Geiger and proportional counters a single initial ion is converted into an electron avalanche when the first electron is accelerated towards a positively charged wire and creates further ions as it collides with other atoms.

electron neutrino A type of neutrino associated with the electron, often created in particle interactions involving electrons (such as beta decay). The electron neutrino has a mass of zero or nearly zero, and has no electric charge.

electron reflectometer An instrument that measures the pitch angle of electrons reflected off a planetary surface. The cut-off pitch angle above which no electrons are seen is related to the magnetic field at the surface.

encryption Translation of data into a secret code for purposes of security, particularly to prevent unauthorized persons from reading it.

equity investor A person or entity that owns a risk interest or ownership right in property (property in this case includes stocks or other capital, not merely real estate).

exoplanets Planets existing outside of Earth's solar system.

extreme ultraviolet Electromagnetic radiation between wavelengths 121 nm to 10 nm.

extremely high frequency (EHF) Electromagnetic radiation between frequencies 30 to 300 gigahertz.

extrusion To shape (as metal or plastic) by forcing through a die.

etymological An account of the history of a particular word or element of a word.

flare star A variable star that undergoes unpredictable dramatic increases in brightness for a few minutes, probably caused by stellar flares.

fluorescence spectrometer An instrument that shines light onto a sample and excites electrons of some sample substances, which then emit electromagnetic radiation—fluorescence, often visible light, to obtain a spectrum, from which certain sample substance(s) can be identified.

flywheel A heavy rotating disk that stores angular momentum; on spacecraft, typically used to control its pointing direction (attitude).

focal length The distance of a focus from the surface of a lens or concave mirror.

focus A point at which rays (such as light, heat, or sound) converge or from which they diverge or appear to diverge.

Frame-readout camera A device that takes a photograph through an optical lens, then scans and records the image electronically for later radio transmission to a receiving station.

gamma ray Electromagnetic radiation with wavelengths shorter than about 0.01 nm (or with E > 0.1 MeV).

gas chromatograph mass spectrometer An instrument that separates chemicals by heating the sample to high temperature to produce a gas, and then carrying the gas past a stationary fluid or solid to differentially separate compounds according to mass and charge.

gas imaging spectrometer An instrument that detects high-energy electromagnetic radiation (usually X rays) by collision of the high-energy radiation with a gas, which produces photons that are detected and amplified by the instrument.

gas scintillation proportional counter An X-ray detector consisting of a chamber filled with a noble (i.e., unreactive) gas such as argon or xenon. Ionization of the gas by X-ray photons produces an avalanche of electrons. Atoms of gas excited by the avalanche emit an ultraviolet photon, producing a flash whose intensity is proportional to the energy of the X-ray.

Geiger counter A device for detecting X rays, gamma rays, cosmic rays and radioactive emissions. The high-energy particle or photon causes an ion within an enclosed chamber. The electron from the ion is then accelerated by a strong electric charge towards a central anode. On its way it collides with further atoms, producing an electron avalanche. The electron avalanche is picked up by the anode and the resulting pulse of current detected or fed to a loudspeaker.

geodesy The scientific discipline concerned with the determination of the size and shape of the Earth and the exact positions of points on its surface and with description of variations in its gravity field.

geoid The surface within or around the Earth that is everywhere normal to the direction of gravity and coincides with the mean sea level in the oceans.

geopolitics The study of the relationship among politics and geography, demography, and economics, especially with respect to the foreign policy of a nation.

geostationary Of or pertaining to a satellite traveling in an orbit 22,300 miles (35,900 km) above the earth's equator at this altitude, the satellite's period of rotation, 24 hours, matches the earth's period of rotation and the satellite always remains in the same spot over the Earth.

geosynchronous Of or having an orbit with a fixed period of 24 hours (although the position in the orbit may not be fixed with respect to the Earth).

globular cluster A gravitationally bound, very old, nearly spherical group of stars.

grain The interior shape of solid propellant inside a solid rocket motor case.

gravity gradient stabilization A means of controlling a spacecraft's pointing direction (attitude) by using variations in the local gravity field.

grazing incidence X-ray mirrors A series of mirrors that allow the focusing of X rays in an X-ray telescope. X rays will penetrate regular mirrors, but do not penetrate mirrors if the incident ray is nearly parallel to the surface of the mirror.

grism A combination of a prism and grating arranged to keep light at a chosen central wavelength undeviated as it passes through.

gyro *See* gyroscope.

gyroscope A wheel or disk mounted to spin rapidly about an axis and also free to rotate about one or both of two axes perpendicular to each other and to the axis of spin so that a rotation of one of the two mutually perpendicular axes results from application of torque to the other when the wheel is spinning and so that the entire apparatus offers considerable opposition depending on the angular momentum to any torque that would change the direction of the axis of spin.

gyrostat A type of gyroscope consisting of a rotating wheel fixed inside a rigid case.

halo orbit An orbit around the L1, L2, or L3 Lagrange point that is roughly perpendicular to the line between the two mutually orbiting bodies.

ham A licensed operator of an amateur radio station.

hard X ray *See* X ray.

heat exchanger A device for transferring heat from one substance or component to another substance or component without allowing the two substances or components to mix.

heliosheath The outer region of the heliosphere (the region of space in which the Sun's solar wind exists) beyond the termination shock (where the solar wind dramatically slows as it encounters the interstellar wind) but prior to the heliopause (where the solar wind and interstellar wind are of equal pressure).

horn antenna A horn-shaped antenna for reception and/or transmission of microwave signals.

hydrazine A colorless fuming corrosive strongly reducing liquid base N_2H_4 used especially as propellants for rocket engines.

hypergolic engine An engine that uses propellants that ignite upon contact without external aid.

hyperspectral Refers to techniques to create images using hundreds to thousands of (usually) contiguous wavebands.

image intensifier A device for producing brighter optical and ultraviolet images. Designs vary but most have a photo-emitter coated onto the rear of a window that forms part of a vacuum chamber. The original image is focused onto this window and electrons, in numbers proportional to the intensities within the image, are emitted into the vacuum chamber. A strong electric field then accelerates the electrons through the chamber to collide with a phosphorescent material coated onto a second window. Their energy is converted into a visible light image that is much brighter than the original and which can be detected.

impact crater A crater caused by the impact of an object.

inclination (A) The angle between the orbital plane and a reference plane for an orbiting body. For an object orbiting the sun the reference plane is the ecliptic. (B) The angle between a body's spin axis and a reference plane, which is usually the body's orbital plane. (C) The angle between the direction of a magnetic field and the local horizontal for a planet or moon.

inertia A property of matter by which it remains at rest or in uniform motion in a straight line unless acted upon by some external force.

inertial measurement unit A self-contained device that measures changes to inertial forces. These typically consist of gyroscopes to measure rotations and/or accelerometers to measure accelerations.

infrared Electromagnetic radiation with a wavelength between about 0.8 and 300 microns.

in situ In the natural or original position. In space applications, in situ often refers to the use of materials at their original location, such as use of lunar regolith to locally manufacture materials to build a base.

interstellar medium The environment that exists between stars in a galaxy. Typically this consists of an extremely dilute mixture of ions, neutral atoms, molecules, dust, cosmic rays, and magnetic fields.

ion An atom that carries a positive or negative electric charge as a result of having lost or gained one or more electrons.

ion chamber A device that detects the presence of electromagnetic radiation and particles. The electromagnetic radiation and charged particles collide with and ionize the gas in a chamber. The resulting ions are then collected by charged plates (anodes and cathodes) to produce a detectable electrical current.

ionosphere A region of a planetary atmosphere where the atoms and molecules are ionized.

ionization Gases are said to be ionized when their atoms or molecules lose electrons by collisions with ionizing radiation in the form of photons or elementary particles. The higher the energy of the photons (i.e., the shorter the wavelength), the more likely are they to ionize a gas.

isotope Any of two or more species of atoms of a chemical element with the same atomic number and position in the periodic table and nearly identical chemical behavior but with differing atomic mass or mass number and different physical properties.

Kevlar A trademark for a relatively lightweight, synthetic fiber with high tensile strength that was created in 1975 by the DuPont Company and is now used as a reinforcing agent in a wide variety of composite-type applications that require a high strength-to-weight ratio.

kieserite A mineral ($MgSO_4H_2O$) that is a white hydrous magnesium sulfate.

L1 Lagrangian Point The Lagrangian point directly in between two mutually orbiting objects.

Lagrangian Point Points in the orbital plane of two objects that orbit around their common center of mass, where a particle of negligible mass can remain in equilibrium. There are five such points for two bodies in circular orbits, three of which are unstable to small perturbations. The other two, which are 60° in front of and behind the less massive body, and in the same orbit, are stable.

laser An acronym for Light Amplification by Stimulated Emission of Radiation; a device that produces a coherent, high-energy light beam.

latitude An imaginary line around Earth or another heavenly body parallel to its equator and measured as angular distance in degrees north or south of the equator, the equator being 0° and the poles being 90°.

lepton A family of elemental particles that includes the electron, the muon, the tau, and their associated neutrinos. A lepton is a fermion, meaning it has a quantum spin of ½, and it is subjected to the electromagnetic, gravitational, and weak forces, but not the strong force (which binds nuclei).

leucitic basalt Igneous rocks with significant content of leucite, which is a white or gray mineral ($KAlSi_2O_6$) consisting of a potassium aluminum silicate.

libration The effect of being able to see more than 50 percent of the Moon's surface from Earth, even though the Moon's spin rate is the same as its orbital period. Libration is mainly due to the fact that the Moon's orbit is not circular (producing a libration in longitude) and its spin axis is not perpendicular to its orbital plane (producing a libration in latitude). There are also smaller librations due to the fact that the Moon's spin axis is not perpendicular to the ecliptic, and its spin is not absolutely regular about a fixed axis. There is also a diurnal libration caused by observing the Moon from either end of the Earth's diameter at 12 hourly intervals.

libration point *See* Lagrangian Point.

lithosphere The rigid outer layer of a planet or moon, including the crust and part of the upper mantle.

longitude An angular measure of relative position east or west on the surface Earth or another heavenly body, given in degrees or hours, minutes, and seconds from a specific meridian. For Earth, this is usually the prime meridian at Greenwich, England, which has a longitude of 0°.

low Earth orbit (LEO) A satellite orbits above the Earth at altitudes between (very roughly) 200 miles and 1,000 miles.

Lyman-alpha telescope A telescope that views images in the Lyman-alpha wavelength of 1,216 angstroms. Lyman-alpha refers to the difference between the lowest energy and next-lowest energy levels of a neutral hydrogen atom.

Mach number The object speed divided by the speed of sound.

magnetic damping (A) Restraining vibratory motion by using electric eddy currents induced in either a coil or an aluminum plate attached to the vibrating object that passes between the poles of a magnet thereby converting the energy of motion to heat. (B) Using the local magnetic field to damp out undesired rotations and movements.

magnetic dipole The pair of north and south magnetic poles, separated by a short distance, that makes up all magnets, equivalent to a flow of electric charge around a loop.

magnetic dipole moment (sometimes called **magnetic moment**) The maximum amount of torque caused by magnetic force on a dipole that arises per unit value of surrounding magnetic field in vacuum.

magnetometer An instrument used to detect the presence of a metallic object or to measure the intensity of a magnetic field.

magnetopause The boundary of the magnetosphere.

magnetosphere The region around a planet in which its magnetic field is constrained by the solar wind.

magnetotail The elongated extension of a planet's magnetosphere on the side facing away from the Sun. The magnetotail is shaped by the pressure of the solar wind as it streams around the magnetosphere, compressing it on the side facing the Sun and stretching it on the opposite side into a long, tail-like shape trailing far into interplanetary space.

maria (plural of mare) The extensive dark areas on the Moon that were originally thought to be seas of water. They are now known to be solidified lava.

mascon An area of "mass concentration" or high density on the Moon.

mass spectrometer An instrument to identifying the chemical constitution for a substance by means of the separation of gaseous ions according to their differing mass and charge.

meridian A great circle or half circle on the surface of a sphere passing through the poles, often numbered for longitude on a map or globe.

microgravity A very low-gravity (or gravitational) environment.

microlensing As predicted by Einstein, when a gravitationally strong foreground object is exactly aligned with a background object, the light from the background object forms either a ring of light or two images as it is bent by the gravity of the foreground object.

morphology In geology, the study of the characteristics and configuration and evolution of rocks and land forms.

multilateral Involving more than two nations or parties.

multi-path propagation The phenomenon whereby electromagnetic signals propagate along multiple paths, which reach a given location at different times depending on refraction and reflection even though they originated from the same source.

multispectral Refers to imaging techniques that use tens to hundreds of sometimes non-contiguous spectral wavebands.

muon An unstable lepton that is common in the cosmic radiation near the Earth's surface, has a mass about 207 times the mass of the electron, and exists in negative and positive forms.

muon neutrino The type of neutrino associated with the muon particle.

nanotechnology Technology built or operating on a scale of 100 nanometers.

nebula Any of many immense bodies of highly rarefied gas or dust in interstellar space.

neutral hydrogen Hydrogen atoms with no charge (not ionized).

neutrino An uncharged elementary particle that has three forms (electron, muon, and tau), with very small but non-zero mass, and that interacts very weakly after being created as a result of particle decay.

neutron star The stellar remnant of stars with masses of from 1.4 to 3.0 solar masses, when they have stopped producing energy by thermonuclear fusion. A neutron star consists almost entirely of neutrons, and is prevented from further collapse by neutron degeneracy pressure, which is a quantum mechanical effect analogous to electron degeneracy pressure in white dwarfs.

nova A star that sudenly increases its light output tremendously and then fades away to its former obscurity in a few months or years.

nutation Oscillating movement of the axis of a rotating body.

occultation The passage of one astronomical object in front of another so that an observer can no longer see the more distant object.

olivine A magnesium or iron silicate, $(Mg,Fe)_2SiO_4$.

orthostatic intolerance A disorder of the autonomic nervous system that occurs when a person stands up.

otolith A calcareous concretion in the internal ear of a vertebrate or in the otocyst of an invertebrate. The otolith organs sense gravity and linear acceleration such as from due to initiation of movement in a straight line.

ozone hole A region in the Earth's upper atmosphere in which little ozone exists.

pallet A portable platform for handling, storing, or moving materials and packages. For in-space systems, these often refer to platforms to hold experiments, exposed to the external space environment.

panchromatic Sensitive to light of all colors in the visible spectrum.

parabolic reflector A concave device used to focus radiated or received energy into a desired beam pattern

parafoil A nonrigid airfoil with an aerodynamic cell structure which is inflated by the wind. Ram-air inflation forces the parafoil into a classic wing cross-section.

payload The load that is carried by a spacecraft and consists of things (as passengers or instruments) that relate directly to the purpose of the flight as opposed to things that are necessary for operation.

perigee The point in the orbit of the Moon or Earth-orbiting spacecraft where it is closest to the Earth.

perigee kick motor A rocket motor that fires at orbital perigee to change a spacecraft's orbit.

periapsis The point in the path of an orbiting body at which it is nearest to the primary.

phased-array radar An antenna with an arrangement of dipoles, the phase of each dipole being controlled by a computer that enables the radiated beam to scan an area very rapidly.

phenolic epoxy resin A usually thermosetting resin or plastic containing epoxy made by condensation of a phenol with an aldehyde and used especially for molding and insulating and in coatings and adhesives. Epoxy containing oxygen attached to two different atoms already united in some other way; specifically containing a 3-membered ring consisting of one oxygen and two carbon atoms.

photoelectron emitter A material that emits electrons when it is exposed to light. Different materials are used at different wavelengths—for example, gallium arsenide (wavelength limit 1,000 nm) and caesium antimonide (wavelength limit about 650 nm).

photometer An instrument for measuring luminous intensity, luminous flux, illumination, or brightness.

photometry A branch of science that deals with measurement of the intensity of light.

photon A quantum of radiant energy.

photopolarimeter An instrument for determining the amount of polarization of light or the proportion of polarized light in a partially polarized ray, combined with a telescope for producing an image (as of a planet) by means of polarized light.

photosphere The visible surface of the Sun at a temperature of about 6,000 K. Sunspots and faculae are in the photosphere.

pixel Any of the small discrete elements that together constitute a digital image.

plagioclase A feldspar of sodium or calcium aluminum silicate.

planetarium A building with seats for an audience that houses a device for producing a projecting images of celestial bodies and other astronomical phenomena onto the inner surface of a spherical dome.

planetesimal A body of rock and/or ice, typically 100 meters to 10 kilometers or so in diameter, that was formed in the early solar nebula.

plasma An ionized gas consisting of electrons and ionized atoms.

polar orbit An orbit in which a satellite passes above or nearly above both poles of a body (such as a planet, moon or Sun).

polarimeter An instrument for determining the amount of polarization of light or proportion of polarized light in a partially polarized ray.

polyhydrated sulfate A mineral with a sulfate compound and several water molecules in the crystal structure.

potable water Water suitable for drinking.

precession A comparatively slow gyration of the rotation axis of a spinning body about another line intersecting it so as to describe a cone caused by the application of a torque tending to change the direction of the rotation axis.

prograde Orbital or axial motion of a body in the solar system that is counterclockwise as seen from north of the ecliptic. The term is also used to describe the motion of an object on the celestial sphere in a west-east direction.

prophylactic Tending to ward off or prevent disease.

proportional counter A device to count particles of ionizing radiation and measure their energy.

pulsar A rotating neutron star which emits a beam or beams of radio and/or other waves. These are detected from Earth as pulses at the rotational frequency of the star, usually of the order of seconds, but sometimes of milliseconds. Neutron stars are ultra-dense stars with a typical diameter of only about 10 kilometers.

quasar A very bright extragalactic object with a high red-shift. It is generally thought to be the most luminous type of nucleus of an active galaxy.

quasi stellar object *See* quasar.

radar altimetry Determination of altitude by use of radar.

radiation budget The balance between energy arriving at a location and energy traveling away from that location.

radio galaxy A galaxy that is very luminous in radio wavelengths.

radioactive label A substance that contains a radioisotope.

radioisotope An atom with an unstable nucleus that emits electromagnetic radiation or subatomic particles (protons, neutrons, electrons, etc.).

radioisotope thermoelectric generators A device that generates electricity from the heat produced from the emissions of a radioisotope.

radiometer An instrument for measuring the intensity of electromagnetic radiation.

rectifying To make an alternating current unidirectional.

red dwarf A small, dim, type M or late K spectral type main sequence star. Red dwarves have low mass (.1 to .5 solar mass) and low surface temperature (2,500 to 3,500 K) that gives them a reddish appearance.

reflection grating spectrometer A device to measure spectra generated from a mirror with tiny grooves to spread incoming light according to wavelength.

refractory metal A metal capable of enduring very high temperatures.

regenerative cooling A process to cool a rocket nozzle in which low temperature fuel or oxidizer is channeled through tubes or jackets around the nozzle to absorb heat, after which the warmed fuel or oxidizer is fed into a gas generator or into the combustion chamber for combustion.

regolith The top layer of a solid planetary or similar body (i.e., the Moon) covering the crust and composed of dust and loose fragments of rock.

repeater A device, frequently called a transponder, that receives uplink signals from the ground, removes unwanted noise from the incoming signals, and retransmits them at the same or different frequencies to destination locations.

resonance line The line of longest wavelength associated with a transition between the ground state and an excited state of an atom or molecule.

resonance orbit An orbit in which the orbital period of one body is a simple multiple of that of another body which is orbiting the same primary body.

resonator mirror An arrangement of optical components that allow a light beam to circulate and, in the process, experience various physical effects that change its spatial distribution so that certain patterns or frequencies of radiation are sustained and others suppressed.

retrograde Orbital or axial motion of a body in the solar system that is clockwise as seen from north of the ecliptic. The term is also used to describe the motion of an object on the celestial sphere in an east-west direction.

robotic For spacecraft, refers to non-crewed vehicles.

RP-1 fuel A highly refined form of kerosene used as a rocket fuel.

scalar magnetometer A device to measure the strength (but not direction) of a magnetic field.

scanning telephotometer A telescopic device to measure the intensity of electromagnetic radiation across a range of wavelengths.

scattered disk object An object scattered out of the plane of the solar system by gravitational interactions with an outer planet, usually Neptune.

scatterometer An instrument to transmit microwave radiation towards the ocean surface and measure their return signal after being scattered by capillary waves on the ocean surface.

scintillation A flash of light produced in a phosphor by an ionizing event.

scramjet A supersonic combustion ramjet, which uses the very high speed of a supersonic aircraft to compress incoming air, without need for a rotary compressor.

search radar A radar that employs fan beams, usually wider in the vertical than in the horizontal, to scan a large amount of space quickly for early detection of targets.

secretariat The officials or office entrusted with administrative duties, maintaining records, and overseeing or performing secretarial duties, esp. for an international organization.

seismic station A facility consisting of an instrument for measuring ground movement, a clock for determining time, and a recorder for collecting data, with multiple facilities at different sites being required to locate events accurately and determine their nature.

seismometer A device to measure and record ground vibrations.

servomechanism An automatic control system in which the output is constantly or intermittently compared with the input through feedback so that the error or difference between the two quantities can be used to bring about the desired amount of control.

Seyfert galaxies A type of active galaxy with a brilliant point-like nucleus and inconspicuous spiral arms. It has a broad emission-line spectrum.

shock-heating The heating caused by a compression wave formed when the speed of a body or fluid relative to a medium exceeds that at which the medium can transmit sound.

shutter control A policy by which a government has the authority to restrict a commercial remote sensing company from taking images of certain politically or militarily sensitive regions.

slush hydrogen A mixture of solid and liquid hydrogen used to lower the volume of hydrogen and hence the mass and size of the tanks needed to carry it.

soft X ray *See* X ray.

solar array An array of photovoltaic cells to convert sunlight into electrical power.

solar constant The amount of energy received at Earth from the Sun per unit area per unit time.

solar cycle The approximately 11-year repeating period in which the Sun's energy output oscillates from a maximum when there are many sunspots and solar storms to a minimum in which few storms or sunspots exist.

solar maximum The period of greatest solar activity (most solar storms) in the Sun's solar cycle.

solar minimum The period of least activity (least solar storms) in the Sun's solar cycle.

solar wind A stream of particles, primarily protons and electrons, flowing from the Sun.

solid-state Referring to electronic components, devices, and systems based entirely on the semiconductor. Historically this term was used in the 1950s to 1970s during the transition period away from vacuum tubes to semiconductors.

sorbent bed A chunk of absorbing material.

sounding Measurement of atmospheric and oceanic conditions at various heights and depths.

sounding rocket A suborbital rocket used to measure atmospheric conditions at various heights, and for astronomical observations.

spectral line A dark or bright line in an otherwise uniform and continuous spectrum, resulting from a deficiency or excess of photons in a narrow frequency range, compared with the nearby frequencies. Each line is specific to a given element, and the brightness, width and position of a line can be used to determine the physical conditions of the absorbing or emitting source.

spectral signature A unique observed pattern of spectral lines.

spectrograph An instrument for dispersing radiation (as in electromagnetic radiation or sound waves) into a spectrum and photographing or mapping the spectrum.

spectroheliograph A device that creates images of the Sun that is made by monochromatic (single frequency) light and shows the Sun's faculae and prominences.

spectrometer A device that measures the spectra resulting from a spectrograph.

spectrozonal Films that combine one panchromatic and one infrachromatic layer on a common basis. During the developing process color ingredients are introduced in both of these layers and tincture the image in various colors.

spin stabilization A means of pointing one axis of a spacecraft in a fixed direction by spinning the spacecraft around a rotational axis.

spread-spectrum voice link A wireless communication medium in which the frequency of the transmitted signal is deliberately varied to diminish the effects of catastrophic interference or unauthorized interference.

star types Classifications of stars based on spectral characteristics.

stationary plasma thruster A Hall-effect or ion thruster in which the propellant is ionized and accelerated by an electric field.

stellar wind A flow of neutral or charged gas ejected from the upper atmosphere of a star.

store-dump satellites Orbiting platforms that pick up and save data on board for later transmission when a ground receiving station comes within line of sight.

strap-on booster A rocket engine strapped to the side of an existing rocket to provide additional thrust.

stratigraphy Geology that deals with the origin, composition, distribution, and succession of layers of differing types of rock or earth.

sublimation To cause to pass directly from the solid to the vapor state, or to condense back to solid form.

substorm A process by which plasma in the Earth's magnetotail becomes quickly energised, flowing towards the Earth and producing bright aurorae for periods of up to an hour or so. A substorm can cause spacecraft charging, resulting in spacecraft malfunctions.

Sun-synchronous orbit An orbit in which a satellite crosses the equator of a planet at the same local time of day on each orbit.

superfluid A state of matter cooled to near absolute zero and characterized by apparently frictionless flow.

superhigh frequency (SHF) Electromagnetic radiation between frequencies 3 to 30 gigahertz.

supernova A massive stellar explosion. There are two types of supernovas. A type I supernova is the explosion of a white dwarf in a binary whose mass has exceeded the Chandrasekhar limit of 1.4 solar masses. Prior to the explosion the white dwarf has been gaining mass from its red giant binary companion. A type II supernova is an explosion produced when a main sequence star of more than eight solar masses stops producing energy by thermonuclear fusion.

superoxide The univalent anion O_2- (O_2 with a single negative charge) or a compound containing it.

synodic period The mean interval for a planet between successive inferior or superior conjunctions with the Earth and Sun, or for a planetary satellite between successive conjunctions of the satellite with the Sun as observed from its parent planet.

synthetic aperture radar A space or airborne system that emits microwave pulses and uses special signal processing to produce high-resolution images of the Earth's surface or other objects while moving a considerable distance along a straight path.

tachocline The abrupt transition zone between the differentially rotating convection zone and the uniformly rotating radiative interior of the Sun (and presumably other stars).

tau neutrino The type of neutrino associated with the tau particle.

telemetry The automatic measurement and transmission of data over a distance, such as between a ground station and an artificial satellite or space probe, to record information, operate guidance apparatus, etc.

telephony The use or operation of an apparatus for transmission of sounds between widely removed points with or without connecting wires.

telephotometer An instrument to measure the amount of light received from a distant source; also called a transmissometer or visibility meter.

teletype A printing device resembling a typewriter that is used to send and receive telephonic signals.

thermonuclear warhead A weapon characterized by extremely destructive explosive power derived from the fusion of atomic nuclei at high temperatures.

tholeiitic basalt A subalkaline igneous rock with relatively less sodium and potassium than alkali basalt.

throttleability The ability to change the magnitude of a rocket engine's thrust.

tracking radar A radar that locks onto a strong signal reflected from a target object and automatically follows it, regardless of sudden changes in the target's direction.

transponder A radio or radar set that upon receiving a designated signal emits a radio signal of its own and that is used for the detection, identification, and location of objects, and to re-transmit communications signals.

three-axis stabilization A mechanism to hold a spacecraft's pointing direction stable in all three axes. Spin stabilization, by contrast, can only keep the single rotational axis stable.

thruster A small rocket engine that produces a relatively low-magnitude thrust, generally used for attitude (pointing) control or orbit control.

triangulation A trigonometric operation to find a position or location by means of bearings from two fixed points a known distance apart.

TNT—trinitrotoluene The explosive yield of this compound is a standard measure of explosive strength.

triode tube A vacuum tube with three electrodes—a cathode, a wire-mesh control grid, and a plate (anode)—for electrical amplification.

Trojan points The L4 and L5 Lagrange points, which lie at the third corners of the two equilateral triangles in the plane of orbit whose common base is the line between the centers of the two orbiting masses, such that the point lies behind (L_5) or ahead of (L_4) the smaller mass with regard to its orbit around the larger mass.

truss An assemblage of members (as beams) forming a rigid framework.

turboprop An airplane powered by turbo-propeller engines, which are jet engines having a turbine-driven propeller, providing thrust by means of a propeller but usually augmented by thrust obtained by the hot exhaust gases which issue in a jet.

ullage motor Small rocket engines used to accelerate a vehicle in zero gravity so as to create pressure that forces propellant from the propellant tank(s) to the primary liquid rocket motor(s) to enable it/them to ignite.

ultrahigh frequency (UHF) Electromagnetic radiation with frequencies between 300 to 3000 megahertz.

ultraviolet Electromagnetic radiation with a wavelength between about 9 and 380 nm.

utopian Founded upon or involving idealized perfection.

variable stars Stars that vary in luminosity over time.

vector magnetometer An instrument to measure the component of the magnetic field in a particular direction.

vernier rockets Any of two or more small supplementary rocket engines for making fine adjustments in the speed or course or controlling the attitude.

very high frequency (VHF) Electromagnetic radiation with frequencies between 30 MHz to 300 MHz.

Very Small Aperture Terminal A small radio antenna and associated electronics to receive or send electromagnetic signals.

very long baseline interferometry A method combine observations of an object that are made simultaneously by many optical or radio telescopes a great distance apart, emulating a telescope with a size equal to the maximum separation between the telescopes.

vidicon A camera tube using the principle of photoconductivity.

visible light Electromagnetic radiation with wavelengths from about 380 nm (blue) to about 780 nm (red).

white dwarf The stellar remnant of stars with a maximum mass of about 1.4 solar masses when they stop producing energy by thermonuclear fusion. A white dwarf consists of atomic nuclei and electrons that have been completely stripped from atoms. White dwarfs are prevented from further collapse by so-called electron degeneracy pressure, which is a quantum mechanical effect.

wind scatterometer An instrument to measure wind speed and direction at the sea surface, by measuring changes in radar reflectivity.

Wolf-Rayet star Very hot, O-type stars producing a spectrum with strong, broad emission lines on a continuum background.

X ray Photons of very short wavelengths. The X-ray waveband is from about 0.01 to 9 nm (100 to 0.1 keV). Those X rays at the short wavelength (high energy) end are called hard X rays, and those at the long wavelength (low energy) end are called soft X rays.

X-ray jet A directional flux of X rays produced by matter falling onto a neutron star or black hole in an X-ray binary star system.

yaw Movement of a satellite left or right about its vertical axis.

zeolite Open, crystalline structures that can be used as absorbents and filters.

Units

Å	Angstrom
arcsec	arcsecond
arcmin	arcminute
AU	Astronomical Unit
bar	bar
bps	bits per second
C	centigrade
cm	centimeter
eV	electron-Volt
ft	feet
G	gauss
Gbps	gigabits per second
GHz	gigahertz
h	hour
Hz	Hertz
in	inch
J	Joule
K	Kelvin
keV	kilo electron-Volt
kg	kilogram
km	kilometer
l	liter
lb	pounds
m	meter
mb	millibar
mm	millimeters
Mbps	megabits per second
MHz	megahertz
MeV	mega electron-Volt
min	minutes

mph	miles per hour
μm	micrometer, micron
nm	nanometer
Oe	Oersteds
R_E	Earth radii
rpm	revolutions per minute
μs	microsecond
T	Tesla
V	Volts
W	Watts
We	Watts of electricity

metric ton	1,000 kilograms

Index

Bold page numbers indicate the main article about the topic.